The Emerging Quantum

Luis de la Peña · Ana María Cetto
Andrea Valdés Hernández

The Emerging Quantum

The Physics Behind Quantum Mechanics

Springer

Luis de la Peña
Instituto de Física
Universidad Nacional Autónoma de México
Mexico, D.F.
Mexico

Andrea Valdés Hernández
Instituto de Física
Universidad Nacional Autónoma de México
Mexico, D.F.
Mexico

Ana María Cetto
Instituto de Física
Universidad Nacional Autónoma de México
Mexico, D.F.
Mexico

ISBN 978-3-319-07892-2 ISBN 978-3-319-07893-9 (eBook)
DOI 10.1007/978-3-319-07893-9
Springer Cham Heidelberg New York Dordrecht London

Library of Congress Control Number: 2014941916

Printed on acid-free paper

Springer is part of Springer Science+Business Media (www.springer.com)

Preface

Fifty years ago—in 1963, to be precise—the British physicist Trevor Marshall published a paper in the *Proceedings of the Royal Society* under the short title *Random Electrodynamics*—an intriguing title, at that time. To date this paper has received just over four citations per year, which means it is alive, but not as present as it could be, considering the perspectives it opened for theoretical physics. Shortly thereafter a related paper was published by a young US physicist, Timothy Boyer, under the longer title *Quantum Electromagnetic Zero-Point Energy and Retarded Dispersion Forces.* Boyer does not cite Marshall's paper (although he does so in his third paper, which is followed by a productive 50-year long work in solitary), but instead he refers to the work of David Kershaw and Edward Nelson on stochastic quantum mechanics. All these papers share a central feature: they are based on conceiving quantum mechanics as a stochastic process. Marshall mentions explicitly the existence of a real, space-filling radiation zero-point field as the source of stochasticity. Boyer sees a deep truth in this, and in a note added to his manuscript he comments that "…in this sense, quantum motions are experimental evidence for zero-point radiation."

From a historical perspective, we recall that nearly 50 years earlier—in 1916, to be precise—Nernst had proposed to consider atomic stability as experimental evidence for Planck's recently discovered zero-point radiation. This visionary idea was largely ignored by the founders of quantum mechanics, the only (brief) exception being the Einstein and Stern paper of 1913; such is history. Both Marshall and Boyer succeed in demonstrating that some quantum phenomena can indeed be understood by the simple expedient of adding this random zero-point field to the corresponding classical description. Their pioneering work was soon followed by that of other colleagues, moved by the conviction that the random zero-point field has something important to tell us about quantum mechanics. Many other results have been obtained during this period, which constitute the essence of the theory largely known under the name of *stochastic electrodynamics*. At the same time, other researchers, notably Nelson, dedicated their efforts to develop the phenomenological stochastic theory of quantum mechanics. The perception that quantumness and stochasticity are but two different aspects of a reality, started to gain support from several sides.

So here we are, 50 years later. In the mean time, quantum mechanics has continued to develop; the new applications derived from it only serve to reaffirm it as a powerful theory. Along with its success, however, comes an increasing recognition that its old foundational problems have not found convincing solution. Recall the birth of quantum theory: Bohr's model of the hydrogen atom was supported on a postulate that implied a fundamental violation of electrodynamics. Truly, such postulate was necessary at its moment, but urgent necessity does not restore physical consistency. Then came the mysterious matrix mechanics, and the no less mysterious de Broglie wavelength. Such obscure premises served as foundations for the interpretative apparatus of quantum theory. And obscurity and vagueness followed, along with a formidable mathematical apparatus. From this perspective, one easily concludes that better supporting and supported principles are required. More recent efforts from a number of authors attest to the conviction that quantum mechanics, and more generally quantum theory, is in need of an alternative that helps to explain the underlying physics and to solve the conundrums that have puzzled many a physicist, from de Broglie and Schrödinger to Einstein and Bell, among many others. Common to most of the recent efforts in search of an alternative is precisely the idea that the quantum description emerges from a deeper level.

Quantum mechanics constitutes usually both, the point of departure and the final reference, for all inquiries about the meaning of the theory itself. Its conceptual problems are therefore looked at from inside, which provides limited space for rationalization, and even in some instances creates a kind of circular reasoning of scant utility, as is amply testified by the unending discussions on these matters. Experience evinces that an external and wider approach is indeed required to grasp the meaning of quantum theory and get a clear, physically understandable, and preferably objective, realistic, causal, local picture of the portion of the world that it scrutinizes.

The main purpose of this book is to show that such alternative exists, and that it is tightly linked to the stochastic zero-point radiation field. This is a fluctuating field, solution of the classical Maxwell equations, yet by having a nonzero mean energy at zero temperature it is foreign to classical physics. The fundamental hypothesis of the theory here developed is that any material system is an open system permanently shaken by this field; the ensuing interaction turns out to be ultimately responsible for quantization. In other words, rather than being an intrinsic property of matter and the (photonic) radiation field, quantization emerges from a deeper stochastic process. A physically coherent way to understand quantum mechanics and go beyond it is thus offered, confirming the notion of emergence—the coming forth of properties of a compound system, which no one of its parts possesses.

The theory here presented has been developed along the years in an effort to find answers to some of the most relevant conceptual puzzles of quantum mechanics, by providing a physical foundation for it. It is thus not one more interpretation of quantum mechanics, but constitutes a comprehensive and self-consistent theoretical framework, based on well-defined first principles in line with a realistic viewpoint of Nature. There is neither the opportunity nor the need to resort to ad hoc tenets or philosophical considerations, to assign physical meaning to the elements of the theory and interpret its results.

As the formalism of quantum mechanics is successfully reproduced, some may argue about the value of redoing what is already well known. However, the usual theory, with its interpretations included, seems to tell us more about our knowledge and our way of thinking about Nature, than about Nature itself. A good part of what really happens out there remains hidden, waiting to be disclosed. With this volume, our intention is to contribute to this disclosure and to share the fascinating experience of discovering some of the quantum mysteries and intricacies along the process. Moreover, a door is opened to further explorations that may unravel new physics. As the reader will appreciate, this chapter is not closed; there is much that remains unexamined, awaiting future investigations.

This book has been prepared for an audience that is conversant with at least the most basic ideas and results of quantum mechanics. More specifically, it is intended to address those readers who (either secretly or openly) seek a remedy to the apocalyptic statement by Feynman, that "nobody understands quantum mechanics." Its contents should be of value to researchers, graduate students and teachers of theoretical, mathematical and experimental physics, quantum chemistry, foundations and philosophy of physics, as well as other scholars interested in the foundations of modern physics.

Throughout this volume, frequent reference is made to *The Quantum Dice. An Introduction to Stochastic Electrodynamics* (*The Dice*), a precursor containing many ideas and results that have survived the test of time and others that have been superseded or improved here. *The Dice* and the present book differ in at least two central aspects. First, the version of stochastic electrodynamics discussed in the former was essentially limited to linear problems and failed to properly address the more general nonlinear case; this limitation is successfully lifted in the present book. Secondly, in addition to applying the Fokker-Planck method (already contained in *The Dice*) with success, particularly in Chaps. 4 and 6, new procedures are developed and crucial physical demands (as e.g., the balance of energy, and ergocidity) are identified, which converge into a theoretical framework that is clearer, richer and more unified than the former one. Further to facilitating a smooth and fruitful incursion into the territories of quantum mechanics and quantum electrodynamics, the new developments result in an expansion of the aims of the theory, for example by including the study of composite systems or by opening the door to future analysis of the system before the attainment of the quantum regime.

In addition to the bibliography at the end of the chapters, a list of suggested references (not cited in the chapters) appears at the end of the volume. In the bibliography, the items marked * refer to stochastic electrodynamics (some of them including stochastic optics) and those marked ** are general or topical reviews on stochastic electrodynamics; papers marked ‡ are overtly critical about stochastic (quantum) mechanics; those marked ‡‡ contribute to the development of that theory, but may express some important criticism about it. Some few abbreviations are used in the text, all of them easy to spell out: QM, QED, SED,

LSED, ZPF, FPE, GFPE for quantum mechanics, quantum electrodynamics, stochastic electrodynamics, linear stochastic electrodynamics, zero-point field, Fokker-Planck equation, and generalized Fokker-Planck equation, respectively. In Chap. 1—and occasionally elsewhere—CI and EI are used for the Copenhagen and ensemble interpretations of quantum mechanics, respectively.

The authors acknowledge numerous valuable observations and suggestions received during the elaboration of the manuscript. We are particularly grateful to Pier Mello, Theo Nieuwenhuizen, Vaclav Spika, and Gerhard Grössing for their support and critical comments. Draft versions of the various chapters were shared with some of our students; special thanks go to David Theurel and Eleazar Bello for their useful comments. Further, we wish to thank the Dirección General de Asuntos del Personal Académico (UNAM) and its Director General, Dante Morán, for the support received for the preparation of this volume, under contracts Numbers IN106412 and IN112714. A special word of appreciation goes to Alwyn van der Merwe for his relentless support as editor of the Springer series, and to the reviewers of Springer for their valuable comments and suggestions. Our thanks go also to Aldo Rampioni, Kirsten Theunissen and the Springer staff for their support and attentions. Finally, we wish to acknowledge the facilities provided to us throughout the years by the Instituto de Física, UNAM.

Mexico, March 2014

Luis de la Peña
Ana María Cetto
Andrea Valdés Hernández

Contents

Chapter 1
Quantum Mechanics: Some Questions

...[quantum-mechanical] vagueness, subjectivity, and indeterminism, are not forced on us by experimental facts, but by deliberate theoretical choice.

Bell (1987, page 160)

... that today there is no interpretation of quantum mechanics that does not have serious flaws, and that we ought to take seriously the possibility of finding some more satisfactory other theory, to which quantum mechanics is merely a good approximation .

Weinberg (2013, page 95)

1.1 On Being Principled... At Least on Sundays

Tied to our microscopic place in the immensities of the Cosmos, we are beginning to unfold its mysteries with remarkable precision. Being as gigantic as we are compared to the atomic and subatomic worlds, we have been able nevertheless to uncover an important fraction of its workings. We do not know yet what an electron is made of, but we know already many of its secrets (see e.g. Wilczek 2002).

The remarkable scientific, technological, philosophical, and even economic success of quantum mechanics is only the beginning. No physicist on Earth would question the numerically fitting description that quantum mechanics offers of the part of the world that pertains to its domains, which extend much beyond the atomic scale the theory originally was intended to cover, both towards the macroscopic and the ultramicroscopic. However, a nonnegligible portion of the practicing physicists would also acknowledge, either openly or reluctantly, that the mysteries of the quantum world have not been satisfactorily cleared or explained, after more than eighty years of successful existence of this most basic theory.

Such acknowledgment depends of course on what is meant by *explanation*. A historical example of what we have in mind follows from the Newtonian theory of

L. de la Peña et al., *The Emerging Quantum*,
DOI: 10.1007/978-3-319-07893-9_1,

gravitation: the clarity, universality, simplicity and high precision of this theory made of it a grandiose paradigm; the theory reigned undisputed for over two centuries and became the ideological pedestal that supported the European Enlightenment. The universal gravitational force became the pivotal element to understand innumerable terrestrial and celestial facts, and a central element in the construction of a whole philosophy of nature. This occurred despite the known shortcomings of the theory in more than one essential aspect. Not only did it rest on the ageing concept of action at a distance, but the specific form of the force was selected ad hoc to lead to the Keplerian ellipses, introduced as a mere patch into the Newtonian system of mechanics, with no theoretical support or physical mechanism that would lead to it or explain it. From this more exacting point of view, one could say that the classical theory gives a precise and simple *description* of the facts, sufficiently good *f*or *a*ll *p*ractical *p*urposes (FAPP); but it hardly constitutes an *explanation* of what is going on in the real world. To find such an explanation the whole edifice of general relativity had to be put forth, allowing us to dispense with ad hoc elements or actions at a distance, and providing us instead with a causal rule. Indeed, general relativity *explains* the Newtonian theory.

Today we can calculate atomic transition frequencies to within a billionth part, and use refined applications of the quantum properties of matter and the radiation field to construct marvelous and powerful devices that have become emblematic of our civilization. However, have we really got an understanding of what is happening deep-down in the quantum world? A glance at the quantum literature dedicated to the discussion of its fundamental aspects is sufficient to reveal the vast spread of meanings and uncertainties that beset current quantum knowledge. Of course, if the number predicted by the theory, or the use that is made of it, is taken as its test, just as was the case with Newtonian gravitation and the extended pragmatic viewpoint it prompted, the conclusion is that there is no problem at all. But we may be a bit more demanding and ask, for instance, for the physical (rather than formal) explanation ofatomic stability, the origin of uncertainty or the quantum fluctuations. Again, are wave-particle duality and quantum nonlocalities the final word? Do superluminal influences really exist?[1] In short: the quantum formalism describes its portion of Nature astonishingly well and we do not know why. It would be difficult to express this kind of feelings about the status of present-day quantum theory more lucidly than Bell did in 1976: quantum mechanics is a FAPP theory. And Maxwell (1992) rightly asks: what is beyond FAPP?

Since the creation of quantum mechanics (QM) there has been a flood of papers and essays discussing these and similar or deeper questions, and almost any conceivable (or inconceivable) argument or answer has been advanced, both from within physics and from the philosophy of science, ranging from a complete accord with quantum orthodoxy to a radical departure from it. Such extended and deep rumination has not been the endeavor of idle physicists and philosophers, since names such as Bohr, de

[1] In statements about superluminal influences, it is difficult to know which kind of influences are being considered. Anyhow, detailed analysis shows that special relativity and quantum mechanics have still a peaceful coexistence (see e.g. Shimony 1978; Redhead 1983, 1987).

Broglie, Dirac, Einstein, Heisenberg, Landé, Popper, Schrödinger, do honor to an unending list of active participants.

Let us listen to some few big voices to get a better feeling of the magnitude of the quantum muddle, as Popper (1959) calls it. Feynman writes:

> I think I can safely say that nobody understands quantum mechanics,

and goes on speaking of the [unsolved] mysteries of QM (Feynman et al. 1965). Referring to matter diffraction he asserts:

> A phenomenon which is impossible, absolutely impossible, to explain in any classical way, and which has in it the heart of quantum mechanics. In reality it contains the only mystery...
>
> How does it really work? What machinery is actually producing this thing? Nobody knows any machinery. Nobody can give you a deeper explanation of this phenomenon than I have given; that is, a description of it.

Gell-Mann (1981) in his turn qualifies:

> In elementary particle theory one assumes the validity of three principles that appear to be exactly correct.
>
> (1) Quantum mechanics, that mysterious, confusing discipline, which none of us really understands but which we know how to use. It works perfectly, as far as we can tell, in describing physical reality, but it is a 'counter-intuitive discipline', as social scientists would say. Quantum mechanics is not a theory, but rather a framework, within which we believe any correct theory must fit. (2) Relativity. (3) Causality.

In his turn Dyson (1958) observes:

> ...the student says to himself: 'I understand QM' or rather he says: 'I understand now that there isn't anything to be understood...' .

And speaking about himself, he adds (Dyson 2007)

> ...the important thing about quantum mechanics is the equations, the mathematics. If you want to understand quantum mechanics, just do the math. All the words that are spun around it don't mean very much.

Despite the hundreds of books and of international conferences discussing both physical and philosophical problems of QM, the basic conundrums remain alive and as unresolved as they were eight decades ago. Fortunately nobody (to our knowledge) has blamed Bell of having been unable to understand QM, as was said about Einstein. He, Bell, solved the matter his own way: at the time of some lectures he explained that during the week he used the handy FAPP theory. The weekends however he would regain his principles and search for something better (quoted in Gisin 2002).

Experience shows that so far, neither physical nor philosophical arguments have been effective to get us out of the muddle. For the normal practicing physicist the philosophical arguments, when they have a meaning for science, are little more than an abstraction, an ethereal generalization of the truths already discovered by science. But if along its lines of reasoning, science has been unable to set foot on the profundities of the quantum world, we cannot expect philosophy to unfold them for us. Something of revealing importance can thus be extracted from these persistent discussions: as long as the issues are debated and the differing points of view defended from *inside* quantum theory, no definite conclusion can be reached. What is required then is to gain a look onto QM from *outside* it, to get a wider and clearer perspective. The work presented here represents precisely a systematic attempt to look onto QM from outside it, with the help of a deeper physical theory. This provides us with the possibility of getting answers from a wider perspective than that obtained by just interpreting (or reinterpreting, or misinterpreting) the formalism.

In fact, many of the difficulties with QM arise as a result of the interpretation ascribed to its formalism. Though there have been claims that QM does not need interpretation,[2] the truth is that in no other place of physics do the theory and its formal content elicit such diverse and even contradictory meanings as in QM (see Sect. 1.2). And indeed, the formal apparatus of a theory is in general not enough to interpret it.[3] If "nobody understands quantum theory" it is difficult to hold that the theory speaks for itself. Apart from the immediate problem that represents the lack of consensus on the interpretation of QM, the critical point is that many interpretations of it, particularly the dominant one, jeopardize (when not simply do away with) some principles that have been pillars of the whole edifice of physics. Even if—or precisely because—the principles of scientific philosophy are a distillate of the most fundamental discoveries of science, *if* QM demonstrates that Nature (not a certain description of it) is incompatible with some of those principles, as might be realism, determinism, locality or objectivism, then the philosophical framework must of course be modified accordingly, instead of forcing us to attune physics to worn presuppositions. It could be that the advances of science demand a revision of what is taken at a given moment for a firmly established general outlook; history is full of experiences of this nature. The central concerns and theories of the philosophy of science should be consistent with scientific discovery, and are therefore subject to revision, just as happens with science itself. When the scientific case is clear, science philosophy must adapt to what science tells us. But that requires an absolutely convincing demonstration, since principles as realism, say, are just that, general principles extracted from a huge plurality of cases and circumstances, so their generality, universality, solidity and soundness are utterly confirmed. Convincing demonstrations, not a mere interpretation of the formal apparatus of QM, are thus required to abandon these solid principles.[4]

In the following section we present and comment on some of the most basic issues that beset QM, which originate when adopting a certain interpretation of the theory.

---

[2] See e.g. Fuchs and Peres (2000), or Omnés (1994). Compare with, e.g. Bunge (1956), de Witt and Graham (1974), and Marchildon (2004).

[3] For example, a given system of linear differential equations can represent a mechanical, an acoustical, an electrical or an electromagnetic system, or even an analog computer as well. There is ample conceptual space to accommodate the interpretation.

[4] Virtually all science philosophers have received with approval the philosophical conclusions arrived at from (orthodox) quantum mechanics, despite its nonrealistic (even antirealistic) and subjective trends. Far from helping to drive quantum physics towards a more realistic conception, this of course has contributed to reinforce such trends.

By the same token, in this introductory chapter there is no attempt to resolve these issues or give answers to them. It is along the subsequent chapters, as we develop the theory, that we will be finding answers. This will allow us to summarize, in the final chapter, the insights afforded by the theory and discuss its outlook.

1.1.1 The Sins of Quantum Mechanics

Let us point out in brief some of the sins of QM—some venial, others capital—that are readily found and discussed in the scientific literature, particularly the one written under the spell of the orthodox interpretation. It may seem amazing that two discussions on the subject written by physicists (one of whom later became a recognized philosopher of science) published almost half a century apart (Bunge 1956; Laloë 2002), touch essentially upon the same fundamental questions, of course with an emphasis that corresponds to the given moment.

- QM is an indeterministic theory. Indeed, though the quantum dynamic laws evolve deterministically, the theory is unable to predict individual events. The most the theory can offer are probabilistic predictions, whence the specific outcome of an experiment cannot be determined in advance. In itself, indeterminism is not a regrettable property of a physical theory. The statistical theories of classical physics are indeterministic (or, for some people, they obey statistical determinism) and this is not considered a shortcoming. The reason is that in such cases the origin of such indeterminacy is clear. Recall for instance the statistical description of a classical gas; there is a distribution of velocities of the molecules that calls for a statistical description with no practical alternative. The distribution of velocities of the molecules is a direct consequence of the fact that there is a myriad of microstates compatible with the macroscopic state under scrutiny, all of them having equivalent possibilities corresponding to the initial conditions. In other words, the indeterminacy is a feature of the description, not of the system itself. By contrast, in the usual rendering of QM we have no more explanation for the statistical indeterminism than the indeterminism of the theory. For some this means quantum indeterminism is irreducible.[5,6]

---

[5] Determinism must be clearly distinguished from causality, the latter referring to an ontological property of the system. The notion of indeterminism wavers in the literature from ontological to epistemic connotations, and from objective to subjective meanings. In this book we understand by (physical) determinism a property of the *description* of a physical system, not of the system itself, and thus of epistemological nature. Although many different meanings are ascribed also to causality, this term refers to a direct *genetic* connection among the elements of the description, i.e. to an ontological property of the underlying physical reality. We could say that causality refers to the hardware of nature, determinism to our software about it.

[6] Whether the indeterminism is ontic or merely manifests itself at the observational or descriptive level is a controversial issue, to which every decoder adds his own preferred interpretation (see Bunge 1956 for examples). Still, the attempts to construct a fundamental and deeper *deterministic* theory from which QM could emerge through an appropriate mechanism to generate indeterminism,

- QM has intrinsic limitations to its predictive power. As stated above, the predictions of QM are only probabilistic. The specific reading of the meter is beyond what QM can predict, yet Nature gives in each instance a well-defined unique answer; we are therefore faced with two possibilities: (a) the predictions of QM are incomplete, or (b) the predictions are complete and God plays dice.
- QM is a noncausal theory. One of the most conspicuous examples of noncausality in QM (which is also a towering manifestation of indeterminism) are the Heisenberg inequalities, which imply the existence of unavoidable (quantum) fluctuations. The cause for such fluctuations is alien to the theory (assuming that a cause must indeed exist), or is simply inexistent at all (assuming that no property of Nature escapes to the quantum description). There is a long list of schools and subschools, with different views on whether the Heisenberg inequalities refer to uncertainties (a measure of our ignorance), to (objective or ontic) indeterminacies, or to something else.[7,8] In any case, the widespread attitude is that no cause for quantum fluctuations is considered to be required, and even less, investigated; they can happily remain 'spontaneous'.
- QM is not a legitimate probabilistic theory. Though the predictions of QM deal with probabilities, no formulation of QM is fully consistent with a genuine probabilistic interpretation (in the classical sense). The use of probability amplitudes instead of probabilities implies a distinctive probability theory by itself. For example, negative probabilities appear in QM not only in connection with phase-space distributions, but also as a result of the superposition principle. The amplitudes can interfere destructively and give rise to negative contributions to the probability densities, of a nonclassical nature. These results have led to a widespread acceptance of negative probabilities as a necessary trait of quantum theory.[9]

---

speak to the existing conviction in some circles that quantum indeterminism demands explanation. For example, t'Hooft has envisioned a process of local information loss leading to equivalence classes that correspond to the quantum states ('t Hooft 2002, 2005, 2006).

[7] The textbook (and historical) explanation of the Heisenberg inequalities as a result of the perturbation of, say, the electron by the observation cannot be taken as the last word, at least because the inequalities follow (as a theorem) from the formalism without introducing observers and measuring apparatus.

Within the statistical interpretation of QM (see Sect. 1.2.2) they indeed refer to the product of the (objective) variances of two noncommuting dynamic variables in a given state (see e.g. Ballentine 1998, Sect. 8.4).

[8] The interpretative difficulties are even greater with the energy-time inequality, because this inequality (in its usual form) does not belong to the customary formal apparatus of the theory. There are of course various proposals to replace it (see e.g. Bunge 1970; Jammer 1974, Sect. 5.4). Also the introduction of a time operator has been explored by several authors (see e.g. Muga et al. 2008, in particular the contribution by P. Busch; see also Hilgevoord and Atkinson 2011).

[9] The acceptance of negative probabilities implies a fundamental change in the axioms of probability theory. Since "they are well-defined concepts mathematically, which like a negative sum of money ...should be considered simply as things which do not appear in experimental results" (Dirac 1942; see also Feynman 1982, 1987; d'Espagnat 1995, 1999; and the detailed discussion in Mückenheim et al. 1986, where they are called *extended probabilities*), they tend to be pragmatically accepted, even if this renders the meaning of probability obscure. Once this door is open, anything may step

- QM is a nonlocal theory. Nonlocality is a major issue for quantum physics. It is inherent to the structure of the theory, although subject to quite different connotations, some of which lead to the notion of action at a distance. Locality is a most fundamental physical demand; it pertains to the conceptual framework upon which theoretical physics is founded, yet it is apparently contravened by *all* quantum systems, not only multipartite ones, in which the entanglement introduces the well-known nonlocal correlations between the subsystems. Thus, to understand the origin and meaning of quantum nonlocality is a major task for a deeper understanding of present-day physics, one that has been put aside in favour of the development and expansions of its applications.
- QM is a theory of observables, not of *beables*. According to the more extended interpretation of QM, it is meaningless to speak of the value of a certain variable of a physical system until the corresponding measurement has been performed. Therefore the theory refers to measured variables (observables) and not to preexisting, objective, individual properties of the system (beables). This is clearly a shortcoming from a realist point of view.
- QM is a contextual theory. In quantum theory (Bell's) contextuality means that the result of measuring an observable A depends both on the state of the system *and* the whole experimental context. In particular, it depends on the result obtained in a previous (or simultaneous) measurement of another, *commuting* observable B. Thus the value attributed to A depends on the whole context.[10]
- QM requires a measurement theory. The pure states of the microworld are not realized in our everyday world. We need some means to reduce the former to mixtures when passing to the macroscopic level. Traditionally the assumed agent is the observation (measurement); thus the observer and his proxy break actively into the description in order to produce results.[11] It would not be an overstatement to say that the notion of measurement in QM raises more conceptual problems than those it is intended to solve.
- QM postulates a nonunitary evolution foreign to its formalism. In its usual interpretation, QM demands the collapse of the vector state (the projection onto a subspace associated with the observable under measurement) as a means to reduce all the possibilities encoded in the state into a single one, to account for the measurement process.[12] It is thus the observer who does the dirty task of suspending the uni-

---

in; thus, for instance, imaginary probabilities have been considered to reconcile quantum theory with locality (Ivanović 1978).

In Khrennikov (2009) the probabilistic machinery of quantum mechanics is extended within a realist point of view, to the description of any kind of contextual contingencies, which leads to a theory that finds application in several fields of inquiry, including economics and psychology.

[10] We are referring to the use of the term 'contextuality' as e.g. in Bell (1985) or Svozil (2005). In particular, this property of a quantum systems is at the base of the response of (Bohr 1935) to the EPR 1935 argument (see Einstein et al. 1935).

[11] One should add that a theory of measurement (i.e., of *our* methods to interrogate nature) cannot be part of a *fundamental* (thus general) description of nature, because the former must be quite specific and detailed in every instance to have any predictive capacity.

[12] The notion of reduction or collapse of the wave function was introduced as a quantum postulate by von Neumann (1932) and Pauli (1933). There is no clear definition of the qualities of the perturbation

tary and causal evolution law to allow for the (nonunitary) collapse of the wave function.[13]

- QM risks becoming subjective with the entry into scene of the observer. The observer is an active intruder, the element that transforms the potential into the real; however, he/she is not part of the libretto. For some people this is an opportunity to add subjective elements to the interpretation.[14]
- QM requires a boundary between the observed and the observer, but the theory cannot define it. To avoid an infinite regression, the measuring instrument must be classical. Thus a part of the world is not described by QM, despite the fact that it is considered to be a fundamental theory, one that should apply to everything.[15] Since quantum theory should lead to the description of the macroscopic world as a limiting process, in principle it cannot refer to elements of the latter in its foundations; yet it does precisely that.
- QM deals with objects of undefined nature. The theory does not embody an objective strict rule of demarcation that distinguishes between corpuscular and wave entities. Worse, even: whether these objects exhibit a corpuscle- or a wavelike behaviour is controlled by the free undertakings of the observer. There is room for three quarks within a proton, but an electron may occupy the whole interferometer before hitting a single point on the screen.
- QM lacks of a space-time description. In particular, the notion of trajectory is foreign to QM, presumably prevented by the Heisenberg inequalities. Thus, QM describes what the atomic electrons do in the abstract Hilbert space, but says nothing about what they do in common three-dimensional space.[16]
- QM is a nonrealist theory. The usual quantum description averts realism from several sides, through the lack of a space-time description, incomplete causality,

---

of the physical system that demarcate the two ways of evolution (the causal one and the collapse). Thus, "[T]he observed system is required to be isolated in order to be defined, yet interacting to be observed" (Stapp 1971). Within the single-system interpretation the collapse is avoided by means of the 'many-worlds interpretation' (or 'relative-state formulation') of QM (Everett 1957, from Everett's thesis 1956), according to which the world splits into as many independent worlds as different results of the measurement can occur. We will not discuss here this (extreme, even if logical) interpretation.

[13] It is of course possible in principle to include the measurement apparatus in the Hamiltonian; a well known example of this is Bohm's theory (see Chap. 8). This helps to express the measurement problem in more realistic terms. Another well-known example is van Kampen (1988).

[14] An argument against the observer, aimed at recovering objectivity in the quantum 'potentialities', has been advanced from cosmology. According to inflationary theory, the early classical inhomogenities in the cosmic microwave background originated in earlier quantum fluctuations. This quantum-to-classical transition took place much before even galaxies existed. It follows that the measurement problem in cosmology is of a different kind (Perez et al. 2006; Valentini 2008).

[15] It is even applied to the universe as a whole; see e.g. Hartle and Hawking (1983). A well-grounded critique of the boundary, for the general public, is contained in Wick (1995).

[16] However, the possibility to construct quantum trajectories (by considering additional elements into the usual quantum description) has received special attention since the times of de Broglie. The best known example of quantum trajectory is perhaps the one afforded by Bohm's theory (discussed in Chap. 8).

unexplained indeterminism, nonlocality... (see Sect. 1.3 for a discussion on realism and quantum mechanics).

1.2 The Two Basic Readings of the Quantum Formalism

1.2.1 The Need for an Interpretation

The pure theoretical skeleton of a physical theory, its *formalism*, says nothing about the world; it is devoid of empirical meaning. To attribute physical meaning to the abstract mathematical apparatus, a set of *semantic rules*, collectively known as *the interpretation*, is required. The interpretation assigns a concrete empirical meaning to the nonlogical terms in the theoretical model (such as mass, force, charge, electric field, and so on). Physically, the model normally does not resemble what it models; the conformity resides in the functioning.

Which is the meaning we should ascribe to the different elements in the quantum formalism, e.g, the wave function, solution of the Schrödinger equation for a given problem? The answer is left to our ingenuity. And this is where the real problem starts... It is not difficult to count a dozen different interpretations of the *same* theory: Copenhagen interpretation (Bohr, Heisenberg, etc., from 1926 on); ensemble interpretation (Einstein, etc., from 1926 on); de Broglie–Bohm theory (de Broglie 1927; Bohm 1952a, b); quantum logic (Birkhoff and von Neumann 1936); many worlds (Everett 1957); stochastic electrodynamics (Marshall 1936); stochastic mechanics (Nelson 1966); modal interpretations (van Fraassen 1972); propensities of smearons (Maxwell 1982); consistent histories (Griffiths 1984); quantum information (Wheeler 1983); transactional interpretation (Cramer 1986); zitterbewegung interpretation (Hestenes 1990); no-signaling plus some nolocality (Popescu and Rohrlich 1994); relational quantum mechanics (Rovelli 1996); and so on. According to other authors, QM does not require an interpretation at all (Peres (2000)), or on the contrary, there is only one legitimate interpretation (Omnès 1994), or even any interpretation goes (Feyerabend 1978). We are further told that the description does not really describe the system, but merely our knowledge (or information) about it (Heisenberg 1958a, b, but see Marchildon 2004; Jaeger 2009); or that the theory is about measurements and observables and not about beables (see Bell 1976, 1985); or that the awareness of our knowledge 'actualizes' the wave function, thus promoting us from external passive bystanders into active (although involuntary) participators (Patton and Wheeler 1975), without being included however in the formal structure. A recent trend is to say that QM refers not to matter, but to bits of information (see e.g. Vedral 2010). And so forth...

Thus we have a nice formal description of the quantum world, empirically adequate for our purposes, but we still lack of a real understanding of that world. No wonder that there are expressed recognitions of the need of a fundamental and deep amendment of our present quantum image (see e.g. Delta Scan 2008; Stenger 2010).

1.2.2 A Single System, or an Ensemble of Them?

A most basic and crucial question for any interpretation of QM relates to the meaning of the wave function: does it describe the dynamics of a single particle, or does it instead refer to an ensemble of similarly prepared particles? The answer to this question distinguishes between the two mainstreams of the interpretation of quantum theory, the Copenhagen and the ensemble interpretations.[17]

The usual textbook standpoint on QM is based on some variant of the Copenhagen (or orthodox) interpretation (CI).[18] It might also be called the *customary*, *mainstream* or *regular* interpretation, although it is not so clear that the present-day practicing physicists (and physical and quantum chemists) adhere to it in their daily endeavours as tightly as such names may fancy. The founding fathers of the CI are of course Heisenberg (1930) and Bohr (1934), who were joined almost from the start by physicists like Pauli, Dirac (1930), Born (1971), von Neumann (1932), and Landau. One should bear in mind, however, that the name CI does not refer to a sharp set of precepts, since a wide range of tenets with respect to some of the central interpretative issues can be distinguished among its practitioners. Thus it encompasses a collection of variants of interpretation rather than a tight doctrine. In a broad sense one refers normally (but not necessarily) to any of the members of such collection as the *conventional interpretation*. The basic tenet of the CI of QM is that a pure state provides a description *as complete and exhaustive as possible of an individual system*. So, QM goes as far as is possible in the knowledge of Nature, and physicists must renounce once and for all the hope for a more detailed description of the individual; Nature imposes upon us a limitation to our knowledge. This assumption has enormous consequences, some of which will be discussed in the following section.

A very different outlook ensues from the ensemble (orstatistical) interpretation (EI) of QM. According to this interpretation the wave function refers to a (theoretical) ensemble of similarly prepared systems, rather than to a single one. The earliest attempts to formulate an ensemble interpretation of QM are found in Slater (1929), Schrödinger (1932) and Fürth (1933). Other early advocates of this interpretation were Langevin (1934), Popper (1959), Einstein (1936, 1949), Landé (1955, 1965), Blokhintsev (1964, 1965) (the original Russian version of 1949 was the first systematic treatment of the ensemble interpretation of QM).[19] Being an intrinsically

---

[17] An early introductory account of the different interpretations of QM and their variants can be found in Bunge (1956). More advanced expositions, also by professional philosophers of science, are found, among others, in Bunge (1973) and Redhead (1987). A more recent monograph by a physicist is Auletta (2000).

[18] Since this interpretation (as indeed all interpretations) contains in an essential way Born's (1926) probabilistic notion of the wave function, and in addition it was strongly influenced by Heisenberg, it would be more properly called *Copenhagen-Göttingen interpretation*. Wigner (1963) proposed to apply the term 'orthodox' more specifically to the view adopted by von Neumann, as reshaped by London and Bauer (1939).

[19] More recent advocates are Margenau (1958, 1978), Sokolov et al. (1962), Mott (1964), Marshall (1965), Lamb (1969, 1978), Belinfante (1975), Newton (1980), Santos (1991), de Muynck (2002), Laughlin (2005), Khrennikov (2009), Nieuwenhuizen (2005) (in Adenier et al. 2006), etc. For an

statistical description, for the advocates of the EI the description afforded by the wave function ψ *is neither complete nor exhaustive of the individual systems* that conform the ensemble (which in its turn gives significance to the different probabilities encoded in ψ). Chance enters into the picture in a fundamental way; the wave function does not "represent things themselves, but merely the probability of their occurrence" (Einstein 1933, slightly adapted).

1.3 Is Realism Still Alive?

"Quantum mechanics demolishes the view that the universe exists out there" (Wheeler 1979).

Quantum mechanics, or a certain interpretation of it?

Such a view of QM is clearly nonrealist. This may not mean much to some, to others it may be unimportant, but to still others it may be of high significance, because philosophical realism is not a capricious free invention. As mentioned earlier, philosophers arrived at the notion of realism by distilling the works of creative scientists (and philosophers) along the centuries, and recognizing and extracting the essence of their diverse procedures. They have thus discovered that there are realist scientists, nonrealist scientists and anti-realist scientists, and that the large majority of creative natural scientists are (spontaneously or consciously) realist and work under the assumption (or conviction) that the world they are studying is not an illusion, but exists by itself. This is the essence of scientific realism: the belief in a real world, external to us, independent of our attention to it, a world in which we act, which acts upon us, and upon which we act to know more about it. A nonrealist negates either the reality of the external world or its independence from us, or both; an antirealist is more extreme and believes that the world is a result of our mental activity.[20] Along the centuries, science, with its remarkable development, has nourished and reinforced realism. Shortly stated, realism is a synthetic result of the scientific venture.

Further to the general defining attributes of scientific realism—external reality, independent from our deeds, and the possibility to know the world—realism in physics embodies other demands of general validity. An obvious one is causality, which lies at the basis of physical science. Another is the recognition that the phenomena occur in space and time, and thus should admit a space-time description. A

important defense of the ensemble interpretation of QM see the old paper by Ballentine (1970), or his more recent books (1989, 1998); Ballentine takes, however, an indeterministic view. Home and Whitaker (1992) contains a detailed discussion, from a realist point of view, of the different versions of the ensemble interpretation of QM. Further, an interesting analysis is that of Rylov (1995) who demonstrates on general arguments that QM (including Dirac's theory) necessarily refers to an ensemble of particles.

[20] It is not too difficult to find openly antirealistic views nourished by the conventional interpretation of QM. See e.g. Rigden (1986), Adler (1989). There are also some researchers that go as far as to consider that the universe itself is not real; see e.g. Henry (2005).

third one is that the causal relations are local, which means that there are no actions at a distance.[21,22]

Let us look at some of the features of QM as seen from the CI and the EI, to make clear the position of these interpretations with regard to realism. In doing so, we will touch upon some of the difficulties encountered in Sect. 1.1.1 and discusse them more at length.

As stated above, a most distictive quality of QM is its indeterminism, which in some instances is taken as noncausality. In a situation commonly considered, a given observation can lead to one of a miscellany of possible results (e.g. a specific eigenvalue among a set of values). Which is the outcome is a matter of chance, and the CI grants that nothing, except chance, determines the result. The example of the decay of a single radioactive nucleus is illustrative: quantum theory can correctly assign a mean lifetime to the nucleus, but it cannot predict the precise moment or direction of the decay products. However, a nearby detector shows that such moment and such directions exist. The precise prediction escapes quantum theory. By considering the quantum description to provide the most complete attainable information about a given system, not unusually the CI declares that precise values of the physical variables cannot be predicted by QM simply because such variables do not have preexistent values; they do not exist until a measurement is performed, until a precise value is recorded).[23] Thus, for example, for the conventional school, the position of the particle is materialized or brought into being, as it were, as a result of its measurement. The values of the dynamical variables are thus objectively undetermined prior to their measurement, and only probable values can be assigned to them; probabilities become irreducible. Since the nonexistent cannot be measured, it is the measurement itself which fixes the measured value, giving reality to it. It is here that the observer (or the observer's proxy) slips into the description; the realist fundamental principle that physics should refer to the world rather than to our knowledge of it (or information about it) is eroded, and with it the no less fundamental demand of a strictly objective rendering of the physical world. All this was clearly recognized

---

[21] We are using here the term *realism* with the meaning of *gnoseologic realism* (Bunge 1985), i.e. ontologically as the belief in an external world, independent of our theories and observations, and epistemologically as the conviction that it is possible to know that world, part by part. However, in some places we use a restricted notion of physical realism which originates in the famous EPR 1935 paper, namely that if a value can be determined for a variable without disturbing the individual system, there exists an element of reality associated with it, even prior to the measurement. According to this notion, the individual systems are at all times in objectively real states (Deltete and Guy 1990), even if unknown, and should in principle be amenable to a space-time description.

[22] An introductory discussion of scientific realism by a realist can be seen in Boyd (1983). The author shows, in particular, how the educated (expressly in science) common sense is a good guide towards scientific realism.

[23] A word of caution is needed here. The measured value may or may not preexist, it suffices to consider that some feature or property related to the measured value preexists. The clearest example is perhaps the measurement of a spin with a Stern-Gerlach apparatus, which obviously may reorient the spin. Thus, a realist theory is compatible with both possibilities; it all depends on the nature of the measured variable. See Allahverdyan et al. (2013).

(and accepted) by Bohr (1928) in his famous Como Lecture of September 1927, a characteristic sentence of which says:

> ...the finite interaction between the object and the measuring devices... implies... the necessity to renounce the classical idea of causality, and a radical revision of our attitude toward the problem of physical reality,

and by Heisenberg in denying the existence of an underlying quantum realm (Heisenberg 1958a, page 129):

> ...the idea of an objective real world whose smallest parts exist objectively in the same sense as stones or trees exist, independently of whether or not we observe them ... is impossible...

or further (Heisenberg 1958b, page 15):

> ... the natural laws formulated mathematically in quantum theory no longer deal with the elementary particles themselves but with our knowledge of them. Nor is it any longer possible to ask whether or not these particles exist in space and time objectively...

The role of the observer is not limited to bringing out a real physical variable out of a mere potentiality, it includes determining the very nature of the system. For instance, in an electron diffraction experiment the electron suffers a series of transformations from being a (more or less) localized entity (with corpuscle-like properties) to becoming a structure that fills a macroscopic volume (with wavelike properties) and vice versa. It seems difficult to bring to terms this series of transmutations with the idea of a reality independent of our undertakings.[24]

Along with the observer, a radical form of nonlocality is introduced into the theoretical framework: the collapse of the wave function—instantaneous over the whole space—determined by a local measurement. Indeed, the collapse, which is the theoretical counterpart of the changes on the individual system brought about by the active observer, becomes the inevitable mechanism by which a specific result is selected from among the various possibilities. The collapse disrupts the orderly causal development described by the evolution equation, introducing an abrupt fall to a lawlessly established state of a certain statistical mixture (these are the spooky actions at a distance, mentioned by Einstein to Born; see Born 1971). Thus two forms of evolution compete within the theory, and it is the observer —the ineluctable intruder—who determines with his actions which of them should operate. Of course, interpreting the collapse as merely a theoretical tool, without ascribing to it a sense of reality, becomes an acceptable pragmatic procedure. But this is not its usual grasp.[25]

---

[24] In a letter to Physics Today by Henry (2004, p. 14) discussing why physics understanding is so poor in the United States, the author ends by saying: "We know from quantum mechanics that nothing is real, except for the observations themselves." Another typical example reads: "one cannot consider quantum properties as being 'real,' in the sense of 'objective reality" (Paul 2008).

[25] As is the case with other quantum paradoxes, the collapse of the wave function becomes understandable within the ensemble interpretation. The fact that an *individual* observation is made does not change the (original) ensemble, it only changes our knowledge by giving us an extra piece of information. We add this information to construct a *new* ensemble that corresponds to the updated situation, a quite normal statistical procedure. The 'collapsed' state vector describes the new situation.

Since according to the spirit of the Copenhagen interpretation it is meaningless to attribute any existence to a certain physical variable until it is measured,[26] the quantum variables have been transformed into *observables*. Hence, the standard adumbration of QM demands from us to assume that the theory is not about existing objects of nature, but about our measurements and observations on them. Bohr states it clearly (as reported by Petersen 1963):

> There is no quantum world. There is only an abstract quantum mechanical description. It is wrong to think that the task of physics is to find out how Nature *is*. Physics concerns what we can say about Nature.

Heisenberg goes even farther (Heisenberg 1958b), by negating the reality of his very object of study:

> ...the atoms or the elementary particles are not as real [as any phenomena in daily life]; they form a world of potentialities or possibilities rather than one of things and facts.

Out of the frying pan into the fire, today we see a modern version of this idealistic vision of the world swiftly extending in connection with information, which argues that the building blocks that constitute the world are not matter and energy, but ... bits of information (see e.g. Vedral 2010; Boriboje and Brukner 2011, and references therein). A most fashionable formula for this was introduced by Wheeler (1990): "It from bit", where 'bit' stands for the unit of information; according to this dictum, the material world emerges from the (qu)bits of quantum information, not conversely.

As for the possibility to construct a space-time description of quantum systems, the very idea was firmly negated by Heisenberg, Bohr and other founders of QM, who declared the quantum world to be nonvisualizable. Thus, the concept of trajectory was taken as untenable in quantum theory since it is contrary to Heisenberg inequalities (and to the wavelike properties, many would add).[27] The view of a nonvisualizable world helped to do away with the need to explain some of the quantum paradoxes (Jones 2008, particularly Chap. 16). By 1927 quantum trajectories were so insistently negated—with the exception of de Broglie and Einstein[28] (Bacciagaluppi and Valentini 2009)—that at the closure of the Solvay 1927 Congress Lorentz felt obliged to make a declaration of principles:

> ... I should like to preserve this ideal of the past, to describe everything that happens in the world with distinct images. I am ready to accept other theories, on condition that one is able to re-express them in terms of clear and distinct images.

---

[26] The dictum "No elementary phenomenon is a phenomenon until it is a registered phenomenon" (Wheeler 1978, 1983) is a transparent revelation of the positivism that permeates usual quantum theory.

[27] We find trajectories in Feynman's method of path integrals, but they are virtual and attain arbitrary velocities, and besides all possible trajectories are considered with equal amplitude, not only those (unknown) related to the actual motion followed by a given electron travelling from point A to point B.

[28] This was precisely one of the persistent arguments put forward byEinstein against the Copenhagen interpretation.

We are not longing for a past full of clear images, if that past is gone for ever. But, is it really gone? As Lorentz put it, we should be ready to accept the new theories, on condition that they are the result of transparent and definitive knowledge, free of free elections. Yet, by embracing the Copenhagen interpretation, we forsake the possibility not only of making precise predictions about individual trajectories, but entertaining that very notion.

The widespread conclusion that the violation of the Bell inequalities by QM demonstrates Nature's nonlocality represents one more argument against realism. As an example, van Fraassen (1989) contends that scientific realism is invalidated at the microlevel by the violation of Bell's inequalities, and therefore it cannot be valid more generally.[29] In fact, there is no need of these inequalities or any of their variants to demonstrate that QM corresponds indeed to a nonlocal *description*, as follows, for example, from Bohm's interpretation of QM. The point is that we must carefully distinguish between Nature being intrinsically nonlocal and a nonlocal rendering of the relevant portion of Nature.

To maintain a realist view of physics, either the definition of realism must be changed to accommodate for the new situation, or we must accept that QM cannot be the final tale. The standard lore purports the first alternative, which leads to consider that our current notion of realism is incompatible with science.[30] For example Stapp (1972) writes "If the statistical predictions of quantum theory are true, an objective universe is incompatible with the law of local causes." It is interesting to compare this with Einsteins contention (in Born 1971, page 221):

> I cannot seriously believe in [quantum mechanics] because it cannot be reconciled with the idea that physics should represent a reality in time and space, free from spooky actions at a distance.

Clearly Einstein opted for the second alternative above, namely to admit that QM is not the final tale. As he expressed in Einstein (1949):

---

[29] By contrast, Shimony (1989) contends that the formalism of QM may have to be modified so that the theory meets certain metaphysical constraints. He even suggests the need to modify QM to save physical realism. By way of example he points out a possible modification of the topology of space-time at a subquantum scale. He alerts the reader, remarking that "[t]his proposal is the antithesis of [his] attempt to draw philosophical consequences from scientific results, for it indicates rather a reliance on philosophical considerations to supply the heuristics for a scientific investigation." (page 34).

As can be surmised, the conceptual problems associated with the violation of the Bell inequalities have led some authors to even question QM as a fundamental theory of nature [see e.g. Howard (1989)].

[30] More precisely, that local realism and quantum theory are incompatible. This can be argued, as summarized by Ferrero (1987), as follows: It is possible to demonstrate that the following four statements are incompatible:

a) Realism; b) Locality; c-EPR) Quantum mechanics is a complete theory; c-Bell) Quantum mechanics accepts hidden variables (it is not a complete theory); d) Quantum mechanics is a valid theory of Nature.

a, b, d and c-EPR are the assumptions in the EPR paper;

a, b, d and c-Bell are the assumptions in the early derivation of Bell's theorem.

Thus, independently of the completeness of QM (i.e., of c-EPR or c-Bell), a, b and d are incompatible. In Bell 1971 the demand c-Bell was eliminated.

> If in quantum mechanics we consider the psi-function as (in principle) a complete description of a real physical situation, we thereby imply the hypothesis of action-at-distance, a hypothesis that is hardly acceptable. If, on the other hand, we consider the psi-function as an incomplete description of a real physical situation, then it is hardly to be believed that, for this incomplete description, strict laws of temporal dependence hold.

By assuming that QM goes as far as possible in the knowledge of Nature, the CI forces us to admit a nonrealistic, irreductibly indeterministic, nonlocal and noncausal world. In contrast, once we concede that the quantum description is incomplete, the possibility of going beyond QM without having to renounce to realism opens in principle. A means to recover realism is thus offered by adhering to the ensemble interpretation. In particular, by recognizing that quantum theory is statistical and as such incomplete, the ensemble school allows for the possibility of understanding the indeterminism as due to such incompleteness, without necessarily assigning to it a more fundamental meaning, as could be that of an ontological property, or, perhaps, an irreducible indeterminism at the observational level. This leaves the door open to further studies at a deeper level, for the identification of the source of the indeterministic (or stochastic) behavior characteristic of quantum systems. For those who profess this credo this is a most important alternative. For a hard realist, who believes that each individual system has always a real state (may be unknown), and that among the tasks of physics an important one is to discover such real states, an *essentially* statistical theory cannot be taken as complete.

In an extended variant of the EI (also here there are variants, of course) the particle is assumed to have at each moment a set of well-defined, objectively real properties, even if these properties are not simultaneously described by the wave function.[31,32] Thus for example, one thing is to say that the values of two variables associated with noncommuting operators cannot be simultaneously ascertained by resorting to ψ, and another one is to say that such values are not simultaneously defined, or simultaneously existent, even if distributed and unknown. Preexisting values thus may exist (Deltete and Deltete and Guy 1990), yet the wave function ψ—a catalog of all the different possible outcomes—can only assign to each of them a certain probability. In the example of the decay of a single radioactive nucleus the fact that

[31] There exists a widespread belief that if two quantities cannot be measured simultaneously, they do not exist simultaneously. This (positivist) identification of existing (being) and being observed (measured) is of course merely a point of view; it is not part of the postulates of QM.

[32] A simple example may be illustrative of the ambiguity of the quantum description. Consider the state vector of two spin 1/2 particles in the singlet state (referred to a certain direction z)

$$|00\rangle_z = \frac{1}{\sqrt{2}}\left(|\uparrow\rangle\,|\downarrow\rangle - |\downarrow\rangle\,|\uparrow\rangle\right).$$

A rotation of the system of reference to an arbitrary direction $\hat{n}$ transforms this description into

$$|00\rangle_{\hat{n}} = \frac{1}{\sqrt{2}}\left(|\hat{n}_+\rangle\,|\hat{n}_-\rangle - |\hat{n}_-\rangle\,|\hat{n}_+\rangle\right).$$

Now the spins are referred to the arbitrary direction $\hat{n}$. Thus, the spins may be aligned in any direction whatsoever. In other words, the state vector gives absolutely no indication of the actual direction of the spins. From the ensemble point of view, the individual spin pairs are distributed uniformly in all directions.

a precise prediction escapes to QM, does not mean by necessity that there are no precise (although unknown) factors precisely determining the result. Thus, the EI advocate distinguishes between the capabilities of our theories or descriptions, and what happens in the real world, at the ontological level. A particular, but immediate consequence of this is that the notion of trajectory, though recognized as foreign to the quantum description, is not forbidden in principle.

From an ontological point of view, what the EI and CI schools claim is the preexistence or not of features that lead to the observed value (see, however, footnote 23). Thus, referring to the observables of the CI, Bell contends: observables are not beables (Bell 1987, particularly articles number 5 and 7).[33] The transition from beables to observables—from preexisting values to undefined or nonexisting values—is one most important issue of quantum theory, which remains nevertheless unstudied. Out of the blue the observer enters the scene, although the quantum-mechanical formalism does not provide tools to establish where that boundary between the observed and the observer lies, leaving room for an ambiguity and cloudiness that is totally strange to theoretical physics. Bell (1987, article 20) refers to this in unequivocal terms: "It is the toleration of such an ambiguity, not merely provisionally but permanently, and at the most fundamental level, that is the real break with the classical ideal. It is this rather than the failure of any particular concept such as 'particle' or 'determinism'."[34]

The pictures provided by the CI and the EI differ so widely—they in fact exclude each other—that at first glance it should be a simple matter to empirically demonstrate the fallacies behind one or the other. But almost eighty years have elapsed since the advent of quantum theory and the dichotomy remains, notwithstanding the endless discussions and enlightened studies on the subject.[35] The root of the difficulties is that the problem is deeply influenced by the personal philosophical stance. There coexist several general outlooks about the world, and each one of us adopts one or another, consciously or unconsciously to different degrees. This is an (apparently) free personal selection, more or less as (apparently) free is the selection of a religious credo. Add to that the characteristic positivistic standpoint that pervades textbooks, entangled with their scientific content. The physics student is normally unprepared to recognize the presence of this mixture, and less so to disentangle it, so that he ends up assimilating as established knowledge what is far from that.

---

[33] Not surprisingly, other terms equivalent to beable have been proposed in the literature, such as 'being' or 'existent' (Shimony 1978; d'Espagnat 1984). Bell (1987, article 19) adds 'beer' as another one, personally suggested to him by Zumino.

[34] A strong contention against the pragmatic and nonrealist views associated with the observer and his (hers in his language) measurements, reigns in the whole little (big) book of Bell on the foundations of quantum mechanics (Bell 1987). He even says that there are words that should not belong to the lingo of theoretical physics and should be banned from it, such as 'measurement', 'observation', 'observer'.

[35] Reviews or reprints of important work expressing differing views, as well as ample lists of references to papers dealing with this subject, can be found in de Witt and Graham (1974); Belinfante (1973); Jammer (1974); Nilson (1976); Wheeler and Zurek (1983); Cushing and McMullin (1989); Ballentine (1989, 1998); Omnès (1994, 1999); Home (1997); Auletta (2000); Bertlmann and Zeilinger (2002), etc. The list is endless.

For a realist the CI is implausible, to say it mildly (other more belicose terms have been used), while a moderate orthodox considers the EI full of unnecessary metaphysics Fuchs and Peres (2000), or just dogmatic. For a more radical orthodox, the EI lacks the space needed to accommodate other elements demanded by his world view, such as the observer and perhaps his mind. The pragmatic (FAPP) physicist argues that the Copenhagen theory has been used successfully for many years without a single failure, which is a proof of its correctness, so we should derive from it our vision of the world and not the other way round. He therefore expects us to renounce our basic principles of physical thought in order to be able to understand physics (Tambakis 1994) on the basis of a 'quantum syllogism', an attitude similar in nature to that required to give theological support to the theory of the epicycles, as Jaynes (1993) put it. Further, not few physicists add that QM describes what can be described, and that importing into the quantum domain knowledge that originated in the classical world leads to contradictions and paradoxes (see e.g. Lévy-Leblond 1973), as Bohr alerted us since 1935.

It should be noted that, much as the strength of the EI lies in its essentially statistical nature, in it lies also its weakness. Indeed, the EI (as expounded e.g. in Ballentine 1970, 1989, 1998) is far from being free of difficulties on a very fundamental level. An immediate one is that the quantum-mechanical description is a very particular sort of statistical description, in terms, not of probabilities, but of amplitudes of probability, which have the peculiarity that they interfere among themselves. This is fundamental for QM; it is the basis for quantum interference and entanglement, two most important and characteristic features of the quantum systems. This superposition of amplitudes has at least two implications that go counter to the usual theory of probability: the occurrence of probabilities that depend on the context (contextuality, for short), and of negative probabilities, as remarked in Sect. 1.1.1. Moreover, and connected to the latter, the quantum description does not allow for a joint distribution for noncommuting variables, so it lacks of a true phase-space distribution of general applicability. The fact that joint probability distributions do not exist for noncommuting variables puts into question the very definition of correlations between them. It should therefore not be surprising to find results such as those of Gleason (1957), Bell (1966), Kochen and Specker (1967),[36] showing that even if each observable is considered as a classical random variable, two incompatible observables (noncommuting operators) cannot be viewed simultaneously as classical random variables defined on the same space of events, with *independence* from the specific context. The consequence of this is the nonexistence of a (context-independent) joint distribution of such variables (Suppes and Zanotti 1981). A particular sequel of such theorems is that any hidden-variables theory of QM is necessarily contextual.

Of course, such problems as negative probabilites and the lack of a phase-space description, being characteristic of the quantum formalism, are common to all inter-

[36] The latter is the name by which the theorem of these authors is commonly known, although a similar result was presented somewhat earlier in Bell (1966). For this reason some authors refer to it under the fairer acronym BKS. There are not so many instances in which an almost simultaneous discovery by several authors is duly recognized—more often, science seems to have become a one-hundred meter steeplechase race.

pretations of QM. However, the problem becomes more accute for the EI, precisely because it sees QM as a statistical theory. The widespread lack of clarity about this topic has led to a series of objections against the ensemble interpretation of QM, with some authors claiming with conviction that such a formulation has been empirically disproved. About this there is still much to say.

1.4 What is this Book About?

Through the following eight chapters, a fundamental theory for quantum mechanics is constructed from first physical principles, disclosing quantization as an emergent phenomenon arising from a deeper stochastic process. The elements that sustain the pillars of the quantum-mechanical formalism are identified; hallmarks such as the mechanism responsible for atomic stability, the nature of quantum fluctuations, the origin and meaning of quantum nonlocalities, as well as other central features of quantum theory, are elucidated. All this is carried out within a comprehensive and self-consistent theoretical framework that reaffirms fundamental scientific principles such as realism, causality, locality, and objectivity. Thus, the theory developed in the present monograph hopefully may serve to show that those principles can survive their apparently unsurmountable adversities.

If one lesson can be drawn from the persistent but inconclusive enlightened studies on the meaning of the quantum laws, it is that the analysis of quantum theory from its inside leads to nowhere. Such studies may add richness, deepness and erudition to an interpretation, but the essentials remain the same. The virtue of the theory presented here is that it offers a perspective on the quantum world from outside it; one arrives at the quantum formalism from a distance, with a well-defined physical perspective. The interpretation comes from the physics, not the physics from the interpretation.

1.4.1 The Underlying Hypothesis

The fundamental hypothesis that is put to test and developed at length in this book is that every material system is an open stochastic system in permanent contact with the random zero-point radiation field (ZPF). The existence of an all-pervading ZPF follows quite naturally from the (classical) Maxwell equations, yet it is foreign to the classical realm, which graciously assigns zero energy to the field oscillators at zero temperature. The ZPF is taken here as the athermal component of the radiation field, as real as any other solution of the Maxwell equations.

The most significant conclusion drawn from the present theory is that the quantum phenomenon, rather than being an intrinsic property of matter or the radiation field, emerges from their interaction. A key element is found in the fluctuations of the ZPF, which correspond to the 'vacuum fluctuations' of quantum electrodynamics (QED). Vacuum fluctuations are commonplace in modern quantum theory, though some of their consequences seem not to be fully appreciated. The fluctuations of the best known vacuum field, the electromagnetic radiation field, are commonly considered to be (totally or partially) responsible for several physical phenomena, such as spontaneous radiation from excited systems (see e.g. Dalibard et al. 1982), the Casimir forces (see, e.g., Davydov 1965; Ballentine 1989), and the Lamb shift (see e.g. Sokolov et al. 1962; Milonni 1994). But apart from serving to explain these quantum *corrections*, the vacuum field is mostly viewed as a nuisance, because it is responsible for several of the infinities that spoil the otherwise smooth quantum calculations.[37] Thus it is swept under the carpet as soon as possible (only to reenter through the back door) and reduced to a merely virtual field. In the theory presented here, rather than being a nuisance, the ZPF becomes central for the understanding of the behavior of atomic matter. Thus, far from being considered as merely the origin of some small corrections or effects to be added on top of the quantum pattern of matter, the ZPF is seen as the source of the quantum behavior of matter. This is the central premise of stochastic electrodynamics (SED), at least from the point of view of the present authors.

Naturally, since all vacuum fields may contribute in principle to the universal background noise, in line with our approach all of them could contribute to the fundamental stochastic behavior of matter on the microscopic level. However, at the scales to which QM is most frequently applied, or for systems basically of an electrodynamic nature, it is the electromagnetic vacuum that plays the pivotal role. At deeper levels or for systems of another sort, it may well be that other vacua become relevant; one can even speculate that all vacuum fields have similar statistical properties, so that a kind of *universality* holds, in the sense that the essential stochasticity of matter is basically independent of the nature of the dominant background field. One could also consider that the required random field is just a construct to simulate the effects of random fluctuations of the metric, and take these as the ultimate origin of the quantum phenomenon (a first heuristic approach to this idea has been given in Santos 2006).

1.4.2 The System Under Investigation

Our system of study is composed of a material charged particle (rather, an ensemble of them) embedded in the ZPF and having a dynamics that is initially described by a classical (stochastic) equation of motion. Due to the randomess of the system, the theory is statistical in essence. The system is then left to evolve. When, and if, it reaches a reversible regime in which detailed energy balance (i.e., at each frequency of the field) is attained in the mean between the field and matter, the radiative terms in

[37] Interestingly, at present the zero-point fields are seen as possible sources of the conjectured dark energy. Even if for the moment this is not much more than a speculation (which carries its own problems), it brings to the fore the possible importance of zero-point fields (see e.g. Saunders and Brown 1991).

the dynamical equations for the mechanical subsystem become mere corrections that can be neglected in a first approximation. Under these conditions the evolution turns out to be controlled by the quantum equations. Two independent and complementary derivations of this fundamental result are presented, one in Chap. 4 (leading to the Schrödinger description), and another in Chap. 5 (leading to the Heisenberg formalism). The ensuing classical-to-quantum transition could in a way evoke the usual textbook derivations in quantum field theory that start from a classical field theory and at some point incorporate an extra-classical (quantum) demand. Of course the converse transition, from quantum to classical, is theoretical commonplace—although not always based on conclusive arguments. Yet our procedure differs profoundly in essence and scope from such formal methods; here no quantum demand is introduced (neither a priori nor a posteriori). The ZPF is the extra-classical physical entity that ultimately endows the system with its quantum properties, and in addition guarantees the internal consistency of the theory. The quantum is not the means, but the consequence.

The present theory should not be confused with a semiclassical theory, which treats matter quantum-mechanically but the field classically, or conversely (see e.g. Sokolov and Tumanov 1956). Quite the contrary, here we deal with an initially continuous radiation field (classical, but with its zero-point component) and a particle that initially satisfies classical equations of motion, and show that both end up being quantized.

As a prelude to the derivations in Chaps. 4 and 5, the phenomenological description of QM as a stochastic theory is discussed in Chap. 2, with the purpose of introducing the reader to some of the (old) methods that succeed in showing that it makes sense indeed to understand QM as a stochastic theory. In Chap. 3 we initiate the testing of our hypothesis, by analyzing the consequences of allowing for a zero-point contribution in the equilibrium radiation field. There it is shown that the ZPF has a decisive role in leading to the Planck distribution for the radiation in thermal equilibrium, and to the quantized spectrum for the oscillators of the field.

The treatment of matter and field as inseparable elements of a whole system makes it possible for the theory to go *beyond* QM in the most natural way. It provides the elements to study the radiation and absorption terms—a matter that is normally considered to belong to the domain of QED—which here appear as radiative corrections (neglected in the previous approximation) to the quantum-mechanical description. In Chap. 6 it is shown that indeed, these terms are responsible for the finite lifetimes of excited atomic states, as well as for the absolute stability of the ground state in the sole presence of the ZPF. A further radiative correction that appears quite naturally gives the Lamb shift for isolated atoms, and the corresponding shifts in more complex situations. Of particular interest is the discussion, in the same Chap. 6, related to the origin of the electron spin from the present perspective, as another consequence of the fluctuations imposed on the particle by the field, in this case, those that give rise to rotational motions. We are thus faced with one more element that cannot be predicted from within the Schrödinger realm, but can be unfolded by recognizing the presence and action of the ZPF. Moreover, being the spin of the charged particle the support for its magnetic moment, it becomes clear that along with it, the theory determines the spin g-factor of the electron, predicting its correct value of 2.

When the theory is generalized to include systems of two particles, which is the subject of Chap. 7, a phenomenon expected in the present treatment appears, namely the emergence of correlations between (even otherwise noninteracting) nearby particles through common relevant modes of the vacuum field. The correlated motions of the particles attest to their entanglement, induced by the ZPF. Therefore, just as the ZPF may be capable of generating decoherence of the system, it also stands as the most important source of coherence in a significant class of bipartite systems. In particular, when the particles are identical and subject to the same external potential, our results disclose the mechanism underlying the Pauli exclusion principle. More generally, the vacuum field is exhibited as an important source of nonlocality: when this field is ignored, the consequences of its action appear as nonlocal. Nonlocality is further studied in Chap. 8, both for the single-particle case and for a pair of correlated (entangled) particles; these studies unfold the important role played by the so-called diffusive velocity, just the one due to the quantum fluctuations, in providing the quantum system with its characteristic nonlocal descriptive features. In addition, in Chap. 8 we make a brief detour to the causal interpretation of QM, which among interesting features provides an opportunity to glance at a hidden-variables description and to take a fresh look at quantum nonlocality.

Attention is paid in Chap. 9 to the undulatory properties of matter; the de Broglie wave is constructed and shown to originate in the radiation field around the moving particle. A well-defined physical wave is thus naturally associated to the moving corpuscle, yet both entities (particle and wave) are clearly distinguished from each other at all times. Further, a brief discussion is presented regarding the diffraction of electrons, which is explained by arguing that the electron diffraction pattern is but a trace of the pattern produced by the diffracted ZPF. A final section is devoted to a discussion on the relationship between atomic and cosmological constants, with the ZPF, of cosmic presence, acting as the bridge between these two realms of Nature. The final Chap. 10 contains an overview of the main results and implications for QM of the theory developed in the previous chapters. It further provides a brief account of several of its limitations and possible extensions, and ends with a brief discussion of SED in the broader context of theories of space-time metric fluctuations.

It should indeed be noted from the start that the treatment given here to the quantum problem corresponds to a restricted theory in several senses. An obvious one is that the entire discussion is nonrelativistic. Further, the dynamics that takes place during the transition from the original classical state—in which the system is far from equilibrium—to the final state—the quantum regime, controlled by the detailed balance of energy—still needs to be worked out in detail; surely such studies will reveal a rich physics that so far remains hidden. Moreover, the entire treatment is limited here to the description of the dynamics of the material part of the system, while the field is considered as basically (though not entirely!) unperturbed. This excludes by construction the possibility of a full quantum-electrodynamic description. Consequently, the calculation of those phenomena that correspond to QED is everywhere limited in this volume to the lowest significative order of approximation. Within these limitations, nevertheless, the results derived are always the correct ones, appropriately coinciding with the corresponding predictions of either (nonrelativistic) QM or QED.

By looking at quantum theory from the perspective offered here, we hope that the reader will find a satisfactory explanation or answer to a number of the issues and puzzles mentioned in this chapter, and to others that may be boggling his mind. On the other hand, as discussed in the final chapter, it is clear that there are still many fundamental (and treacherous) facets to learn about the quantum world and its intriguing machinery. QM is a marvelous theory. Just because it is marvelous, it deserves to be better understood.

In concluding, we should note that the theory developed in this volume is an alternative, more advanced, complete and elaborate version of the previously developed theory of SED.[38] When it is necessary to distinguish between the traditional theory and the present version, the latter will be designated with LSED (the *l* stands for *linear*; see the explanation in Sect. 5.2). The theory offers substantial answer to a fundamental question posed by T. H. Boyer,[39] namely: which quantum problems can be explained using classical physics plus the ZPF? A large collection of papers published in the past half century by different authors (by Boyer himself, P. Claverie, D. C. Cole, H. M. França, T. W. Marshall, A. Rueda, E. Santos, ourselves and several others) provided the ground for the construction of the present version and anticipated some of the results derived here. Recent results obtained by some of these authors and others serve to legitimate or reinforce the ones presented here. We therefore wish, through the present work, to pay tribute to all those colleagues who have joined us in this exciting endeavour with the shared conviction that the quantum puzzle *can* be solved, and that the ZPF is a central part of the solution.

Appendix A: The Ensemble Meaning of Probability

Considering that probability is a somewhat obscure subject, about which all sorts of debates have taken place, the following observations—due in essence to Brody (1975, 1993)—may be appreciated by some of our readers. The point is that several notions of probability coexist and are used in the physical literature, with their respective caveats. It would not be an overstatement to say that the personal grasp of the notion of probability plays an important role in the espousal of one or the other interpretation of QM. It therefore seems appropriate to give some precision to the meaning given to it in the present work.[40]

[38] A comprehensive account of the results obtained in SED up to 1995 is contained in the book The Quantum Dice, by L. de la Peña and A.M. Cetto (1996), hereafter referred to as *The Dice.*

[39] We attribute this question to Boyer by inferring it from his papers. In a private communication he has expressed himself in similar terms. See however Boyer (2011).

[40] Among the many different perspectives on the subject within physics, the following cover a wide range of possibilities: Bunge (1970); Lucas (1970); Gillies (1973); Rédei and Szegedi (1989); Home

Apart from the formal or axiomatic (Kolmogorovian) probabilities and the subjective interpretation of probability,[41] there are two interpretations of probability popular among the practitioners of physics. One of them is the *frequentist* or *objective (empirical) interpretation*. According to this interpretation, proposed by Venn (1880), and developed by Reichenbach (1949) and von Mises (1957), among others, a series of observations is made and the relative frequency of an event is thus determined; its probability is taken as the value attained in the limit when the number of cases in the series tends to infinity. Here we are dealing with events (not with propositions as in the formal rendering, or with opinions or beliefs as is the case with the subjective interpretation), and the determination of the relative frequency is an empirical, objective (although approximate) process. There are however some problems that hamper a strict formulation of this probability: if experimental frequencies are used, the infinite limit is unattainable; if the relative frequency is a theoretical estimate, then the limit is probabilistic and the frequentist definition becomes circular. Again, the existence of the limit value should be assumed. Moreover, the theoretical structure lacks an experimental counterpart: why should the experimental relative frequencies correspond to the theoretical estimates? Notwithstanding such difficulties, this interpretation constitutes a widely used practical tool. As Bunge (1970) puts it: "All we have is a frequency *evaluation* of probability".

Let us turn our attention to another important view on probability, much extended among physicists, namely the *ensemble interpretation*. We follow here the discussion on the subject by Brody (1975, 1975), particularly Chap.10), and start by recalling the usual concept of *ensemble* . Each theoretical model of reality should be in principle applicable to all cases of the same kind, i.e., to all cases where the properties of the system considered by the model are equal; the factors neglected by the model may vary or fluctuate freely, but in consistency with the applicable physical laws. The set of all these cases constitutes the ensemble of interest. The notion of ensemble as a set of theoretical constructs can thus be established without recourse to the concept of probability, and can be structured so as to possess a measure, which is then used to define averages over the ensemble. The ensemble concept of probability can then be introduced as follows. Let A be a property of interest and let χ_A be the indicator function of A, i.e., $\chi_A(\omega) = 1$ if the member ω of the ensemble has the property A,

---

and Whitaker (1992). See also *Interpretations of Probability* in the online Stanford Encyclopedia of Philosophy.

[41] The most extended subjective views of probability are the *individual* degree of acceptability of a proposition (de Fenetti 1974), or its Bayesian version (Jeffreys (1939); Jaynes (1995); Caticha (2008) as a measure of the informed personal opinion. According to the Bayesian views, any evaluation of a probability is conditional to some evidence that partially entails it; thus, Keynes (1921) asserts that "the probability of the same statement varies with the evidence presented". By contrast, the probability of decay of an atomic nucleus depends on the internal physical situation of the constituent nucleons, and is entirely independent of any personal information. This illustrates the different use of the concept of probability in physics and in other fields of knowledge. It should be considered that even if an assigned numerical probability is taken as depending on our degree of rational belief (or our degree of partial entailment), it contains some logical elements, since it is limited by rational constraints that ensure the possibility of using a mathematical apparatus (see Gillies (1973), Introduction).

$\chi_A(\omega) = 0$ otherwise. Then the probability of A is the expectation over the ensemble of $\chi_A(\omega)$,

$$\Pr(A) = \int_{\Omega} \chi_A(\omega) d\mu(\omega), \tag{A.1}$$

where $\mu(\omega)$ is the measure function for the ensemble, usually normalized over Ω, the range of the events ω. It is possible to show that this definition satisfies all the axioms of Kolmogorov (1956), so that indeed the ensemble can become the basic tool for probabilistic theorization.

The experimental counterpart of this probability is the relative frequency as measured in an actual (and of course finite) series of experiments. If the relative frequencies thus measured do not correspond to the theoretical estimates, the ensemble (the measure) should be redefined until agreement is reached through the appropriate research work. Here there is no global recipe. Of course, as is the case with any other physical quantity, theoretical probabilities and their experimental values need not necessarily be exactly the same.

The ensemble definition of probability does not allow the application of the notion of probability to a singular case (there is no ensemble). Thus, for example, the philosophical problem of the probability of a given theory being true, becomes meaningless. To give meaning to the assertion about the probability of a single event, it must be translated into a statement about its relative frequency.

The most interesting aspect of the ensemble notion of probability is its direct correspondence with the concept used by physicists in their daily undertakings, so that we adhere to it in the present work, even though it is not entirely free of conceptual and philosophical problems—as any other interpretation of probability.

References

Adenier, G., Khrennikov, A.Yu., Nieuwenhuizen, Th.M. (eds.): Quantum theory: reconsideration of foundations 3. AIP Conference Proceedings, p. 810 (2006)

Adler, C.G.: Realism and/or physics. Am. J. Phys. **57**, 878 (1989)

Allahverdyan, A.E., Balian, R., Nieuwenhuizen, Th M.: Understanding quantum measurement from the solution of dynamical models. Phys. Rep. **525**, 1 (2013)

Auletta, G.: Foundations and Interpretation of Quantum Mechanics. In the Light of a Critical-Historical Analysis of the Problems and of a Synthesis of the Results. World Scientific, Singapore (2000)

Bacciagaluppi, G., Valentini, A.: Quantum Theory at the Crossroads. Reconsidering the 1927 Solvay Conference. Cambridge University Press, Cambridge (2009)

Ballentine, L.E.: The statistical interpretation of quantum mechanics. Rev. Mod. Phys. **42**, 358 (1970)

Ballentine, L.E.: Quantum Mechanics. Prentice-Hall, New York (1989)

Ballentine, L.E.: Quantum Mechanics. A Modern Development. World Scientific, Singapore (1998)

Belinfante, F.J.: A Survey of Hidden-Variables Theories. Pergamon, Oxford (1973)

Belinfante, F.J.: Measurement and Time Reversal in Objective Quantum Theory. Pergamon, Oxford (1975)

Bell, J.S.: On the problem of hidden variables in quantum mechanics. Rev. Mod. Phys. **38**, 447 (1966). Reprinted in Wheeler and Zurek 1983 and in Bell 1987
Bell, J.S.: Introduction to the hidden-variable question. In: Proceedingsof the International School of Physics 'Enrico Fermi', course IL. AcademicPress, New York (1971). Reprinted in Bell 1987 as number 4
Bell, J.S.: The theory of local beables. Epistemological Lett. **9** (1976). Reprinted in Bell 1987 as number 7
Bell, J.S.: The theory of local beables. Dialectica **39**, 86 (1985). Reprinted in Bell 1987
Bell, J.S.: Speakable and Unspeakable in Quantum Mechanics. Princeton University Press, Princeton (1987)
Bertlmann, R.A., Zeilinger, A.: Quantum [Un]speakables. From Bell to Quantum Information. Springer, Berlin (2002)
Birkhoff, G., von Neumann, J.: The Logic of Quantum Mechanics **37** (1936)
Blokhintsev, D.I.: Fundamentals of Quantum Mechanics. Reidel, Dordrecht (1964). English translation from the Russian edition of 1949 (Leningrad)
Blokhintsev, D.I.: The Philosophy of Quantum Mechanics. Reidel, Dordrecht (1965)
Bohm, D.: A suggested interpretation of the quantum theory in terms of "Hidden" variables. I. Phys. Rev. **85**, 166 (1952a). Reprinted in Wheeler and Zurek 1983
Bohm, D.: A suggested interpretation of the quantum theory in terms of "Hidden" variables. II. Phys. Rev. **85**, 180 (1952b). Reprinted in Wheeler and Zurek 1983
Bohr, N.: Nature **121**, 580 (1928). Reprinted in Électrons et photons. Rapports et discussions du Cinquième Conseil de Physique. Gauthier-Villars, Paris (1928)
Bohr, N.: Atomic Theory and the Description of Nature. Cambridge University Press, Cambridge (1934). Reprinted in The Philosophical Writings of Niels Bohr, vol. I
Bohr, N.: Can quantum-mechanical description of physical reality be considered complete? Phys. Rev. **48**, 696 (1935)
Boriboje, D., Brukner, Č.: Quantum theory and beyond: is entanglement special?. In: Halvorson, H. (ed.) Deep Beauty: Understanding the Quantum World Through Mathematical Innovation. Cambridge University Press, Cambridge (2011)
Born, M.: Zur Quantenmechanik der Stossvorgänge. Zeit. Phys. **37**, 863 (1926)
Born, M. (ed.): The Born-Einstein Letters. Macmillan, London (1971)
Boyd, N.: The current status of the issue of scientific realism. Erkenntnis **19**, 45 (1983)
Boyer, H.: Any classical description of nature requires classical electromagnetic zero-point radiation. Am. J. Phys. **79**, 1163 (2011)
Brody, A.: Probability: a new look at old ideas. Rev. Mex. Fís. **24**, 25 (1975)
Brody, A.: The Philosophy Behind Quantum Mechanics. In: De la Peña, L., Hodgson, P. (eds.). Springer, Berlin (1993)
Bunge, M.: Survey of the interpretations of quantum mechanics. Am. J. Phys. **24**, 272 (1956)
Bunge, M.: Problems concerning intertheory relations. In: Weingartner, P., Zecha, G. (eds.) Induction, Physics and Ethics, p. 285. Reidel, Dordrecht (1970)
Bunge, M.: Philosophy of Physics. Reidel, Dordrecht (1973)
Bunge, M.: Racionalidad y Realismo. Alianza Universidad, Madrid (1985)
Caticha, A.: Lectures on Probability, Entropy and Statistical Physics. arXiv: 0808.0012v1 [physics.data-an] (2008)
Cramer, J.G.: The transactional interpretation of quantum mechanics. Rev. Mod. Phys. **58**, 647 (1986)
Cushing, J.T., McMullin, E. (eds.): Philosophical Consequences of Quantum Theory. University of Notre Dame Press, Indiana (1989)
Dalibard, J., Dupont-Roc, J., Cohen-Tannoudji, C.: Vacuum fluctuations and radiation reaction: identification of their respective contributions. J. Phys. (Paris) **43**, 1617 (1982)
Davydov, A.S.: Quantum Mechanics. Addison-Wesley, Reading (1965). English translation with additions by D. ter Haar from the Russian edition (Moscow, 1963)

de Broglie, L.: La Mécanique ondulatoire et la structure atomique de la matière et du rayonnement. J. Phys. Rad. **8**, 225 (1927)
Delta Scan: The future of Science and Technology 2005–2055. Better Understanding of Quantum Theory. http://humanitieslab.stanford.edu/2/365 (2008)
de Finetti, B.: Theory of Probability. Wiley, London (1974)
**de la Peña, L., Cetto, A.M.: The Quantum Dice. An Introduction to Stochastic Electrodynamics. Kluwer, Dordrecht (1996) (Referred to in the book as *The Dice*)
Deltete, R., Guy, R.: Einstein's opposition to the quantum theory. Am. J. Phys. **58**, 673 (1990)
de Muynck, W.M.: Foundations of Quantum Mechanics, an Empiricist Approach. Kluwer, Dordrecht (2002)
d'Espagnat, B.: Nonseparability and the tentative descriptions of reality. Phys. Rep. **110**, 201 (1984)
d'Espagnat, B.: Veiled Reality. An Analysis of Present-day Quantum Mechanical Concepts. Addison-Wesley, Reading (1995)
d'Espagnat, B.: Conceptual Foundations of Quantum Mechanics. Perseus Books, Reading (1999)
de Witt, B.S., Graham, R.N.: Resorce Letter IQM-I on the interpretation of quantum mechanics. Am. J. Phys. **39**, 724 (1974)
Dirac, P.A.M.: The Principles of Quantum Mechanics. Clarendon, Oxford (1930 (1947,1958))
Dirac, P.A.M.: The physical interpretation of quantum mechanics. Proc. Roy. Soc. London A **180**, 1 (1942)
Dyson, F.J.: Innovation in Physics. Sci. Am. **199**(9), 74 (1958). Quoted in Landé 1965, p. 148, and requoted in Selleri, Quantum Paradoxes, p. 2
Dyson, F.J.: Interview with Onnesha Roychoudhuri, Sep 29, 2007 (in Atoms & Eden)
Einstein, A.: On the Method of Theoretical Physics. The Herbert Spencer Lecture. Clarendon, Oxford (1933). Reprinted in A. Einstein: Ideas and Opinions. Crown, New York (1954)
Einstein, A.: Physik und Realität. J. Franklin Inst. **221**, 313 (1936). English translation: Physics and Reality, ibid, 349. Reprinted in A. Einstein, Out of my Later Years. Thames & Hudson, London (1950)
Einstein, A.: Autobiographical notes and Einstein's reply in P. A. Schilpp, Albert Einstein, Philosopher-Scientist, Library of Living Philosophers (Evanston Ill) (1949)
Einstein, A., Podolsky, B., Rosen, N.: Can quantum-mechanical description of physical reality be considered complete?. Phys. Rev. **47**, 777 (1935). Reprinted in Wheeler and Zurek 1983
Everett, H.: Relative state formulation of quantum mechanics. Rev. Modern Phys. **29**, 454 (1957)
Ferrero, M.: El Teorema de EPR y las desigualdades de Bell. In: Fundamentos de la Física Cuántica, (notes for a graduate course by several authors, Universidad de Cantabria) (1987)
Feyerabend, K.: Against Method. Verso, London (1978)
Feynman, R.P.: Simulating physics with computers. Int. J. Theor. Phys. **21**, 467 (1982)
Feynman, R.P.: Negative probability. In: Peat, F.D., Hiley, B. (eds.) Quantum Implications: Essays in Honour of David Bohm, p. 235. Routledge & Kegan Paul, London (1987)
Feynman, R.P., Leighton, R.B., Sands, M.: The Feynman Lectures on Physics, vol. III. Addison-Wesley, Reading, Mass (1965)
Fuchs, C.A., Peres, A.: Quantum theory needs no interpretation. Zeit. Phys. Today **53**, 70 (2000). Discussion in the September issue
Fürth, R.: Über einige Beziehungen zwischen klassischer Statistik und Quantenmechanik. Zeit. Phys. **81**, 143 (1933)
Gell-Mann, M.: Questions for the future. Series Wolfson College lectures, 1980. Oxford University Press, Oxford (1981). Also in the collection The Nature of Matter, Wolfson College Lectures 1980. J. H. Mulvey, ed. (Clarendon Press, Oxford, 1981)
Gillies, A.: An Objective Theory of Probability. Methuen & Co., London (1973)
Gisin, N.: Sundays in a Quantum Engineers's Life, in Bertlmann and Zeilinger (2002)
Gleason, A.M.: Measures on the closed subspaces of a Hilbert space. J. Math. Mech. **6**, 885 (1957). Reprinted in The logico-algebraic approach to quantum mechanics, vol. I. In: Hooker, C.A. (ed.). Reidel, Dordrecht (1975)
Griffiths, R.B.: Preferred consistent history sets. J. Stat. Phys. **36**, 219 (1984)

Hartle, J.B., Hawking, S.W.: Wave function of the Universe. Phys. Rev. D **28**, 2960 (1983)
Heisenberg, W.: The Physical Principles of the Quantum Theory. Dover, New York (1930)
Heisenberg, W. Physics and Philosophy. The Revolution in Modern Science. Harper and Row, New York (1958a)
Heisenberg, W.: The Physicist's Conception of Nature. Harcourt Brace, New York (1958b)
Henry, R.C.: A Private Universe, Letter to Physics Today. Feb 2004, p. 14
Henry, R.C.: The mental universe. Nature **436**, 29 (2005)
Hestenes, D.: The zitterbewegung interpretation of quantum mechanics. Found. Phys. **20**, 1213 (1990)
Hilgevoord, J., Atkinson, D.: Time in Quantum Mechanics. Clarendon Press, Oxford (2011)
Home, D.: Conceptual Foundations of Quantum Physics. An Overview from Modern Perspectives. Plenum, New York (1997)
Home, D., Whitaker, M.A.B.: Ensemble interpretations of quantum mechanics. A modern perspective. Phys. Rep. **210**, 223 (1992)
Howard, D.: Holism, separability, and the metaphysical implications of the Bell experiments. In: Cushing, J.T., McMullin, E. (eds.) Philosophical Consequences of Quantum Theory. University of Notre Dame Press (1989)
Ivanović, I.D.: On complex Bell's inequality. Lett. Nuovo Cim. (1978)
Jaeger, G.: Entanglement. Information and the Interpretation of Quantum Mechanics. Springer, Berlin (2009)
Jammer, M.: The Philosophy of Quantum Mechanics: The Interpretation of Quantum Mechanics in Historical Perspective. Wiley, New York (1974)
Jaynes, T. In: Mandel, L., Wolf, E. (eds.) Coherence and Quantum Optics. Plenum, New York (1993)
Jaynes, T.: Probability Theory: The Logic of Science (personal edition online) (1995)
Jeffreys, H.: Theory of Probability. Clarendon Press, Oxford (1939)
Jones, S.: The Quantum Ten. Oxford University Press, Oxford (2008)
Keynes, M.: A Treatise on Probability. Macmillan, London (1921)
Khrennikov, A.: Contextual Approach to Quantum Formalism. Springer, Berlin (2009)
Kochen, S., Specker, E.P.: The problem of hidden variables in quantum mechanics. J. Math. Mech. **17**, 59 (1967)
Kolmogorov, N.: Foundations of the Theory of Probability. Chelsea, New York (1956)
Laloë, F.: Do We Really Understand Quantum Mechanics? Strange Correlations, Paradoxes and Theorems (2002). arXiv http://arxiv.org/abs/quant-ph/0209123v2quant-ph/0209123v2
Lamb Jr, W.E.: An operational interpretation of nonrelativistic quantum mechanics. Zeit. Phys. Today **22**(4), 23 (1969)
Lamb, W.E. Jr. In: Fujita, S. (ed.) The Ta-You Wu Festschrift. Gordon and Breach, London (1978)
Landé, A.: Foundations of Quantum Theory. Yale University Press, New Haven (1955)
Landé, A.: New Foundations of Quantum Mechanics. Cambridge University Press, Cambridge (1965)
Langevin, P.: La Notion de Corpuscle et d'Atome. Hermann, Paris (1934)
Laughlin, B.: A Different Universe. Basic Books, Cambridge (2005)
Lévy-Leblond, J.-M.: Les inegalités de Heisenberg, p. 14. Bull. Soc. Française de Physique (1973)
London, F., Bauer, E.: La théorie de l'observation en mècanique quantique. Hermann, Paris (1939)
Lucas, R.: The Concept of Probability. Clarendon, Oxford (1970)
Marchildon, L.: Why should we interpret quantum mechanics? Found. Phys. **34**, 1453 (2004). http://arxiv.org/abs/quant-ph/0405126arXiv:quant-ph/0405126, v2 30 Jul 2004
Margenau, H.: Philosophical problems concerning the meaning of measurement in physics. Phil. Sci. **25**, 23 (1958)
Margenau, H.: Physics and Philosophy: Selected Essays. Reidel, Dordrecht (1978)
*Marshall, T.W.: Random electrodynamics. Proc. Roy. Soc. A **276**, 475 (1963)
*Marshall, T.W.: Statistical electrodynamics. Proc. Camb. Phil. Soc. **61**, 537 (1965)
Maxwell, N.: Instead of particles and fields: a micro realistic quantum "Smearon" theory. Found. Phys. **12**, 607 (1982)

Maxwell, N.: Beyond "FAPP": three approaches to improving orthodox quantum heory and an experimental test. In: van der Merwe, A., et al. (eds.) Bell's Theorem and the Foundations of Modern Physics, p. 362. World Scientific, Singapore (1992)
Milonni, W.: The Quantum Vacuum. An Introduction to Quantum Electrodynamics. Academic Press, San Diego (1994)
Mott, F.: On teaching quantum phenomena. Contemp. Phys. **5**, 401 (1964)
Mückenheim, W., Ludwig, G., Dewdney, C., Holland, P.R., Kyprianidis, A., Vigier, J.P., Cufaro-Petroni, N., Bartlett, M.S., Jaynes, E.T.: A review of extended probabilities. Phys. Rep. **133**(6), 337 (1986)
Muga, J.G., Sala Mayato, R., Egusquiza, I.L. (eds.): Time in Quantum Mechanics. Lecture Notes in Physics, vol. 1, p. 734. Springer, Berlin (2008)
Nelson, E.: Derivation of the Schrödinger equation from Newtonian mechanics. Phys. Rev. **150**, 1079 (1966)
Newton, G.: How physics confronts reality: Einstein was correct, but Bohr won the game. Am. J. Phys. **48**, 1029 (1980)
Nilson, R.: Bibliography of the history and philosophy of quantum mechanics. In: Suppes, P. (ed.) Logic and Probability in Quantum Mechanics. Reidel, Dordrecht (1976)
Omnès, R.: The Interpretation of Quantum Mechanics. Princeton University Press, Princeton (1994a)
Omnès, R.: Quantum Philosophy. Princeton University Press, Princeton (1999b)
Patton, C.M., Wheeler, J.A. In: Isham, C.J., Penrose, P., Sciama, D.W. (eds.): Quantum Gravity. Clarendon, Oxford (1975)
Paul, H.: Quantum Theory. Cambridge University Press, Cambridge (2008)
Pauli, W.: Die allgemeinen Prinzipien der Wellenmechanik. In: Geiger, H., Scheel, K. (eds.) Handbuch der Physik, vol. 24. Springer, Berlin (1933). English translation: General Principles of Quantum Mechanics
Perez, A., Sahlmann, H., Sudarsky, D.: On the quantum origin of the seeds of cosmic structure. Class. Quantum Grav. **23**, 2317 (2006)
Petersen, A.: The philosophy of Niels Bohr. Bull. Atomic Scientist **19**, 8 (1963)
Popescu, S., Rohrlich, D.: Quantum nonlocality as an axiom (1994)
Popper, K.: Logik der Forschung, p. 1959. Springer, Vienna (1935). English translation: The Logic of Scientific Discovery. Basic Books, New York (1959)
Rédei, E., Szegedi, P. In: Bitsakis, E.I., Nicolaides, C.A. (eds.): The Concept of Probability. Kluwer, Dordrecht (1989)
Swinburne, R. (ed.): Nonlocality and peaceful coexistence. In: Space, Time, and Causality. Reidel, Dordrecht (1983)
Redhead, M.L.G.: Incompleteness, Nonlocality and Realism: A Prolegomenon to the Philosophy of Quantum Mechanics. Clarendon Press, Oxford (1987)
Reichenbach, H.: The Theory of Probability. University of California, Berkeley (1949)
Rigden, J.: Editorial comment. Am. J. Phys. **54**, 387 (1986)
Rovelli, C.: Relational quantum mechanics. Int. J. Theor. Phys. **35**, 1637 (1996)
Rylov, Y.A.: Spacetime distortion as a reason for quantum stochasticity. In: Garbaczewski, P., Wolf, M., Weron, A. (eds.) Chaos-The Interplay Between Stochastic and Deterministic Behaviour. Lecture Notes in Physics, vol. 457, p. 523. Springer, Berlin (1995). See also Quantum mechanics as a dynamic construction, http://hypress.hypres.com/rylov/yrylov.htm
*Santos, E.: Interpretation of the quantum formalism and Bell's theorem. Found. Phys. **21**, 221 (1991)
Santos, E.: On a heuristic point of view concerning the motion of matter: from random metric to Schrödinger equation. Phys. Lett. A **352**, 49 (2006)
Saunders, S., Brown, H.R.: The Philosophy of Vacuum. Clarendon Press, Oxford (1991)
Schrödinger, E.: Sur la théorie relativiste de l'électron et l'interpretation de la mécanique quantique. Ann. Inst. Henri Poincaré **2**, 269 (1932)

Shimony, A.: Metaphysical problems in the foundations of quantum mechanics. International Philosophical Quarterly **18**, 3 (1978)
Shimony, A.: Search for a worldview which can accommodate our knowledge of microphysics. In: Cushing, J.T., McMullin, E. (eds.) (1989)
Slater, C.: Physical meaning of wave mechanics. J. Franklin Inst. **207**, 449 (1929)
Sokolov, A.A., Tumanov, V.S.: The uncertainty relation and fluctuation theory. Soviet Phys. JETP **30**, 802 (1956, Russian version). **3**, 958 (1957, English translation)
Sokolov, A.A., Loskutov, Y.M., Ternov, I.M.: Qvantovaya mekhanika. MVO, Moscow (1962). English translation: Quantum Mechanics. Holt, Rinehart and Winston, New York (1966) (All sections related to the ensemble interpretation of quantum mechanics were omitted in the English translation)
Stapp, P.: S-matrix interpretation of quantum theory. Phys. Rev. D **3**, 1303 (1971)
Stapp, P.: The Copenhagen interpretation. Am. J. Phys. **40**, 1098 (1972)
Stenger, J.: The Unconscious Quantum: Metaphysics in Modern Physics and Cosmology. Prometheus Books, New York (2010)
Suppes, P., Zanotti, M.: When are probabilistic explanations possible? Synthèse **48**, 191 (1981)
Svozil, K.: On counterfactuals and contextuality. In: Adenier, G., Khrennikov, A.Y. (eds.) Foundations of Probability and Physics, vol. 3. CP 750, AIP, Melville (2005)
Tambakis, A.: On the empirical law of epistemology: physics as an artifact of mathematics. In: Ferrero, M., van der Merwe, A. (eds.) Proceedings of the International Symposium on Fundamental Problems in Quantum Physics. Kluwer, Dordrecht (1994)
't Hooft, G.: The obstinate reductionist's point of view on the laws of physics. In: Proceedings of Europa—Vision und Wirklichkeit, Europe (2002)
't Hooft, G.: Determinism beneath quantum mechanics. In: Elitzur, A., et al. (eds.) Quo Vadis Quantum Mechanics?, p. 99. Springer, Berlin (2005). http://arxiv.org/abs/quant-ph/0212095v1arXiv: quant-ph/0212095v1
't Hooft, G.: The mathematical basis for deterministic quantum mechanics. In: Nieuwenhuizen, T.M. et al. (eds.) Beyond the Quantum, vol. 3. World Scientific, Singapore. arXiv: http://arxiv.org/abs/quant-ph/0604008v2quant-ph/0604008v2
Valentini, A.: Inflationary cosmology as a probe of primordial quantum mechanics. Phys. Rev. D **82**, 063613 (2008). arXiv:hep-th/0805.0163
van Fraassen, B.: A formal approach to the philosophy of science. In: Colodny, R. (ed.) Paradigms and Paradoxes: The Philosophical Challenge of the Quantum Domain, vol. 303. University of Pittsburgh Press (1972)
van Fraassen, B.: The charybdis of realism: epistemological implications of Bell's inequality. In: Cushing, J.T., McMullin, E. (eds.) (1989)
van Kampen, N.G.: Ten theorems about quantum mechanical measurements. Phys. A **153**, 97 (1988)
Vedral, V.: Decoding Reality: The Universe as Quantum Information. Oxford University Press, Oxford (2010)
Venn, J.: On the diagrammatic and mechanical representation of propositions and reasonings. Philos. Mag. J. Sci., Ser. 5 **10**(59) (1880)
von Mises, R.: Probability, Statistics and Truth. George Allen and Unwin, London (1957)
von Neumann, J.: Mathematische Grundlagen der Quantenmechanik. Springer, Berlin (1932). English Translation: Mathematical Methods of Quantum Mechanics. Princeton University Press, Princeton (1955)
Weinberg, S.: Lectures on Quantum Mechanics. Cambridge University Press, Cambridge (2013)
Wheeler, A.: The 'Past' and the 'Delayed-Choice Double-Slit Experiment'. In: Marlow, A.R. (ed.) Mathematical Foundations of Quantum Theory, vol. 9. Academic Press, New York (1978)
Wheeler, J.A.: The quantum and the universe. In: Pantaleo, M., de Finis, F. (eds.) Relativity, Quanta and Cosmology II. Johnson Reprint Corporation, New York (1979)
Wheeler, A.: Law without law. In: Wheeler and Zurek, vol. 182 (1983)
Wheeler, A.: Information, physics, quantum: the search for links (1990)

Wheeler, J.A., Zurek, W.H. (eds.) Quantum Theory and Measurement. Princeton University Press, Princeton (1983)

Wick, D.: The Infamous Boundary. Seven Decades of Controversy in Quantum Physics. Birkhäuser, Boston (1995)

Wigner, P.: The problem of measurement. Am. J. Phys. **31**, 6 (1963)

Wilczek, F.: Four Big Questions with Pretty Good Answers, Palestra dada num Simpósio em Honra do 100° Aniversário de Heisenberg, Munique, Dez. (2001). http://arxiv.org/abs/hep-ph/0201222hep-ph/0201222

Chapter 2
The Phenomenological Stochastic Approach: A Short Route to Quantum Mechanics

> *Some physicists, among them myself, cannot believe that we must abandon, actually and forever, the idea of direct representation of physical reality in space and time...*
>
> A. Einstein (1954)

2.1 Why a Phenomenological Approach to Quantum Mechanics?

Independently of the interpretation embraced, randomness enters into the quantum description as a central and ever present ingredient.[1] Therefore, and leaving aside for the moment the problem of identifying the source of the randomness, a direct analysis of QM as a stochastic theory seems befitting as an introduction to the subject and in preparation for the chapters that follow. The phenomenological approach presented in this chapter is particularly suitable for such purposes, and in spite of its limitations, it reinforces the notion that a stochastic process underlies QM.[2]

The natural procedure to deal with a stochastic problem in physics involves a statistical treatment. However, any direct stochastic interpretation of QM faces immediately a fundamental difficulty, since QM is not a genuine statistical theory, as mentioned

[1] Not *quite independently* of the interpretation, strictly speaking. For example, Bell (1987, article 19) argues that "...the reversibility of the Schrödinger equation strongly suggests that quantum mechanics is not fundamentally stochastic in nature." This sentence sounds tempting... At this stage, how would you respond to it? (An answer is given at the end of the chapter.)

[2] Randomness plays an important role in several interpretations of QM, in addition to the one developed in this book. A case of major interest is Griffiths' theory of consistent histories, according to which most of the evolution is due to randomness. See Omnès (1994, 1999a, b), Griffiths (1996), Griffiths and Omnés (1999).

L. de la Peña et al., *The Emerging Quantum*,
DOI: 10.1007/978-3-319-07893-9_2,

in Chap. 1. This presents us with a quite discomforting alternative: (i) either a true statistical description of quantum systems is achieved, but then it cannot faithfully reproduce all of QM, or (ii) the stochastic model is constructed so as to reproduce the quantum behavior in detail, but then it will be endowed with some bizarre properties. The derivation of standard QM from a genuine stochastic theory, as developed in later chapters, will allow us to fully appreciate the importance of this assertion, since several approximations have to be made along the way, which result in a sui generis statistical description.

The stochastic theory presented in this chapter is based on a phenomenological model that goes along the second alternative and hence is completely in line with QM. As will become clear, this model has the advantage of its intuitive appeal and its simplicity. The fact that it is expressed in terms of dynamical variables that are not part of the standard quantum formalism, contributes to enrich the description by looking at the quantum phenomenon from a different angle, yet the phenomenological model also has some important shortcomings.

In developing the stochastic approach to QM a stochastic physical source with certain simple statistical properties is normally assumed to exist, but not identified. This lack of definition has given way to a variety of most dissimilar proposals, and even to the assumption that the hidden source does not exist at all and that the stochasticity is spontaneous, which of course leaves things as noncausal as they are without the need of another theory. In the context of the present chapter, it is a small sin to leave the source of stochasticity unspecified, since the rest of the book is devoted, to a large extent, to identify such source and to extract the consequences of its presence.

An important point to be made here is that nearly all forms of phenomenological approach to QM consider the stochastic process as a kind of Brownian motion. As will become evident, this is incorrect. Quantum stochasticity does *not* mean Brownian motion; at the phenomenologic level quantum and classical stochastic particles follow their own dynamical rules.

2.2 The Stochastic Description of Quantum Mechanics

Among the profuse arguments given in support of the notion of a stochastic process underlying QM, a couple of them go as follows (see also Comisar 1965; Hall and Collins 1971; Vasudevan et al. 2008). The first one, merely formal, is based on the analogy first observed by Schrödinger (1931, 1932) (and later by 1933) between his equation and the diffusion equation, which are related with one another by analytical continuation into imaginary times(see Sect. 4.4.5). The seed planted by Schrödinger expanded much later into a fuller theory in terms of Bernstein processes (Blanchard and Garbaczewski 1994; Jamison 1974; Zambrini 1986; see also Cramer 1986; Garbaczewski 1990, 1992, 1992, 1993a, 1993b, 1994, 1995.

A second interesting argument is that if a quantum particle is considered to follow a stochastic process in configuration space, the resulting Hausdorff fractal dimension of such process turns out to be the same as that of a Brownian particle (Abbott and Wise 1981). This fact is in itself interesting, although one can think with no less legitimacy of a phase-space description, and then the analogy breaks down. Many other particular reasons have been given in support of a stochastic approach to the quantum problem; for example, de Broglie (1967) felt compelled to introduce it in order to make the particle switch at random from one guiding wave to another (the guiding wave proposed by de Broglie is touched upon in Chap. 8).

The first relatively accomplished stochastic theory of the quantum process was proposed by the Hungarian physicist Fényes (1946, 1952) [which was strongly criticized by Nicholson (1954)], and further developed by Kershaw (1964) and Bess (1973), among others.[3] Féynes' theory is based on an ad hoc Lagrangian within a Brownian context; the author went as far as to recover a good part of the Hilbert-space formalism and concluded that QM describes an inherently stochastic phenomenon. But perhaps the most widely known theory of this kind, based on a non dissipative Markov process, is *stochastic mechanics*, initiated by Nelson in (1966) (and subsequent papers 1967–2013), and further developed by Guerra (1981, 1984, 1985, 1988), Guerra and Marra (1983), Guerra and Morato (1983), and Davidson (1978, 1979a, b, c, 1981, 2007). A common characteristic of this collection of works, at least during the earlier stages of development, was the identification of the underlying process as classical and of a Brownian nature. This led Jammer (1974, p. 418) to the statement: "The main objective of the stochastic interpretation of quantum mechanics has been to show that quantum theory is fundamentally a classical theory of probabilities or stochastic processes, and as such conceptually of the same structure as, say, the Einstein-von Smoluchowski theory of Brownian motion..." in our own words (de la Peña and Cetto 1982), "this sounds as astonishing and implausible as the complementary assertion would sound, namely, that Brownian motion is fundamentally a quantum theory conceptually of the same structure as the Schrödinger theory of the electron". And indeed, the need for a clear conceptual distinction between these two stochastic processes gave rise to a different branch of research, whose scope was also the development of a possible stochastic interpretation of QM, but on the basis that the quantum stochasticity is distinctly nonclassical, i.e., essentially different from Brownian motion. This theory, which has been called *stochastic quantum mechanics*, is the one that will be presented here.[4]

[3] At least two other stochastic proposals were made almost simultaneously to Féynes' work, by Novobátzky (1951) and Takabayasi (1952). A detailed account of the first developments of the stochastic approach to QM can be seen in Jammer (1974), Chap. 9.

[4] As noted earlier, Nelson calls the theory simply *stochastic mechanics*. His work is that of a mathematician and should be of major interest to the more mathematically inclined readers. There is another entirely different theory that goes under the same name *stochastic quantum mechanics*, pioneered by Prugovečki (1984, 1995) [see also Ali and Engliš (2005); Ali and Prugovečki (1986)]. It represents an attempt to unify physics into a rigorous quantum structure that considers a quantum spacetime and a universe which on the microscopic level follows a stochastic rather than deterministic evolution. Further, it should be noted that some authors speak of stochastic *quantum mechanics* while referring to Bohm's theory (see e.g. Feligioni et al. 2005).

A general feature of all these stochastic theories is their phenomenological nature; since they are aimed at reproducing QM, whether the process is considered classical or not, they are in principle unable to go *beyond* QM. In general, no specific assumption is made about the nature of the stochastic force, although, as indicated above, one can find the most varied suggestions in this regard, ranging from collisions with vacuum particles or *zerons*, interactions with a diversity of vacuum fields or even neutrinos, or a universal action reservoir (Lisi 2006), to fluctuations of the space-time metric (Santos 2006). The description may even be made compatible with the idea of an indeterministic electron, which is far from the realistic and causal persuasion that inspires the whole enterprise. Their phenomenological character is perhaps the strongest objection that can be made to these models, but taken at their face value they can be and indeed have been useful, at least because of the picture they suggest, and for several other reasons that will become evident in what follows.

2.3 Stochastic Quantum Mechanics

Our first task is to construct a theory of stochastic processes in configuration space that is sufficiently general (within the proper limits of the theory), so as to accommodate the quantum processes, assuming such a description is feasible. The (rather informal) exposition that follows, essentially based on de la Peña (1969), de la Peña and Cetto (1975, 1982, 1991, 1996) (we follow more closely the exposition in this last work), and Santos (1973), starts with the formulation of the appropriate kinematics. Different or complementary discussions can be seen in Nelson (1966, 2012), Guerra (1981), Blanchard et al. (1987), Kyprianidis (1992) and references therein.

2.3.1 Kinematics

Consider a particle undergoing a stochastic motion, so that its position $\boldsymbol{x}(t)$ constitutes a stochastic process. Thus, for each possible event (or rather, for each realization of the source of randomness, if any) a specific trajectory is followed, starting from the initial conditions. Assume that at a certain time t the particle is located at a point $\boldsymbol{x}$; at a slightly earlier time $t' = t - \Delta t$ it had a different position denoted by $\boldsymbol{x}'$, and similarly, at a slightly later time $t'' = t + \Delta t$ it will occupy the position $\boldsymbol{x}''$. For an arbitrary C_∞-function g of the stochastic variable $\boldsymbol{x}$ a Taylor series expansion, with

$$\Delta_+ \boldsymbol{x} = \boldsymbol{x}'' - \boldsymbol{x}, \quad \Delta_- \boldsymbol{x} = \boldsymbol{x} - \boldsymbol{x}', \tag{2.1}$$

gives the expression (a sum over repeated indices is understood)

$$\frac{g(\boldsymbol{x}'') - g(\boldsymbol{x}')}{2\Delta t} = \frac{\partial g}{\partial x_i}\frac{\Delta_+ x_i + \Delta_- x_i}{2\Delta t} + \frac{\partial^2 g}{\partial x_i \partial x_j}\frac{\Delta_+ x_i \Delta_+ x_j - \Delta_- x_i \Delta_- x_j}{4\Delta t} + \cdots \tag{2.2}$$

For a smooth (sure or nonstochastic) motion we can take the limit $\Delta t \to 0$, when this expression reduces to $dg(\boldsymbol{x})/dt = (\nabla g) \cdot (d\boldsymbol{x}/dt)$.[5] However, in the presence of stochasticity Δt cannot be taken arbitrarily small. The reason is that at the time scale of the 'instantaneous' description (i.e., according to the available experimental time resolution) the components $\Delta_\pm x_i$ for a given member of the ensemble may happen to be non differentiable (or changing very fast), due to abrupt kicks impressed by the random source. On the other hand, since $\Delta_\pm x_i(t)$ refers also to a stochastic variable, also the 'derivative' defined above becomes a random function. Nevertheless, it is possible to construct an approximate or phenomenological derivative, for small Δt, as follows.

The first change consists in averaging over all the possible events (or realizations of the background randomness), or rather over the ensemble of particles that reproduce all the possible trajectories. This operation is denoted by $\langle \cdot \rangle$, so that instead of $g(\boldsymbol{x}'') - g(\boldsymbol{x}')$ we will consider $\langle g(\boldsymbol{x}'') - g(\boldsymbol{x}') \rangle$. Next, the problem of taking the limit $\Delta t \to 0$ is solved by performing a moving averaging of the function $\boldsymbol{x}(t)$ during a 'small' time Δt,[6] much smaller than the characteristic time T_0 of the systematic (relevant) motions, but long enough for the particle to feel the effects of many blows from the stochastic source, so as to effectively smoothen out the most rapid changes in the instantaneous position. For example, in the case of Brownian motion the particle is so large compared with the solvent molecules that it receives a large number of molecular impacts during the time interval Δt, thus effectively averaging them into a (much) smoother function of time. These averaged quantities are the ones that obey the diffusion laws. Thus we choose

$$T_0 \gg \Delta t \gg t_c, \tag{2.3}$$

where t_c is appropiately selected so as to embrace many of the most closely spaced violent changes in each particular 'instantaneous' $\boldsymbol{x}(t)$. The resulting (coarse-time-scale) average time derivative or *systematic derivative* is denoted by the symbol $\mathcal{D}_c$; hence,

---

[5] By writing

$$\frac{g(\boldsymbol{x};t'') - g(\boldsymbol{x};t')}{2\Delta t} = \frac{1}{2\Delta t}\int_{t'}^{t'+2\Delta t} \frac{\partial g(\boldsymbol{x};s)}{\partial s} ds$$

it becomes clear that this expression is a coarse-grained time-derivative obtained by time-averaging the derivative $\partial g/\partial t$. This procedure mimicks the time smoothing produced by an observation, which is always extended in time. Such smoothing is particularly appropriate to deal with highly irregular (and even non-differentiable) functions.

[6] The moving average $x_{\Delta t}(t)$ of $x(t)$ is defined as $x_{\Delta t}(t) = (1/\Delta t)\int_t^{t+\Delta t} x(\tau) d\tau$.

$$\mathcal{D}_c g(\boldsymbol{x}) = \frac{\langle g(\boldsymbol{x}'') - g(\boldsymbol{x}')\rangle}{2\Delta t}, \tag{2.4}$$

with Δt such that Eq. (2.3) holds.

At this point it is necessary to make several assumptions about the properties of the stochastic motion. Considering the desired generality of the treatment and the lack of a specific model, these properties are unknown in principle; but for the cases of interest here it proves sufficient to assume that the stochasticity is due to a stationary, isotropic and homogeneous source; the second moments of $\Delta_{\pm}\boldsymbol{x}$ are then independent of the sign for equal signs, $\langle\Delta_+ x_i \Delta_+ x_j\rangle = \langle\Delta_- x_i \Delta_- x_j\rangle$ (up to terms of order Δt). Further, the fluctuations are assumed to be statistically independent, $\langle\Delta_+ x_i \Delta_- x_j\rangle = 0$ for all i and j. Each surviving second moment may have a contribution of order Δt due to the randomness of the motion, plus higher-order contributions,

$$\langle\Delta_+ x_i \Delta_+ x_j\rangle = 2D_{ij}(\boldsymbol{x}, t)\Delta t \ + \cdots, \quad \langle\Delta_- x_i \Delta_- x_j\rangle = 2D_{ij}(\boldsymbol{x}, t)\Delta t \ + \cdots \tag{2.5}$$

These expressions define (to zero order in Δt) the elements D_{ij} of the diffusion tensor as

$$D_{ij}(\boldsymbol{x}, t) = \frac{\langle\Delta_{\pm} x_i \Delta_{\pm} x_j\rangle}{2\Delta t}. \tag{2.6}$$

The difference $\langle\Delta_+ x_i \Delta_+ x_j\rangle - \langle\Delta_- x_i \Delta_- x_j\rangle$ is therefore of order higher than the first in Δt, and from Eqs. (2.2) and (2.4) we have to zero order in Δt (adding the contribution that may come from a possible explicit time dependence of g)

$$\mathcal{D}_c g(\boldsymbol{x}, t) = \left(\frac{\partial}{\partial t} + \boldsymbol{v} \cdot \nabla\right) g(\boldsymbol{x}, t), \tag{2.7}$$

with the *flux* or *systematic* (*convective*) *velocity* $\boldsymbol{v}(\boldsymbol{x}, t)$ given by

$$\boldsymbol{v}(\boldsymbol{x}, t) = \frac{\langle\Delta_+\boldsymbol{x} + \Delta_-\boldsymbol{x}\rangle}{2\Delta t} = \frac{\langle\boldsymbol{x}'' - \boldsymbol{x}'\rangle}{2\Delta t}. \tag{2.8}$$

This $\mathcal{D}_c$ coincides with the total time derivative of hydrodynamics. Note that application of Eq. (2.7) to each of the components x_i gives

$$\boldsymbol{v}(\boldsymbol{x}, t) = \mathcal{D}_c \boldsymbol{x}. \tag{2.9}$$

The systematic derivative defined above is only one of an infinite number of possible derivatives. Another one, equally important for what follows, is obtained by considering not the difference in (2.2), but the sum

$$\frac{g(\boldsymbol{x}'')+g(\boldsymbol{x}')}{2\Delta t} = \frac{2g(\boldsymbol{x})}{2\Delta t} + \frac{\partial g}{\partial x_i}\frac{\Delta_+ x_i - \Delta_- x_i}{2\Delta t} + \frac{\partial^2 g}{\partial x_i \partial x_j}\frac{\Delta_+ x_i \Delta_+ x_j + \Delta_- x_i \Delta_- x_j}{4\Delta t} + \cdots \tag{2.10}$$

This leads to the definition of the *stochastic derivative* of the function $g(\boldsymbol{x}, t)$, namely,

$$\mathcal{D}_s g(\boldsymbol{x}, t) = \frac{\langle g(\boldsymbol{x}'') + g(\boldsymbol{x}') - 2g(\boldsymbol{x})\rangle}{2\Delta t} \tag{2.11}$$

or

$$\mathcal{D}_s g(\boldsymbol{x}, t) = \frac{\partial g}{\partial x_i}\frac{\langle \Delta_+ x_i - \Delta_- x_i\rangle}{2\Delta t} + \frac{\partial^2 g}{\partial x_i \partial x_j}\frac{\langle \Delta_+ x_i \Delta_+ x_j + \Delta_- x_i \Delta_- x_j\rangle}{4\Delta t} + \cdots . \tag{2.12}$$

As for the first moments of the deviations of the coordinates entering into the expression for $\mathcal{D}_s g$, note that for a smooth motion the difference $(\Delta_+\boldsymbol{x} - \Delta_-\boldsymbol{x})$ is of order $(\Delta t)^2$; however, if there is a 'diffusion pressure', i.e., if the distribution of the instantaneous motions is inhomogeneous (there are more impacts per unit time from one side than from the other), the average of this difference may contain a term of order Δt. Therefore, we write

$$\boldsymbol{u} = \frac{\langle \Delta_+\boldsymbol{x} - \Delta_-\boldsymbol{x}\rangle}{2\Delta t} = \frac{\langle \boldsymbol{x}'' + \boldsymbol{x}' - 2\boldsymbol{x}\rangle}{2\Delta t} \tag{2.13}$$

and call $\boldsymbol{u}$ the *diffusive*, *stochastic*, or *osmotic velocity*. Collecting results, and neglecting again all higher-order terms, we get

$$\mathcal{D}_s g(\boldsymbol{x}, t) = u_i \frac{\partial g}{\partial x_i} + D_{ij}\frac{\partial^2 g}{\partial x_i \partial x_j}, \tag{2.14}$$

an equation that applied to $\boldsymbol{x}$ gives

$$\boldsymbol{u} = \mathcal{D}_s \boldsymbol{x}. \tag{2.15}$$

In what follows we consider the simple case of a diagonal, isotropic and uniform diffusion tensor $D_{ij} = D\delta_{ij}$, with D constant, so that Eq. (2.14) reduces to

$$\mathcal{D}_s g(\boldsymbol{x}, t) = \left(\boldsymbol{u}\cdot\nabla + D\nabla^2\right) g(\boldsymbol{x}, t). \tag{2.16}$$

The Markovian approximation made above (here *Markovian* means only retention of terms up to and including second-order moments) is by no means trivial and in each specific application its validity should be verified. However, it will prove sufficient and appropriate for the reproduction of the quantum description.

Note that neither $\boldsymbol{u}$ nor $\mathcal{D}_s$ exist in the *Newtonian limit*, i.e., for smooth motions in the absence of stochasticity. This allows us to define the Newtonian limit through

$$\text{Newtonian limit:} \quad \mathcal{D}_s \to 0 \quad \text{and} \quad \boldsymbol{u} \to 0. \tag{2.17}$$

In this limit of course $\mathcal{D}_c \to d/dt$, with the derivative taken along the flux of particles. As is now evident, by considering a sequence of time intervals previous to t' and following t'' it becomes possible to define as many different velocities as desired, and each additional one renders a more complete (but less local) statistical description of the motion. However, for the present purposes the two velocities $\boldsymbol{v}$ and $\boldsymbol{u}$ defined above happen to be sufficient. Yet certain linear combinations of them, as well as of the operators of time derivation $\mathcal{D}_c$ and $\mathcal{D}_s$, are particularly useful. Specifically, we have the *exit* and *access* combinations, denoted by the indices e and a, respectively [also called *forward* (+) and *backward* (−)]. The velocities and operators of interest are summarized as follows,

$$\mathcal{D}_e = \mathcal{D}_c + \mathcal{D}_s, \quad \mathcal{D}_a = \mathcal{D}_c - \mathcal{D}_s; \tag{2.18a}$$

$$\mathcal{D}_c = \frac{\partial}{\partial t} + \boldsymbol{v} \cdot \nabla, \quad \mathcal{D}_s = \boldsymbol{u} \cdot \nabla + D\nabla^2; \tag{2.18b}$$

$$\mathcal{D}_e = \frac{\partial}{\partial t} + \boldsymbol{v}_e \cdot \nabla + D\nabla^2, \quad \mathcal{D}_a = \frac{\partial}{\partial t} + \boldsymbol{v}_a \cdot \nabla - D\nabla^2; \tag{2.18c}$$

$$\boldsymbol{v}_e = \frac{\langle \boldsymbol{x}'' - \boldsymbol{x} \rangle}{\Delta t} = \boldsymbol{v} + \boldsymbol{u} = \mathcal{D}_e \boldsymbol{x}, \quad \boldsymbol{v}_a = \frac{\langle \boldsymbol{x} - \boldsymbol{x}' \rangle}{\Delta t} = \boldsymbol{v} - \boldsymbol{u} = \mathcal{D}_a \boldsymbol{x}; \tag{2.18d}$$

$$\boldsymbol{v} = \tfrac{1}{2} (\boldsymbol{v}_e + \boldsymbol{v}_a) = \mathcal{D}_c \boldsymbol{x}, \quad \boldsymbol{u} = \tfrac{1}{2} (\boldsymbol{v}_e - \boldsymbol{v}_a) = \mathcal{D}_s \boldsymbol{x}. \tag{2.18e}$$

Equation (2.18d) exhibits the access (exit) velocity $\boldsymbol{v}_a$ ($\boldsymbol{v}_e$) as the local velocity of the particles reaching (leaving) point $\boldsymbol{x}$ at time t. However, these refer to the (local) average values of $\Delta \boldsymbol{x}_{\pm}$; to get instantaneous expressions it is required to add to each increment the corresponding instantaneous deviation from their respective local mean value, i.e.,

$$\Delta_+ \boldsymbol{x} = \boldsymbol{v}_e \Delta t + \delta \boldsymbol{x}_+, \quad \Delta_- \boldsymbol{x}_i = \boldsymbol{v}_a \Delta t + \delta \boldsymbol{x}_-, \tag{2.19}$$

with $\delta \boldsymbol{x}_+$ and $\delta \boldsymbol{x}_-$ independent stochastic vector variables that average to zero.[7] In the absence of diffusion, $\boldsymbol{v}_a = \boldsymbol{v}_e$; but if there is diffusion, there may be more

[7] To reproduce the above results it is required that the second moment $\langle (\delta \boldsymbol{x}_+)^2 \rangle$ be proportional to Δt, so that $\langle (\Delta_+ \boldsymbol{x})^2 \rangle / \Delta t$ acquires a finite value [as demanded by Eq. (2.6)]. This is a characteristic

(or fewer) particles leaving than entering the neighborhood of $\boldsymbol{x}$ in a given small time interval, the difference $2\boldsymbol{u}$ being then a measure of the intensity of the diffusion [see Eq. (2.27) below].

An important feature of the velocities is their different behavior with respect to time reversal. A time-reversal operation $\hat{T}$ interchanges t' and t'', and thus also the points $\boldsymbol{x}'$ and $\boldsymbol{x}''$:

$$\hat{T}\boldsymbol{x}'' = \boldsymbol{x}', \quad \hat{T}\boldsymbol{x}' = \boldsymbol{x}''. \tag{2.20}$$

It follows from Eqs. (2.18d) and (2.18e) that

$$\begin{aligned} \hat{T}\boldsymbol{v}_e &= -\boldsymbol{v}_a; & \hat{T}\boldsymbol{v}_a &= -\boldsymbol{v}_e, \\ \hat{T}\boldsymbol{v} &= -\boldsymbol{v}; & \hat{T}\boldsymbol{u} &= \boldsymbol{u}, \end{aligned} \tag{2.21}$$

and similarly for the derivative operators,

$$\begin{aligned} \hat{T}\mathcal{D}_e &= -\mathcal{D}_a; & \hat{T}\mathcal{D}_a &= -\mathcal{D}_e, \\ \hat{T}\mathcal{D}_c &= -\mathcal{D}_c; & \hat{T}\mathcal{D}_s &= \mathcal{D}_s. \end{aligned} \tag{2.22}$$

The next step is to construct appropriate expressions for the acceleration; this can be readily achieved by applying a time derivation to a velocity. We have at our disposal two velocities and two time derivatives, which can be combined into four different accelerations. These accelerations and their corresponding behavior under time reversal are

$$\begin{aligned} \boldsymbol{a}_{cc} &= \mathcal{D}_c\boldsymbol{v} = \mathcal{D}_c\mathcal{D}_c\boldsymbol{x}; & \hat{T}\boldsymbol{a}_{cc} &= +\boldsymbol{a}_{cc}; \\ \boldsymbol{a}_{ss} &= \mathcal{D}_s\boldsymbol{u} = \mathcal{D}_s\mathcal{D}_s\boldsymbol{x}; & \hat{T}\boldsymbol{a}_{ss} &= +\boldsymbol{a}_{ss}; \\ \boldsymbol{a}_{cs} &= \mathcal{D}_c\boldsymbol{u} = \mathcal{D}_c\mathcal{D}_s\boldsymbol{x}; & \hat{T}\boldsymbol{a}_{cs} &= -\boldsymbol{a}_{cs}; \\ \boldsymbol{a}_{sc} &= \mathcal{D}_s\boldsymbol{v} = \mathcal{D}_s\mathcal{D}_c\boldsymbol{x}; & \hat{T}\boldsymbol{a}_{sc} &= -\boldsymbol{a}_{sc}. \end{aligned} \tag{2.23}$$

2.3.2 Spatial Probability Density and Diffusive Velocity

Let $\rho(\boldsymbol{x}, t)$ denote the probability density of particles in configuration space. This function satisfies the forward (exit) Fokker-Planck Eq. (Risken 1984),

$$\frac{\partial\rho}{\partial t} + \nabla\cdot\rho\boldsymbol{v}_e - D\nabla^2\rho = 0. \tag{2.24a}$$

(Footnote 7 continued)
feature of Brownian motion (or rather, of a white noise), and explains the extended reference to theories as the present one as 'Brownian-motion theories'.

The backward (access) Fokker-Planck equation can be obtained from (2.24a) by performing the substitution $t - \Delta t \to t + \Delta t$. This amounts to change the sign of the temporal derivate, and transforms $\boldsymbol{v}_e$ into $-\boldsymbol{v}_a$ [see the first line in Eq. (2.21)]. The resulting equation is

$$\frac{\partial \rho}{\partial t} + \nabla \cdot \rho \boldsymbol{v}_a + D\nabla^2 \rho = 0. \tag{2.24b}$$

By combining Eqs. (2.24a) and (2.24b) and using (2.18e) one is led to

$$\frac{\partial \rho}{\partial t} + \nabla \cdot \rho \boldsymbol{v} = 0, \tag{2.25}$$

$$\nabla \cdot \rho \boldsymbol{u} = D\nabla^2 \rho. \tag{2.26}$$

The first of these equations is the continuity equation expressing the local conservation of particles. The second one can be rewritten as $\nabla \cdot (\rho \boldsymbol{u} - D\nabla \rho) = 0$, which integrates into $\rho \boldsymbol{u} = D\nabla \rho + \nabla \times \boldsymbol{G}$, with $\boldsymbol{G}$ an arbitrary vector; however, by considering the balance of particles that go into and out of any small volume around a point $\boldsymbol{x}$ in space, it can be seen that one should take $\boldsymbol{G} = 0$ in general (de la Peña and Cetto 1969), which leads to the important formula[8]

$$\boldsymbol{u} = D\frac{\nabla \rho}{\rho} = D\nabla \ln \rho, \tag{2.27}$$

confirming that the motions described by $\boldsymbol{u}$ are due to diffusion, as discussed in relation with Eq. (2.13). This observation substantiates the selection $\boldsymbol{G} = 0$, since then $\boldsymbol{u}$ is due exclusively to the spatial changes in ρ. Notice that $\boldsymbol{u}$ can be rewritten alternatively in the form

$$\boldsymbol{u} = D\nabla \ln \left(\frac{\rho}{\rho_{\text{ref}}}\right), \tag{2.28}$$

where ρ_{ref} is any arbitrary reference constant value. Thus, the diffusive velocity does not depend on the scale of the density $\rho(\boldsymbol{x}, t)$. In Chap. 4 we will relate the coefficient D in Eq. (2.27) to the source of the fluctuations (namely the zero-point radiation field) and in doing so we will endow $\boldsymbol{u}$ with a deeper physical meaning.

---

[8] In the literature it is possible to find the velocity $\boldsymbol{u}$ defined with the sign reversed. Equation (2.27) can be recast into the form $\boldsymbol{j}_{\text{diff}} \equiv \boldsymbol{u}\rho = D\nabla \rho$, known as Fick's law (with due allowance for the reversed sign).

2.3.3 Dynamics

The lack of a specific model requires that we use a few basic arguments for the construction of the dynamics of stochastic mechanics. The best way is to opt for the most general law consistent with several obvious requirements. In the first place, one should expect the relationship between the (coarse grained) accelerations and the forces to be linear. The acceleration $\boldsymbol{a}$ must then be expressible as a linear combination of the previous four accelerations, Eq. (2.23),

$$\boldsymbol{a} = \lambda_1 \boldsymbol{a}_{cc} + \lambda_2 \boldsymbol{a}_{ss} + \lambda_3 \boldsymbol{a}_{cs} + \lambda_4 \boldsymbol{a}_{sc}, \tag{2.29}$$

where the λ's are constant parameters to be determined. Notice that this expression is not time-reversal invariant, since upon time inversion the last two terms reverse their sign, whereas the first two remain unchanged.

The total force acting on the particles can be represented as the sum of the external force and a stochastic force. In its turn this latter can be decomposed into two terms, namely the dissipative force (which embodies the systematic effects of the stochasticity on the particle), and the purely random force. The effective (locally averaged) force is thus composed of the external force plus a coarse-grained friction term. A force that depends only on the position should remain invariant with respect to time reversal, whereas velocity-dependent forces may change their sign under such operation. Therefore, if we decompose the net force $\boldsymbol{f}$ in the general form $\boldsymbol{f} = \boldsymbol{f}_+ + \boldsymbol{f}_-$, where $\hat{T}\boldsymbol{f}_\pm = \pm \boldsymbol{f}_\pm$, and assume a linear relation between forces and accelerations, it follows that the most general equations of motion acquire the form

$$\begin{aligned} m\,(\lambda_1 \boldsymbol{a}_{cc} + \lambda_2 \boldsymbol{a}_{ss}) &= \boldsymbol{f}_+, \\ m\,(\lambda_3 \boldsymbol{a}_{cs} + \lambda_4 \boldsymbol{a}_{sc}) &= \boldsymbol{f}_-, \end{aligned} \tag{2.30}$$

with m the mass of the particle. The parameters in these equations can be selected so as to adjust the theory to different purposes (see, e.g., Davidson 1978, 1979b; Nassar 1986a). In particular, in order to reproduce the quantum-mechanical description any friction term must be taken as zero (or considered negligible), since the dynamics as described by the Schrödinger equation is reversible. This situation differs substantially from the corresponding one in classical stochastic problems (say, of the Brownian-motion family), where the dissipative effects never cease and the long-term motions are purely stochastic.[9] Therefore, in the quantum case the source of stochasticity must be different from a white noise, since the total lack of coherence of the latter makes it unsuitable to sustain a systematic (mean) motion.

Considering first the conservative problem $\boldsymbol{f} = -\nabla V(\boldsymbol{x})$, for which $\boldsymbol{f}_- = 0$ (the case $\boldsymbol{f}_- \neq 0$ is straightforward and is considered in the next section), the second equation in (2.30) gives

[9] The hipothesis of a Brownian process without friction is just the most characteristic feature of Nelson's (1966, 1985a, 2012) theory.

$$\mathcal{D}_c \boldsymbol{u} + \gamma \mathcal{D}_s \boldsymbol{v} = 0, \tag{2.31}$$

where the two last expressions in Eq. (2.23) were used and we put $\gamma = \lambda_4/\lambda_3$. On the other hand, by taking the gradient of the continuity Eq. (2.25) and using Eqs. (2.7) and (2.16), one arrives after some simplifications at

$$\frac{\partial \boldsymbol{u}}{\partial t} + \nabla (\boldsymbol{u} \cdot \boldsymbol{v}) + D\nabla^2 \boldsymbol{v} = -D\nabla \times (\nabla \times \boldsymbol{v}). \tag{2.32}$$

From (2.27) it follows that $\nabla \times \boldsymbol{u} = 0$; further, in the conservative problem it is reasonable to reduce the description to the case $\nabla \times \boldsymbol{v} = 0$ (in the next section the more general problem with $\nabla \times \boldsymbol{v} \neq 0$ is reviewed). The above equation transforms thus into

$$\frac{\partial \boldsymbol{u}}{\partial t} + (\boldsymbol{u} \cdot \nabla) \boldsymbol{v} + (\boldsymbol{v} \cdot \nabla) \boldsymbol{u} + D\nabla^2 \boldsymbol{v} = 0, \tag{2.33}$$

which can be rewritten as

$$\mathcal{D}_c \boldsymbol{u} + \mathcal{D}_s \boldsymbol{v} = 0. \tag{2.34}$$

This equation is an alternative form of the continuity equation for the conservative problem. Comparison with Eq. (2.31) gives

$$\gamma = \frac{\lambda_4}{\lambda_3} = 1. \tag{2.35}$$

The results allow us to identify the second of Eq. (2.30) as a constraint on the system rather than a dynamical relation. It follows also that for $\nabla \times \boldsymbol{v} \neq 0$ the continuity equation will determine, via the second equation in (2.30), the appropriate expression for the force $\boldsymbol{f}_-$, which will no longer be null [see, for example, Eq. (2.50)].

The first of Eq. (2.30), on the other hand, is a true dynamical law, which can be recast into the form

$$m\lambda_1 (\boldsymbol{a}_{cc} - \lambda \boldsymbol{a}_{ss}) = \boldsymbol{f}_+, \tag{2.36}$$

with $\lambda = -\lambda_2/\lambda_1$. In the Newtonian limit, $\boldsymbol{a}_{ss} \to 0$ and $\boldsymbol{a}_{cc} = \mathcal{D}_c^2 \boldsymbol{x} \to d^2\boldsymbol{x}/dt^2$; therefore, to recover the correct classical limit one must take $\lambda_1 = 1$, and the equation of motion becomes

$$m (\mathcal{D}_c \boldsymbol{v} - \lambda \mathcal{D}_s \boldsymbol{u}) = \boldsymbol{f}_+. \tag{2.37}$$

This is the most general dynamical law allowed by the theory, under the principles adopted. Note that it contains a single free parameter, whose value will be discussed below. Notice also that for $\lambda \neq 0$ Eq. (2.37) differs from the classical equation of motion.

With the derivatives given by (2.18b), Eq. (2.37) reads explicitly

$$m\left[\frac{\partial \boldsymbol{v}}{\partial t}+(\boldsymbol{v}\cdot\nabla)\,\boldsymbol{v}-\lambda\,(\boldsymbol{u}\cdot\nabla)\,\boldsymbol{u}-\lambda D\nabla^2\boldsymbol{u}\right]=\boldsymbol{f}_+. \tag{2.38}$$

This equation was first proposed with the specific value $\lambda = 1$ in Nelson 1966. It can also be cast in the form

$$m\frac{d\boldsymbol{v}}{dt}=\boldsymbol{f}_+ + \boldsymbol{f}_{\text{diff}}, \tag{2.39}$$

with the time derivative taken along the mean motion (with respect to the flux velocity $\boldsymbol{v}$). The term $\boldsymbol{f}_{\text{diff}}$ stands for a force (additional to the external one) originating in the diffusive velocity,

$$\boldsymbol{f}_{\text{diff}} = m\lambda\boldsymbol{a}_{ss} = -\nabla V_{\text{diff}}, \tag{2.40}$$

with

$$V_{\text{diff}} = -\lambda\left(\frac{1}{2}m\boldsymbol{u}^2 + mD\nabla\cdot\boldsymbol{u}\right) = -\lambda\left(2mD^2\frac{\nabla^2\sqrt{\rho}}{\sqrt{\rho}}\right), \tag{2.41}$$

where Eq. (2.27) was used to write the second equality.

It should be borne in mind that the equation of motion (2.39) refers just to mean values in $\boldsymbol{x}$-space. For a description of the instantaneous motion of a (single) particle it is necessary to resort to the Langevin equation for the problem. In Nelson's (1967) theory such equation is proposed to correspond to a frictionless Brownian motion, which (in one dimension, for simplicity) reads

$$dx = (v+u)\,dt + \sqrt{2D}\Delta W(t), \tag{2.42}$$

where ΔW is taken as a Gaussian white noise (or Wiener process), $\langle\Delta W(t)\rangle = 0$, $\left\langle(\Delta W(t))^2\right\rangle = dt$ [compare with Eq. (2.19)]. The trajectories are now nowhere differentiable with probability one, so that some appropriate procedure is required to integrate the equation of motion. In Nelson's formulation the Ito definition of the time integral is used, which means that the increment $\Delta W(t)$ is interpreted as equal to $\Delta W(t) = W(t+dt) - W(t)$ for $dt > 0$. Details can be seen e.g. in Nelson's works; Gardiner (1983), and Vasudevan et al. (2008).

2.3.4 Integrating the Equation of Motion

By taking $\lambda_1 = 1$ and $\lambda_3 = \lambda_4$ in Eq. (2.30), and further observing that λ_3 can be absorbed in f_- by an appropriate redefinition (which is equivalent to taking $\lambda_3 = 1$), we write the law of motion and the constraint in the form

$$\begin{aligned} m\left(\mathcal{D}_c \boldsymbol{v} - \lambda \mathcal{D}_s \boldsymbol{u}\right) &= \boldsymbol{f}_+, \\ m\left(\mathcal{D}_c \boldsymbol{u} + \mathcal{D}_s \boldsymbol{v}\right) &= \boldsymbol{f}_-. \end{aligned} \tag{2.43}$$

Equations (2.43) with the derivatives given by (2.18b) look impressive: they form a system of coupled, nonlinear partial differential equations involving $\boldsymbol{v}$ and $\boldsymbol{u}$. However, this system has the remarkable property that it can be integrated (once), uncoupled, and linearized if expressed in terms of appropriate functions. This is achieved in several steps as follows.

From Eq. (2.38) and writing $\boldsymbol{f}_+$ generically as

$$\boldsymbol{f}_+ = -\nabla V + \boldsymbol{F}, \tag{2.44}$$

the first equation in (2.43) reads

$$m\left[\frac{\partial \boldsymbol{v}}{\partial t} + (\boldsymbol{v}\cdot\nabla)\,\boldsymbol{v} - \lambda\,(\boldsymbol{u}\cdot\nabla)\,\boldsymbol{u} - \lambda D\nabla^2\boldsymbol{u}\right] = -\nabla V + \boldsymbol{F}. \tag{2.45}$$

We now decompose $\boldsymbol{v}$ in the general form

$$\boldsymbol{v} = 2D\nabla S + \boldsymbol{b}, \tag{2.46}$$

with $S = S(\boldsymbol{x}, t)$ a (dimensionless) real function and $\boldsymbol{b}$ a vector containing any possible rotational contribution to $\boldsymbol{v}$. With the help of the identities

$$\begin{aligned} \frac{1}{2}\nabla \boldsymbol{w}^2 &= (\boldsymbol{w}\cdot\nabla)\,\boldsymbol{w} + \boldsymbol{w}\times(\nabla\times\boldsymbol{w}), \\ \nabla^2\boldsymbol{w} &= \nabla\,(\nabla\cdot\boldsymbol{w}) - \nabla\times(\nabla\times\boldsymbol{w}), \end{aligned}$$

Equation (2.45) rewrites as

$$\nabla\left[2mD\frac{\partial S}{\partial t} + \frac{1}{2}m\boldsymbol{v}^2 + V_{\text{diff}} + V\right] = \boldsymbol{F} - m\frac{\partial \boldsymbol{b}}{\partial t} + m\boldsymbol{v}\times(\nabla\times\boldsymbol{v}). \tag{2.47}$$

As for the second equation in (2.43), it reads explicitly

$$m\left[\frac{\partial \boldsymbol{u}}{\partial t} + (\boldsymbol{v}\cdot\nabla)\,\boldsymbol{u} + (\boldsymbol{u}\cdot\nabla)\,\boldsymbol{v} + D\nabla^2\boldsymbol{v}\right] = \boldsymbol{f}_-. \tag{2.48}$$

The identity

$$\nabla(\boldsymbol{u}\cdot\boldsymbol{v}) = (\boldsymbol{u}\cdot\nabla)\,\boldsymbol{v} + (\boldsymbol{v}\cdot\nabla)\,\boldsymbol{u} + \boldsymbol{u}\times(\nabla\times\boldsymbol{v}) + \boldsymbol{v}\times(\nabla\times\boldsymbol{u}),$$

together with Eq. (2.27), gives

$$mD\nabla\left[\frac{1}{\rho}\left(\frac{\partial\rho}{\partial t} + \nabla\cdot\rho\boldsymbol{v}\right)\right] = \boldsymbol{f}_- + m\boldsymbol{u}\times(\nabla\times\boldsymbol{v}) + mD\nabla\times(\nabla\times\boldsymbol{v}). \quad (2.49)$$

We thus see that, as expected, the equation $m\,(\mathcal{D}_c\boldsymbol{u} + \mathcal{D}_s\boldsymbol{v}) = \boldsymbol{f}_-$ imposes constraints via the continuity equation, which causes the term within square brackets to vanish. This fixes the force $\boldsymbol{f}_-$ as

$$\boldsymbol{f}_- = -m\boldsymbol{u}\times(\nabla\times\boldsymbol{v}) - mD\nabla\times(\nabla\times\boldsymbol{v}). \quad (2.50)$$

Now we come back to the dynamical Eq. (2.47) and assume that the additional force $\boldsymbol{F}$ in Eq. (2.44) refers to a Lorentz force due to an external electromagnetic potential $\boldsymbol{A}$,

$$\boldsymbol{F} = -\frac{e}{c}\frac{\partial\boldsymbol{A}}{\partial t} + \frac{e}{c}\boldsymbol{v}\times(\nabla\times\boldsymbol{A}). \quad (2.51)$$

In this case the flow velocity becomes

$$\boldsymbol{v} = 2D\nabla S - \frac{e}{mc}\boldsymbol{A}, \quad (2.52)$$

thus fixing $\boldsymbol{b} = -(e/mc)\boldsymbol{A}$. This value for $\boldsymbol{b}$, together with Eq. (2.51), implies that the right-hand side of Eq. (2.47) vanishes, so the equation reduces to

$$\nabla\left[2mD\frac{\partial S}{\partial t} + \frac{1}{2}m\boldsymbol{v}^2 + V_{\text{diff}} + V\right] = 0. \quad (2.53)$$

Further, introduction of Eq. (2.52) into (2.50) gives a diffusion-dependent Lorentz-force term that changes sign under time reversal. In the Coulomb gauge (i.e., taking $\nabla\cdot\boldsymbol{A} = 0$) this force is given by

$$\boldsymbol{f}_- = \frac{e}{c}\boldsymbol{u}\times(\nabla\times\boldsymbol{A}) + D\frac{e}{c}\nabla^2\boldsymbol{A}. \quad (2.54)$$

Now we are in position to integrate the dynamical Eq. (2.53). The result, after absorbing into S the arbitrary function of time that arises from the integration, is

$$2mD\frac{\partial S}{\partial t} + \frac{1}{2}m\boldsymbol{v}^2 - \lambda 2mD^2\frac{\nabla^2\sqrt{\rho}}{\sqrt{\rho}} + V = 0. \quad (2.55a)$$

This, together with the continuity equation

$$\frac{\partial \rho}{\partial t} + \nabla \cdot \rho \boldsymbol{v} = 0, \tag{2.55b}$$

with $\boldsymbol{v}$ given by (2.52), constitutes a pair of differential equations for the variables ρ and S that is equivalent to the original pair (2.43), and bears the dynamical information for an ensemble of particles subject to the conservative force $-\nabla V$ and immersed in an external electromagnetic field represented by the vector potential $\boldsymbol{A}$. Equations (2.55a) and (2.55b) uncouple if we perform a transformation that takes ρ and S to new variables w_+ and w_-, such that

$$\begin{aligned} w_+ + w_- &= \ln \rho, \\ w_+ - w_- &= 2\frac{S}{\sqrt{-\lambda}}. \end{aligned} \tag{2.56}$$

Direct substitution leads, after some algebra, to the pair of separated equations

$$\begin{aligned} -2m\zeta\frac{\partial w_+}{\partial t} &= -2m\zeta^2\left[(\nabla w_+)^2 + \nabla^2 w_+\right] + V - \\ &-\zeta\frac{e}{c}\left(2\boldsymbol{A}\cdot\nabla w_+ + \nabla\cdot\boldsymbol{A}\right) + \frac{e^2}{2mc^2}\boldsymbol{A}^2; \end{aligned} \tag{2.57}$$

$$\begin{aligned} +2m\zeta\frac{\partial w_-}{\partial t} &= -2m\zeta^2\left[(\nabla w_-)^2 + \nabla^2 w_-\right] + V + \\ &+\zeta\frac{e}{c}\left(2\boldsymbol{A}\cdot\nabla w_- + \nabla\cdot\boldsymbol{A}\right) + \frac{e^2}{2mc^2}\boldsymbol{A}^2, \end{aligned} \tag{2.58}$$

with $\zeta = D\sqrt{-\lambda}$. As a final step, this system of equations can be linearized by introducing the further change of functions

$$\psi_+ = \exp w_+ = \exp\left(\ln\sqrt{\rho} + \frac{S}{\sqrt{-\lambda}}\right) = \sqrt{\rho}\exp\left(\frac{S}{\sqrt{-\lambda}}\right), \tag{2.59a}$$

$$\psi_- = \exp w_- = \exp\left(\ln\sqrt{\rho} - \frac{S}{\sqrt{-\lambda}}\right) = \sqrt{\rho}\exp\left(-\frac{S}{\sqrt{-\lambda}}\right), \tag{2.59b}$$

thus obtaining

$$-2mD\sqrt{-\lambda}\frac{\partial \psi_+}{\partial t} = \frac{1}{2m}\left(2mD\sqrt{-\lambda}\nabla - \frac{e}{c}\boldsymbol{A}\right)^2\psi_+ + V\psi_+, \tag{2.60a}$$

$$2mD\sqrt{-\lambda}\frac{\partial \psi_-}{\partial t} = \frac{1}{2m}\left(-2mD\sqrt{-\lambda}\nabla - \frac{e}{c}\boldsymbol{A}\right)^2\psi_- + V\psi_-. \tag{2.60b}$$

Further, Eqs. (2.59a, 2.59b) give

$$\rho = \psi_+ \psi_-, \tag{2.61}$$

and the velocities $\boldsymbol{v}$ and $\boldsymbol{u}$ rewrite as

$$\boldsymbol{v} = D\sqrt{-\lambda}\nabla \ln \frac{\psi_+}{\psi_-} - \frac{e}{mc}\boldsymbol{A} = D\sqrt{-\lambda}\left(\frac{\nabla\psi_+}{\psi_+} - \frac{\nabla\psi_-}{\psi_-}\right) - \frac{e}{mc}\boldsymbol{A}, \tag{2.62}$$

$$\boldsymbol{u} = D\nabla \ \ln \psi_+\psi_- = D\left(\frac{\nabla\psi_+}{\psi_+} + \frac{\nabla\psi_-}{\psi_-}\right). \tag{2.63}$$

In the absence of external electromagnetic field ($\boldsymbol{A} = 0$), $\boldsymbol{v}$ reduces to

$$\boldsymbol{v} = 2D\nabla S.$$

The function $S(\boldsymbol{x}, t)$ represents therefore a velocity potential, or a kind of statistical action function, the gradient of which gives the momentum associated with the systematic (mean local) velocity.

2.3.5 Quantum and Classical Stochastic Processes

Equations (2.60a, 2.60b) apply to any system that can be described by the present stochastic treatment, subject to the free (though nontrivial) choice of D and λ. This exhibits at once the strength and the weakness of the procedure. For on the one hand, an appropriate selection of the parameters leads to a Schrödinger-like description of the stochastic system; but on the other hand, the equation thus obtained is quite unspecific and the 'appropriate' selection of the parameters seems quite arbitrary. A complete theory should allow for an unambiguous derivation of both λ and the coefficients D_{ij} (not necessarily constant nor diagonal in the more general case), from first principles. This is beyond reach for the present phenomenological approach, due first and foremost to the nonspecificity of the random field. In Sect. 2.4 an argument is given that helps to perceive the generality (and arbitrariness) of Schrödinger-like equations, and to realize that the selection of the parameters is a matter of no minor importance. At this stage we just briefly explore the possible applications of the results just derived.[10]

Notice that in the integrated Eqs. (2.60a, 2.60b) the free parameter is the product $D\sqrt{-\lambda}$, not each factor separately. One may therefore consider that D takes care of

[10] In (Davidson 1979b and 2001) an interesting, slightly different selection of the parameters is discussed, which reproduces the *classical* nonlinear Schrödinger equation, derived in Sect. 4.5.5 of this book. See also (Bacciagaluppi 2011).

the scale, whereas the relevant property of λ is its sign, so that one can take $\lambda = \pm 1$. This leads to essentially two different theories, according to the sign of λ.

2.3.5.1 The Parabolic Solution

Take first $\lambda = -1$. In this case Eqs. (2.60a, 2.60b) are parabolic, the functions ψ_+ and ψ_- are both real and the process described by them is irreversible. This theory can be used to describe classical Markov processes, if due allowance is made for the unbalanced friction force, which can be introduced via an expression such as $f_{\rm f} = -\beta \boldsymbol{v}_e$ with $\boldsymbol{v}_e = 2D(\nabla\psi_+/\psi_+)$, or the like (see e.g. Cetto 1972). However, with the introduction of such a term the theory ceases to be linear and it is then simpler to go back to the (linear) Fokker-Planck equation. Moreover, it happens that the values of both the friction parameter β and the diffusion coefficient D are problem-specific; the single way out of this situation is the fluctuation-dissipation relation, when it is at hand. As is well known, the presence of the friction force, together with the incoherence of the noisy background, leads to purely noisy solutions for $t \to \infty$. With all these drawbacks, a procedure as the present one seems to be of little help, if any, for such problems. Further elaborations can be seen in de la Peña and Cetto (1975), Skagerstam (1977), Nassar (1986a, b) and references therein.

2.3.5.2 The Hyperbolic Solution

Take now $\lambda = 1$. In this case Eqs. (2.59a, 2.59b) give $\psi_- = \sqrt{\rho}\exp(iS) = \psi_+^*$, and Eqs. (2.60a, 2.60b) become hyperbolic and each other's complex conjugate. The process is therefore reversible. The Schrödinger equation is obtained with the selection

$$D = \frac{\hbar}{2m} \tag{2.64}$$

for the diffusion coefficient. It acquires the nature of a wave equation thanks to the factor i in front of the first derivative with respect to time, which mimics a second-order time derivative, as discussed in Sect. 4.4.5.[11] It seems reasonable to ask why the theory should predict an undulatory behavior, when the whole treatment has been made in terms of particles that follow (deterministic) trajectories. These matters will be briefly discussed in Sect. 2.5.1 below (see also Chap. 9). In any case, we see that $\lambda = +1$ is the back door through which undulatory aspects enter the theory.

It should be stressed that the selection (2.64) is far from obvious or natural; there is not an a priori reason to assume that the tensor mD has a universal value, independent of the specific problem (this point is discussed in Sect. 2.4). Given the phenomenological character of the present theory, this formula enters as an empirical selection,

[11] Although giving rise to some bizarre wave phenomena; for comments and examples of this see e.g. Ballentine 1990, 1998.

although there have been of course some attempts to justify it from fundamental considerations (see e.g. de Broglie 1967) so as to ground the theory on a more solid basis. Since the problem of identifying the noise source behind the assumed stochastic process is left open in this kind of approach, that of *deriving* the detailed form and value of the diffusion tensor and the constant λ remains open as well.

A key point of the present stochastic approach is its adequacy to distinguish classical from quantum stochastic processes, which have become described by essentially different equations, even if at first sight the corresponding Langevin equations seem to be quite similar. One should therefore speak not of a Brownian analog of QM (as is so frequently done), but of a quantum stochastic process in itself. Insufficient attention to this crucial point is the cause of much confusion in the literature. More specifically, according to Eq. (2.37), the accelerations $\boldsymbol{a}_B$ for the classical (Brownian) case ($\lambda = -1$) and $\boldsymbol{a}_Q$ for the quantum system ($\lambda = 1$) are, respectively,

$$\boldsymbol{a}_B = \mathcal{D}_c \boldsymbol{v} + \mathcal{D}_s \boldsymbol{u}, \quad \boldsymbol{a}_Q = \mathcal{D}_c \boldsymbol{v} - \mathcal{D}_s \boldsymbol{u}. \tag{2.65}$$

Similarly, Eqs. (2.39)–(2.41) show that the sign of the extra potential V_{diff} is essential in determining the different dynamics. In the quantum case, V_{diff} (with $D = \hbar/2m$) becomes the so-called *quantum potential* V_Q, and $\boldsymbol{f}_{\text{diff}} = \boldsymbol{f}_Q$ can then be interpreted (in the language proper of Bohm's theory discussed in Chaps. 4 and 8) as a quantum force. From Eq. (2.41) we see that this extra potential is intimately related to the diffusion. Further, being due solely to the spatial variations in the density of particles ρ, it introduces a nonlocal ingredient into the description, since the probability density ρ contains information about the entire setup. This point will reappear in several of the following chapters, particularly in Chap. 8.

2.4 On Schrödinger-Like Equations

According to the exposition in Sect. 2.3.5, deriving a Schrödinger-like equation would seem to be quite an easy matter. However, as noted above, the proper selection of the parameters λ and D—which is crucial to obtain quantization—is by no means trivial. Thus a *true* derivation of a quantum equation of motion requires more than arriving at a Schrödinger-*like* equation, it requires also deriving the value of the parameters involved. To make this point clear, suffice it to recall the following alternative way of 'deriving' the Schrödinger equation, based merely on general arguments of a statistical nature. The sole intention of this example is to elaborate on the general relationship between the Schrödinger equation and a simple stochastic description in configuration space.

The starting point is the continuity Eq. (2.25) for the density of particles. Assume the flux to be laminar and write the drift (systematic) velocity $\boldsymbol{v}$ in terms of a velocity potential (an 'action' S) according to

$$\boldsymbol{v} = \frac{a}{m} \nabla S, \tag{2.66}$$

with the parameter a so selected as to make S a dimensionless function of $\boldsymbol{x}$ and t. A change of variables from ρ, S to a new complex pair ψ, ψ^*, defined as

$$\psi = \sqrt{\rho} e^{iS}, \quad \psi^* = \sqrt{\rho} e^{-iS}, \tag{2.67}$$

gives

$$\rho = \psi^* \psi, \quad \boldsymbol{v} = \frac{ia}{2m} \nabla \left(\ln \psi^* - \ln \psi \right). \tag{2.68}$$

The continuity equation thus transforms into

$$\psi^* \left(i \frac{\partial \psi}{\partial t} + \frac{a}{2m} \nabla^2 \psi \right) - \psi \left(-i \frac{\partial \psi^*}{\partial t} + \frac{a}{2m} \nabla^2 \psi^* \right) = 0. \tag{2.69}$$

At this point a separating real function U is introduced so that

$$\psi^* \left(i \frac{\partial \psi}{\partial t} + \frac{a}{2m} \nabla^2 \psi \right) = \psi \left(-i \frac{\partial \psi^*}{\partial t} + \frac{a}{2m} \nabla^2 \psi^* \right) = U \psi \psi^*. \tag{2.70}$$

As a result, Eq. (2.69) becomes separated into

$$i \frac{\partial \psi}{\partial t} = -\frac{a}{2m} \nabla^2 \psi + U \psi \tag{2.71}$$

and its complex conjugate. The procedure can be applied to any mechanical system obeying the continuity equation (with a laminar flow), and hence to classical or quantum particles alike (de la Peña 1967). However, two problems remain, namely the determination of the parameter a and the function U, which may depend on $\boldsymbol{x}$, t *and even* on ψ and ψ^*. In what follows we proceed to determine the function U. See also Kracklauer (1992), and de la Peña and Cetto (1993) for other determinations of U.

We start by combining the expression (2.68) for $\boldsymbol{v}$ with Eq. (2.71), to obtain

$$\frac{\partial \boldsymbol{v}}{\partial t} = \frac{a^2}{4m^2} \nabla \left(\frac{1}{\psi^*} \nabla^2 \psi^* + \frac{1}{\psi} \nabla^2 \psi \right) - \frac{a}{m} \nabla U. \tag{2.72}$$

On the other hand, $\nabla \ln \psi = (1/2) \nabla \ln \rho + i \nabla S$, whence

$$\frac{\nabla^2 \psi}{\psi} = \frac{\nabla^2 \sqrt{\rho}}{\sqrt{\rho}} + i \nabla^2 S + i \nabla S \cdot \frac{\nabla \rho}{\rho} - (\nabla S)^2. \tag{2.73}$$

This latter equation, when introduced into (2.72), leads to

$$\frac{\partial \boldsymbol{v}}{\partial t} = \nabla \left(\frac{a}{2m} \nabla \cdot \boldsymbol{u} + \frac{1}{2} \boldsymbol{u}^2 \right) - \frac{1}{2} \nabla \boldsymbol{v}^2 - \frac{a}{m} \nabla U, \tag{2.74}$$

with

$$\boldsymbol{u} = \frac{a}{2m} \nabla \ln \rho. \tag{2.75}$$

Since $\nabla \times \boldsymbol{v} = 0$, it follows that $\nabla \boldsymbol{v}^2 = 2 (\boldsymbol{v} \cdot \nabla) \boldsymbol{v}$, which allows to rewrite Eq. (2.73) in terms of the total time derivative along the trajectory $d\boldsymbol{v}/dt = (\partial \boldsymbol{v}/\partial t) + (\boldsymbol{v} \cdot \nabla) \boldsymbol{v}$, as

$$m \frac{d\boldsymbol{v}}{dt} = -\nabla \left(-\frac{a}{2} \nabla \cdot \boldsymbol{u} - \frac{1}{2} m \boldsymbol{u}^2 \right) - a \nabla U. \tag{2.76}$$

On the other hand, combining Eqs. (2.39)–(2.41) [with $D = a/2m$, in accordance with Eqs. (2.27) and (2.75)] one obtains

$$m \frac{d\boldsymbol{v}}{dt} = -\lambda \nabla \left(-\frac{a}{2} \nabla \cdot \boldsymbol{u} - \frac{1}{2} m \boldsymbol{u}^2 \right) + \boldsymbol{f}_+. \tag{2.77}$$

For the conservative case ($\boldsymbol{f}_+ = -\nabla V$), comparison of the last two equations implies that

$$(1 - \lambda) \nabla \left(-\frac{a}{2} \nabla \cdot \boldsymbol{u} - \frac{1}{2} m \boldsymbol{u}^2 \right) = \nabla (V - aU). \tag{2.78}$$

Integration of this expression gives the solution

$$aU = V + (1 - \lambda) V_a + h(t), \tag{2.79}$$

with $h(t)$ an arbitrary function of time that can be taken as zero without loss of generality, and

$$V_a = \left(\frac{1}{2} m \boldsymbol{u}^2 + \frac{1}{2} a \nabla \cdot \boldsymbol{u} \right) = \frac{a^2}{2m} \frac{\nabla^2 \sqrt{\rho}}{\sqrt{\rho}}. \tag{2.80}$$

Notice that according to Eq. (2.41), $V_{\text{diff}} = -\lambda V_a$, since $D = a/2m$.

We now introduce Eq. (2.79) into (2.71), thus obtaining

$$ia \frac{\partial \psi}{\partial t} = -\frac{a^2}{2m} \nabla^2 \psi + V\psi + (1 - \lambda) V_a \psi. \tag{2.81}$$

The result just obtained evinces the distinctive nature of the case $\lambda = 1$: this is the single value of λ that linearizes Eq. (2.81), and transforms it into a Schrödinger equation,

$$ia\frac{\partial \psi}{\partial t} = -\frac{a^2}{2m}\nabla^2\psi + V\psi. \tag{2.82}$$

The value $\lambda = -1$, corresponding to the Brownian case as discussed above, leads to a Schrödinger-like equation but with a total potential $V + 2V_a$. Further, the value $\lambda = 0$ gives a net potential $V + V_a$. This case corresponds to a classical equation of motion devoid of stochasticity, but allows distributed velocities, since, according to the discussion following Eq. (2.36), the condition $\lambda = 0$ is equivalent to taking the Newtonian limit. As discussed in Sect. 4.5.5, Eq. (2.81) with $\lambda = 0$ is formally equivalent to a field theory for ψ with a classical Lagrangian.

Leaving aside the problem of justifying the selection $\lambda = 1$ to arrive at (2.82), we observe that the parameter a fixes the scale of the action aS, and is therefore, in principle, problem-dependent. Hence the a priori identification of Eq. (2.82) with the Schrödinger equation containing a universal constant a, is *not* warranted. This observation explains the singular role played by Schrödinger's equation in quantum theory, and therein resides a specific feature of quantum systems. Whereas in the classical case the value of the action integrals is determined by the initial conditions (whence a becomes highly arbitrary), in the quantum case this parameter becomes fixed in a more fundamental way: it is the initial conditions what are conditioned by the parameter, and determined so as to comply with an energy-balance condition, as will be shown in Chap. 4.[12]

Notwithstanding its importance, this discussion is frequently overlooked in the literature, characteristically in many published attempts to present variants of the above procedure as *bona fide* derivations of the Schrödinger equation from classical arguments. Still, doubts have been cast on the phenomenological stochastic theory, in particular on the legitimacy of the demand of single-valuedness on the wave solution, as discussed in Sect. 2.6.

Some 15 years ago, Yves Couder and his colleagues discovered the *bouncer*, a macroscopic particle (a small drop of silicon oil) that can be made to dance over the surface of a vertically vibrated bath of the same fluid. By increasing the peak acceleration of the vibrations, the droplet can be made to self-propel with constant speed. With this arrangement they have observed a variety of behaviors of the droplet that have a striking similarity with the wavelike behavior of quantum particles (see e.g. Couder and Fort 2006; Couder et al. 2005; Wind-Willassen et al. 2013). A detailed study by Brady and Anderson (2013) has revealed that this macroscopic hydrodynamic system can indeed be described by Eq. (2.82), with the parameter a appropriately selected for the specific system (and hence *not universal*), so that it can be taken as a close mechanical model of the quantum behavior. Here the vibrating oil bath is representing the substratum, which by its interplay with the particle generates a quantum-like dynamics, including quantization of orbital motions!

---

[12] For example, when solving the Heisenberg equations of motion for x and p, the initial conditions are given by matrices, which guarantees that the Heisenberg inequalities are satisfied starting from $t = 0$.

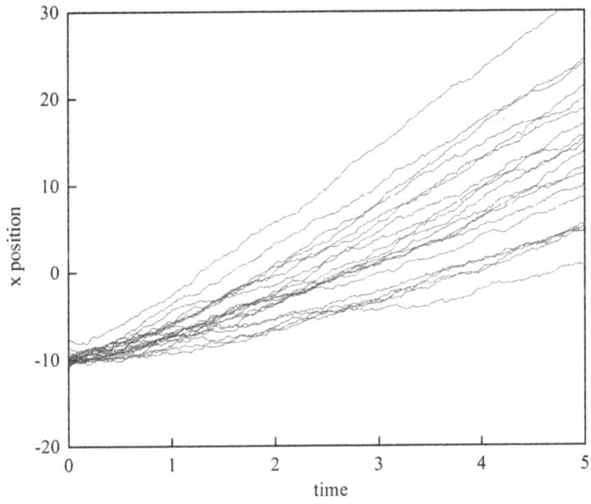

Fig. 2.1 Trajectories of a Gaussian wave packet made of 20 free particles with fixed energy, according to Eq. (2.42). Obtained by numerical simulation. Courtesy of Bárbara Moreno Munguía)

2.5 Stochastic Quantum Trajectories

Let us return to stochastic quantum mechanics. In this phenomenological approach, the ensemble is composed of particles (localized entities), although their statistical description appears encoded in Eq. (2.82), which has formally acquired the nature of a wave equation. One can therefore go back to the equation of motion (2.42) and use it to find individual trajectories, by means of numerical simulation. Compared to the corresponding calculations usually carried out in the framework of Bohm's theory (see Chap. 8), which resort to the *mean local* velocity $\boldsymbol{v}(\boldsymbol{x}, t)$ (though in that approach they are seen as referring to the actual velocity of a single particle), these ones are somewhat more elaborate. The trajectories here obtained follow more closely the *instantaneous* motions, therefore they show more detail and provide extra information about the quantum dynamics. Figures 2.1 and 2.2 (similar to those in Moreno Murguía 2006), illustrate the results obtained for a Gaussian wave packet containing 20 particles, in the first case moving freely, in the second one impinging upon a narrow semitransparent barrier centered at the origin. The presence of rapid fluctuations—absent in Bohm's description and merely implicit in the quantum description—is conspicuous.

An analysis of Fig. 2.2 reveals several interesting aspects of the dynamics. The majority of the particles are reflected by the barrier, although an important fraction of them cross it and some remain inside for a relatively long time, going to and fro, until they escape in one direction or the other. This is particularly interesting because it shows that it is legitimate to speak of real particles in motion 'inside' the barrier. A most remarkable peculiarity displayed by the trajectories is the nonlocality of their behavior, as is further discussed in Chap. 8. It is clear from Fig. 2.2 that long before reaching the barrier, the particles already 'feel' its presence and start modifying their energy, either losing or gaining some, even enough to 'jump over' the barrier in some

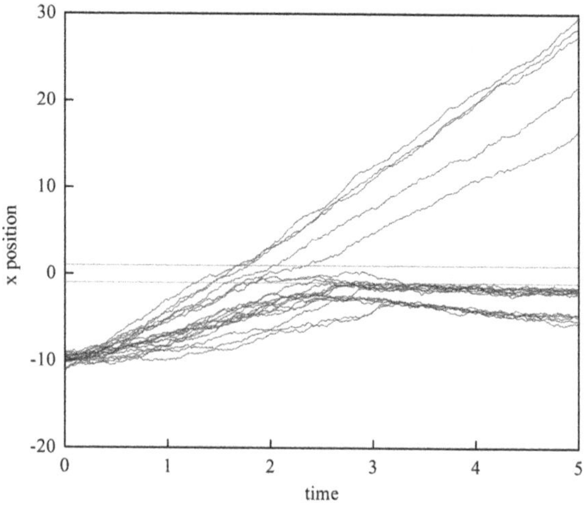

Fig. 2.2 The same packet as in Fig. 2.1, now in the presence of a barrier of width 1 and height 10 (arbitrary units), represented by a pair of fine *horizontal lines*. The energy of incidence is 1. Courtesy of Bárbara Moreno Murguía

cases (constant energy corresponds to a constant slope of the trajectories). Beyond the barrier the energy (hence the velocity) of the particles remains stable in statistical terms, and close to its original value. This applies also to the reflected particles, which rapidly tend to move as free particles. Both figures also clearly show that typically the trajectories intersect, due to the presence of stochasticity, in contrast with the prediction derived from Bohm's description (see e.g. Holland 1993). Figure 2.2 also hints at the effects of the interference of the incident and the reflected packets, which gives rise to diffraction.

2.5.1 Wavelike Patterns

A remarkable property of quantum particles is of course their wavelike behavior. At first it might seem counterintuitive to expect from a stochastic mechanical formulation as the one developed here, to reproduce the undulatory behavior of particles. On the other hand, as said below Eq. 2.64, we have arrived at a wave equation for describing the dynamics of the ensemble. In addition, a stochastic model that reproduces QM must account for the wavelike features. That this is so has been confirmed with the help of various numerical simulations, similar to the ones carried out for the semitransparent barrier mentioned in the previous section. One such example is presented in Fig. 2.3, taken from McClendon and Rabitz (1988). This figure shows the fringe pattern obtained for a 'wave packet' of several thousand particles emerging from two Gaussian slits, obtained by numerical integration within stochastic mechanics. The result compares well, statistically speaking, with that obtained from a quantum-mechanical calculation. A most important observation made with this

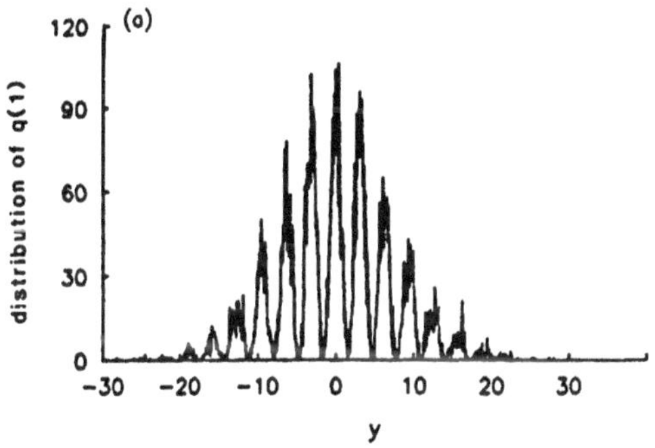

Fig. 2.3 Fringe pattern for a Gaussian wave packet containing 10,000 particles crossing a screen with two Gaussian slits. Obtained by numerical simulation. Reprinted with permission from McClendon and Rabitz (1988). Copyright 1988 by the American Physical Society

numerical experiment is that it demonstrates that each electron comes from just one slit.

The same conclusion is obtained in Webb (2011) using an event-based model for particles emitted one-at-a-time in the two-slit experiment. The results of the numerical simulations confirm that the fringe patterns should be interpreted in terms of the aggregate behavior of individual particles. Other interesting related quantum simulations are discussed in Michielsen et al. (2010) and De Raedt and Michielsen (2012). This gives a clear answer to a much raised question about the behavior of the electrons in such case. Of course such numerical calculations cannot give an explanation of the physics behind the diffraction pattern, since they are based on phenomenological models. In Chap. 9 we reconsider this problem from the point of view of the theory developed in the next chapters.

2.6 Extensions of the Theory, Some Brief Comments, and Assessment

Stochastic quantum mechanics can be extended along several important directions. For example—and this is perhaps one of its most remarkable outcomes—conditional *probabilities* (not amplitudes) have been constructed describing interference phenomena and the like (Petroni 1989). Variational methods and path-integral procedures have been introduced. In particular, by following a variational approach it has been possible to show that for stationary states $\boldsymbol{v}$ is irrotational wherever the density is different from zero, while in the nodes at $\rho = 0$ the vorticity tensor $\boldsymbol{\Omega}_{ij} = (\nabla \times \boldsymbol{v})_{ij}$ can be different from zero. Using a generalization to mean velocities that are not irrotational, the process is shown to relax towards a standard (irrotational) solution, which can be seen as an attractor for the extended family of stochastic solutions.[13]

[13] The idea that the quantum stationary states are some kind of attractors within an appropriate set of solutions has been arrived at from other, complementary points of view; see e.g. de la Peña and Cetto (1995), 't Hooft (2006).

Extensions of the theory to a wider range of problems include, among others, the electron spin (de la Peña 1971), the description of mixtures (Guerra 1984), radiative corrections (de la Peña and Cetto 1971), the relativistic case (de la Peña 1970; Hakim 1968; Morato 1992), and gravity (Smolin 1986). Spinning and relativistic particles have been studied by Dohrn et al. (1979), and a statistical description that can accommodate relativity and spin in a natural way has been proposed by Tiwari (1988). An independent, interesting development in a similar direction is the treatment of the Dirac equation in terms of a dichotomic (telegraph) stochastic process (Gaveau et al. 1984). Systems composed of several particles have been considered, dramatically exhibiting the characteristic nonlocalities of the description (de la Peña and Cetto 1969; Loffredo and Morato 2007). Further, a quantum field theory has been developed within Nelson's framework, as well as a procedure for stochastic quantization and a full study of quantum coherent states. The theory has also received close attention from the point of view of Bernstein processes (see e.g. the works of Garbaczewski). Dissipation in quantum systems is highly amenable to treatment with the stochastic methods (Marra 1987). Another noteworthy result is the sub-quantum H-theorem in Valentini (1991a, b). Further, stochastic quantum mechanics has been of some value in the study of stochastic chaos in Brownian systems obeying a Fokker-Planck equation that is formally analogous to the Schrödinger equation (see e.g. Alpatov and Reichl 1994).[14] A somewhat different and interesting realist and objective formulation of the stochastic approach to the quantum phenomenon has been developed in recent years by Budiyono (2012a, b, c, 2013a, b).

The stochastic theory also helps to gain some intuition on specific quantum problems, notably the (anti)symmetrization of the wave function (Nelson 1985a, Sect. 20; see also Loffredo and Morato 1987; Petroni and Morato 2000). Its application to the tunnel effect (Jona-Lasinio et al. 1981; Yasue 1981) is convenient for addressing aspects related to quantum trajectories, such as arrival times, first hitting time, sojourn times, and so on, and provides an illustration of typical (one-particle) quantum nonlocality, as shown in Sect. 2.5. The analysis of particle trajectories represents undoubtedly a valuable plus of the stochastic approach to QM (see McClendon and Rabitz 1988; Moreno Murguía 2006).

An additional contribution of the theory is that it discloses the link between the quantum potential V_Q [see paragraph following Eq. (2.65)] and the diffusive velocity $\boldsymbol{u}$, as shown in Eq. (2.41). This helps to assign a kinetic nature to V_Q, a point that will be revisited in detail in Chap. 8, where the relation between $\boldsymbol{u}$ and the nonlocal properties of the quantum system will become clear. An illustration of this can be seen in the expression for the acceleration $\boldsymbol{a}_Q$, Eq. (2.65), which is a function of the diffusive velocity and thus of the (changes of the) density of particles ρ: the essentially nonlocal nature of $\boldsymbol{u}$ is conveyed to the acceleration through $\lambda \neq 0$.[15]

[14] Some of these matters are discussed in Vasudevan et al. (2008). For the relativistic case see also Ramanathan (1997). Extensive and complementary lists of references to earlier work can be found in Jammer (1974), Guerra (1981, 1984, 1988), Blanchard et al. (1987), de la Peña and Cetto (1991), and *The Dice*.

[15] The formula for the acceleration $\boldsymbol{a}_B$ for classical (Brownian) particles is of course as nonlocal as the quantum acceleration, but nobody denies the usefulness of the Brownian-motion theory of

Despite its advantages in providing an alternative route for the understanding of QM, the theory has also its downsides, the most obvious one being its phenomenological nature, as has been stressed already. This may not perturb those who argue that thermodynamics is also a phenomenological theory and yet nobody quarrels about that. The point is that in the present description a most fundamental element is missing: the *physical* cause of the fluctuations, so the physical elements that determine the parameters in Schrödinger-like equations remains unidentified. This leaves things more or less as in QM itself: we face again the unexplained (noncausal) fluctuations, and the universality of a in Eq. (2.71) must be assumed a priori.

The theory has received further criticisms from a diversity of standpoints (see e.g. Ghirardi et al. 1978; Gillespie 1995; Grabert et al. 1979; Mielnik and Tengstrand 1980). One in particular, is that not a single stochastic process, but an infinity of them can be associated to a quantum state (Davidson 1979b). This is a peculiarity of the description in terms of a Schrödinger-type equation involving the sole product $D\sqrt{-\lambda}$, as discussed above, rather than a problem for its stochastic interpretation. This difficulty is solved by determining D on physical grounds, as is done in Chap. 4. It is further argued that, contrary to what happens with classical diffusions, the quantum stochastic process cannot be separated into 'subprocesses' satisfying a given set of initial conditions (Grabert et al. 1979); this means that the trajectory of a given particle depends nonlocally on all other trajectories that it *could* have followed, which is of course unrealistic and unacceptable. However, these (and other) bizarre peculiarities (see e.g. Ghirardi et al. 1978) are a manifestation of the quantum behavior; they constitute an integral part of quantum theory, even if some of them remain normally hidden. In other words, bizarre quantum properties manifest themselves as bizarre stochastic properties. Accepting quantum theory implies accepting them. We have become accostumed with time to accept the former, but are still very sensitive to the latter. What stochastic QM does is to expose them for further analysis.

The nonlocality problem in stochastic quantum mechanics has been strongly—and rightly—criticized by Nelson (1985a, b, 2005, Sect. 23) on the ground that any fundamental physical theory that violates locality is untenable. It is noteworthy that Nelson decided to abandon his succesful efforts in the development of his stochastic mechanics—which to a large extent is the one discussed in the present chapter—for a reason of principle, namely, when he discovered its nonlocal nature. Now, it is clear that a theory designed to reproduce QM will reproduce the niceties but also the quandaries of QM. And the nonlocality of Nelson's theory is a mere rebound of the quantum nonlocalities —yet nobody renounces QM by rejecting its nonlocalities. Quite the contrary: today it is fashionable to happily speak of quantum nonlocalities; a look at the literature around the Bell inequalities serves to attest this. It seems that the problem has two facets. For on the one hand it is important to understand why

(Footnote 15 continued)
Einstein and Smoluchowski within its domain of applicability. It even played a most important and historic role in the empirical demonstration of the reality of molecules at the beginning of the 20th century! Such description of the Brownian case is admittedly not a fundamental one. In the quantum case a problem arises when interpreting it as a fundamental theory, since a fundamental expression for the acceleration *must* be local.

QM implies a nonlocal description, and on the other hand, it is important to find the theory that supersedes this trait which, as Nelson put it, is untenable.[16]

Probably the most extended criticism towards stochastic quantum mechanics (or stochastic mechanics) is the one raised by Wallstrom (1989, 1994) in a frequently cited work (see also Goldstein 1987; Takabayasi 1952). In essence it asserts that the transition from the couple of Eq. (2.43) [or Eq. (2.39) and the continuity equation] to the Schrödinger equation may be unbefitting due to the fact that in the construction for $\psi \sim e^{iS}$ the function S may be many-valued, so that there is no reason to assume that ψ is single-valued, it being a mere mathematical object. Detailed rebuttals of Wallstrom's argument have been given in Smolin (2006) in a significant contribution to Nelson's theory, and by Fritsche and Haugk (2009) (and 2003), this latter offering a proof that the single-valuedness of the wave function ensues from the conservation of its normalization at all times. It is important to insist on these rebuttals because Wallstrom's work has been considered by many as the definitive blow against Nelson's and similar theories. An unfortunate example is Wick's (1995) book, an excellent and highly advisable book for the wide public, which contains a careful discussion of several of the conceptual problems of QM, particularly the 'infamous boundary' between the observed and the observer. Given the book's realistic and objective approach to the subject, one would expect it to pay serious attention to the stochastic theory—which it does not. In fact, the author confesses that he used to be appreciative of Nelson's theory, but was forced to change his point of view by Wallstrom's paper. The replies provided by Smolin and by Fritsche and Hangk hopefully help restore confidence in the stochastic theories of QM—within their natural limitations. In Chap. 4 we come back to this point.

To put things in the proper perspective we should bear in mind that as a phenomenological theory, stochastic quantum mechanics is not to be doomed for its properties or shortcomings. The error would lie in taking such a limited description as the accomplished theory. The parameter λ of the stochastic description of QM is *selected* so as to reproduce the latter, with all virtues and limitations of such selection. Difficulties appear due to the poorness of the configuration-space description: it is too restricted to hold the richness of the real stochastic phenomena.

Generally speaking, the critics of stochastic (quantum) mechanics are formally correct in their criticisms, although their objections normally relate in the last instance (and unknowingly to the critics) to the peculiarities of quantum systems rather than to the stochastic approach itself. What in reality many of the critics of the stochastic description of QM are doing is contribute to the catalog of the most relevant differences between classical and quantum stochastic processes. The differences are so substantial that one should not be surprised to find that the required stochastic quantum description falls far from the corresponding classical one.

---

16 Recently, Nelson has attempted to apply stochastic mechanics to relativistic fields, hoping to avoid the above mentioned nonlocality features, and aiming to develop useful technical tools in constructive field theory (see Nelson 2013).

2.6.1 A Summing Up

The material of this chapter has hopefully served its purpose to assess the value of stochastic quantum mechanics in the present context. One first advantage of the theory is that it neatly discloses the stochastic nature of the quantum system, highlighting essential similarities and differences between QM and classical, Brownian-type stochastic processes. Another is that it leads to QM through a simple, phenomenological approach which, not being part of usual QM, enriches it by offering a complementary, intuitive picture of some important aspects of the theory.

The stochastic approach provides with relative simplicity a way to arrive at QM from a realist and objective physical picture; however, it leaves us with the feeling that the real thing continues to be hidden behind the phenomenological curtain. Something more fundamental is required. The search for it is the subject matter of the following chapters.

Answer to the quiz: The reversibility of the Schrödinger equation means that this equation describes an average behavior after any (subquantum) irreversible process that could exist, has ceased to be active. The meaning of this answer will become clearer as we proceed.

References

Abbott, L.F., Wise, M.B.: Dimension of a quantum-mechanical path. Am. J. Phys. **49**, 37 (1981)

Ali, T., Engliš, M.: Quantization methods: a guide for physicists and analysts. Rev. Math. Phys. **17**, 391 (2005)

Ali, S.T., Prugovečki, E.: Mathematical problems of stochastic quantum mechanics. Acta Appl. Math. **6**, 1 (1986)

Alpatov, P., Reichl, L.E.: Spectral properties of a time-periodic Fokker-Planck equation. Phys. Rev. E **49**, 2630 (1994)

Bacciagaluppi, G.: Non-equilibrium in Stochastic Mechanics. J. Phys. Conf. Series **361**, 012017 (2011)

Ballentine, L.E.: Quantum Mechanics. Prentice Hall, NJ (1990)

Ballentine, L.: Quantum Mechanics. A Modern Development. World Scientific, Singapore (1998)

Bell, J.S.: Speakable and Unspeakable in Quantum Mechanics. Princeton, Princeton U. P (1987)

Bess, L.: A diffusion model for the Schrödinger equation. Progr. Theor. Phys. **49**, 1889 (1973)

Blanchard, P., Combe, P., Zheng, W.: Mathematical and Physical Aspects of Stochastic Mechanics. Lecture Notes in Physics, vol. 281. Springer, Berlin (1987)

Blanchard, P., Garbaczewski, P.: Natural boundaries for the Smoluchowski equation and affiliated diffusion processes. Phys. Rev. E **49**, 3815 (1994)

Brady, R., Anderson R.: Analogue Physics. A Student's Guide to Waves in An Ideal Fluid (Cambridge University Computer Center, 2013)

Budiyono, A.: Possible corrections to quantum mechanical predictions in a hidden variable model. Physica A **391**, 3081 (2012a)

Budiyono, A.: Quantum fluctuations as deviation from classical dynamics of ensemble of trajectories parameterized by unbiased hidden random variable. Physica A **391**, 3102 (2012b)

Budiyono, A.: Quantization from Hamilton-Jacobi theory with a random constraint. Physica A **391**, 4583 (2012c)

Budiyono, A.: Objective uncertainty relation with classical background in a statistical model. Physica A **392**, 43 (2013a)
Budiyono, A.: Quantization from an exponential distribution of infinitesimal action. Physica A **392**, 307 (2013b)
Cetto, A.M.: Investigaciones sobre una teoría estocástica de la mecánica cuántica. Ph.D. thesis, UNAM, Mexico, 1972
Comisar, G.G.: Brownian-motion model of nonrelativistic quantum mechanics. Phys. Rev. B **138**, 1332 (1965)
Couder, Y., Fort, E.: Single-particle diffraction and interference at a macroscopic scale. Phy. Rev. Lett. **97**, 154101 (2006)
Couder, Y., Protière, S., Fort, E., Boudaoud, A.: Dynamical phenomena: Walking and orbiting droplets. Nature **437**(7056), 208 (2005)
Cramer, J.G.: The transactional interpretation of quantum mechanics. Rev. Mod. Phys. **58**, 647 (1986)
Davidson, M.P.: A classical realization of quantum mechanics. Found. Phys. **8**, 481 (1978)
Davidson, M.P.: A dynamical theory of markovian diffusion. Physica A **96**, 465 (1979a)
Davidson, M.P.: A generalization of the Fenyes-Nelson stochastic model of quantum mechanics, Lett. Math. Phys. **3**, 271 (1979b), arXiv: quant-ph/0112063
Davidson, M.P.: A model for the stochastic origins of Schrodinger's equation. J. Math. Phys. **20**, 1865 (1979c)
Davidson, M.P.: Momentum in stochastic quantum mechanics, Lett. Math. Phys. **5**, 23 (1981), arXiv: quant-ph/0112114v1
Davidson, M.P.: Stochastic models of quantum mechanics—A perspective. In: Adenier, G., Khrennikov, A.Y., Fuchs, C.A. (eds.) Foundations of Probability and Physics 4. American Institute of Physics Inc., New York, p. 106 (2007) (AIP Conference Proceedings 889)
de Broglie, L.: Le mouvement brownien d'une particule dans son onde. C. R. Acad. Sci. Paris B **264**, 1041 (1967)
de la Peña, L.: A simple derivation of the Schrödinger equation from the theory of Markov processes. Phys. Lett. A **24**, 603 (1967)
de la Peña, L.: New formulation of stochastic theory and quantum mechanics. J. Math. Phys. **10**, 1620 (1969)
de la Peña, L: New formulation of stochastic theory and quantum mechanics. V: The relativistic spinless particle. Rev. Mex. Fís. 19, 133 (1970)
de la Peña, L.: Stochastic theory of quantum mechanics for particles with spin. J. Math. Phys. **12**, 453 (1971)
de la Peña, L., Cetto, A.M.: New formulation of stochastic theory and quantum mechanics. IV: The two-particle system. Rev. Mex. Fís. **18**, 323 (1969)
de la Peña, L., Cetto, A.M.: Self-interaction corrections in a nonrelativistic stochastic theory of quantum mechanics. Phys. Rev. D **3**, 795 (1971)
‡‡de la Peña, L., Cetto, A.M.: Stochastic theory for classical and quantum mechanical systems. Found. Phys. **5**, 355 (1975)
‡‡de la Peña, L., Cetto, A.M.: Does quantum mechanics accept a stochastic support? Found. Phys. **12**, 1017 (1982) (Reprinted in Quantum Space and Time: The Quest Continues, A. O. Barut, A. van der Merwe and J.-P. Vigier, eds., (Cambridge University Press 1984).)
*de la Peña, L., Cetto, A.M.: Electromagnetic vacuum-particle interaction as the source of quantum phenomena. Found. Phys. Lett. **12**, 476 (1991)
*de la Peña, L., Cetto, A.M.: Role of the zeropoint field in wave mechanics. In: Courants, Amers, Écueils en Microphysique, p. 311. Fondation Louis de Broglie, Paris (1993)
*de la Peña, L., Cetto, A.M.: Is quantum mechanics a limit cycle theory? In: Ferrero, M., van der Merwe, A. (eds.) Fundamental Problems in Quantum Physics. Kluwer, Dordrecht (1995)
**de la Peña, L., Cetto, A.M.: The Quantum Dice: An Introduction to Stochastic Electrodynamics. Kluwer, Dordrecht (1996) (Referred to in the book as *The Dice*)

De Raedt, H., Michielsen, K.: Event-by-event simulation of quantum phenomena. Ann. d. Phys. **524**, 393 (2012)

‡Dohrn, D., Guerra, F., Ruggiero, P.: Spinning particles and relativistic particles in the framework of Nelson's stochastic mechanics. In: Albeverio, S., et al. (eds.) Feynman Path Integrals. Springer Lecture Notes in Physics, vol. 106, pp. 165–181. Springer, Berlin (1979)

Einstein, A.: On Quantum Physics (1954)

L. Feligioni, O. Panella, Y.N. Srivastava and A. Widom: Two-time correlation functions: stochastic and conventional quantum mechanics. Eur. Phys. J. B **48**, 233 (2005)

Fényes, I.: A deduction of Schrödinger equation. Acta Bolyaina **1**, 5 (1946)

Féynes, I.: Eine wahrscheinlichkeitstheoretische Begründung und Interpretation der Quantenmechanik. Zeit. Phys. **132**, 81 (1952)

Fritsche, L., Haugk, M.: A new look at the derivation of the Schrödinger equation from Newtonian mechanics. Ann. Phys. (Leipzig) **12**, 371 (2003)

Fritsche, L., Haugk, M.: Stochastic Foundation of Quantum Mechanics and the Origin of Particle Spin (2009), arXiv: abs/0912.3442v1 [physics.gen-ph]

Fürth, R.: Über einige Beziehungen zwischen klassischer Statistik und Quantummechanik. Zeit. Phys. **81**, 143 (1933)

Garbaczewski, P.: Random versus deterministic paths in stochastic mechanics. Phys. Lett. A **143**, 85 (1990)

Garbaczewski, P.: Relativistic problem of random flights and Nelson's stochastic mechanics. Phys. Lett. A **164**, 6 (1992)

Garbaczewski, P.: Derivation of the quantum potential from realistic Brownian particle motions. Phys. Lett. A **162**, 129 (1992)

Garbaczewski, P.: Physical significance of the Nelson-Newton laws. Phys. Lett. A **172**, 208 (1993a)

Garbaczewski, P.: On the statistical origins of the de Broglie-Bohm quantum potential: Brownian motion in a field of force as Bernstein diffusion. Phys. Lett. A **178**, 7 (1993b)

Garbaczewski, P.: Dynamics of coherent states as the diffusion process: Schrödinger's system, Bernstein processes and quantum evolution. In: Feng, D.H., Klauder, J.R., Strayer, M.R. (eds.) Coherent States: Past, Present and Future. World Scientific, Singapore (1994)

Garbaczewski, P., Klauder, J.R., Olkiewicz, R.: Schrödinger problem, Lévy processes, and noise in relativistic quantum mechanics. Phys. Rev. E **51**, 4114 (1995)

Gardiner, C.W.: Handbook of Stochastic Methods. Springer, Berlin (1983)

Gaveau, B., Jacobson, T., Kac, M., Schulman, L.S.: Relativistic extension of the analogy between quantum mechanics and brownian motion. Phys. Rev. Lett. **53**, 419 (1984)

‡Ghirardi, G.C., Omero, C., Rimini, A., Weber, T.: The stochastic interpretation of quantum mechanics: a critical review. Riv. Nuovo Cim. **1**, 1 (1978)

‡Gillespie, D.T.: Incompatibility of the Schrödinger equation with Langevin and Fokker-Planck equations. Found. Phys. **25**, 1041 (1995)

‡‡Goldstein, S.: Stochastic mechanics and quantum theory. J. Stat. Phys. **47**, 645 (1987)

‡Grabert, H., Hänggi, P., Talkner, P.: Is quantum mechanics equivalent to a classical process? Phys. Rev. A **19**, 2440 (1979)

Griffiths, R.B.: Consistent histories and quantum reasoning. Phys. Rev. A **54**, 2759 (1996)

Griffiths, R.B., Omnés, R.: Consistent histories and quantum measurements. Phys. Today **52**(8), 26 (1999)

Guerra, F.: Structural aspects of stochastic mechanics and stochastic field theory. Phys. Rep. **77**, 263 (1981)

Guerra, F.: Probability and quantum mechanics. The conceptual foundations of stochastic mechanics. In: Accardi, L., Waldenfels, W.V. (eds.) Quantum Probability and Applications I. Lecture Notes in Mathematics, vol. 1055. Springer, Berlin (1984)

Guerra, F.: Probability and quantum mechanics. The conceptual foundations of stochastic mechanics. In: Accardi, L., Waldenfels, W.V. (eds.) Quantum Probability and Applications II. Lecture Notes in Mathematics, vol. 1136. Springer, Berlin (1985)

Guerra, F.: Stochastic variational principles and quantum mechanics. Ann. Inst. Henri Poincaré **49**, 314 (1988)

Guerra, F., Marra, R.: Origin of the quantum observable operator algebra in the frame of stochastic mechanics. Phys. Rev. D **28**, 1916 (1983)

Guerra, F., Morato, L.M.: Quantization of dynamical systems and stochastic control theory. Phys. Rev. D **27**, 1774 (1983)

Hakim, R.: Relativistic stochastic processes. J. Math. Phys. **9**, 1085 (1968)

Hall, F.G., Collins, R.E.: Stochastic processes and their representations in Hilbert space. J. Math. Phys. **12**, 100 (1971)

Holland, P.: The Quantum Theory of Motion. Cambridge University Press, Cambridge (1993)

Jamison, B.: Reciprocal processes. Z. Wahrsch. Verw. Gebiete **30**, 65A (1974)

Jammer, M.: The Philosophy of Quantum Mechanics: The Interpretation of Quantum Mechanics in Historical Perspective. Wiley, New York (1974)

Jona-Lasinio, G., Martinelli, F., Scoppola, E.: The semiclassical limit of quantum mechanics: a qualitative theory via stochastic mechanics. Phys. Rep. **77**, 313 (1981)

Kershaw, D.: Theory of hidden variables. Phys. Rev. B **136**, 1850 (1964)

Kracklauer, A.F.: An intuitive paradigm for quantum mechanics. Phys. Essays **5**, 226 (1992)

Kyprianidis, A.: The principles of a stochastic formulation of quantum theory. Found. Phys. **22**, 1449 (1992)

Lisi, A.G.: Quantum mechanics from a universal action reservoir (2006), arXiv: physics/0605068 [physics.pop-ph]

Loffredo, M.I., Morato, L.M.: Selfconsistent hydrodynamical model for He II near the absolute zero in the framework of stochastic mechanics. Phys. Rev. B **35**, 1742–1747 (1987)

Loffredo, M.I., Morato, L.M.: Stochastic quantization for a system of N identical interacting Bose particles. J. Physica A Math. Theor. **40**, 8709 (2007)

Marra, R.: Variational principles for conservative and dissipative diffusions. Phys. Rev. D **36**, 1724 (1987)

McClendon, M., Rabitz, H.: Numerical simulations in stochastic mechanics. Phys. Rev. A **37**, 3479 (1988)

‡Mielnik, B., Tengstrand, G.: Nelson-Brown motion: some question marks. Int. J. Theor. Phys. **19**, 239 (1980)

Michielsen, K., Yuan, S., Zhao, S., Jin, F., De Raedt, H.: Coexistence of full which-path information and interference in Wheeler's delayed-choice experiment with photons. Physica E **42**, 348 (2010)

Morato, L.M.: A new approach to the problem of relativistic kinematics in stochastic mechanics and a derivation of Klein-Gordon equation. In: Guerra, F., Loffredo, M.I., Marchioro C. (eds.) Proceedings of the International Workshop on "Probabilistic Methods in Mathematical Physics". World Scientific, Singapore (1992)

Moreno Murguía B.: Simulación del efecto túnel en el formalismo de la mecánica cuántica estocástica. Professional thesis, Universidad Nacional Autónoma de México, Mexico (2006)

Nassar, A.B.: Ermakov and non-Ermakov systems in quantum dissipative models. J. Math. Phys. **27**, 755 (1986a)

Nassar, A.B.: Logarithmic nonlinearities within the framework of stochastic mechanics. Phys. Rev. A **33**, 2134 (1986b)

Nelson, E.: Derivation of the Schrödinger equation from Newtonian mechanics. Phys. Rev. **150**, 1079 (1966)

Nelson, E.: Dynamical Theories of Brownian Motion. Mathematical Notes. Princeton University Press, New Jersey (1967)

Nelson, E.: Construction of quantum fields from Markoff fields. J. Funct. Anal. **12**, 97 (1973)

Nelson, E.. In: Brittin, W.E., Gustafson, K.E., Wyss, W. (eds.) Mathematical Physics VII, Proceedings of the Seventeenth International Congress on Mathematical Physics, p. 509. North-Holland, Amsterdam (1984)

Nelson, E.: Quantum Fluctuations. Princeton Series in Physics. Princeton University Press, New Jersey (1985a)

Nelson, E.: A survey of stochastic mechanics. In: Proceedings of the International Conference on Quantum Statistics and Foundational Problems on Quantum Mechanics, Hadronic Journal Supplement, vol. 1, p. 401 (1985b)
Nelson, E.: What is stochastic mechanics? In: Mathématiques Finitaires et Analyse non-Standard, Journées Société Mathématique de France, vol. 1. CIRM, Luminy (1985c)
Nelson, E.. In: Albeverio, S., Casati, G., Merlini, D. (eds.) Stochastic Processes in Classical and Quantum Systems. Lecture Notes in Physics, vol. 262. Springer, Berlin (1986)
‡‡Nelson, E.: The mystery of stochastic mechanics. manuscript of a talk delivered on 22 Nov 2005, http://www.math.princeton.edu/nelson/papers/talk.pdf (2005)
Nelson, E.: Review of stochastic mechanics. J. Phys. Conf. Ser. **361**, 012011 (2012)
Nelson, E.: Stochastic mechanics of relativistic fields. J. Phys. Conf. Ser. **504**, 012013 (2014)
Nicholson, A.F.: On a theory due to I. Fenyes, Australian. J. Phys. **7**, 14 (1954)
Novobátzky, K.: Das klassische modell der quantentheorie. Ann. d. Phys. **444**, 406 (1951)
Omnès, R.: The Interpretation of Quantum Mechanics. Princeton University Press, New Jersey (1994)
Omnès, R.: Understanding Quantum Mechanics. Princeton University Press, New Jersey (1999a)
Omnès, R.: Quantum Philosophy. Princeton University Press, New Jersey (1999b)
Petroni, N.C.: Stochastic mechanics and quantum interference. Phys. Lett. A **141**, 370 (1989)
Petroni, N.C., Morato, L.: Entangled states in stochastic mechanics. J. Physica A Math. Gen. **33**, 5833 (2000)
Prugovečki, E.: Stochastic Quantum Mechanics and Quantum Spacetime. Reidel, Dordrecht (1984)
Prugovečki, E.: Principles of Quantum General Relativity. World Scientific, Singapore (1995). (The main results are summarized in the paper Quantum Geometry and Gravity, in Quantum Gravity: International School of Cosmology and Gravitation, XIV Course in Erice 1995, P. G. Bergmann, V. de Sabbata and H.-D. Treder, eds. (World Scientific, Singapore, 1996))
Ramanathan, R.: The Present Status of the Quantum Theory of Light. Kluwer, Dordrecht (1997)
Risken, H.: The Fokker-Planck equation. Methods of Solution and Applications. Springer, Berlin (1984)
Santos, E.: Brownian motion and the Stochastic theory of quantum mechanics. In: Biel, J., Rae, J. (eds.) Irreversibility in the Many Body Problem, p. 457. Plenum Press, New York (1973)
Santos, E.: On a heuristic point of view concerning the motion of matter: from random metric to Schrödinger equation. Phys. Lett. A **352**, 49 (2006)
Schrödinger, E.: Sitzungsber. Preuss. Akad. Wissen. Berlin, Phys. Math. Klasse 144 (1931)
Schrödinger, E.: Sur la théorie relativiste de l'électron et l'interpretation de la mécanique quantique. Ann. Inst. Henri Poincaré **2**, 269 (1932)
Skagerstam, B.S.: Stochastic mechanics and dissipative forces. J. Math. Phys. **18**, 308 (1977)
Smolin, L.: Stochastic mechanics, hidden variables, and gravity. In: Penrose, R., Isham, C.J. (eds.) Quantum Concepts in Space and Time. Clarendon, Oxford (1986)
Smolin, L.: Could quantum mechanics be an approximation to another theory? (2006), arXiv: quant-ph/0609109v1
‡‡Takabayasi, T.: On the formulation of quantum mechanics associated with classical pictures. Progr. Theor. Phys. **8**, 143 (1952)
Tiwari, S.C.: Derivation of the Hamiltonian form of the Klein-Gordon equation from Schrödinger-Furth quantum diffusion theory: Comments. Phys. Lett. A **133**, 279 (1988)
't Hooft, G.: The mathematical basis for deterministic quantum mechanics. In: Nieuwenhuizen T.M. et al. (eds.) Beyond the Quantum. World Scientific, Singapore (2006), arXiv: quant-ph/0604008v2
Valentini, A.: Signal-locality, uncertainty, and the subquantum H-theorem. I. Phys. Lett. A **156**, 5 (1991a)
Valentini, A.: Signal-locality, uncertainty, and the subquantum H-theorem. II. Phys. Lett. A **158**, 1 (1991b)
Vasudevan, R., Parthasarathy, K.V., Ramanathan, R.: Quantum Mechanics. A Stochastic Approach. Alpha Science International, Oxford (2008) (See also Stochastic Quantum Mechanics by R Vasudevan, in Stochastic processes and their applications, M. J. Beckmann, M. N. Gopalan and

R. Subramanian, eds., Lecture Notes in Economics and Mathematical Systems 370 (Springer, Berlin 1991))

‡Wallstrom, T.C.: On the derivation of the Schrödinger equation from stochastic mechanics. Found. Phys. Lett. **2**, 113 (1989)

‡Wallstrom, T.C.: Inequivalence between the Schrödinger equation and the Madelung hydrodynamic equations. Phys. Rev. A **49**, 1613 (1994)

Webb, G.F.: Event-based interpretation of Schrödinger's equation for the two-slit experiment. Int. J. Theor. Phys. **50**, 3571 (2011)

Wick, D.: The Infamous Boundary. Copernicus, Springer, New York (1995)

Wind-Willassen, O., Moláček, J., Harris, D.M., Bush, J.W.: Exotic states of bouncing and walking droplets. Phys. Fluids **25**, 082002 (2013)

Yasue, K.: Quantum mechanics and stochastic control theory. J. Math. Phys. **22**, 1010 (1981)

Zambrini, J.-C.: Stochastic mechanics according to E. Schrödinger. Phys. Rev. A **33**, 1532 (1986)

Chapter 3
The Planck Distribution, a Necessary Consequence of the Fluctuating Zero-Point Field

With this chapter we initiate our analysis of the implications of considering the fluctuating zero-point radiation field (ZPF) as a fundamental constituent of an otherwise classical system. As announced in the introductory chapter, our journey starts with a fresh look at a simple though physically (and historically) relevant system, namely the electromagnetic radiation field in equilibrium with matter at temperature T . The blackbody problem, the one that gave birth to quantum mechanics, is thus revisited, taking into account the ZPF. The mere existence of this nonthermal field is shown to have far-reaching consequences. In particular, by performing a thermodynamic *and* statistical analysis of an ensemble of harmonic oscillators of frequency ω representing the modes of the radiation field of the respective frequency, we find that Planck's law, as well as irreducible (quantum) fluctuations, arise as *necessary* consequences of allowing for the presence of the pervarsive ZPF, without any assumption of discreteness.

3.1 Thermodynamics of the Harmonic Oscillator

Let us start by considering a one-dimensional harmonic oscillator of frequency ω, with the Hamiltonian given by[1]

$$H = (p^2 + \omega^2 q^2)/2. \tag{3.1}$$

For a material oscillator of mass $m = 1$, q and p stand for the oscillator's position and momentum, respectively. Now, of relevance for our purposes is that a monochromatic mode of frequency ω of the radiation field is equivalent to a harmonic oscillator of that same frequency. In this case H refers to the energy of such mode, and q and p represent its quadratures.

[1] This first part of the exposition borrows from the work of Boyer (1969b, 2003), who has contributed substantially to the analysis of the Planck distribution from a perspective akin to the one developed here. See also Boyer (1969a, 1976, 1983, 1984, 2010a, b, 2012), Marshall (1965), and Theimer (1971).

L. de la Peña et al., *The Emerging Quantum*,
DOI: 10.1007/978-3-319-07893-9_3,

Several basic properties of the harmonic oscillator can be derived from the structure of (3.1), and thus hold irrespective of the oscillator's nature. In particular, for a given constant energy U, the trajectory in phase space is the ellipse

$$p^2 + \omega^2 q^2 = 2U, \tag{3.2}$$

and its area gives the action

$$J = \frac{1}{2\pi} \oint p dq = \frac{1}{2\pi\omega} \oint \sqrt{2U - \omega^2 q^2}\, d(\omega q) = \frac{U}{\omega}. \tag{3.3}$$

The action J is an adiabatic invariant of the harmonic oscillator (see e.g. Landau and Lifshitz (1976), Sect. 49; José and Saletan (1998), Sect. 6.4), which means that it remains constant under a slow change of the frequency. Therefore, the change dU in the energy concomitant with the slow change $d\omega$ is given by

$$dU = J d\omega = \frac{U}{\omega} d\omega, \tag{3.4}$$

so that the work dW done *by* the system on the external device effecting the change of frequency is

$$dW = -\frac{U}{\omega} d\omega. \tag{3.5}$$

From here it follows that if $S(T, \omega)$ stands for the entropy of the system when this latter is in thermodynamic equilibrium at temperature T, for a reversible process one may write

$$T dS(T, \omega) = dU(T, \omega) + dW = dU(T, \omega) - \frac{U}{\omega} d\omega, \tag{3.6}$$

consequently

$$T \left(\frac{\partial S}{\partial T}\right)_\omega dT + T \left(\frac{\partial S}{\partial \omega}\right)_T d\omega = \left(\frac{\partial U}{\partial T}\right)_\omega dT + \left[\left(\frac{\partial U}{\partial \omega}\right)_T - \frac{U}{\omega}\right] d\omega. \tag{3.7}$$

Since the changes in the variables T and ω are independent, this relation naturally splits into the pair of equations

$$T \left(\frac{\partial S}{\partial \omega}\right)_T = \left(\frac{\partial U}{\partial \omega}\right)_T - \frac{U}{\omega}, \tag{3.8}$$

$$T \left(\frac{\partial S}{\partial T}\right)_\omega = \left(\frac{\partial U}{\partial T}\right)_\omega. \tag{3.9}$$

We utilize these two relations by taking the partial derivative of the first one with respect to T and of the second one with respect to ω, and combine the results to get

$$\left(\frac{\partial S}{\partial \omega}\right)_T = -\frac{1}{\omega}\left(\frac{\partial U}{\partial T}\right)_\omega. \tag{3.10}$$

Substitution into Eq. (3.8) gives

$$\left(\frac{\partial U}{\partial \omega}\right)_T - \frac{U}{\omega} = -\frac{T}{\omega}\left(\frac{\partial U}{\partial T}\right)_\omega. \tag{3.11}$$

The solution of this equation can be found by writing $U = \omega f(T, \omega)$ to cancel the term U/ω, whence

$$\frac{\omega}{T}\left(\frac{\partial f}{\partial \omega}\right)_T = -\left(\frac{\partial f}{\partial T}\right)_\omega. \tag{3.12}$$

This equation holds for any function f of the single variable ω/T, as can be easily verified; hence Eq. (3.11) admits the general solution

$$U = \omega f(\omega/T). \tag{3.13}$$

Equation (3.13) is indeed a very important result: it is Wien's law, which establishes the general form of the mean energy U of any harmonic oscillator as a function of its frequency ω and the temperature T. This law will be at the basis of our considerations below.[2,3]

We now present some additional results concerning the thermodynamics of the harmonic oscillator that will be useful below. The Helmholtz free energy F takes the form

$$F(T, \omega) = -k_B T \phi(\omega/T), \tag{3.14}$$

where k_B is Boltzmann's constant and ϕ is a thermodynamic potential from which the thermodynamic functions of the oscillator can be determined. In particular the mean equilibrium energy becomes

---

[2] Wien's law is a fundamental law of physics, since only simple and very general principles are required for its derivation. It is valid in classical as well as in quantum physics, and is even consistent with relativity, so it was the appropriate law to herald the 20th century. To get a better feeling of its fundamental nature, a derivation based solely on dimensionality arguments can be found in Sommerfeld's classical book on Thermodynamics (Sommerfeld 1956). Simple clear discussions of Wien's law can be seen in two highly pedagogical papers: Piña and de la Selva (2010), and del Río-Correa (2010).

[3] We recall that for the derivation of his law, Wien studied the Doppler effect of the modes of an adiabatically disturbed *radiation* field in thermal equilibrium (see e.g. Milonni 1994).

$$U(T,\omega) = k_B T^2 \left(\frac{\partial \phi}{\partial T}\right)_\omega = -k_B \omega \frac{d\phi(z)}{dz}, \tag{3.15}$$

with $z = \omega/T$. Comparison with Wien's law gives

$$f(z) = -k_B \frac{d\phi}{dz}. \tag{3.16}$$

Finally, the entropy is also a function of the variable z,

$$S(z) = k_B \phi(z) + z f(z). \tag{3.17}$$

These results suffice for our purposes.

3.1.1 Unfolding the Zero-Point Energy

In the low-temperature limit $T \to 0$, Eqs. (3.13) and (3.15) give for the mean energy

$$\mathcal{E}_0 \equiv U(0,\omega) = \omega f(\infty) = -k_B \omega \frac{d\phi}{dz}(\infty) = A\omega, \tag{3.18}$$

so that the zero-point energy $\mathcal{E}_0$—the mean energy of the oscillator at absolute temperature $T = 0$—is determined by the value that the function $f(z)$ (or $d\phi/dz$) attains at infinity.[4] In the usual thermodynamic analysis the value of the constant $A = f(\infty)$ is arbitrarily chosen as zero, so there is no athermal energy. However, the more general (and more natural) solution corresponds to a nonnull value of A. In the case of the radiation field oscillators, this represents a physically more reasonable choice than a vacuum that is completely devoid of electromagnetic phenomena. By taking A to be nonzero we attest the existence of a zero-point energy that fills the whole space and is proportional to the frequency of the oscillator,[5,6]

---

[4] Some textbook demonstrations of Wien's law cast doubt about extending its validity to the limit $T = 0$. That Eq. (3.13) holds also at $T = 0$ is explicitly demonstrated in Cole (1990).

[5] That the only spectrum consistent with relativity (and hence with electromagnetic theory) corresponds to $\mathcal{E}_0(\omega) \sim \omega$, has been demonstrated independently by several authors. The earliest of such demonstrations are those in Marshall (1963), Santos (1968), and Boyer (1969b). See also Cole (1990), Milonni (1994), Chap. 2; and *The Dice*, Chap. 4. The present thermodynamic calculation leads to the same expression, Eq. (3.18). Further, the Schrödinger equation provides a similar prediction for the ground-state energy of a particle in a harmonic oscillator potential. Here we have a vivid example of the intrinsic unity of physics, reinforcing the idea that it refers to different aspects of a single reality.

[6] Taking $A = 0$ is equivalent to putting the boundary condition for the solutions of Maxwell's equations at infinity in the past equal to zero, i. e. no radiation. The choice $A \neq 0$ replaces this unnatural boundary condition by a zero-point field at infinity, simultaneously restoring time-reversal symmetry in electrodynamics.

$$\mathcal{E}_0 = A\omega = \tfrac{1}{2}\hbar\omega. \tag{3.19}$$

The value of A (with dimensions of action) must be universal because it determines the equilibrium spectrum at $T = 0$, which, according to Kirchhoff's law, has a universal character. We have put it equal to $\hbar/2$ in order to establish contact with present-day knowledge. However, it must be stressed that the presence of the Planck constant here does not imply any quantum connotation. In addition, it should be noticed that many of the results to be obtained in the present chapter do not depend on the precise value of $\mathcal{E}_0$, the only requirement being in such instances that it be different from zero.

A nonnull value of A means a violation of energy equipartition among the oscillators, since the equilibrium energy becomes now a function of the oscillator frequency. Though at this stage such violation can strictly be assured only at $T = 0$, the result suggests that the physics ensuing from the existence of $\mathcal{E}_0 \neq 0$ necessarily transcends classical physics. This opens up interesting possibilities that will be explored along this chapter.

In concluding this section, let us note that the existence of a zero-point energy provides a natural energy scale, which, along with $k_B T$, suggests to introduce the dimensionless quantity

$$\mathring{z} = \frac{2\mathcal{E}_0}{k_B T} = \frac{\hbar}{k_B} z = \frac{\hbar\omega}{k_B T}. \tag{3.20}$$

This will be the natural dimensionless variable of the thermodynamic functions, since the potential ϕ in Eq. (3.14) is a dimensionless function of z and can therefore be expressed as a function of $\mathring{z}$.

3.2 General Thermodynamic Equilibrium Distribution

Our aim is to find the average energy U per oscillator in an ensemble of such systems when equilibrium has been reached at a fixed temperature T. For this purpose we first follow the standard description of a canonical ensemble (Pathria 1996). In this case, the probability that a member of the ensemble is in a state with energy between $\mathcal{E}$ and $\mathcal{E} + d\mathcal{E}$ can be written in the general form

$$W_g(\mathcal{E})d\mathcal{E} = \frac{1}{Z_g(\beta)} g(\mathcal{E}) e^{-\beta\mathcal{E}} d\mathcal{E}, \tag{3.21a}$$

$$Z_g(\beta) = \int g(\mathcal{E}) e^{-\beta\mathcal{E}} d\mathcal{E}, \tag{3.21b}$$

where $\beta = 1/(k_B T)$, $Z_g(\beta)$ is the partition function that normalizes $W_g(\mathcal{E})$ to unity, and $g(\mathcal{E})$ is a weight function representing the density of states with energy $\mathcal{E}$. The mean value $\langle f(\mathcal{E})\rangle$ of any function $f(\mathcal{E})$ is thus

$$\langle f(\mathcal{E})\rangle = \int W_g(\mathcal{E}) f(\mathcal{E}) d\mathcal{E}. \tag{3.22}$$

For $f(\mathcal{E}) = \mathcal{E}$, (3.22) gives the mean energy

$$U = \langle \mathcal{E} \rangle = \int \mathcal{E} W_g(\mathcal{E}) d\mathcal{E}. \tag{3.23}$$

Equation (3.21a) constitutes the general form of a Boltzmann distribution.[7] In particular, the corresponding classical distribution for the harmonic oscillator is obtained from (3.21a) with $g(\mathcal{E})$ given by Pathria (1996)

$$g_{\text{classic}}(\mathcal{E}) = \frac{1}{s\omega}, \tag{3.24}$$

where s is a constant with dimensions of action, so g has the dimension of (energy)$^{-1}$. In this case one gets from the above equations

$$W_{\text{cl}}(\mathcal{E}) = W_{g_{\text{cl}}}(\mathcal{E}) = \frac{e^{-\beta \mathcal{E}}}{\int e^{-\beta \mathcal{E}} d\mathcal{E}}; \tag{3.25a}$$

$$Z_{\text{cl}}(\beta) = \int g_{\text{cl}}(\mathcal{E}) e^{-\beta \mathcal{E}} d\mathcal{E} = \frac{1}{s\beta\omega}; \tag{3.25b}$$

$$\langle \mathcal{E} \rangle = U = -\frac{1}{Z_{\text{cl}}} \frac{dZ_{\text{cl}}}{d\beta} = \frac{1}{\beta} = k_B T. \tag{3.25c}$$

From the last equation it follows that $U(T = 0) = 0$. This means that to allow for a zero-point energy, a form for $g(\mathcal{E})$ different from that given by Eq. (3.24) must be used. The specific structure of this $g(\mathcal{E})$ consistent with a zero-point energy for the harmonic oscillator will be determined below.

3.2.1 Thermal Fluctuations of the Energy

Equations (3.21a, 3.21b) and (3.22) lead to a series of important and general results. With $f(\mathcal{E}) = \mathcal{E}^r$, r a positive integer, it follows that (the prime indicates derivative with respect to β)[8]

---

[7] This form of writing $W_g(\mathcal{E})$ was used, for example, by Einstein (1907) in his early work on the specific heat of solids. He considered the distribution in (3.21a) assuming from the start a form for the function $g(\mathcal{E})$ equivalent to (3.82) below, as was dictated by the quantization discovered by Planck. Here we proceed in the opposite sense, by allowing the theory to determine $g(\mathcal{E})$.

[8] The present discussion draws closely from de la Peña and Cetto (2002), de la Peña et al. (2008, 2010a, b), Valdés-Hernández et al. (2010), Valdés-Hernández (2010).

$$\left\langle \mathcal{E}^{r}\right\rangle' = -\frac{Z_g'}{Z_g}\left\langle \mathcal{E}^{r}\right\rangle - \frac{1}{Z_g}\int \mathcal{E}^{r+1} g(\mathcal{E}) e^{-\beta\mathcal{E}} d\mathcal{E} = -\frac{Z_g'}{Z_g}\left\langle \mathcal{E}^{r}\right\rangle - \left\langle \mathcal{E}^{r+1}\right\rangle, \tag{3.26}$$

and further, from (3.21b),

$$\langle \mathcal{E}\rangle = U = \frac{1}{Z_g}\int \mathcal{E} g(\mathcal{E}) e^{-\beta\mathcal{E}} d\mathcal{E} = -\frac{Z_g'}{Z_g}. \tag{3.27}$$

These two expressions combined give the recurrence relation

$$\left\langle \mathcal{E}^{r+1}\right\rangle = U\left\langle \mathcal{E}^{r}\right\rangle - \left\langle \mathcal{E}^{r}\right\rangle', \tag{3.28}$$

which can be extended to any continuous function $h(\mathcal{E})$ to obtain

$$-\left\langle h(\mathcal{E})\right\rangle' = \left\langle \mathcal{E} h(\mathcal{E})\right\rangle - U\left\langle h(\mathcal{E})\right\rangle. \tag{3.29}$$

Thus $-\left\langle h(\mathcal{E})\right\rangle'$ is given in general by the covariance of $h(\mathcal{E})$ and $\mathcal{E}$.

Equation (3.28) with $r = 1$ gives a most important expression for the energy variance,

$$\sigma_{\mathcal{E}}^2 \equiv \left\langle (\mathcal{E}-U)^2\right\rangle = \left\langle \mathcal{E}^2\right\rangle - U^2 = -\frac{dU}{d\beta}, \tag{3.30}$$

which can be rewritten as the well-known relation (Mandl 1988)

$$\sigma_{\mathcal{E}}^2 = -\frac{dU}{d\beta} = k_B T^2\left(\frac{\partial U}{\partial T}\right)_{\omega} = k_B T^2 C_{\omega} \tag{3.31}$$

in terms of the specific heat (or heat capacity) C_{ω}.[9] Because C_{ω} is surely finite at low temperatures, the right-hand side of this expression is zero at $T = 0$, whence

$$\sigma_{\mathcal{E}}^2(T=0) = 0, \tag{3.32}$$

which shows that the description provided by the distribution W_g does not allow for the dispersion of the energy at zero temperature. The fact that W_g offers a thermodynamic description that admits thermal fluctuations only, and has no room for temperature-independent fluctuations, is an important shortcoming, as is clear when we consider a collection of harmonic oscillators (such as those of the electromagnetic field in equilibrium inside a cavity) which are endowed with a zero-point energy given by (3.19). Indeed, for such system the distribution W_g leaves out the fluctuations of the nonzero nonthermal component of the energy. We continue to work here

---

[9] C_{ω} coincides with the specific heat at constant volume, so the usual notation in this context is C_V. Still, we employ the subindex ω since we are considering ω to be a fixed parameter.

with the thermodynamic description, but later on we shall introduce a full-fledged statistical description that overcomes this limitation.

3.2.2 Some Consequences of the Recurrence Relation

The recurrence relation (3.28) and the Wien law can be recast into other interesting forms as follows. First we observe that the equation

$$\langle \mathcal{E}^r \rangle = \frac{1}{Z_g(\beta)} \int \mathcal{E}^r g(\mathcal{E}) e^{-\beta\mathcal{E}} d\mathcal{E}, \tag{3.33}$$

with the substitutions $\mathcal{E} = \mathcal{E}_0 \epsilon$ (ϵ dimensionless), and $\mathring{z} = 2\mathcal{E}_0\beta$, gives

$$\langle \mathcal{E}^r \rangle = \mathcal{E}_0^r f_r(\mathcal{E}_0, \mathring{z}), \tag{3.34}$$

where $f_r(\mathcal{E}_0, \mathring{z})$ is defined as

$$f_r(\mathcal{E}_0, \mathring{z}) = \frac{\int \epsilon^r g(\mathcal{E}_0\epsilon) e^{-\mathring{z}\epsilon/2} d\epsilon}{\int g(\mathcal{E}_0\epsilon) e^{-\mathring{z}\epsilon/2} d\epsilon}.$$

As follows from Eq. (3.34), f_r is an adimensional function, hence it can be expressed as a function of the adimensional parameter $\mathring{z}$ only. For the harmonic oscillator we use Wien's law to write $\mathcal{E}_0 = A\omega$, so that Eq. (3.34) reads

$$\langle \mathcal{E}^r \rangle = \omega^r A^r f_r(\mathring{z}), \tag{3.35}$$

which is a generalization of Wien's law for any power r.

On the other hand, the general recurrence relation between the moments of the energy, Eq. (3.28), can be rewritten as follows, using Eq. (3.31),

$$\left\langle \mathcal{E}^{r+1} \right\rangle = U \langle \mathcal{E}^r \rangle - \frac{d}{d\beta} \langle \mathcal{E}^r \rangle = U \langle \mathcal{E}^r \rangle + \sigma_{\mathcal{E}}^2 \frac{d}{dU} \langle \mathcal{E}^r \rangle, \tag{3.36}$$

$$\text{or} \quad \left\langle \mathcal{E}^{r+1} \right\rangle = \left(U + \sigma_{\mathcal{E}}^2 \frac{d}{dU} \right) \langle \mathcal{E}^r \rangle. \tag{3.37}$$

Successive iterations of this equation yield

$$\langle \mathcal{E}^r \rangle = \left(U + \sigma_{\mathcal{E}}^2 \frac{d}{dU} \right)^{r-1} U. \tag{3.38}$$

This reveals $U + \sigma_{\mathcal{E}}^2 \, (d/dU)$ as a kind of 'raising' operator for the higher moments of the energy, beginning with the first moment $\langle \mathcal{E} \rangle = U$. It is clear that for $\sigma_{\mathcal{E}}^2(U)$ even in U, the moments $\langle \mathcal{E}^r \rangle \, (U)$ have the parity of r.

The second centered moment of the energy is $\sigma_{\mathcal{E}}^2 = \left\langle (\mathcal{E}-U)^2 \right\rangle$; for the third one we obtain

$$\left\langle (\mathcal{E}-U)^3 \right\rangle = \left\langle \mathcal{E}^3 \right\rangle - 3U\sigma_{\mathcal{E}}^2 - U^3 = \sigma_{\mathcal{E}}^2 \frac{d}{dU} \left\langle (\mathcal{E}-U)^2 \right\rangle, \tag{3.39}$$

and by induction it can be seen that this last result generalizes into

$$\left\langle (\mathcal{E}-U)^r \right\rangle = \sigma_{\mathcal{E}}^2 \frac{d}{dU} \left\langle (\mathcal{E}-U)^{r-1} \right\rangle \tag{3.40}$$

for any integer $r \geq 1$. This equation shows that at $T = 0$, all centered moments are zero because of (3.32); hence the energy is exactly $\mathcal{E}_0$, and its distribution function reduces to $\delta(\mathcal{E} - \mathcal{E}_0)$ in this thermodynamic analysis.

3.3 Planck's Law from the Thermostatistics of the Harmonic Oscillator

3.3.1 General Statistical Equilibrium Distribution

It now becomes necessary to extend our description so as to allow for nonthermal fluctuations of the zero-point energy of the field, which are excluded by W_g. This can be achieved by paying attention to the *statistical* distribution of the energy $W_s(\mathcal{E})$. Since for every frequency the field contains a huge number of modes, the central limit theorem applies (Grimmett and Stirzaker 1983; Papoulis 1991) and hence the field amplitude of frequency ω follows a normal distribution. This means that the energy distribution follows the simple law

$$W_s(\mathcal{E}) = \frac{1}{U} e^{-\mathcal{E}/U}, \tag{3.41}$$

with

$$\int W_s(\mathcal{E}) d\mathcal{E} = 1, \qquad \int \mathcal{E} W_s(\mathcal{E}) d\mathcal{E} = U, \tag{3.42}$$

and the corresponding energy dispersion is given by (the subscript s denotes averages taken with respect to W_s, to be distinguished from those calculated with W_g)[10]

---

[10] This is a well-known statistical result, established for the first time by Lorentz for the *thermal* radiation field. A simple demonstration is given in Vedral (2005). Inclusion of the zero-point

$$\left(\sigma_{\mathcal{E}}^2\right)_s = U^2. \tag{3.43}$$

This (exponential) distribution of the energy [subject to the constraints (3.42)] has the property of maximizing the statistical entropy S_s, defined as

$$S_s = -k_B \int W_s(\mathcal{E}) \ln c_s W_s(\mathcal{E}) d\mathcal{E}, \tag{3.44}$$

where c_s is an appropriate constant with dimension of energy. Since the entropy is usually interpreted as a measure of the disorder present in the system (see e.g. Callen 1985; Mandl 1988), the maximal entropy property means maximum disorder, which is the natural demand for a system constituted by a huge number of independent components once equilibrium has been reached.

From Eq. (3.43) we see that W_s allows indeed for zero-point fluctuations, since at $T = 0$

$$(\sigma_{\mathcal{E}}^2)_s \Big|_0 = U^2(T=0) = \mathcal{E}_0^2, \tag{3.45}$$

which means that there is a nonthermal contribution to the energy fluctuations, with variance $\mathcal{E}_0^2$. The thermal contribution $\sigma_{\mathcal{E}_T}^2$ to the energy fluctuations at any temperature is obtained by subtracting from the total ones this nonthermal term $\mathcal{E}_0^2$. This is true because the thermal and nonthermal fluctuations have an entirely different source, so they are statistically independent, with a null correlation [see the discussion following Eq. (3.90)]. That is,

$$\sigma_{\mathcal{E}_T}^2 = (\sigma_{\mathcal{E}}^2)_s - \mathcal{E}_0^2, \tag{3.46}$$

whence

$$\sigma_{\mathcal{E}_T}^2 = U^2 - \mathcal{E}_0^2. \tag{3.47}$$

Recalling that $\sigma_{\mathcal{E}}^2$ in Eq. (3.31) stands for the *thermal* fluctuations of the energy, we can combine this latter with (3.47) and write (omitting the subindex T)

$$\sigma_{\mathcal{E}}^2 = U^2 - \mathcal{E}_0^2 = -\frac{dU}{d\beta}. \tag{3.48}$$

Before studying the consequences of this relation we observe that the distribution (3.41) leads to recurrence relations incorporating the nonthermal fluctuations. Indeed (3.41) gives for the moments of the energy

(Footnote 10 continued)
component does not modify this statistical property, since the argument to establish it remains in force.

$$\left\langle \mathcal{E}^r \right\rangle_s = r! U^r, \tag{3.49}$$

and making reiterative use of this equation one obtains

$$\left\langle \mathcal{E}^r \right\rangle_s = U \left\langle \mathcal{E}^{r-1} \right\rangle_s + (\sigma_{\mathcal{E}}^2)_s \frac{d}{dU} \left\langle \mathcal{E}^{r-1} \right\rangle_s . \tag{3.50}$$

Thus a sophisticated form of writing (3.49) in terms of a raising operator is

$$\left\langle \mathcal{E}^r \right\rangle_s = \left(U + (\sigma_{\mathcal{E}}^2)_s \frac{d}{dU} \right) \left\langle \mathcal{E}^{r-1} \right\rangle_s , \tag{3.51}$$

a result analogous to the previous recurrence relation (3.38), but now including the zero-point fluctuations. A much simpler, alternative form of this relation is

$$\left\langle \mathcal{E}^r \right\rangle_s = r U \left\langle \mathcal{E}^{r-1} \right\rangle_s . \tag{3.52}$$

3.3.2 *Mean Energy as Function of Temperature; Planck's Formula*

We note from Eq. (3.48) that knowledge of the variance $\sigma_{\mathcal{E}}^2$ as a function of U is enough to determine $U(\beta)$. Indeed, an integration of this equation—which articulates both thermodynamic and statistical information via Eqs. (3.31) and (3.46), respectively—

$$\frac{dU}{d\beta} = \mathcal{E}_0^2 - U^2(\beta) \tag{3.53}$$

gives the function $U(\beta)$. Subject to the condition $U \to \infty$ as $T \to \infty$, the result is

$$U(\beta) = \begin{cases} \frac{1}{\beta}, & \text{for } \mathcal{E}_0 = 0; \\ \mathcal{E}_0 \coth \mathcal{E}_0 \beta, & \text{for } \mathcal{E}_0 \neq 0. \end{cases} \tag{3.54}$$

Although the case $\mathcal{E}_0 = 0$ can of course be obtained from the last expression in the limit $\mathcal{E}_0 \to 0$, it is more illustrative to treat the two cases separately. As seen from Eq. (3.54), the mean energy as a function of the temperature depends critically on the presence of $\mathcal{E}_0$. For $\mathcal{E}_0 = 0$ the classical energy equipartition is recovered,

$$U_{\text{cl}} = \beta^{-1} = k_B T, \tag{3.55}$$

whereas for $\mathcal{E}_0 = \hbar\omega/2$ Planck's law is obtained,

$$U_{\text{Planck}}(\omega, T) = \tfrac{1}{2}\hbar\omega \coth \tfrac{1}{2}\hbar\omega\beta. \tag{3.56}$$

By taking the limit $T \to 0$, we verify that U_{Planck} includes the zero-point energy,[11]

$$U_{\text{Planck}}(\beta \to \infty) = \tfrac{1}{2}\hbar\omega = \mathcal{E}_0. \tag{3.57}$$

This establishes Planck's law as a physical result whose ultimate meaning—or cause—is the existence of a fluctuating zero-point energy of the field oscillators.

It is important to stress that Planck's law has been obtained without the introduction of any explicit quantum or discontinuity requirement. Equation (3.53) results from a thermostatistical analysis of the field modes, based on the properties of W_g and W_s, together with Wien's law, which opens the door to their zero-point energy $A\omega$. This leads us to conclude that Wien's law with $A \neq 0$ in Eq. (3.18) constitutes an extension of classical physics into the quantum domain—as evidenced by the quantum properties of the harmonic oscillator that ensue from Planck's law (see below). Thus, strictly speaking, Wien's law stands as a precursor of Planck's, and should be considered historically to contain the first quantum law.

The demonstration that the law that gave rise to quantum theory stems from the existence of a fluctuating zero-point energy, brings to the fore the crucial importance of this nonthermal energy for the understanding of quantum mechanics or, more generally, of quantum theory.

A brief comment on the thermal fluctuations of the energy seems in place before ending this section. We have seen that for $\mathcal{E}_0 \neq 0$ the thermal energy dispersion is given by

$$\sigma^2_{\mathcal{E}_T}(U) = U^2 - \mathcal{E}_0{}^2 \quad (U = U_{\text{Planck}}), \tag{3.58}$$

whereas in the classical case ($\mathcal{E}_0 = 0$),

$$\sigma^2_{\mathcal{E}_T}(U) = U^2 \quad (U = U_{\text{cl}}). \tag{3.59}$$

Whilst in the latter case the thermal fluctuations of the oscillator's energy depend solely on its (purely) thermal mean energy, U_{cl}, in the former case Eq. (3.58) relates the thermal fluctuations with the *total* mean energy U_{Planck}, which includes the temperature-independent contribution. The statistical description initiated in Sect. 3.3.1 will be resumed below, in Sect. 3.6.

---

[11] Planck's law without zero-point energy (the first relation derived by Planck) is obtained by fixing the constant of integration precisely as $-\mathcal{E}_0$, so that $U(\beta) = \mathcal{E}_0 \coth \mathcal{E}_0\beta - \mathcal{E}_0$. The existence of the zero-point energy remains hidden with this choice.

An additional comment is in place here. At first sight it would seem plausible to take the constant of integration in the first line of Eq. (3.54) as $\mathcal{E}_0$, so that the resulting function, $U(\omega, T) = k_B T + \mathcal{E}_0$ is apparently consistent with both the existence of a nonthermal energy and Wien's law. However, such choice must be discarded since this U cannot be obtained as a limit case of $\mathcal{E}_0 \neq 0$.

3.4 Planck, Einstein and the Zero-Point Energy

The previous discussion suggests separating the average energy U_{Planck} (which as of now will be denoted simply by U) into a thermal contribution U_T and a temperature-independent part $\mathcal{E}_0$,

$$U = U_T + \mathcal{E}_0, \tag{3.60}$$

so that Eq. (3.58) becomes

$$\sigma_{\mathcal{E}}^2 = U_T^2 + 2\mathcal{E}_0 U_T. \tag{3.61}$$

The first term in Eq. (3.60)

$$U_T = \mathcal{E}_0 \coth \mathcal{E}_0\beta - \mathcal{E}_0 = \frac{2\mathcal{E}_0}{e^{2\mathcal{E}_0\beta} - 1}, \tag{3.62}$$

with $\mathcal{E}_0 = \hbar\omega/2$, is Planck's law without the zero-point energy. At sufficiently low temperatures U_T takes the form

$$U_T(\beta \to \infty) = 2\mathcal{E}_0 e^{-2\mathcal{E}_0\beta}. \tag{3.63}$$

This is the (approximate) expression suggested by Wien at the end of the 19th century, and considered for some time to be the exact law for the blackbody spectral distribution. Equations (3.63) and (3.61) represent the germ of quantum theory, since it is precisely on their basis that Planck and Einstein advanced the notion of the quantum (for the material oscillators and for the radiation field, respectively). The following pages contain a discussion of their respective points of view and of the relations between these and our present notions based on the reality of the zero-point energy. A remarkable relationship will thus be disclosed.

3.4.1 Comments on Planck's Original Analysis

In his initial studies on the radiation field in equilibrium with matter, Planck (1900a, b) used as point of departure the expression for the derivative of the entropy[12]

$$\frac{\partial S}{\partial U} = \frac{1}{T}. \tag{3.64}$$

In line with the views and knowledge of his time, Planck recognized only the thermal energy, so U should be replaced here by U_T. In the high-temperature limit the relation (3.64) led him to write (putting $U_T(T \to \infty) = k_B T$)

---

[12] An early account of the material in this and the following two subsections is presented in de la Peña and Cetto (2002), and de la Peña et al. (2010a, b).

$$\frac{\partial^2 S}{\partial U_T^2} = \frac{\partial}{\partial U_T}\left(\frac{k_B}{U_T}\right) = -\frac{k_B}{U_T^2}. \tag{3.65}$$

For low temperatures Planck used Wien's result (3.63), assuming it to afford an exact description of the properties of the equilibrium field. He thus wrote

$$U_T = 2\mathcal{E}_0 e^{-2\mathcal{E}_0\beta} = 2\mathcal{E}_0 e^{-2\mathcal{E}_0/k_B T} = 2\mathcal{E}_0 e^{-2(\mathcal{E}_0/k_B)(\partial S/\partial U_T)}, \tag{3.66}$$

whence

$$\frac{\partial S}{\partial U_T} = -\frac{k_B}{2\mathcal{E}_0}\ln\frac{U_T}{2\mathcal{E}_0}, \tag{3.67a}$$

$$\frac{\partial^2 S}{\partial U_T^2} = -\frac{k_B}{2\mathcal{E}_0 U_T}. \tag{3.67b}$$

Not surprisingly, Eqs. (3.65) and (3.67b) give different results, since different temperature regimes were used in each case. As the simplest possibility Planck assumed that the description for arbitrary temperatures could be obtained by interpolating Eqs. (3.65) and (3.67b) and consequently he proposed the relation

$$\frac{\partial^2 S}{\partial U_T^2} = -\frac{k_B}{U_T^2 + 2\mathcal{E}_0 U_T}. \tag{3.68}$$

This equation leads directly to Planck's law without the zero-point term [Eqs. (3.62)], a result that Planck (against his will) interpreted, as is well known, as due to the quantization of the energy exchanged between the material oscillators of the cavity and the equilibrium radiation field.[13]

---

[13] The rationale behind Planck's reading of his formula is the following. If the system composed by the walls of the cavity (represented by a collection of material oscillators) and the enclosed radiation field exchanges energy not continuously but by lumps (which he called quanta) of value $n\hbar\omega$ ($n = 1, 2, 3, \ldots$), then the mean equilibrium energy is

$$U = \frac{\sum_{n=0}^{\infty} n\hbar\omega e^{-\beta n\hbar\omega}}{\sum_{n=0}^{\infty} e^{-\beta n\hbar\omega}}.$$

Performing the summations with the aid of the relation $\sum_{n=0}^{\infty} x^n = 1/(1-x)$, one gets

$$U = \frac{\hbar\omega}{e^{\hbar\omega\beta} - 1},$$

which is just the U_T in Planck's theory. If by contrast a continuous exchange of energy is assumed instead of a discrete one, the sum above must be replaced by an integral from 0 to ∞. The reader can easily check that in this case the result is the classical formula $U_T = 1/\beta = k_B T$.

3.4.2 Einstein's Revolutionary Step

A few years later, Einstein argued that even though Eq. (3.68) was empirically confirmed (through Planck's law), its full meaning remained to be clarified. For this purpose Einstein chose also to take Eq. (3.64) as a safe point of departure, whence he wrote

$$\frac{\partial^2 S}{\partial U_T^2} = \frac{\partial}{\partial U_T}\frac{1}{T} = -\frac{1}{T^2 C_\omega}, \tag{3.69}$$

or

$$k_B T^2 C_\omega = -k_B \left(\frac{\partial^2 S}{\partial U_T^2}\right)^{-1}. \tag{3.70}$$

Equation (3.70) combined with (3.68) and (3.31) gives

$$k_B T^2 C_\omega = -\frac{dU_T}{d\beta} = \sigma_{\mathcal{E}}^2 = U_T^2 + 2\mathcal{E}_0 U_T, \tag{3.71}$$

which is the same as (3.61). Einstein recognized the disagreement between this result and the classical expression $\sigma_{\mathcal{E}}^2 = U_T^2$. As is well known, it is here where he made his most—according to him (Rigden 2005), his *only*—revolutionary step in physics. He interpreted the first term on the right-hand side of (3.71) as due to the fluctuations of the thermal field produced by the interference among its modes of a given frequency. This interpretation follows from considering the limit of (3.71) at high temperatures, at which $U_T \gg \mathcal{E}_0$ and therefore $\sigma_{\mathcal{E}}^2 = U_T^2$, as was predicted by Lorentz on the basis of Maxwell's equations and is discussed in relation with Eq. (3.43). Einstein thus saw in this term a direct manifestation of the wavelike nature of light.

As for the second term in (3.71), which in the context of classical thermodynamics is completely unexpected, the fact that it leads to the quantum theory of Planck led Einstein to interpret it in terms of light quanta (Einstein 1905a, b), seeing in the expression $2\mathcal{E}_0 U_T$ a manifestation of discrete properties of the radiation field, as follows. According to Planck, the average energy exchanged between n material oscillators (representing the walls of the cavity) of frequency ω and the radiation field is $\Delta U = \hbar\omega\langle n\rangle$, and contributes with $\sigma_{\Delta U}^2 = 2\mathcal{E}_0\Delta U = \hbar^2\omega^2\langle n\rangle$ to the fluctuations of the field, as follows from (3.71). For Einstein, the linearity of the variance in $\langle n\rangle$ suggested a Poisson distribution of n independent events, each corresponding to an exchange of energy equal to $\hbar\omega = 2\mathcal{E}_0$.[14] It is the interpretation by Einstein of the linear term as representing a discrete or 'corpuscular' contribution, with each

[14] A Poisson distribution refers to the probability of n independent discrete events taking place simultaneously, and has the form

$$P_a(n) = e^{-a}\frac{a^n}{n!}.$$

It is easy to verify that for this distribution the mean of n is $\langle n\rangle = a$ and its variance is precisely $\sigma_n^2 = \langle n\rangle$.

corpuscle being an independent packet of energy $\hbar\omega$, what gave birth to the notion of the photon (see Vedral 2005 for a simple derivation). It is clear from Eq. (3.71) that the discrete structure of the field will manifest itself only at very low temperatures, when the linear term dominates over the quadratic, wavelike one. However, it is important to stress, as Einstein did as of 1909, that the two terms coexist at all temperatures, and thus, both particle and wave manifestations of light coexist at all temperatures (Einstein 1909). This observation is sometimes ignored to argue that they are mutually exclusive, although there exist both theoretical arguments and experiments that demonstrate the possible coexistence of the two aspects of the behaviour of light.[15]

3.4.3 Disclosing the Zero-Point Field

It is important to note that no zero-point energy was considered by either Planck or Einstein in their analysis of Eqs. (3.68) and (3.71), respectively. Instead, as stated above, Planck interpreted the term $2\mathcal{E}_0 U_T$ in Eq. (3.68) as a result of the discontinuities in the processess of energy exchange between matter and field (more specifically in the emissions, as of 1912). Einstein in his turn saw in $2\mathcal{E}_0 U_T$ a manifestation of the corpuscular nature of the field, and thus pointed to it as the key to Planck's law. Now, from the point of view proposed here the consideration of the zero-point energy gives rise to a third understanding of Eq. (3.71) that does not depend on the notion of quanta. The elucidation of U_T^2 as the result of the interference of the modes of frequency ω of the thermal field suggests to interpret $2\mathcal{E}_0 U_T$ as due to additional interferences, now between the thermal field and a *zero-point radiation field* of mean energy $\mathcal{E}_0$ (per mode of frequency ω) that is present at all temperatures. As is by now clear, Eq. (3.71) lacks the extra term $\mathcal{E}_0^2$ representing the nonthermal fluctuations, just because the thermodynamic description has no room for them; this shortcoming has been overcome with the introduction of the distribution W_s, Eq. (3.41).

From this new perspective the notion of intrinsic discontinuities in the energy exchange or in the field itself is unnecessary to explain either Planck's law or the linear term in Eq. (3.71); it is the existence of a (fluctuating) zero-point radiation field (ZPF) what accounts for that law. This could of course not be Planck's or Einstein's interpretation because the zero-point energy (and more so the zero-point field) was still unknown at that time, even though their results were consistent with its existence.

The concept of a zero-point energy of the radiation field appeared for the first time in 1912, in a work where Planck attempted another derivation of his law, motivated by his well-known uneasiness with the idea of introducing discontinuities in

---

[15] Graded realizations of complementarity relations (wave-like or particle-like behavior) have been under close scrutiny during the last decades; see e.g. Jaeger et al. (1995), Englert (1996), Engert and Bergou (2000), Liu et al. (2009), Flores and de Tata (2010) (see also Ghose and Home 1996). The general validity of Einstein's fluctuation formula (3.71) had been verified experimentally since earlier times; see Aldemade et al. (1966), Kattke and van der Ziel (1970). The authors are grateful to M. D. Godfrey for drawing their attention to these references.

our theoretical descriptions (Planck 1912). Some time thereafter Einstein and Stern (1913) used the idea of a zero-point energy, although applied to molecules, i.e., to mechanical oscillators. Unfortunately the authors were obliged to use the (incorrect) value $\hbar\omega$ for this energy; this along with other difficulties led Einstein to abandon such line of research.[16] Shortly thereafter the notion of a zero-point field was born anew, when Nernst made his visionary proposal (Nernst 1916), as briefly mentioned in the preface.

3.5 Continuous Versus Discrete

We have just seen how three alternative approaches provide three quite different readings of the same quantity, $U_T^2 + 2\mathcal{E}_0 U_T$. In these approaches, either the zero-point energy (of a continuous field) or the energy quantization is identified as the notion underlying the Planck spectral energy distribution. Therefore the next logical step is to inquire about the relation between the zero-point energy and quantization. Is quantization inevitably linked to Planck's law, or is it merely the result of a point of view, of a voluntary but dispensable choice?

3.5.1 The Partition Function

An answer to the above question is found from an analysis of the partition function obtained from (3.54). As follows from Eq. (3.27), $Z_g(\beta)$ can be determined by direct integration of

$$U = -\frac{d \ln Z_g(\beta)}{d\beta}, \tag{3.72}$$

with $U(\beta)$ given by the second of Eq. (3.54). The result is

$$Z_g = \frac{C}{\sin h\mathcal{E}_0\beta}, \tag{3.73}$$

where C is a numerical constant whose value is determined by requiring the classical result $Z_g = (s\beta\omega)^{-1}$ [Eq. (3.25b)] to be recovered in the limit $T \to \infty$. This leads to $C = \mathcal{E}_0/s\omega = \hbar/2s$, so that

$$Z_g(\beta) = \frac{\mathcal{E}_0}{s\omega \sinh \mathcal{E}_0\beta}. \tag{3.74}$$

[16] More detailed discussions of these points, from a modern perspective, are presented in Milonni (1994); see also Boyer (1969a) and Jiménez et al. (1980).

On the other hand, from Eqs. (3.15) and (3.72) the thermodynamic potential ϕ can be written in the form

$$\phi = \ln Z_g. \tag{3.75}$$

This along with Eq. (3.17) gives for the entropy (up to an additive constant, and writing $S = S_g$)

$$\begin{aligned} \S_g &= k_B \ln Z_g + \frac{U}{T} \\ &= k_B \ln \tfrac{\hbar}{s} - k_B \ln(2 \sinh \mathcal{E}_0 \beta) + k_B \beta U, \end{aligned} \tag{3.76}$$

which in the zero-temperature limit reduces to

$$S_g(\beta \to \infty) = k_B \ln \frac{\hbar}{s}. \tag{3.77}$$

To set the origin of the entropy at $T = 0$ one must take $s = \hbar$,[17] hence the partition function takes the form

$$Z_g(\beta) = \frac{1}{2 \sin h \mathcal{E}_0 \beta}. \tag{3.78}$$

3.5.2 The Origin of Discreteness

Once we have determined the partition function Z_g we are in position to discuss the discontinuities characteristic of the quantum theory, which are hidden in the continuous description given by the distribution W_g. To this end we expand Eq. (3.78) and write (see Santos 1975; Theimer 1976; Landsberg 1981 for related discussions)

$$Z_g = \frac{1}{2 \sinh \mathcal{E}_0 \beta} = \frac{e^{-\beta \mathcal{E}_0}}{1 - e^{-2\beta \mathcal{E}_0}} = \sum_{n=0}^{\infty} e^{-\beta \mathcal{E}_0 (2n+1)} = \sum_{n=0}^{\infty} e^{-\beta \mathcal{E}_n}, \tag{3.79}$$

where

[17] This is a most significant quantum result. In the quantum statistical description the finite quantity $\hbar^3$ plays the role of a minimal element of volume in phase space. This idea was introduced formally for the first time by Planck in his early studies of the blackbody spectrum (Planck 1900a, b). Later, in 1924, Bose assumed that two or more distributions of microstates that differ only in the permutation of phase points within a subregion of phase space of volume $\hbar^3$, are to be regarded as identical, which already corresponds to the Bose-Einstein statistics. In the classical description the volume of such elementary cells is taken to tend to zero in order to recover the continuity of the phase space. It is remarkable that, already in his classical statistical studies, Boltzmann introduced formally the idea of a discrete phase space (see e.g., Jones 2008, Chap. 3).

$$\mathcal{E}_n \equiv (2n+1)\mathcal{E}_0 = \hbar\omega n + \tfrac{1}{2}\hbar\omega. \tag{3.80}$$

Equation (3.79) allows now the determination of the function $g(\mathcal{E})$ by means of (3.21b),

$$Z_g(\beta) = \int g(\mathcal{E})e^{-\beta\mathcal{E}}d\mathcal{E} = \sum_{n=0}^{\infty} e^{-\beta\mathcal{E}_n} = \int_0^{\infty}\sum_{n=0}^{\infty}\delta(\mathcal{E}-\mathcal{E}_n)e^{-\beta\mathcal{E}}d\mathcal{E}, \tag{3.81}$$

whence

$$g(\mathcal{E}) = \sum_{n=0}^{\infty}\delta(\mathcal{E}-\mathcal{E}_n). \tag{3.82}$$

The substitution of (3.82) into Eq. (3.21a) finally determines the probability density $W_g(\mathcal{E})$,

$$W_g(\mathcal{E}) = \frac{1}{Z_g}\sum_{n=0}^{\infty}\delta(\mathcal{E}-\mathcal{E}_n)e^{-\beta\mathcal{E}}. \tag{3.83}$$

This distribution gives for the mean value of any function $f(\mathcal{E})$

$$\langle f(\mathcal{E})\rangle = \int W_g(\mathcal{E})f(\mathcal{E})d\mathcal{E} = \frac{1}{Z_g}\sum_{n=0}^{\infty} f(\mathcal{E}_n)e^{-\beta\mathcal{E}_n} = \sum_{n=0}^{\infty} w_n f(\mathcal{E}_n), \tag{3.84}$$

with the weights w_n given by

$$w_n = \frac{e^{-\beta\mathcal{E}_n}}{Z_g} = \frac{e^{-\beta\mathcal{E}_n}}{\sum_{n=0}^{\infty} e^{-\beta\mathcal{E}_n}}. \tag{3.85}$$

The final form of $W_g(\mathcal{E})$, Eq. (3.83), identifies $\{\mathcal{E}_n = \hbar\omega(n+1/2)\}$ with the set of discrete energy levels accesible to the oscillators. Such discreteness, seemingly excluding all other values of the energy, is due to the highly pathological distribution $g(\mathcal{E})$, Eq. (3.82). As a result, (3.84) shows that the mean value of a function of the *continuous* variable $\mathcal{E}$ calculated with the distribution $W_g(\mathcal{E})$, can be obtained equivalently by averaging over the set of *discrete* indices (or *states*) n, with respective weights w_n. Thus, although both averages are formally equivalent, their descriptions are essentially different: one refers to the continuous energy $\mathcal{E}$, the other one to discrete states (levels) with energy $\mathcal{E}_n$. As this latter is completely characterized by the state n, it is natural to interpret the last equality in Eq. (3.84) as a manifestation of the discrete (quantized) nature of the energy. Indeed, the last equality in Eq. (3.84) can be recognized as the description afforded by the density matrix for a canonical ensemble of quantum oscillators at temperature T, with the weights w_n given by (3.85) (see e.g. Cohen-Tannoudji et al. 1977).

The above discussion points to the fundamental role played by the zero-point energy in explaining quantization, by putting it at the root of Eq. (3.79) and hence of Eq. (3.82). From the present point of view, and contrary to the usual credo, the radiation field is not intrinsically quantized, but it becomes so when attaining equilibrium through its interaction with matter. In other words, quantization is here exhibited as an emergent property of matter and field in interaction, an idea that is closely examined from several angles in the following chapters, becoming thus the leitmotiv of the book.

3.6 A Quantum Statistical Distribution

The thermostatistical analysis of a canonical ensemble of oscillators has led to the conclusion that although $\mathcal{E}$ is a continuous variable, its equilibrium distribution possesses extremely peaked values. In other words, the energies that conform to the thermal equilibrium state described by the distribution W_g belong, roughly speaking, to a discrete spectrum. This explains why the mean value $\langle f(\mathcal{E})\rangle$, which corresponds to an equilibrium state, involves only the discrete set $\mathcal{E}_n$. However, the energy still fluctuates and in doing so tends to fill the interspaces between its discrete values.[18] Thus we find that temperature-independent fluctuations appear as a characteristic trait of quantum systems. A closer study of this property allows to establish contact with one of the most frequently used distributions in quantum statistics.

3.6.1 Total Energy Fluctuations

The appropriate statistical distribution that includes all (thermal as well as nonthermal) fluctuations is given by Eq. (3.41),

$$W_s(\mathcal{E}) = \frac{1}{U}e^{-\mathcal{E}/U}, \tag{3.86}$$

and the variance of the energy at all temperatures (including $T = 0$) is $(\sigma^2_{\mathcal{E}})_s = U^2$. Using the decomposition (3.60) we may write for the total energy fluctuations

$$(\sigma^2_{\mathcal{E}})_s = U^2 = (U_T + \mathcal{E}_0)^2 = U_T^2 + 2\mathcal{E}_0 U_T + \mathcal{E}_0^2. \tag{3.87}$$

[18] The existence of energy fluctuations associated with the natural linewidth and other processes (see e.g. Schiff 1955; Louisell 1973), effectively dilutes this discrete distribution of energies into a somewhat smoothened-out distribution acquiring a more continuous shape. Thus $g(\mathcal{E})$ should be seen as a theoretical limiting distribution.

This result generalizes Eq. (3.59) to include both thermal and nonthermal energy fluctuations. In conformity with the present discussion, the total energy can be written in terms of its thermal and nonthermal *fluctuating* parts,

$$\mathcal{E} = \mathcal{E}_T + \mathcal{E}_0. \tag{3.88}$$

The total energy fluctuations are then given by

$$(\sigma^2_{\mathcal{E}})_s = \sigma^2_{\mathcal{E}_T} + \sigma^2_{\mathcal{E}_0} + 2\Gamma(\mathcal{E}_T, \mathcal{E}_0), \tag{3.89}$$

where $\Gamma(\mathcal{E}_T, \mathcal{E}_0)$ is the covariance

$$\Gamma(\mathcal{E}_T, \mathcal{E}_0) = \langle \mathcal{E}_T \mathcal{E}_0 \rangle - \langle \mathcal{E}_T \rangle \langle \mathcal{E}_0 \rangle. \tag{3.90}$$

Comparing Eqs. (3.89) and (3.87), and identifying the temperature-dependent part of the fluctuations of the whole field $U_T^2 + 2\mathcal{E}_0 U_T$ with $\sigma^2_{\mathcal{E}_T}$ and $\mathcal{E}_0^2$ with $\sigma^2_{\mathcal{E}_0}$, we verify that $\Gamma(\mathcal{E}_T, \mathcal{E}_0) = 0$, as was expected considering that the fluctuations of $\mathcal{E}_T$ and $\mathcal{E}_0$ are statistically independent, due to the independence of their sources.

The entropy S_s follows from Eqs. (3.44) and (3.86),

$$S_s = -k_B \int W_s(\mathcal{E}) \ln c_s W_s(\mathcal{E}) d\mathcal{E} = k_B \ln c_s^{-1} U + k_B, \tag{3.91}$$

whence

$$\frac{\partial S_s}{\partial U} = \frac{k_B}{U}. \tag{3.92}$$

A comparison with the thermodynamic entropy, which satisfies

$$\frac{\partial S_g}{\partial U} = \frac{1}{T}, \tag{3.93}$$

shows that these two entropies coincide only when $\mathcal{E}_0 = 0$, i.e., for $U = k_B T$.

3.6.2 Quantum Fluctuations and Zero-Point Fluctuations

Let us now investigate how the nonthermal fluctuations become manifest in the statistical properties of the ensemble of oscillators. The value of the energy of the harmonic oscillator [(cf. Eq. (3.1)]

$$\mathcal{E} = (p^2 + \omega^2 q^2)/2 \tag{3.94}$$

can be used as a starting point to perform a transformation from the energy distribution $W_s(\mathcal{E})$ to a distribution $w_s(p, q)$ defined in the oscillator's phase space (p, q). To this end we introduce the pair of variables $(\mathcal{E}, \theta)$ related to the couple (p, q) by[19]

$$p = \sqrt{2\mathcal{E}} \cos\theta, \tag{3.95a}$$

$$q = \sqrt{\frac{2\mathcal{E}}{\omega^2}} \sin\theta, \tag{3.95b}$$

so that $w_s(p, q)$ is given by (Papoulis 1991; Birnbaum 1961)[20]

$$w_s(p, q) = W_s(\mathcal{E}(p, q), \theta(p, q)) \left| \frac{\partial(\mathcal{E}, \theta)}{\partial(p, q)} \right|, \tag{3.96}$$

with the Jacobian of the transformation

$$\frac{\partial(p, q)}{\partial(\mathcal{E}, \theta)} = \left| \frac{\partial(\mathcal{E}, \theta)}{\partial(p, q)} \right|^{-1} = \frac{1}{\omega}. \tag{3.97}$$

Now, $W_s(\mathcal{E})$ is a marginal probability density that can be obtained from $W_s(\mathcal{E}, \theta)$ by integrating over the variable θ, so that

$$W_s(\mathcal{E}) = \int_0^{2\pi} W_s(\mathcal{E}, \theta) d\theta. \tag{3.98}$$

For a system of harmonic oscillators in equilibrium, the trajectories (in general, the surfaces) of constant energy do not depend on θ, so all values of θ are equally probable, which means that

$$W_s(\mathcal{E}, \theta) = \frac{1}{2\pi} W_s(\mathcal{E}). \tag{3.99}$$

Using Eqs. (3.86), (3.94) and (3.96) we thus obtain for the distribution in phase space:

$$w_s(p, q) = \frac{\omega}{2\pi} W_s(\mathcal{E}(p, q)) = \frac{\omega}{2\pi U} \exp\left(- \frac{p^2 + \omega^2 q^2}{2U} \right). \tag{3.100}$$

This expression, which is known in quantum theory as the Wigner function for the harmonic oscillators (Hillery et al. 1984), can be factorized as a product of two normal distributions,

---

[19] The transformation defined by (3.95a) and (3.95b) is an extended canonical transformation (Goldstein 1980), which differs from a canonical one—from the action and angle variables (J, θ), with $J = \mathcal{E}/\omega$, to the phase space variables (p, q)—only by a constant factor ω. Of course $\theta = \omega t$.

[20] When a probability $P(x)$ is expressed in terms of a new variable $y(x)$ as $W(y)$, the equality $P(x)dx = W(y)dy$ holds. Equation (3.96) is simply the generalization of this result to a two-dimensional space. See Papoulis (1991), Chap. 6 for a detailed derivation.

$$w_s(p,q) = w(p)w(q) = \frac{1}{\sqrt{2\pi\sigma_p^2}} e^{-p^2/2\sigma_p^2} \cdot \frac{1}{\sqrt{2\pi\sigma_q^2}} e^{-q^2/2\sigma_q^2}, \qquad (3.101)$$

where $\sigma_p^2 = U$ and $\sigma_q^2 = U/\omega^2$. The product of these dispersions gives

$$\sigma_q^2 \sigma_p^2 = \frac{U^2}{\omega^2} = \frac{\mathcal{E}_0^2}{\omega^2} + \frac{\sigma_{\mathcal{E}_T}^2}{\omega^2} \geq \frac{\mathcal{E}_0^2}{\omega^2} = \frac{\hbar^2}{4}, \qquad (3.102)$$

where Eq. (3.58) was used to write the second equality and the value $\mathcal{E}_0 = \hbar\omega/2$ was introduced into the last one.

Equation (3.102) points to the fluctuating zero-point energy as the ultimate (and irreducible) source of the so-called quantum fluctuations. Indeed, the magnitude of $\sigma_q^2\sigma_p^2$ is bounded from below because of the nonthermal energy fluctuations; the minimum value $\hbar^2/4$ is reached when all thermal fluctuations have been suppressed, which means $T = 0$. Therefore, descriptions afforded by purely thermal distributions such as W_g cannot account for the meaning of these inequalities. This result stresses again the fact that once a zero-point energy has been introduced into the theory, new distributions (specifically statistical rather than thermodynamic) are needed to include its fluctuations and to obtain the corresponding quantum statistical properties. Though here we have arrived at the Heisenberg inequality (3.102) by considering a system of harmonic oscillators, later on (particularly in Chap. 5) we will derive it for an arbitrary system, and again the presence of the ZPF will turn out to be decisive in reaching the result. Finally, note that the Heisenberg inequalities should be understood as referring to statistical variances, due to the statistical nature of (3.102).

3.6.3 Comments on the Reality of the Zero-Point Fluctuations

As mentioned earlier, the concept of a zero-point energy of the radiation field entered into scene as early as 1912, with Planck's second derivation of the blackbody spectrum. Yet further to the frustrated attempt by Einstein and Stern (1913), and despite the suggestive proposal made by Nernst (1916) to consider the ZPF as responsible for atomic stability, little or no attention was paid to its existence as a real physical entity that could have a role in the newly developing quantum mechanics.[21] Interestingly, it was the crystallographers who, prompted by Debye's theoretical work, set out to measure the spectroscopic effects of the zero-point energy through X-ray analysis and thereby seemengly verified its existence (James et al. 1928; Wollan 1931).

[21] The value of the deep insight of Nernst will be substantiated in Chaps. 4–7. This conceptualization is in vivid contrast with the notion of 'virtual' usually applied to the fluctuating vacuum field.

As mentioned in Sect. 1.4.1, today it is well accepted that the fluctuations of the electromagnetic vacuum are responsible for important observable physical phenomena. Perhaps their best known manifestations, within the atomic domain, are the Lamb shift of energy levels (see e.g. Milonni 1994) and their contribution to the spontaneous transitions of the excited states to the ground state. They are known to contribute one half of the Einstein A-coefficient for 'spontaneous' transitions, the other half being due to radiation reaction (see e.g. Milonni 1994; Davydov 1965).[22] By far the most accepted evidence of the reality of the ZPF is the Casimir effect, that is, the force between two parallel neutral metallic plates resulting from the modification of the field by the boundaries (see e.g. Boyer 1970; Bordag et al. 2009). The existence of the ZPF vcan therefore be considered a reasonably well established physical fact.[23] In the following chapters we will have occasion to study in depth the essential role played more broadly by this random field in its interaction with matter at the atomic level.

References

Aldemade, C.Th.J, Bolwijn, P.T., van der Veer, J.H.C.: Single-beam measurement of Bose-Einstein fluctuations in a natural Gaussian radiation field. Phys. Lett. **22**, 70 (1966)

Birnbaum, Z.W.: Introduction to probability and mathematical statistics. Harper & Row, Tokyo (Chap. 5) (1961)

Bordag, M., Klimchitskaya, G.L., Mohideen, U., Mostepanenko, V.M.: Advances in the Casimir Effect. Oxford University Press, Oxford (2009)

Bose, S.N.: Plancks Gesetz und Lichtquanten-hypothese. Zeit. Phys. **26**, 178 (1924)

Boyer, T.H.: Derivation of the blackbody radiation spectrum without quantum assumptions. Phys. Rev. **182**, 1374 (1969a). A more comprehensive list of references on the role played by the zpf in connection with Planck's law, particularly of the earlier extensive work by T. H. Boyer on the subject, can be seen in *The Dice*

Boyer, T.H.: Classical statistical thermodynamics and electromagnetic zero-point radiation, Phys. Rev. **186**, 1304 (1969b). Boyer uses the terms 'caloric' and 'probabilistic' entropies instead of the terms 'thermal' and 'statistical', respectively, used here

Boyer, T.H.: Quantum zero-point energy and long-range forces. Ann. Phys. **56**, 474 (1970)

*Boyer, T.H.: Equilibrium of random classical electromagnetic radiation in the presence of a nonrelativistic nonlinear electric dipole oscillator. Phys. Rev. D **13**, 2832 (1976)

Boyer, T.H.: Derivation of the Planck radiation spectrum as an interpolation formula in classical electrodynamics with classical electromagnetic zero-point radiation. Phys. Rev. D **27**, 2906 (1983)

---

[22] This point is discussed in Chap. 6, in connection with the derivation of the Einstein A and B coefficients.

[23] There is an extended stance against the assumption of the reality of the fluctuating ZPF, based mainly on the argument that it does not activate photon detectors. Another frequent argument refers to the unobserved tremendous gravitational effects that such field should produce. The first objection has been answered by offering models of photon counters compatible with the reality of the zero-point fluctuations (Santos 2002a, b). The gravitational puzzle is related to the problem of the cosmological constant, and represents an age-old unsolved fundamental problem that besets a broad parcel of physics (see e.g. Weinberg 1989).

Boyer, T.H.: Derivation of the blackbody radiation spectrum from the equivalence principle in classical physics with classical electromagnetic zero-point radiation. Phys. Rev. D **29**, 1096 (1984)
Boyer, T.H.: Thermodynamics of the harmonic oscillator: Wien's displacement law and the Planck spectrum. Am. J. Phys. **71**, 866 (2003)
Boyer, T.H.: Blackbody radiation and the scaling symmetry of relativistic classical electron theory with classical electromagnetic zero-point radiation. Found. Phys. **40**, 1102 (2010a)
Boyer, T.H.: Derivation of the Planck spectrum for relativistic classical scalar radiation from thermal equilibrium in an accelerating frame. Phys. Rev. D **81**, 105024 (2010b)
Boyer, T.H.: The blackbody radiation spectrum follows from zero-point radiation and the structure of relativistic spacetime in classical physics. Found. Phys. **42**, 595 (2012)
Callen, H.B.: Thermodynamics and an introduction to thermostatistics, 2nd edn. Wiley, New York (1985) (Chap.17)
Cohen-Tannoudji, C., Diu, B., Laloë, F.: Quantum Mechanics, vol. I. Wiley, New York (1977) (Chap. 3)
Cole, D.C.: Derivation of the classical electromagnetic zero-point radiation spectrum via a classical thermodynamic operation involving van der Waals forces. Phys. Rev. A **42**, 1847 (1990)
Davydov, S.: Quantum Mechanics. Addison-Wesley, Oxford (1965)
de la Peña, L., Cetto, A.M.: The Quantum Dice, An Introduction to Stochastic Electrodynamics. Kluwer, Dordrecht (1996) (Referred to in the book as *The Dice*)
de la Peña, L., Cetto, A.M.: Planck's law as a consequence of the zero-point radiation field. Rev. Mex. Fís. **48** (Suppl. 1), 1 (2002)
de la Peña, L., Valdés-Hernández, A., Cetto, A.M.: Statistical consequences of the zero-point energy of the harmonic oscillator. Am. J. Phys. **76**, 947 (2008)
de la Peña, L., Valdés-Hernández, A., Cetto, A.M.: La ley de Planck, notable e inevitable consecuencia del campo de punto cero. in: García-Colín Scherer, L., del Río-Correa, J.L., Uriarte Rivera, H. (eds.) Max Planck, a ciento cincuenta años de su nacimiento. El Colegio Nacional, Mexico (2010a)
de la Peña, L., Valdés-Hernández, A., Cetto, A.M.: Wien's Law with Zero-Point Energy Implies Planck's Law Unequivocally. In: New Trends in Statistical Physics. Festschrift in Honor of Leopoldo García-Colín's 80th Birthday. World Scientific, Singapore (2010b)
del Río-Correa, J.L.: Las diferentes facetas de la ley de Wien, in: Peralta, J.A., Ramírez, D., Mendez, A.F. (eds.) Memorias de la XV Reunión Nacional de Física y Matemáticas 2010, p. 274. ESFM, IPN, México (2010)
Einstein, A.: Über einen die Erzeugung und Verwandlung des Lichtes betreffenden heuristischen Gesichtspunkt, Ann. d. Phys. **17**, 132 (1905a). http://onlinelibrary.wiley.com/doi/10.1002/andp.19053220607/pdf. English translation in A. B. Arons and M. B. Peppard 1965, Einstein's proposal of the photon concept: A translation of the Annalen der Physik paper of 1905, Am. J. Phys. **33**, 367;http://astro1.panet.utoledo.edu/ljc/PE_eng.pdf. Also in The Collected Papers of Albert Einstein (Princeton University Press, Princeton), Vol. 3
Einstein, A.: Letter to Conrad Habicht, May 18 or 25. English translation in The Collected Papers of Albert Einstein, vol. 5, 31. An interesting discussion of this point is presented in Rigden 2005 (1905b)
Einstein, A.: Die Plancksche Theorie der Strahlung und die Theorie der spezifischen Wärme, Ann. d. Phys. **22**, 180 (1907). English translation in The Collected Papers of Albert Einstein
Einstein, A.: Zum gegenwärtigen Stand des Strahlungsproblems, Zeit. Phys. **10**, 185 (1909). English translation in The Collected Papers of Albert Einstein
Einstein, A., Stern, O.: Einige Argumente für die Annahme einer molekularen Agitation beim absoluten Nullpunkt, Ann. d. Phys. **40**, 551 (1913). English translation in The Collected Papers of Albert Einstein
Englert, B.-G.: Fringe visibility and which-way information: an inequality. Phys. Rev. Lett. **77**, 2154 (1996)
Engert, B.-G., Bergou, J.A.: Quantitative quantum erasure. Opt. Commun. **179**, 337 (2000)

Flores, E.V., de Tata, J.M.: Complementarity paradox solved: surprising consequences. Found. Phys. **40**, 1731 (2010)
Ghose, P., Home, D.: The two-prism experiment and wave-particle duality of light. Found. Phys. **26**, 943 (1996)
Goldstein, H.: Classical Mechanics, 2nd edn. Addison Wesley, Reading (1980)
Grimmett, G.R., Stirzaker, D.R.: Probability and Random Processes. Clarendon Press, Oxford (1983) (Chap. 4)
Hillery, M., O'Connell, R.F., Scully, M.O., Wigner, E.P.: Distribution functions in physics: fundamentals. Phys. Rep. **106**, 121 (1984)
Jaeger, G., Shimony, A., Vaidman, L.: Two interferometric complementarities. Phys. Rev. A **51**, 54 (1995)
James, R.W., Waller, T., Hartree, D.R.: An investigation into the existence of zero-point energy in the rock-salt lattice by an X-ray diffraction method. Proc. Roy. Soc. A **118**, 334 (1928)
Jiménez, J.L., de la Peña, L., Brody, T.: Zero-point term in cavity radiation. Am. J. Phys. **48**, 840 (1980)
Jones, S.: The Quantum-Ten. Oxford University Press, New York (2008)
José, J.V., Saletan, E.J.: Classical Dynamics. A Contemporary Approach. Cambridge University Press, Cambridge (1998)
Kattke, G.W., van der Ziel, A.: Verification of Einstein's formula for fluctuations in thermal radiation. Physica **49**, 461 (1970)
Landau, L.D., Lifshitz, E.M.: Mechanics, 3rd edn. Course of Theoretical Physics, vol. 1. Butterworth-Heinemann, Oxford (1976)
Landsberg, P.T.: Einstein and statistical thermodynamics II. Oscillator quantisation. Eur. J. Phys. **2**, 208 (1981)
Liu, N.-L., Li, L., Yu, S., Chen, Z.-B.: Duality relation and joint measurement in a Mach-Zehnder interferometer. Phys. Rev. A **79**, 052108 (2009)
Louisell, W.H.: Quantum Statistical Properties of Radiation. Wiley, New York (1973) (Chap. 5)
Mandl, F.: Statistical Physics. Wiley, New York (1988)
Marshall, T.W.: Random electrodynamics. Proc. Roy. Soc. A **276**, 475 (1963)
Marshall, T.W.: A classical treatment of blackbody radiation. Nuovo Cim. **38**, 206 (1965)
Milonni, P.W.: The Quantum Vacuum. Academic Press, New York (1994)
Nernst, W.: Uber einen Versuch, von quantentheoretischen Betrachtungen zur Annahme stetiger Energieänderungen zurückzukehren. Verh. Deutsch. Phys. Ges. **18**, 83 (1916)
Papoulis, A.: Probability, Random Variables, and Stochastic Processes. McGraw-Hill, Boston (1991) (Chap. 6)
Pathria, R.K.: Statistical Mechanics. Butterworth-Heinemann, Oxford (1996)
Piña, E., de la Selva, S.M.T.: Thermodynamics of radiation modes. Eur. J. Phys. **31**, 393 (2010)
Planck, M.: Über eine Verbesserung der Wienschen Spektralgleichung, Verh. Deutsch. Phys. Ges. **2**, 202 (1900a). English translation in ter Haar 1967, 79, On an Improvement of Wien's Equation for the Spectrum
Planck, M.: Zur Theorie des Gesetzes der Energieverteilung im Normalspektrum, Verh. Deutsch. Phys. Ges. **2**, 237 (1900b). English translation in D. ter Haar, The Old Quantum Theory (Pergamon Press, Oxford, 1967) 82
Planck, M.: Über die Begründung des Gesetzes der schwarzen Strahlung. Ann. d. Phys. **37**, 642 (1912)
Rigden, J.S.: Einstein's revolutionary paper, Physics World, http://physicsweb.org/articles/world/18/4/2 (2005)
Santos, E.: Is there an electromagnetic background radiation underlying the quantum phenomena?, p. 317. LXIV. An. Real Soc. Esp. Fís. Quím (1968)
Santos, E.: Comment on 'presenting the Planck's relation $E = nh\nu$'. Am. J. Phys. **43**, 743 (1975)
Santos, E.: Can quantum vacuum fluctuations be considered real? arXiv: http://arxiv.org/abs/quant-ph/0206161v1quant-ph/0206161v1 (2002a)

Santos, E.: How photon detectors remove vacuum fluctuations. arXiv: http://arxiv.org/abs/quant-ph/0207073v2quant-ph/0207073v2 Sept 2002 (2002b)
Schiff, L.: Quantum Mechanics. McGraw-Hill, New York (1955)
Sommerfeld, A.: Thermodynamics and Statistical Mechanics. Academic Press, New York (1956)
Theimer, O.: Derivation of the blackbody radiation spectrum by classical statistical mechanics. Phys. Rev. D **4**, 1597 (1971)
Theimer, O.: Blackbody spectrum and the interpretation of the quantum theory. Am. J. Phys. **44**, 183 (1976)
Valdés-Hernández, A.: Investigación del origen del enredamiento cuántico desde la perspectiva de la Electrodinámica Estocástica Lineal, Ph. D. Thesis, Universidad Nacional Autónoma de México, Mexico (2010)
*Valdés-Hernández, A., de la Peña, L., Cetto, A.M.: in: Macías, A., Dagdug, L. (eds.) New Trends in Statistical Physics: Festschrift in Honor of Leopoldo García-Colín's 80th Birthday, p. 113. World Scientific, Singapore (2010)
Vedral, V.: Modern Foundations of Quantum Optics. Imperial College Press, London (2005) (Chap. 3)
Weinberg, S.: The cosmological constant problem. Rev. Mod. Phys. **61**, 1 (1989)
Wollan, E.O.: Experimental electron distributions in atoms of monatomic gases. Phys. Rev. **38**, 15 (1931)

Chapter 4
The Long Journey to the Schrödinger Equation

I am, in fact, rather firmly convinced that the essentially statistical character of contemporary quantum theory is solely to be ascribed to the fact that this [theory] operates with an incomplete description of physical systems.
[In] a complete physical description, the statistical quantum theory would...take an approximately analogous position to the statistical mechanics within the framework of classical mechanics...

A. Einstein (1949)

...I think that we cannot afford to neglect any possible point of view for looking at Quantum Mechanics and in particular its relation to Classical Mechanics. Any point of view which gives us any interesting feature and any novel idea should be closely examined to see whether they suggest any modification or any way of developing the theory along new lines.

P. A. M. Dirac (1951)

An important conclusion was derived from the discussion in Chap. 3: the radiation field in equilibrium with matter acquires a discrete energy distribution, in presence of its zero-point component. This was interpreted to mean that the field appears quantized because of its interaction with matter. Then, what about matter? Could it be that it becomes as much affected as the field by this interaction? In this chapter we investigate this most important question. As our journey progresses it will become clear that indeed, matter is so strongly influenced by the background field, that it ends up behaving 'quantum-mechanically'. Again, quantization is revealed as a phenomenon that emerges as a result of the permanent matter-field interaction.

The approach taken in the present chapter starts with the (nonrelativistic) equation of motion for a particle immersed in the stochastic zero-point radiation field (ZPF) and subject to possible external forces, which are nonlinear in general. The problem is, therefore, a complicated one of electrodynamics with a stochastic field, the exact solution of which is beyond present-day mathematics. A statistical treatment is then carried out, using a standard approach to obtain an equation for the probability

L. de la Peña et al., *The Emerging Quantum*,
DOI: 10.1007/978-3-319-07893-9_4,

density of particles in phase space. The reduction of the description to configuration space leads in the radiationless approximation to the Schrödinger equation, under the condition of energy balance. The main lessons and implications of this important result are discussed in the final part of the chapter.[1]

4.1 Elements of the Dynamics

The main character in this chapter will be a charged particle—typically an electron, the case of neutral particles being briefly discussed in Sect. 4.7—immersed in the ZPF. The particle may be subject in addition to external forces (normally a conservative binding force, on the atomic or molecular scale), and possibly also to some external radiation field in its vicinity. What is important, however, is that the ZPF is always present.

4.1.1 The Equation of Motion

For a particle of mass m and electric charge e subject to the action of an external force $\boldsymbol{f}(\boldsymbol{x})$ and immersed in the pervasive ZPF, the Abraham-Lorentz equation of motion reads[2]

$$m\ddot{\boldsymbol{x}} = \boldsymbol{f}(\boldsymbol{x}) + m\tau\dddot{\boldsymbol{x}} + e\boldsymbol{E}(\boldsymbol{x},t) + \frac{e}{c}\boldsymbol{v} \times \boldsymbol{B}(\boldsymbol{x},t). \tag{4.1}$$

The term $m\tau\dddot{\boldsymbol{x}}$, with $\tau = 2e^2/3mc^3$, represents the radiation reaction force on the particle due to its acceleration (Landau and Lifshitz 1951). For an electron this term is normally small, since τ is of the order of 10^{-23} s; nevertheless, it will prove to be an important ingredient in Eq. (4.1).[3] The fields $\boldsymbol{E}(\boldsymbol{x},t)$ and $\boldsymbol{B}(\boldsymbol{x},t)$ in the Lorentz force

---

[1] This chapter draws to a large extent from previous work, contained in the following references: de la Peña and Cetto (1977a, b, 1995, 1996, 2005, 2006, 2007), de la Peña et al. (2009, 2012a, b), Cetto et al. (1984, 2012).

[2] Instead of starting from the set of Hamilton equations for the entire system (particle plus field), we use as point of departure the (approximate) equation of motion for the particle. A detailed derivation of Eq. (4.1) from the Hamiltonian can be seen in many texts on electrodynamics. A particularly clear discussion is given by Cohen et al. (1989). See also de la Peña (1983), or *The Dice*.

[3] In the usual derivations of Eq. (4.1), a retarded integral over time is written as a series expansion in terms of time derivatives of $\boldsymbol{x}$. The lion's share of this series pertains to the external (Lorentz) force. Then follow terms proportional to $\ddot{\boldsymbol{x}}$ and $\dddot{\boldsymbol{x}}$, and higher time derivatives are neglected, so the equation is approximate.

The term proportional to $\ddot{\boldsymbol{x}}$ adds a 'small' electromagnetic correction δm to the mass. It happens that the integral that expresses this correction is divergent, because of the approximations made. In Eq. (4.1) this poses no problem, since $m = m_0 + \delta m$ is considered to correspond to the experimental (renormalized) mass, m_0 being the mass that appears in the initial Hamiltonian. A formal procedure to solve this problem consists in adding to the initial Hamiltonian a mass *counterterm* that

are the electric and magnetic components of the fluctuating ZPF, hence they must be represented by stochastic variables. This makes Eq. (4.1) practically unsolvable. A number of simplifications and approximations are therefore introduced to go ahead. Firstly, a nonrelativistic description is assumed to be sufficient; this means that under normal conditions ($v/c \ll 1$) the magnetic force becomes negligible compared with the electric force, and (4.1) takes the simpler form

$$m\ddot{x} = f(x) + m\tau\dddot{x} + eE(x,t). \tag{4.2}$$

Secondly, it is anticipated that those modes of the field that are relevant for the dynamics have wavelengths much larger than the characteristic dimensions of the motion, so that one may assume that in the region of space occupied by the particle during its (approximately periodic) motion, the electric field is assumed not to vary appreciably; this is the long-wavelength approximation.[4] The $\boldsymbol{x}$-dependence of $\boldsymbol{E}(\boldsymbol{x},t)$ can then be neglected, and one can write the pair of equations

$$m\dot{\boldsymbol{x}} = \boldsymbol{p}, \quad \dot{\boldsymbol{p}} = \boldsymbol{f}(\boldsymbol{x}) + m\tau\dddot{\boldsymbol{x}} + e\boldsymbol{E}(t). \tag{4.3}$$

4.1.2 Basic Properties of the Zero-Point Field

We need to recall here some basic properties of the field represented by $\boldsymbol{E}(t)$ in Eq. (4.3) that will be extensively used below. Because it refers to the fluctuating vacuum, $\boldsymbol{E}(t)$ must be a stationary random variable with zero mean value,

---

(Footnote 3 continued)
takes into account (with the opposite sign) the contribution to the mass of all neglected terms, and thus eliminates the infinite contribution (of course, it is infinite itself!). This clever cancellation of one infinity with another —a procedure that in QED (and more generally in quantum field theory) gives excellent results—represents a regularization by renormalization.

The radiation-reaction term, being proportional to $\dddot{x}$, transforms the equation of motion into one of third order, thus demanding extra initial (or final) conditions. This term is known to lead to some awkward noncausal effects, such as *preacceleration*, i.e., response in advance to the external force (although in the present case the advanced times are of order $\tau \sim 10^{-23}$ s, so the effect is in practice negligible). It should be clearly understood that this noncausal behaviour is also a result of the neglect of the higher-order terms, since the theory in its closed form is absolutely causal. A similar situation is met with the Lorentz-Dirac equation, which is the relativistic version of the Abraham-Lorentz equation (see Rohrlich 1965). A more extensive discussion of the radiation force and the problems connected with it in the context of the present theory, as well as a causal variant of it, can be seen in *The Dice*, Sect. 3.3. The current approximate form is more convenient in practical terms, provided one bears in mind that its noncausal features are an artifact of the approximation.

Notice that $m\tau\dddot{x}$ represents the electric component of the field radiated by the particle, which is of a nature similar to $e\boldsymbol{E}(t)$; this can be made explicit by writing $m\tau\dddot{x} = e\boldsymbol{E}_{\text{rad}}(\boldsymbol{x},t)$, where $\boldsymbol{E}_{\text{rad}}(\boldsymbol{x},t)$ stands for the radiated field.

[4] As will become clear in Chap. 5, the relevant modes are those associated with the dominant response of the particle to the field (exhibited e.g. in the atomic transitions), which have indeed wavelengths much larger than the atomic dimensions.

$$\overline{\boldsymbol{E}(t)}^{(i)} = 0, \tag{4.4}$$

where $\overline{(\cdot)}^{(i)}$ denotes the average over all realizations (i) of the field. Further, its spectral energy density must correspond to a mean energy $\hbar\omega/2$ per frequency mode. Given that the volume density of modes is equal to[5]

$$\rho_{\text{modes}}(\omega) = \frac{\omega^2}{\pi^2 c^3}, \tag{4.5}$$

the spectral energy density is

$$\rho_0(\omega) = \frac{\omega^2}{\pi^2 c^3}\frac{1}{2}\hbar\omega = \frac{\hbar\omega^3}{2\pi^2 c^3}. \tag{4.6}$$

This expression determines the correlation between the Fourier-transformed spatial components of the electric field. The Fourier transform for the ith-component reads

$$\tilde{E}_i(\omega) = \frac{1}{2\pi}\int_{-\infty}^{+\infty} E_i(t)e^{i\omega t}dt, \tag{4.7}$$

with

$$\overline{\tilde{E}_i(\omega)\tilde{E}_j^*(\omega')}^{(i)} = \delta_{ij}\frac{4\pi}{3}\rho_0(\omega)\delta\left(\omega - \omega'\right). \tag{4.8}$$

Inserting here (4.7) gives the autocorrelation function,

$$\overline{E_i(t)E_j(t')}^{(i)} = \delta_{ij}\varphi(t - t'), \tag{4.9}$$

where

$$\varphi(t - t') = \frac{4\pi}{3}\int_0^{\infty} \rho_0(\omega)\cos\omega(t - t')d\omega, \tag{4.10}$$

since $\rho(\omega) = \rho(|\omega|)$. Equation (4.8) means that the Fourier components pertaining to different frequencies are statistically independent. In the present case the autocorrelation describes a highly colored noise manifested through its strong (cubic) dependence on the frequency. If there is any extra component due to an external field, as could be a thermal equilibrium radiation or any other excitation of the field modes, the corresponding contribution must be added to $\rho_0(\omega)$ in Eq. (4.6).

The total energy density of the vacuum follows from Eq. (4.6),[6]

---

[5] This formula is the result of counting the number of field modes of frequency $\omega = 2\pi c/\lambda$ (with both polarizations) per unit volume within an interval of frequency $\Delta\omega$. Assuming the distribution of radiation to be homogeneous and isotropic, this gives after integrating over the solid angle: $\Delta n = (1/\pi^2 c^3)\int_{\Delta\omega}\omega^2 d\omega$.

[6] As remarked in note 5 of Chap. 3, the spectrum proportional to ω^3 is the single one for which all inertial observers are equivalent. This can be confirmed by calculating the force exerted by a

$$\int_0^\infty \rho_0(\omega)d\omega = \frac{\hbar}{2\pi^2 c^3}\int_0^\infty \omega^3 d\omega. \tag{4.11}$$

Because of the divergence of this integral, strictly speaking $\rho_0(\omega)$ cannot represent a true spectral density, as it would bestow an infinite energy density to the simmering field! Even with a reasonable cutoff in the integral, the result continues to be inadmissibly high. This represents a major open problem for contemporary physics. A mystery besets the huge value of the energy content of the vacuum, particularly in connection with its unobserved supposedly enormous gravitational effects. A speculation (see, i.a., Wesson 1991) is that a perfectly uniformly distributed field filling the whole space would produce equal effects on and from all directions, so that matter would not be perturbed by its gravitational effects; only the possible effects of its fluctuations would reveal its presence. This problem is intimately linked to the mystery of the origin of the cosmological constant. As Weinberg (1989) puts it: "Any solution of the cosmological constant problem is likely to have a much wider impact on other areas of physics or astronomy". Its solution is expected to come some day with a better understanding of the meaning and nature of dark energy. It will be left aside here (just as is done in other important areas of physics where it appears, such as quantum field theory and astrophysics), taking advantage of the fact that the developments of the present theory are not directly affected by it.

As a complement to this brief discussion on the ZPF, the final section of Chap. 9 is devoted to a consideration of the relationship between atomic and cosmological constants, which leads to an interesting expression for Planck's constant in terms of the latter.

4.2 Generalized Fokker-Planck Equation in Phase Space

Solving the exact problem associated with Eq. (4.1) is a hopeless enterprise. However, the detailed description of the stochastic motion of a single particle is not of interest, since it depends on the (unknown) specific realization of the field in a given circumstance. This suggests making a description of an ensemble of similar systems, which represent all (field) realizations compatible with the conditions of the problem. We therefore resort to a statistical treatment, by constructing the corresponding (generalized) Fokker-Planck equation, GFPE.[7]

For this purpose we start by considering the density R of points in the phase space of the particle. For each realization of the field, R satisfies the (approximate) continuity equation (summation over repeated indices is used throughout),

(Footnote 6 continued)
homogeneous and isotropic background field on a dipole moving with velocity v, which is given by $F = -(6/5)\pi^2\tau c[\rho(\omega) - (\omega/3)(d\rho/d\omega)]v$ (see Einstein and Hopf 1910; *The Dice*, Chap. 4). Only for $\rho(\omega) \sim \omega^3$ this force becomes zero.

[7] A Fokker-Planck equation (FPE) is a differential equation of second order that describes the evolution of the probability density for the particle subject to a white noise (an uncorrelated noise, with

$$\frac{\partial R}{\partial t}+\frac{\partial}{\partial x_i}(\dot{x}_i R)+\frac{\partial}{\partial p_i}(\dot{p}_i R)=0, \tag{4.12}$$

where $\dot{x}_i$ and $\dot{p}_i$ are given in accordance with Eq. (4.3).[8] Due to the stochasticity of $\boldsymbol{E}$ (impressed upon $\dot{x}_i$ and $\dot{p}_i$), the density R is still a stochastic variable, which means that it varies from realization to realization of the field. The transition from the description afforded by Eq. (4.12) to an effective one requires focusing not on R, but on its average over the realizations $\{(i)\}$, $Q \equiv \overline{R}^{(i)}$ (Edwards and McComb 1969; Kampen 1976, p. 209). The averaged probability density $Q(\boldsymbol{x}, \boldsymbol{p}, t)$ provides information on how the particles of the ensemble are distributed in the vicinity of a point $(\boldsymbol{x}, \boldsymbol{p})$ at time t. Any reference to the specific realization (i) [or to a specific trajectory that solves Eq. (4.3)] is therefore absent from the description afforded by Q. In this scenario the direct role of the stochasticity of the equation of motion is that of making of the position and momentum distributed variables, even for fixed initial conditions. The statistical (phase-space) treatment requires passing from Eq. (4.12) to the equation that governs the evolution of $Q(\boldsymbol{x}, \boldsymbol{p}, t)$. As shown in Appendix A, such equation reads

$$\frac{\partial Q}{\partial t}+\frac{1}{m}\frac{\partial}{\partial x_i}p_i Q+\frac{\partial}{\partial p_i}f_i Q+m\tau\frac{\partial}{\partial p_i}\dddot{x}_i Q=e^2\frac{\partial}{\partial p_i}\hat{\mathcal{D}}_i Q, \tag{4.13}$$

with the diffusion operator $\hat{\mathcal{D}}(t)$ defined by means of the expression

$$\hat{\mathcal{D}}_i(t)Q=\hat{P}E_i\hat{G}\frac{\partial}{\partial p_j}E_j\sum_{k=0}^{\infty}\left[e\hat{G}\frac{\partial}{\partial p_l}(1-\hat{P})E_l\right]^{2k}Q, \tag{4.14}$$

and the projection operator $\hat{P}$ and the inverse evolution operator $\hat{G}$ given by

---

(Footnote 7 continued)
a flat power spectrum). Given that the present problem involves a colored noise, the corresponding equation for the probability density in phase space is not a true FPE, but a generalization of it that contains memory terms, leading to an integro-differential equation. It is to such equation that we refer as a GFPE. For conceptually rich, early introductions to the FPE for the study of Brownian motion see the papers by S. Chandrasekhar and by Ming Chen Wang and G. E. Uhlenbeck in Wax (1954/1985). For a first-rate presentation of the subject see Stratonovich 1963. For a more recent presentation see Risken (1984); see also Cetto et al. (1984).

[8] The approximation consists in assuming that the field remains essentially unmodified. For a complete description one should write the continuity equation as

$$\frac{\partial R}{\partial t}+\frac{\partial}{\partial x_a}(\dot{x}_a R)+\frac{\partial}{\partial p_a}(\dot{p}_a R)=0,$$

where $R(\{x_a, p_a\}, t)$ stands for the density of points in the entire phase space of the particle plus field system, so that $\{x_a\} = \{x_f, x_i\}$ and $\{p_a\} = \{p_f, p_i\}$, where the index f refers to the field quadratures and i to the particle's variables.

$$\hat{P}A = \overline{A}^{(i)}, \quad \hat{G}A(\boldsymbol{x}, \boldsymbol{p}, t) = \int_{-\infty}^{t} e^{-\hat{L}(t-t')} A(\boldsymbol{x}, \boldsymbol{p}, t')dt', \tag{4.15}$$

where $\hat{L}$ is the Liouville operator for the particle [Eq. (A.17)],

$$\hat{L} = \frac{1}{m}\frac{\partial}{\partial x_i} p_i + \frac{\partial}{\partial p_i}(f_i + m\tau\dddot{x}_i). \tag{4.16}$$

The operator $e^{-\hat{L}(t-t')}$ in Eq. (4.15) acts on all variables to its right and makes them evolve from $\boldsymbol{x}(t')$, $\boldsymbol{p}(t')$ $(t' < t)$ towards $\boldsymbol{x}(t)$, $\boldsymbol{p}(t)$ as final conditions, following a deterministic path. The details are explained in the Appendix A.

The GFPE (4.13) is an integro-differential equation, or equivalently, a differential equation of infinite order. A good part of its complication is due to the memory developed as the system evolves in time according to Eq. (4.15). Nevertheless, as shown in Appendix B, in the time-asymptotic limit, when the Markovian approximation is applicable,[9] the diffusion operator $\hat{\mathcal{D}}$ can be approximately expressed in terms of two diffusion coefficients

$$e^2\hat{\mathcal{D}}_i = D_{ij}^{pp}\frac{\partial}{\partial p_j} + D_{ij}^{px}\frac{\partial}{\partial x_j}, \tag{4.17}$$

which to lowest order in e^2 are given by Eqs. (B.10a), (B.10b), namely

$$D_{ij}^{pp} = e^2\int_{-\infty}^{t} dt'\varphi(t-t')\frac{\partial p_j}{\partial p_i'}, \quad D_{ij}^{px} = e^2\int_{-\infty}^{t} dt'\varphi(t-t')\frac{\partial x_j}{\partial p_i'}, \tag{4.18}$$

with $\varphi(t-t')$ given by Eq. (4.10). In this Markovian approximation the GFPE (4.13) becomes a true FPE (a differential Equation of second order),

$$\frac{\partial Q}{\partial t} + \frac{1}{m}\frac{\partial}{\partial x_i} p_i Q + \frac{\partial}{\partial p_i}(f_i + m\tau\dddot{x}_i)\, Q = \frac{\partial}{\partial p_i} D_{ij}^{pp}\frac{\partial Q}{\partial p_j} + \frac{\partial}{\partial p_i} D_{ij}^{px}\frac{\partial Q}{\partial x_j}. \tag{4.19}$$

This equation will prove to be very useful for actual calculations.

[9] Right after particle and field start to interact, the system is far from equilibrium. In this regime the main effect of the ZPF on the particle is due to the high-frequency modes, which produce violent accelerations and randomize the motion. Eventually, the interplay between the electric field force and radiation reaction is expected to drive the system close to equilibrium; in this (time-reversible) regime the Markovian approximation applies. The duration of the transient period, i.e. the time t_M required by the system to reach the Markovian limit, is determined basically by the effect of the high-frequency modes. Since the particle is assumed to respond to modes of frequency up to $mc^2/\hbar$ (see Chap. 6), t_M is estimated to be of of the order of $\hbar/mc^2 \simeq 10^{-20}$ s for an electron.

4.2.1 Some Important Relations for Average Values

Equation (4.13) contains a wealth of statistical information on the dynamics of the system, which will be of much value for further developments. In the following we present some revealing statistical relations that can be readily derived from it. For this purpose we introduce the (phase-space) average

$$\langle \cdot \rangle = \int (\cdot)\, Q dxdp, \tag{4.20}$$

with the volume element d^3xd^3p written succinctly as $dxdp$. For a general phase function $\mathcal{G}(\boldsymbol{x}, \boldsymbol{p})$ that has no explicit time dependence,

$$\frac{d}{dt}\langle \mathcal{G} \rangle = \int \mathcal{G}\frac{\partial Q}{\partial t} dxdp. \tag{4.21}$$

Equation (4.13) multiplied from the left by $\mathcal{G}$ and integrated over the entire phase space (assuming in the integration by parts that the system is bounded so that Q vanishes at infinity), gives the general relation

$$\frac{d}{dt}\langle \mathcal{G} \rangle = \left\langle \dot{x}_i \frac{\partial \mathcal{G}}{\partial x_i} \right\rangle + \left\langle f_i \frac{\partial \mathcal{G}}{\partial p_i} \right\rangle + m\tau \left\langle \dddot{x}_i \frac{\partial \mathcal{G}}{\partial p_i} \right\rangle - e^2 \left\langle \frac{\partial \mathcal{G}}{\partial p_i} \hat{\mathcal{D}}_i \right\rangle, \tag{4.22}$$

which can be recast in a more illuminating form as

$$\frac{d}{dt}\langle \mathcal{G} \rangle = \left\langle \frac{d\mathcal{G}}{dt} \right\rangle_{\text{nr}} + m\tau \left\langle \dddot{x}_i \frac{\partial \mathcal{G}}{\partial p_i} \right\rangle - e^2 \left\langle \frac{\partial \mathcal{G}}{\partial p_i} \hat{\mathcal{D}}_i \right\rangle. \tag{4.23}$$

The term

$$\left\langle \frac{d\mathcal{G}}{dt} \right\rangle_{\text{nr}} = \left\langle \dot{x}_i \frac{\partial \mathcal{G}}{\partial x_i} + f_i \frac{\partial \mathcal{G}}{\partial p_i} \right\rangle \tag{4.24}$$

represents the nonradiative contribution to $d\langle \mathcal{G} \rangle/dt$, in contrast with the two remaining terms in Eq. (4.23), which originate in the radiation reaction and the fluctuating field. Notice that the latter terms give rise to radiative contributions in the mean only if $\mathcal{G}$ depends on the momentum, since for $\mathcal{G} = \mathcal{G}(\boldsymbol{x})$ Eq. (4.23) reduces to

$$\frac{d}{dt}\langle \mathcal{G}(\boldsymbol{x}) \rangle = \left\langle \frac{d\mathcal{G}(\boldsymbol{x})}{dt} \right\rangle_{\text{nr}}. \tag{4.25}$$

For example, for $\mathcal{G} = x_i$ and $x_i x_j$, respectively,

$$\frac{d}{dt}\langle x_i \rangle = \frac{1}{m}\langle p_i \rangle; \tag{4.26a}$$

$$\frac{d}{dt}\left\langle x_i x_j \right\rangle = \frac{1}{m}\left\langle p_i x_j + p_j x_i \right\rangle. \tag{4.26b}$$

The effect of the radiative terms on the (average) dynamics is disclosed by applying (4.23) to $\mathcal{G} = p_i, x_i p_j, p_i p_j$ successively,

$$\frac{d}{dt}\langle p_i \rangle = \langle f_i \rangle + m\tau \langle \dddot{x}_i \rangle - e^2 \left\langle \hat{\mathcal{D}}_i \right\rangle; \tag{4.27a}$$

$$\frac{d}{dt}\left\langle x_i p_j \right\rangle = \left\langle \frac{1}{m} p_i p_j + x_i f_j \right\rangle + m\tau \left\langle x_i \dddot{x}_j \right\rangle - e^2 \left\langle x_i \hat{\mathcal{D}}_j \right\rangle; \tag{4.27b}$$

$$\frac{d}{dt}\left\langle p_i p_j \right\rangle = \left\langle f_i p_j + p_i f_j \right\rangle + m\tau \left\langle \dddot{x}_i p_j + p_i \dddot{x}_j \right\rangle - e^2 \left\langle p_i \hat{\mathcal{D}}_j + p_j \hat{\mathcal{D}}_i \right\rangle. \tag{4.27c}$$

Equation (4.27a) exhibits the role played by the last term, $-e^2\left\langle \hat{\mathcal{D}}_i \right\rangle$, as a mean effective force due to diffusion, analogous to the osmotic force in the case of Brownian diffusion. As for Eq. (4.27b), its antisymmetric form gives

$$\frac{d}{dt}\langle \boldsymbol{L} \rangle = \langle \boldsymbol{M} \rangle + m\tau \langle \boldsymbol{x} \times \dddot{\boldsymbol{x}} \rangle - e^2 \left\langle \boldsymbol{x} \times \hat{\boldsymbol{\mathcal{D}}} \right\rangle, \tag{4.28}$$

with $\boldsymbol{L}$ and $\boldsymbol{M}$ the angular momentum and the torque due to the external force $\boldsymbol{f}$, respectively. This equation is useful for the study of the angular momentum, as will be shown in Chap. 6. On the other hand, the symmetric form of Eq. (4.27b) (taking $i = j$) leads to

$$\frac{d}{dt}\langle \boldsymbol{x} \cdot \boldsymbol{p} \rangle = \frac{1}{m}\left\langle \boldsymbol{p}^2 \right\rangle + \langle \boldsymbol{x} \cdot \boldsymbol{f} \rangle + m\tau \langle \boldsymbol{x} \cdot \dddot{\boldsymbol{x}} \rangle - e^2 \left\langle \boldsymbol{x} \cdot \hat{\boldsymbol{\mathcal{D}}} \right\rangle, \tag{4.29}$$

which is related to the virial theorem (see Eq. (4.35) below).

Equation (4.27c) gives also (again taking $i = j$)

$$\frac{1}{2m}\frac{d}{dt}\left\langle \boldsymbol{p}^2 \right\rangle = \frac{1}{m}\langle \boldsymbol{p} \cdot \boldsymbol{f} \rangle + \tau \langle \boldsymbol{p} \cdot \dddot{\boldsymbol{x}} \rangle - \frac{1}{m} e^2 \left\langle \boldsymbol{p} \cdot \hat{\boldsymbol{\mathcal{D}}} \right\rangle. \tag{4.30}$$

In terms of

$$H = \frac{1}{2m}\boldsymbol{p}^2 + V(\boldsymbol{x}), \tag{4.31}$$

with $dV/dt = -(1/m)\boldsymbol{p} \cdot \boldsymbol{f}$, Eq. (4.30) becomes

$$\frac{d}{dt}\langle H \rangle = \tau \langle \boldsymbol{p} \cdot \dddot{\boldsymbol{x}} \rangle - \frac{1}{m} e^2 \left\langle \boldsymbol{p} \cdot \hat{\boldsymbol{\mathcal{D}}} \right\rangle. \tag{4.32}$$

This equation is important for the analysis of the mean energy exchange during radiative transitions, as will be seen in Chap. 6. More generally, for any $\mathcal{G} = \xi(\boldsymbol{x}, \boldsymbol{p})$ that represents an integral of the motion of the 'radiationless' problem, (4.24) is zero and one is left with

$$\frac{d}{dt}\langle\xi\rangle = m\tau\left\langle \dddot{x}_i \frac{\partial \xi}{\partial p_i}\right\rangle - e^2\left\langle \frac{\partial \xi}{\partial p_i}\hat{\mathcal{D}}_i\right\rangle, \tag{4.33}$$

which shows that $\langle\xi\rangle$ can in general be affected by both radiation and diffusion. Equation (4.32) is just a particular instance of (4.33) for $\xi = H$.

The above equations are especially useful to obtain expressions for average values of dynamical quantities in the regime in which $\langle\mathcal{G}\rangle$ acquires a constant value, so that $d\langle\mathcal{G}\rangle/dt$ vanishes. This occurs in particular in a stationary situation, since then $(\partial Q/\partial t) = 0$ and the left-hand side of (4.23) is zero for any $\mathcal{G}$. In this case, from Eqs. (4.26a), (4.26b) we have $\langle\boldsymbol{p}\rangle = 0$ and $\langle\boldsymbol{x}\cdot\boldsymbol{p}\rangle = 0$, meaning that $\boldsymbol{x}$ and $\boldsymbol{p}$ are uncorrelated. As for Eq. (4.27a), it reduces to

$$\langle\boldsymbol{f}\rangle = e^2\left\langle\hat{\boldsymbol{\mathcal{D}}}\right\rangle - m\tau\left\langle\dddot{\boldsymbol{x}}\right\rangle, \tag{4.34}$$

which shows that up to terms of order e, the mean value of the external force is zero. Further, under stationarity, Eq. (4.27b) with $i = j$ becomes

$$\frac{1}{2m}\left\langle p^2\right\rangle + \frac{1}{2}\langle\boldsymbol{x}\cdot\boldsymbol{f}\rangle + \frac{1}{2}m\tau\langle\boldsymbol{x}\cdot\dddot{\boldsymbol{x}}\rangle - \frac{1}{2}e^2\left\langle\boldsymbol{x}\cdot\hat{\boldsymbol{\mathcal{D}}}\right\rangle = 0, \tag{4.35}$$

which in the radiationless approximation reduces to the virial theorem (with averages over the ensemble instead of over time). Equation (4.35) represents therefore a more complete form of the latter, indicating the presence of radiative corrections to the mean kinetic energy, of value

$$\langle\delta T\rangle = -\tfrac{1}{2}m\tau\langle\boldsymbol{x}\cdot\dddot{\boldsymbol{x}}\rangle + \tfrac{1}{2}e^2\left\langle\boldsymbol{x}\cdot\hat{\boldsymbol{\mathcal{D}}}\right\rangle. \tag{4.36}$$

As will be shown in Chap. 6, this formula gives the (nonrelativistic) Lamb shift.

Finally, when $\langle H\rangle$ does not evolve anymore in time, Eq. (4.32) reduces to the *energy-balance condition*

$$\tau\langle\dddot{\boldsymbol{x}}\cdot\boldsymbol{p}\rangle = \frac{e^2}{m}\left\langle\boldsymbol{p}\cdot\hat{\boldsymbol{\mathcal{D}}}\right\rangle. \tag{4.37}$$

The left-hand side represents the average power lost by the particle to the field through radiation reaction along its smoothed trajectory, and the right-hand side is the average power gained by the particle from the background field and stored in the momentum fluctuations.

The physical meaning of (4.37) can be made more transparent by resorting to the original stochastic equation of motion (4.3) (note that in contrast with previous expressions, in what follows $\boldsymbol{x}$ and $\boldsymbol{p}$ are stochastic variables that depend on the field realization). Multiplying the second expression in Eq. (4.3) by $\boldsymbol{p}$ allows us to write the time derivative of the (here stochastic) Hamiltonian (4.31) as

$$\frac{dH}{dt} = \tau \boldsymbol{p} \cdot \dddot{\boldsymbol{x}} + \frac{e}{m} \boldsymbol{p} \cdot \boldsymbol{E}, \tag{4.38}$$

or equivalently,

$$\frac{d}{dt}\left(H - \frac{\tau}{m} \dot{\boldsymbol{p}} \cdot \boldsymbol{p}\right) = \frac{d}{dt} H_r = -m\tau \ddot{\boldsymbol{x}}^2 + \frac{e}{m} \boldsymbol{p} \cdot \boldsymbol{E}, \tag{4.39}$$

where H_r differs from the (mechanical, radiationless) Hamiltonian H by a radiative correction,

$$H_r = H - \frac{\tau}{m} \boldsymbol{p} \cdot \dot{\boldsymbol{p}} = \frac{1}{2m} \boldsymbol{p}^2 - \frac{\tau}{m} \boldsymbol{p} \cdot \dot{\boldsymbol{p}} + V. \tag{4.40}$$

The first two terms on the right-hand side coincide with the first two terms of a Taylor series expansion

$$\frac{1}{2m} \boldsymbol{p}^2 (t - \tau) = \frac{1}{2m} \boldsymbol{p}^2 (t) - \frac{\tau}{2m} \frac{d}{dt} \boldsymbol{p}^2 (t) + \cdots ; \tag{4.41}$$

therefore, up to terms of order τ,

$$H_r = \frac{1}{2m} \boldsymbol{p}^2 (t - \tau) + V. \tag{4.42}$$

This expression indicates the presence of preacceleration. However, as mentioned in footnote 3, such noncausal description is merely a result of the approximations made to arrive at the Abraham-Lorentz equation; therefore, taking $H_r \approx H$ above is a means to recover causality. This is legitimate given the smallness of the preacceleration (and the fact that $\dot{\boldsymbol{p}} \cdot \boldsymbol{p}$ averages to zero for periodic motions). Averaging over the field realizations gives for $\overline{H}^{(i)}$ independent of time,

$$m\tau \overline{\ddot{\boldsymbol{x}}^2}^{(i)} = \frac{e}{m} \overline{\boldsymbol{p} \cdot \boldsymbol{E}}^{(i)}. \tag{4.43}$$

This expression is physically equivalent to Eq. (4.37), but has the advantage of disclosing the ZPF as the source of the energy delivered to the particle through the fluctuations impressed on its momentum. As can be expected, the energy-balance equation will play a central role in the development of the theory, specifically from Sect. 4.4.4 on.

In actual applications it is normally more convenient to use the Markovian approximation to determine $d \langle \mathcal{G} \rangle / dt$. This is done by resorting to Eq. (4.17) for the calculation of the last term in Eq. (4.22). After some integrations by parts one arrives at

$$
\begin{aligned}
\frac{d}{dt}\langle \mathcal{G}\rangle = & \left\langle \frac{d\mathcal{G}}{dt}\right\rangle_{\mathrm{nr}} + m\tau\left\langle \dddot{x}_i \frac{\partial \mathcal{G}}{\partial p_i}\right\rangle \\
& + \left\langle D_{ij}^{pp}\frac{\partial^2\mathcal{G}}{\partial p_i \partial p_j}\right\rangle + \left\langle D_{ij}^{px}\frac{\partial^2\mathcal{G}}{\partial p_i \partial x_j}\right\rangle + \left\langle \frac{\partial \mathcal{G}}{\partial p_i}\left(\frac{\partial D_{ij}^{pp}}{\partial p_j} + \frac{\partial D_{ij}^{px}}{\partial x_j}\right)\right\rangle .
\end{aligned}
\tag{4.44}
$$

At the end of Appendix B it is shown that the last term in this expression (containing derivatives of the diffusion coefficients) vanishes, whence (4.44) reduces to

$$
\frac{d}{dt}\langle \mathcal{G}\rangle = \left\langle \frac{d\mathcal{G}}{dt}\right\rangle_{\mathrm{nr}} + m\tau\left\langle \dddot{x}_i \frac{\partial \mathcal{G}}{\partial p_i}\right\rangle + \left\langle D_{ij}^{pp}\frac{\partial^2\mathcal{G}}{\partial p_i \partial p_j}\right\rangle + \left\langle D_{ij}^{px}\frac{\partial^2\mathcal{G}}{\partial p_i \partial x_j}\right\rangle . \tag{4.45}
$$

We shall use either Eq. (4.23) or (4.45) as turns out more convenient.

4.3 Transition to Configuration Space

The results of the previous section are very suggestive of the kind of behavior one can expect for the mean values of relevant dynamical quantities of the material system under the combined action of the ZPF and radiation reaction (in addition to an external force). 'Classical-like' equations are obtained for these mean values, closely reminiscent of the Ehrenfest theorem of quantum mechanics (QM), with additional contributions (corrections of order e^2 and higher) due the radiation terms.

In order to establish the connection between the previous results and the quantum-mechanical laws, we now focus on the GFPE (4.13), and reduce the present (phase-space) description to the configuration space of the particle, which is the space in which the quantum description is usually made in terms of the wave function. An entirely similar procedure can of course be followed to reduce the description to the momentum space of the particle (as is briefly indicated in Sect. 4.3.1). Both processes are equally permissible, but mutually exclusive. For a full description in phase space one would need to return to Eq. (4.13). A concise discussion on why the phase-space description cannot be obtained by going backward from the quantum formalism is left for Sect. 4.6.

The transition to configuration space can be performed in a systematic way by multiplying Eq. (4.13) successively by $p_i^n p_j^m p_k^l$ (with $n, m, l = 0, 1, 2\ldots$) and integrating over the momentum space, to obtain a family of equations containing moments (of increasing order) of the components of $\boldsymbol{p}$. Equivalently, it can be done with the help of the characteristic function $\tilde{Q}$ (the Fourier transform) associated with the density Q,

$$
\tilde{Q}(\boldsymbol{x}, \boldsymbol{z}, t) = \int Q(\boldsymbol{x}, \boldsymbol{p}, t) e^{i\boldsymbol{p}\cdot\boldsymbol{z}} d\boldsymbol{p}. \tag{4.46}
$$

By applying the Fourier operator $\widetilde{(\cdot)} = \int (\cdot)\, e^{i\boldsymbol{p}\cdot\boldsymbol{z}} dp$ to Eq. (4.13) and recalling that all surface terms appearing along the integrations vanish at infinity, one gets

$$\frac{\partial \tilde{Q}}{\partial t} - i\frac{1}{m}\frac{\partial^2 \tilde{Q}}{\partial x_i \partial z_i} - iz_i\,(f_i + m\tau \dddot{x}_i)\,\tilde{Q} = -ie^2 z_i \widetilde{(\hat{\mathcal{D}}_i Q)}. \tag{4.47}$$

According to the equation of motion (4.3), to lowest order in τ the term $m\tau\dddot{x}_i$ can be approximated by $\tau(df_i/dt) = \tau\,(\nabla f_i)\cdot(\boldsymbol{p}/m)$. As will be clear below (Sect. 4.4.1), in the time-asymptotic (radiationless) limit such approximation is legitimate. Moreover, all our calculations will be carried out up to terms of order τ; hence, for convenience we shall as of now make this substitution. In particular, introducing the approximation in (4.13) we get, instead of (4.47), the equation

$$\frac{\partial \tilde{Q}}{\partial t} - i\frac{1}{m}\frac{\partial^2 \tilde{Q}}{\partial x_i \partial z_i} - iz_i f_i \tilde{Q} - \frac{\tau}{m} z_i \frac{\partial f_i}{\partial x_j}\frac{\partial \tilde{Q}}{\partial z_j} = -ie^2 z_i \widetilde{(\hat{\mathcal{D}}_i Q)}. \tag{4.48}$$

For consistency, in the time-asymptotic limit the right-hand side should be calculated in the Markovian approximation.

The probability density $\rho(\boldsymbol{x},t)$ in configuration space is given by the marginal probability

$$\rho(\boldsymbol{x},t) = \int Q(\boldsymbol{x},\boldsymbol{p},t)dp = \tilde{Q}(\boldsymbol{x},0,t), \tag{4.49}$$

and the local average of a generic function $g(\boldsymbol{x},\boldsymbol{p},t)$ is given by the formula

$$\langle g\rangle(\boldsymbol{x}) \equiv \langle g\rangle_{\boldsymbol{x}} = \frac{1}{\rho}\int g(\boldsymbol{x},\boldsymbol{p},t)Q dp. \tag{4.50}$$

It is important to note that $\langle g\rangle_{\boldsymbol{x}}$ represents a *partially* averaged quantity, which is still a function of the position variables, hence the name *local*. The fully averaged quantity is of course

$$\langle g\rangle = \int \langle g\rangle_{\boldsymbol{x}}\, \rho dx. \tag{4.51}$$

For simplicity in the writing, the possible time dependence of both $\langle g\rangle_{\boldsymbol{x}}$ and $\langle g\rangle$ is omitted.

The characteristic function $\tilde{Q}(\boldsymbol{x},z,t)$ is called also *momentum-generating function* because it is possible to derive all (local) moments of p_i from it. For example, the local average of $p_i^n p_j^m p_k^l$ reads

$$\begin{aligned}\left\langle p_i^n p_j^m p_k^l\right\rangle_{\boldsymbol{x}} &= \frac{1}{\rho}\int p_i^n p_j^m p_k^l Q dp \\ &= (-i)^{(n+m+l)}\left(\frac{1}{\tilde{Q}}\frac{\partial^{n+m+l}}{\partial z_k^l \partial z_j^m \partial z_i^n}\tilde{Q}\right)\Bigg|_{z=0}.\end{aligned} \tag{4.52}$$

4.3.1 A Digression: Transition to Momentum Space

Although it will not be required in this book, it is interesting for reference purposes to indicate how a similar transition can be made to the momentum space. This is achieved by applying the Fourier transformation $\int (\cdot)\, e^{i\mathbf{k}\cdot\mathbf{x}} dx$ to equation (4.13). With (in one-dimensional notation, for simplicity)

$$\tilde{P}(p, k, t) = \int Q(x, p, t) e^{ikx} dx, \tag{4.53}$$

and after some simple calculations, one obtains the integro-differential equation

$$\frac{\partial \tilde{P}}{\partial t} - i\frac{p}{m}k\tilde{P} + \frac{1}{2\pi}\frac{\partial}{\partial p}\int \tilde{K}(k-k', p)\tilde{P}(k', p)dk' = e^2\frac{\partial}{\partial p}\widetilde{(\hat{\mathcal{D}}P)}(k, p), \tag{4.54}$$

where $\tilde{K}$ is the Fourier transform of the acting force K,

$$K(x, p) = f(x) + \frac{\tau}{m}\frac{df}{dx}p. \tag{4.55}$$

Note that the last term was again written in its time-asymptotic form. It is clear that things get more complicated in the p-description, because of the integro-differential term on the left-hand side of Eq. (4.54).

4.3.2 A Hierarchy of Coupled Transfer Equations

An alternative procedure to get the most of Eq. (4.48) in the transition to configuration space, consists in expanding the factor $e^{i\mathbf{p}\cdot\mathbf{z}}$ that enters into each term of Eq. (4.48) and separating into powers $z_i^n z_j^m z_k^l$ for $n, l, m = 0, 1, 2, \ldots$ The first three equations thus obtained are

$$\frac{\partial \rho}{\partial t} + \frac{1}{m}\frac{\partial}{\partial x_j}\left(\langle p_j\rangle_x \rho\right) = 0; \tag{4.56a}$$

$$\frac{\partial}{\partial t}\left(\langle p_i\rangle_x \rho\right) + \frac{1}{m}\frac{\partial}{\partial x_j}\left(\langle p_i p_j\rangle_x \rho\right) - f_i\rho - \frac{\tau}{m}\frac{\partial f_i}{\partial x_j}\langle p_j\rangle_x \rho = -e^2\left.\widetilde{(\hat{\mathcal{D}}_i Q)}\right|_{z=0}; \tag{4.56b}$$

$$\begin{aligned}\frac{\partial}{\partial t}\left(\langle p_i p_j\rangle_x \rho\right) &+ \frac{1}{m}\frac{\partial}{\partial x_k}\left(\langle p_i p_j p_k\rangle_x \rho\right) - \left(f_i\langle p_j\rangle_x + f_j\langle p_i\rangle_x\right)\rho \\ &+ \frac{2\tau}{m}\left[\left(\frac{\partial f_i}{\partial x_k}\right)\langle p_j p_k\rangle_x + \left(\frac{\partial f_j}{\partial x_k}\right)\langle p_i p_k\rangle_x\right]\rho = e^2\left.\widetilde{(p_i\hat{\mathcal{D}}_j + p_j\hat{\mathcal{D}}_i)Q}\right|_{z=0}.\end{aligned} \tag{4.56c}$$

The first equation is connected with the second one through the first moments $\langle p_j \rangle_{\boldsymbol{x}}$, the second with the third one through the local correlations $\langle p_l p_j \rangle_{\boldsymbol{x}}$, and so on. The entire set of equations of which (4.56a, 4.56b, 4.56c) are the first three, constitutes thus an infinite hierarchy of coupled nonlinear equations.

The first member of the hierarchy is the continuity equation, which describes the transfer of matter. It follows that the current density (or mean flux of particles) is $\boldsymbol{j}(\boldsymbol{x}) = \rho(\boldsymbol{x}) \langle \boldsymbol{p} \rangle_{\boldsymbol{x}} / m$, hence the flow (or flux) velocity is given by

$$\boldsymbol{v}(\boldsymbol{x}) = \frac{\boldsymbol{j}(\boldsymbol{x})}{\rho(\boldsymbol{x})} = \frac{1}{m} \langle \boldsymbol{p} \rangle_{\boldsymbol{x}} . \tag{4.57}$$

Note that, being proportional to $\langle \boldsymbol{p} \rangle_{\boldsymbol{x}}$, $\boldsymbol{v}(\boldsymbol{x})$ is a *local mean velocity*, an average over the subensemble of particles that at a given time t are localized in the neighborhood of $\boldsymbol{x}$ and move with any velocity $\boldsymbol{p}/m$.

The second equation of the hierarchy describes the transfer of momentum, or equivalently, the evolution of $\boldsymbol{j}(\boldsymbol{x}, t)$. In addition to ρ and first local moments $\langle p_i \rangle_{\boldsymbol{x}}$, it contains second local moments or correlations $\langle p_i p_j \rangle_{\boldsymbol{x}}$, while the transfer of these (and of the kinetic energy, for $i = j$ and up to a factor $1/2m$) is contained in the third equation. This coupling between successive members of the hierarchy (4.56) creates a highly difficult mathematical problem. However, in Sect. 4.4.1 we shall find that under well-defined physical assumptions (and some approximations) a decoupling takes place, having far-reaching consequences.

In order to rewrite the first two equations of the hierarchy in a form that is more convenient for our purposes, we proceed as follows. In terms of the function $\tilde{Q}$ introduced through (4.46), the first moments are [from (4.52)]

$$\langle p_i \rangle_{\boldsymbol{x}} = -i \left(\frac{\partial}{\partial z_i} \ln \tilde{Q} \right) \bigg|_{z=0} , \tag{4.58a}$$

$$\langle p_i p_j \rangle_{\boldsymbol{x}} = - \frac{\partial^2}{\partial z_j \partial z_i} \ln \tilde{Q} \bigg|_{z=0} + \langle p_i \rangle_{\boldsymbol{x}} \langle p_j \rangle_{\boldsymbol{x}} . \tag{4.58b}$$

The structure of these expressions suggests introducing the variables

$$z_+ = \boldsymbol{x} + \eta z, \quad z_- = \boldsymbol{x} - \eta z, \tag{4.59}$$

instead of $\boldsymbol{x}$ and z, and writing $\tilde{Q}$ in the unconditional form

$$\tilde{Q}(z_+, z_-, t) = q_+(z_+, t) q_-(z_-, t) \chi(z_+, z_-, t), \tag{4.60}$$

where η is a real parameter with dimensions of action [as follows from Eq. (4.46)], to be determined in Sect. 4.4.4. The function χ represents the nonfactorizable part of $\tilde{Q}(z_+, z_-, t)$.

Notice from (4.46) that $\tilde{Q}^*(\boldsymbol{x}, z, t) = \tilde{Q}(\boldsymbol{x}, -z, t)$, whence

$$q_+(z_\pm, t) = q_-^*(z_\pm, t), \quad \chi^*(z_+, z_-, t) = \chi(z_-, z_+, t). \tag{4.61}$$

$\tilde{Q}$ can therefore be rewritten in the form

$$\tilde{Q}(z_+, z_-, t) = q(z_+, t) q^*(z_-, t) \chi(z_+, z_-, t), \tag{4.62}$$

$$\text{where} \quad q(z_\pm, t) \equiv q_+(z_\pm, t). \tag{4.63}$$

The whole description can thus be carried out in terms of only the complex function q and the function χ. Further, from (4.49) and (4.62) it follows that

$$\rho(\boldsymbol{x}, t) = \tilde{Q}(\boldsymbol{x}, 0, t) = q^*(\boldsymbol{x}, t) q(\boldsymbol{x}, t) \chi_0(\boldsymbol{x}, t), \tag{4.64}$$

with $\chi_0(\boldsymbol{x}, t) = \chi(z_+, z_-, t)|_{z_+ = z_-}$ a real function that can be taken as a constant, absorbing its possible time and space dependence into the functions $q(\boldsymbol{x}, t)$, $q^*(\boldsymbol{x}, t)$. We therefore write

$$\chi_0(\boldsymbol{x}, t) = 1, \quad \rho(\boldsymbol{x}, t) = q^*(\boldsymbol{x}, t) q(\boldsymbol{x}, t). \tag{4.65}$$

With the above results Eq. (4.58a, 4.58b) thus transform into

$$\langle p_i \rangle_x = -i\eta \frac{\partial}{\partial x_i} \ln \frac{q(\boldsymbol{x}, t)}{q^*(\boldsymbol{x}, t)} + g_i \tag{4.66a}$$

$$\text{with} \quad g_i = i\eta \left(\frac{\partial}{\partial z_{-i}} - \frac{\partial}{\partial z_{+i}} \right) \ln \chi \bigg|_{z_+ = z_-}, \tag{4.66b}$$

and

$$\Gamma_{ij} \equiv \left\langle p_i p_j \right\rangle_{\boldsymbol{x}} - \langle p_i \rangle_{\boldsymbol{x}} \left\langle p_j \right\rangle_{\boldsymbol{x}} = -\eta^2 \frac{\partial^2}{\partial x_i \partial x_j} \ln \rho + \Sigma_{ij} \tag{4.67a}$$

$$\text{with} \quad \Sigma_{ij} = \Sigma_{ji} = 2\eta^2 \left[\left(\frac{\partial^2}{\partial z_{+i} \partial z_{-j}} + \frac{\partial^2}{\partial z_{-i} \partial z_{+j}} \right) \ln \chi \right]_{z_+ = z_-}. \tag{4.67b}$$

Notice that both quantities g_i and Σ_{ij} are determined by the function χ.

Using Eqs. (4.57) and (4.67a), the first two equations of the hierarchy (4.56) become

$$\frac{\partial \rho}{\partial t} + \frac{\partial}{\partial x_j} \left(v_j \rho \right) = 0, \tag{4.68a}$$

$$m\frac{\partial}{\partial t}(v_i\rho)+m\frac{\partial}{\partial x_j}\left(v_iv_j\rho\right)-\frac{\eta^2}{m}\frac{\partial}{\partial x_j}\left(\rho\frac{\partial^2}{\partial x_i\partial x_j}\ln\rho\right)+\frac{1}{m}\frac{\partial}{\partial x_j}\Sigma_{ij}\rho-f_i\rho$$
$$=\tau v_j\frac{\partial f_i}{\partial x_j}\rho-e^2\widetilde{(\hat{\mathcal{D}}Q)_i}\bigg|_{z=0}. \tag{4.68b}$$

4.4 The Schrödinger Equation

Let us now focus on the couple of Eq. (4.68). As shown in Appendix C, a series of algebraic manipulations allows to recast them in the form

$$\nabla\left(\frac{1}{q}\hat{M}q\right)=\boldsymbol{F}_{\text{rad}}+\boldsymbol{F}_{\Sigma}-\frac{\partial \boldsymbol{g}}{\partial t}+\boldsymbol{v}\times(\nabla\times\boldsymbol{g}), \tag{4.69}$$

with $\hat{M}$ the differential operator

$$\hat{M}=-2i\eta\frac{\partial}{\partial t}+\frac{1}{2m}\left(-2i\eta\nabla+\boldsymbol{g}\right)^2+V. \tag{4.70}$$

$\boldsymbol{F}_{\text{rad}}$ and $\boldsymbol{F}_{\Sigma}$ are force vectors with components given by

$$F_{i\text{rad}}=\tau v_j\frac{\partial f_i}{\partial x_j}-\frac{e^2}{\rho}\widetilde{(\hat{\mathcal{D}}Q)_i}\bigg|_{z=0},\quad F_{i\Sigma}=-\frac{1}{m\rho}\frac{\partial}{\partial x_j}\Sigma_{ji}\rho, \tag{4.71}$$

and the functions g_i and Σ_{ij} are defined through Eqs. (4.66b) and (4.67b).

From (4.65) it follows that the operator $\hat{M}$ acts over an *amplitude of probability density*.[10] This, together of course with the structure of $\hat{M}$, strongly suggests to identify (4.69) with a general equation that contains the Schrödinger equation as a particular case, provided $\eta=\hbar/2$. In the following we discuss the conditions under which (4.69) reduces to the Schrödinger equation; as mentioned earlier, the derivation of the value of η is left for Sect. 4.4.4.

4.4.1 The Radiationless Approximation

What is the physical connection between the process described by Eq. (4.68) and the Schrödinger equation? To give an answer to this central question, let us examine the role of the radiative terms appearing in (4.68).

[10] It was Born who introduced the interpretation of $q*q$ as a probability in quantum mechanics, though limited to the description of dispersion states. The proposal of interpreting this quantity as a probability density more generally was put forward by Pauli in (1927). Here, Born's rule ensues from the theory itself.

The first term on the right-hand side of (4.68b), due to the radiation reaction, has a dissipative effect on the motion. In the absence of the second term, this one would be responsible for the decay of the system to its (classical) state of lowest energy. For instance, it would make a hydrogen atom collapse under the attractive force exerted on the (radiating) electron by the nucleus. The second term, on the other hand, exerts a permanent, random action on the motion. In the absence of the first term, this one would lead to an erratic, unstable motion of the electron. Furthermore, this term introduces memory effects since it contains a time integration from the initial moment (at which particle and field were connected) to time t.

Now, in line with the discussion of Sect. 4.2.1 we are particularly interested in those situations (assuming they exist) in which as a result of the combined action of the dissipative and fluctuating forces, the system eventually reaches a regime of energy balance in the mean, or rather, a reversible regime that complies with the condition $d\langle H\rangle/dt = 0$. Probably the hypothesis that such a state is reached is applicable to all (bound) systems, but we leave this point open. What is important is that under such condition, the two terms on the right-hand side of Eq. (4.32) essentially cancel each other, and any residual effect of the radiative terms in (4.68b) leads to minor corrections in the mean. We therefore consider the time-asymptotic limit (see footnote 9) when the two terms on the right-hand side of (4.68b) can be neglected in a first approximation (this is the so-called *radiationless approximation*), i.e.,

$$\frac{\partial \rho}{\partial t} + \frac{\partial}{\partial x_j}\left(v_j \rho\right) = 0, \tag{4.72a}$$

$$m\frac{\partial}{\partial t}\left(v_i \rho\right) + m\frac{\partial}{\partial x_j}\left(v_i v_j \rho\right) - \frac{\eta^2}{m}\frac{\partial}{\partial x_j}\left(\rho\frac{\partial^2}{\partial x_i \partial x_j}\ln\rho\right) + \frac{1}{m}\frac{\partial}{\partial x_j}\Sigma_{ij}\rho - f_i\rho = 0. \tag{4.72b}$$

Further, also $\boldsymbol{F}_{\mathrm{rad}}$ vanishes in this approximation, so that Eq. (4.69) becomes

$$\nabla\left(\frac{1}{q}\hat{M}q\right) = -\frac{1}{m\rho}\nabla\cdot\left(\tilde{\boldsymbol{\Sigma}}\rho\right) - \left[\frac{\partial \boldsymbol{g}}{\partial t} - \boldsymbol{v}\times(\nabla\times\boldsymbol{g})\right], \tag{4.73}$$

where $\tilde{\boldsymbol{\Sigma}}$ is the symmetric tensor with components Σ_{ij}. According to (4.66b) and (4.67b), the forces appearing on the right-hand side of this equation depend on the first and second derivatives of $\chi(z_+, z_-, t)$, evaluated at $z = 0$. As will be confirmed in Sect. 4.6, from the expression $\tilde{Q}(z_+, z_-) = q_+(z_+)q_-(z_-)\chi(z_+, z_-)$ it is clear that a knowledge of the function χ is central for a complete description of the system in phase-space. However, the exact form of χ can only be found by solving the complete hierarchy of equations in configuration space (or equivalently, the full GFPE in phase space), which is beyond present-day possibilities. Nevertheless, even though χ is undetermined at this point, Eq. (4.73) suggests a physical meaning for $\boldsymbol{g}$ and $\tilde{\boldsymbol{\Sigma}}$. This follows from noticing that the term within square brackets has the structure of a Lorentz force $e\boldsymbol{E}_\chi + (e/c)\,\boldsymbol{v}\times\boldsymbol{B}_\chi$, indicating that $\boldsymbol{g}$ plays the role of an effective vector potential,

$$g(x,t) = -\frac{e}{c}A_\chi(x,t), \tag{4.74}$$

so that $E_\chi = -(1/c)\left(\partial A_\chi/\partial t\right)$, and $B_\chi = \nabla \times A_\chi$. This expression for g combined with (4.66a) gives the correct relation between the (local) mechanical momentum $\langle p\rangle_x$ and the (local) canonical momentum $\langle P\rangle_x$,

$$\langle p\rangle_x = -i\eta\nabla \ln \frac{q(x,t)}{q^*(x,t)} + g = \langle P\rangle_x - \frac{e}{c}A_\chi(x,t). \tag{4.75}$$

The structure of the remaining term on the right-hand side of (4.73) suggests assigning to $\tilde{\Sigma}$ the role of an effective stress tensor transmitted through the field A_χ. Therefore, in the radiationless approximation all terms containing g and $\tilde{\Sigma}$ can be neglected separatedly (which amounts to taking $\chi = 1$), and Eq. (4.73) becomes $\nabla(q^{-1}\hat{M}q) = 0$ with $\hat{M}$ is given by (4.70) with $g = 0$; whence

$$\hat{M}q = h(t)q, \tag{4.76}$$

with h an arbitrary function of time. In terms of the function

$$\psi(x,t) = q(x,t)\, e^{-\frac{i}{2\eta}\int^t h(t')dt'}, \tag{4.77}$$

Eq. (4.76) becomes $\hat{M}\psi = 0$, i.e.,

$$-\frac{2\eta^2}{m}\nabla^2\psi + V\psi = 2i\eta\frac{\partial\psi}{\partial t}, \tag{4.78}$$

where

$$\rho(x,t) = \psi(x,t)\psi^*(x,t). \tag{4.79}$$

Without loss of generality one may take $h(t) = 0$, hence $\psi(x,t) = q(x,t)$. Equation (4.78) has precisely the form of the Schrödinger equation for the probability amplitude ψ, in terms of the as yet undetermined parameter η.

We shall occasionally use the polar form of $\psi(x,t)$ (sometimes called the Madelung representation),

$$\psi(x,t) = q(x,t) = \sqrt{\rho(x,t)}e^{iS(x,t)}. \tag{4.80}$$

The local canonical momentum coincides in the radiationless approximation with the mechanical momentum, given according to (4.75) and (4.80) by

$$\langle p\rangle_x = mv = -i\eta\nabla \ln \frac{\psi(x,t)}{\psi^*(x,t)} = 2\eta\nabla S(x,t), \tag{4.81}$$

and the local mean dispersion of the momentum reads (4.67a),

$$\left\langle \boldsymbol{p}^2 \right\rangle_x - \langle \boldsymbol{p} \rangle_x^2 = -\eta^2 \nabla^2 \ln \rho. \tag{4.82}$$

Notice that if we take $\chi \neq 1$ so that the terms containing $\boldsymbol{g}$ are not neglected, but the forces on the right-hand side of (4.73) comply with the relation

$$\frac{\partial \boldsymbol{g}}{\partial t} - \boldsymbol{v} \times (\nabla \times \boldsymbol{g}) = -\frac{1}{m\rho} \nabla \cdot \left(\tilde{\boldsymbol{\Sigma}} \rho \right), \tag{4.83a}$$

Equation (4.73) becomes $\nabla(q^{-1}\hat{M}q) = 0$, where $\hat{M}$ now includes the vector $\boldsymbol{g}$, according to (4.70). Using again (4.77) we get

$$\frac{1}{2m} (-2i\eta\nabla + \boldsymbol{g})^2 \psi + V\psi = 2i\eta \frac{\partial \psi}{\partial t}, \tag{4.84}$$

which has the form of the Schrödinger equation with minimal coupling to the field $\boldsymbol{A}_\chi$.[11]

In the radiationless approximation the first two equations of the hierarchy, (4.72), become an *independent* system of two nonlinear equations for $\rho(\boldsymbol{x}, t)$ and $\boldsymbol{v}(\boldsymbol{x}, t)$,

$$\frac{\partial \rho}{\partial t} + \frac{\partial}{\partial x_j} (v_j \rho) = 0, \tag{4.85a}$$

$$m \frac{\partial}{\partial t} (v_i \rho) + m \frac{\partial}{\partial x_j} (v_i v_j \rho) - \frac{\eta^2}{m} \left(\rho \frac{\partial}{\partial x_i} \nabla^2 \ln \rho + \frac{\partial \rho}{\partial x_j} \frac{\partial^2 \ln \rho}{\partial x_i \partial x_j} \right) - f_i \rho = 0, \tag{4.85b}$$

decoupled from the rest of the hierarchy. By their form, (4.85a, 4.85b) resemble the corresponding pair of equations of kinetic theory (see e.g. Balescu 1975, Sect. 12.4, or Fujita and Godoy (2010), Sect. 8), with the stress tensor replaced by the term proportional to η^2 in Eq. (4.85b). To carry further this analogy (which originated in the early work of Madelung 1926), one would have to treat this term as a kind of stress in a 'quantum fluid'—which is somewhat artificial, as no real fluid is supporting such stress. Actually the term proportional to η^2 does not require any odd interpretation,

[11] More generally, the condition under which Eq. (4.84) holds is

$$\int \boldsymbol{F}_{\boldsymbol{\Sigma}} \cdot d\boldsymbol{x} - \int \boldsymbol{F}_g \cdot d\boldsymbol{x} + h(t) = 0,$$

with $\boldsymbol{F}_g = (\partial \boldsymbol{g}/\partial t) - \boldsymbol{v} \times (\nabla \times \boldsymbol{g})$, and $\boldsymbol{F}_{\boldsymbol{\Sigma}} = -(1/m\rho)\nabla \cdot \left(\tilde{\boldsymbol{\Sigma}} \rho \right)$. The vector field $\boldsymbol{g}$ given by Eq. (4.74) determines the function $\tilde{\boldsymbol{\Sigma}}$ and should contain all electromagnetic contributions: the ZPF, the self-field radiated by the particle, any external field, and any existing excitation. This leads to the territory of QED. Since the radiated field depends on the dynamics, it can be known only by solving the entire (matter-field) problem.

since according to Eqs. (4.67a) and (4.82) it is due to the momentum correlations (transcribed to configuration space by the reduction process). An important property of this term, difficult to overstress, is its nonlocal nature, manifested through its dependence on the distribution of the particles in the entire configuration space. No wonder that this becomes a key term in the theory, as it encapsulates some peculiar features such as quantum fluctuations and quantum nonlocality.[12] In fact, it gives rise to the so-called *quantum* (or *Bohm's*) *potential*

$$V_Q = -\frac{2\eta^2}{m}\frac{\nabla^2\sqrt{\rho}}{\sqrt{\rho}}, \tag{4.86}$$

as follows from Eq. (C.8) in Appendix C, where it is shown that Eq. (4.85a, 4.85b) are equivalent to the Hamilton-Jacobi-type equation found so characteristically in Bohm's theory,

$$2\eta\frac{\partial S}{\partial t} + \frac{1}{2}m\boldsymbol{v}^2 + V + V_Q = 0, \tag{4.87}$$

with the functions ρ, $\boldsymbol{v}$ and S defined through (4.80) and (4.81). This matter will be further discussed in Chap. 8; for the time being suffice it to note that the term proportional to η^2 is the sole element that distinguishes Eq. (4.85a, 4.85b) from those describing an ensemble of classical particles. Its presence signals therefore a definitive departure from classical physics. Section 4.5.1 includes more detailed comments on the subject.

4.4.2 Statistical and Quantum Averages

A remarkable feature of the description constructed above is that it gives the possibility to calculate statistical averages, such as those defined in Sect. 4.2.1, in terms of appropriate operations performed on the probability amplitude $\psi(\boldsymbol{x}, t)$. Thus for instance, according to Eqs. (4.51) and (4.81), the mean momentum is given by

$$\begin{aligned}\langle \boldsymbol{p}\rangle &= \int \langle \boldsymbol{p}\rangle_{\boldsymbol{x}}\,\rho dx = -i\eta\int\left(\psi^*\nabla\psi - \psi\nabla\psi^*\right)dx \\ &= -2i\eta\int\psi^*\nabla\psi dx = \int\psi^*\hat{\boldsymbol{p}}_\eta\psi dx,\end{aligned} \tag{4.88}$$

[12] In Sect. 4.3.1, the transition from phase space to the momentum subspace was shown to lead to an integro-differential equation that is explicitly nonlocal due to the integral transform. The nonlocal character of the reduced description becomes then obvious. A similar situation occurs in the transition to configuration space, but then the nonlocality of the description (even for a single-particle system) is manifested through the term containing η^2, which embodies information of the probability distribution of particles. This point is discussed more at length in Chap. 8.

where an integration by parts was performed in the second line (with ρ vanishing at infinity), and $\hat{\boldsymbol{p}}_\eta$ stands for the differential operator

$$\hat{\boldsymbol{p}}_\eta \equiv -2i\eta\nabla. \tag{4.89}$$

Similarly, from (4.82) one obtains, after some simple algebraic manipulations,

$$\left\langle \boldsymbol{p}^2 \right\rangle = \int \left\langle \boldsymbol{p}^2 \right\rangle_{\boldsymbol{x}} \rho dx = \int \left(\langle \boldsymbol{p} \rangle_{\boldsymbol{x}}^2 - \eta^2 \nabla^2 \ln \rho \right) \rho dx = \int \psi^* \hat{\boldsymbol{p}}_\eta^2 \psi dx. \tag{4.90}$$

On the other hand, since any function of the form $g(\boldsymbol{x}, t)$ coincides with its local mean value $\langle g \rangle_{\boldsymbol{x}}$ [see Eq. (4.50)], one may write

$$\langle g(\boldsymbol{x}) \rangle = \int g(\boldsymbol{x}) \rho dx = \int \psi^* g(\hat{\boldsymbol{x}}) \psi dx, \tag{4.91}$$

with $\hat{\boldsymbol{x}} \equiv \boldsymbol{x}$. As a result, in the present formalism the *statistical* averages $\langle \boldsymbol{p} \rangle$, $\left\langle \boldsymbol{p}^2 \right\rangle$, $\langle g(\boldsymbol{x}) \rangle$ coincide with those defined by the usual quantum rules

$$\left\langle \hat{A} \right\rangle = \int \psi^* \hat{A} \psi dx, \tag{4.92}$$

in terms of an appropriate operator $\hat{A}$ associated with the dynamical variable A. In particular, for $A = H = \left(p^2/2m\right) + V(x)$, Eqs. (4.78) and (4.89) give for the Hamiltonian and the energy operators

$$\hat{H} = \left(\hat{p}_\eta^2/2m\right) + V(\hat{x}), \quad \hat{E} = 2i\eta(\partial/\partial t). \tag{4.93}$$

In the time-asymptotic, radiationless regime, the dynamics of the mechanical subsystem is governed by the Schrödinger-like Eq. (4.78), which means that the quantum rules (still in terms of the parameter η) ensuing from the structure of Eq. (4.78) already operate (more comments on this in the following chapters). Under such circumstances the dynamical variables become represented by their corresponding (usually noncommuting) quantum operators, and the expectation values are to be calculated with the usual quantum tools. In particular, the customary correspondence rule applies, which means replacing the Poisson brackets with the quantum commutators according to

$$[x, p]_{\mathrm{PB}} \to (2i\eta)^{-1} \left[\hat{x}, \hat{p}_\eta\right]. \tag{4.94}$$

However, the correspondence between dynamical variables and operators must be handled carefully when making use of the relations between average values derived in Sect. 4.2.1. Equation (4.22)—from which all such relations are extracted—has been derived directly from the GPFE, and it is a statistical law. In the following Section and in Chap. 6, extensive use is made of Eq. (4.22) to carry out calculations of averages already in the quantum regime. This is a legitimate procedure when the calculations do not involve an ambiguity in the operator ordering (see, e.g., Eq. (4.106) below). Such procedure suffices for our purposes, and corresponds to the regime governed by Eq. (4.78). When convenient we will remind the reader that the relation is to be understood as a quantum expression, by adding a roman subindex Q, i.e., $\left\langle \hat{A} \right\rangle = \langle A \rangle_{\mathrm{Q}}$.

4.4.3 Stationary Schrödinger Equation

We recall from Sect. 4.2.1 that a sufficient condition for $d\,\langle \mathcal{G}(\boldsymbol{x}, \boldsymbol{p}) \rangle /dt = 0$ to hold, is that the system has reached a stationary state, in which $(\partial Q/\partial t) = 0$. Let us now see how Eq. (4.78) can be used to describe such state.

With $(\partial Q/\partial t) = 0$ the function Q becomes $Q(\boldsymbol{x}, \boldsymbol{p})$ and the marginal distribution ρ depends only on $\boldsymbol{x}$. Therefore the local mean velocity does not depend on time [see Eq. (4.52)],

$$\langle \boldsymbol{p} \rangle_{x} = \frac{1}{\rho(\boldsymbol{x})} \int \boldsymbol{p} Q(\boldsymbol{x}, \boldsymbol{p}) d\boldsymbol{p} = m\boldsymbol{v}(\boldsymbol{x}). \tag{4.95}$$

It follows from (4.81) that $\nabla S(\boldsymbol{x}, t)$ is a time-independent function, whence $S(\boldsymbol{x}, t)$ decomposes as $S(\boldsymbol{x}, t) = s(\boldsymbol{x}) + S(t)$, and Eq. (4.80) gives $\psi(\boldsymbol{x}, t) = \varphi(\boldsymbol{x}) e^{iS(t)}$. Application of $\hat{H}_\eta$ to this function thus results in

$$\hat{H}_\eta \psi = -2\eta \frac{dS}{dt} \psi. \tag{4.96}$$

A multiplication from the left by ψ^* gives, after integration,

$$\left\langle \hat{H}_\eta \right\rangle = \langle H \rangle = \frac{1}{2m} \left\langle \boldsymbol{p}^2 \right\rangle + \langle V \rangle = \mathcal{E} = -2\eta \frac{dS}{dt}. \tag{4.97}$$

Integrating the last equality gives finally $S(t) = (-\mathcal{E}/2\eta)t$ so that

$$\psi(\boldsymbol{x}, t) = \varphi(\boldsymbol{x}) e^{-i\mathcal{E}t/2\eta}, \tag{4.98}$$

and Eq. (4.78) becomes the stationary equation,

$$-\frac{2\eta^2}{m} \nabla^2 \psi + V\psi = \mathcal{E}\psi. \tag{4.99}$$

This is an eigenvalue equation (for bound systems), with the eigenvalues representing the possible energies $\mathcal{E}$. The eigenvalue corresponding to the eigenfunction ψ_k will be denoted in general with $\mathcal{E}_k$ $(k \geq 0)$, $\mathcal{E}_0$ representing the lowest energy state. Thus, from Eq. (4.98),

$$\psi_k(\boldsymbol{x}, t) = \varphi_k(\boldsymbol{x}) e^{-i\mathcal{E}_k t/2\eta}. \tag{4.100}$$

4.4.4 Detailed Energy Balance: The Entry Point for Planck's Constant

According to the above results, in the time-asymptotic limit the system reaches a regime in which its statistical behavior is described by the Schrödinger equation—provided the parameter η has the value $\hbar/2$. This description has been obtained by taking the radiationless approximation, in which the ZPF disappeared from the picture, taking Planck's constant along. Now we are in a condition to bring the ZPF back into the picture, by imposing the energy-balance condition (4.37) to the ground state described by Eq. (4.99). We thus assume that the radiation field is in its ground state, with spectral energy density given by (4.6),

$$\rho_0(\omega) = \frac{\hbar\omega^3}{2\pi^2 c^3}, \tag{4.101}$$

and that the mechanical system is in its state of lowest energy ($\mathcal{E} = \mathcal{E}_0$). Therefore we write the energy-balance condition in the form

$$\tau \langle \dddot{\boldsymbol{x}} \cdot \boldsymbol{p} \rangle_0 = \frac{e^2}{m} \left\langle \boldsymbol{p} \cdot \hat{\boldsymbol{\mathcal{D}}} \right\rangle_0 . \tag{4.102}$$

It is possible to give a simpler (equivalent) form to this equation by using the true FPE (4.19) (instead of the more elaborate and exact GFPE), which holds in the time-asymptotic regime. For this purpose we resort to Eq. (4.45) with $\mathcal{G} = H$, thus obtaining the balance condition

$$\tau \langle \dddot{\boldsymbol{x}} \cdot \boldsymbol{p} \rangle_0 = -\frac{1}{m} \left\langle \mathrm{Tr} \boldsymbol{D}^{pp} \right\rangle_0 , \tag{4.103}$$

with $\mathrm{Tr}\boldsymbol{D}^{pp} = \sum_i D_{ii}^{pp}$. To calculate both sides of this equation we apply the usual mathematical methods ensuing from the structure of Eq. (4.78). In particular, resorting to Eq. (4.100) we write (in one dimension, for simplicity)

$$x_{kn}(t) = \int \psi_k^* x \psi_n dx = e^{i\omega_{kn}t} x_{kn}(0), \tag{4.104}$$

where $\omega_{kn} \equiv (\mathcal{E}_k - \mathcal{E}_n)/2\eta$. This gives for the matrix elements of the operators $\hat{p}$ and $\hat{\dddot{x}}$[13]

$$p_{kn} = m\dot{x}_{kn} = im\omega_{kn}x_{kn}, \tag{4.105}$$

$$\dddot{x}_{kn} = -i\omega_{kn}^3 x_{kn}.$$

[13] In the present context these relations appear simply as a result of the calculations. Their physical meaning will become clear when they reappear in Chap. 5, in connection with the derivation of the Heisenberg description of QM.

Now, it is straightforward to show that for any state n,

$$(\widehat{\ddot{x}}\,\hat{p})_{nn} = (\hat{p}\,\widehat{\ddot{x}})_{nn}, \tag{4.106}$$

whence the ordering of these operators is irrelevant for calculating the mean value. Consequently the left-hand side of Eq. (4.103) becomes

$$\tau(\widehat{\ddot{x}}\,\hat{p})_{00} = -m\tau \sum_k \omega_{0k}^4 \left|x_{0k}\right|^2 . \tag{4.107}$$

The calculation of the right-hand side is made in Appendix D and gives

$$-\frac{1}{m}\left\langle D^{pp}\right\rangle_0 = -\frac{\hbar m \tau}{2\eta} \sum_k \omega_{0k}^4 \left|x_{0k}\right|^2 . \tag{4.108}$$

Equating these two results we obtain

$$\eta = \frac{\hbar}{2}. \tag{4.109}$$

In other words, for the ground state to be a truly stationary state, η must be given by this last expression. With this value for η, the general Eq. (4.78) becomes precisely the Schödinger equation,

$$i\hbar\frac{\partial\psi}{\partial t} = -\frac{\hbar^2}{2m}\nabla^2\psi + V\psi. \tag{4.110}$$

The time-asymptotic and radiationless regime, in which the mechanical system is correctly described by the Schrödinger equation, is therefore called the *quantum regime*.[14]

In order to better grasp the significance of the ZPF in connection with Eq. (4.110), let us recall that the generic Eq. (4.78), as well as its stationary form (4.99), were derived neglecting the radiative terms, once they had played their main role in taking

[14] During the development of SED in the late ninetiteenseventies and until the eighties, use was made of the true FPE derived in Appendix B. That equation was applied to several problems, giving some correct results for linear problems, and wrong answers for the rest, particularly the H atom. From what we have just seen, it is clear that a mistake was being made by applying the classical FPE (with *classical* variables) to a system that is already following a nonclassical behavior. In particular, the FPE by itself does not guarantee that the quantum regime has been reached; hand in hand with it, the energy-balance condition must be in force. The erroneous results obtained, characterized by the violation of the energy-balance condition, led unfortunately to a quite extended belief that the stochastic approach to QM based on the ZPF was wrong. This was first shown in Boyer (1976), (1980), and then by several other authors (see e.g. Marshall and Claverie 1980; Alcubierre and Lozano 1988); a detailed exposition and further references are given in *The Dice*). In this context it is relevant to recall the statement by Claverie and Diner in (1977): "The relationship between quantum theory and SED, if it exists, is of a more subtle nature than [a] mere formal equivalence."

the system close to equilibrium, when it had reached a situation of energy balance. Now we have found that the energy-balance condition (4.103) determines Schrödinger's equation *univocally*. It follows that QM contains information about the central connection between the quantum particle and the ZPF, even though this information is contained in a cryptic form. One may therefore conclude that the Schrödinger equation describes more than just the particle; it describes the particle *in interaction* with the field that has taken it to the time-reversible regime.

The result just obtained is of utmost significance for the theory: the fact that η does not depend on the specific problem (e.g. on the external potential V) but only on the ZPF, conveys to the Schrödinger equation its recognized universal validity. In addition to introducing Planck's constant into the picture and furnishing the right value for η, (4.103) confirms the spectrum $\rho \sim \omega^3$ as the *single one* that leads to equilibrium with the ground state of the mechanical system, by guaranteeing that both expressions (4.107) and (4.108) have the same algebraic structure. This means that the energy-balance condition is satisfied not only globally, but term by term; therefore, a situation of *detailed energy balance* between particle and field has been reached.

The energy-balance condition can be imposed to find the specific stationary solution(s) for *any* problem at hand, be it classical or quantum. In the case of a classical stochastic problem, equilibrium is obtained only if the noise is characterized by a Rayleigh-Jeans spectrum, $\rho \sim \omega^2$, as was established by van Vleck almost a century ago (van Vleck 1924; an updated revision is van Vleck and Huber 1977). In the quantum context, by contrast, the lowest-energy stationary solution is that of the Schrödinger equation, and equilibrium is reached with the ZPF with spectrum $\rho \sim \omega^3$, as has just been demonstrated.

We have here the answer to the question of *why* atoms reach and maintain their stability. Without the notion of the ZPF it becomes impossible to explain this stability, since the moving electrons radiate their energy, and eventually (rather quickly) the atom collapses, as mentioned earlier. As will be studied in Chap. 6 (and as is well known), the excited states decay spontaneously, so eventually (if there is no external agent capable of inducing excitations), the excess energy will be radiated. Yet the atom does not collapse, because the stability of the ground state is guaranteed precisely thanks to the action of the field, which counterbalances (in the mean) the tendency of the electrons to radiate. For certain, well-defined orbital states, this mechanism leads to stable situations (strictly, quasi-stable in the case of excited states), which correspond to the quantum states. This is just the mechanism of stability envisioned by Nernst as early as 1916 and supported generally by the authors who have worked on SED along the years. Here we can appreciate the difference between *predicting* the existence of quantum states (as the Schrödinger equation does) and *explaining* them, i.e., exhibiting the mechanism that supports them.

It is appropriate here to contrast the present SED approach with the one that prevailed during the early attempts to develop the SED theory and apply it to the atomic problem (see, e.g., *The Dice* and references therein). It was believed that during the entire evolution of the dynamics, the diffusive terms should play a primary role, on a par with the external forces. Now we have learned that although the random field (in combination with radiation reaction) has an important effect for very short times in taking the system to an equilibrium regime, in the long run its role becomes much more subtle (though no less important): besides maintaining this equilibrium regime it determines the size of the fluctuations and produces radiative corrections. This explains the universal presence of the quantum fluctuations, and assigns to them a causal origin. In Chaps. 5–7 we discuss several other important manifestations of the presence of the ZPF in the quantum description.

A further important remark follows from Eq. (4.33), which under stationarity reduces to the balance condition for any of the 'classical' integrals of motion of the system under study, namely,

$$\left\langle \dddot{x}_i \frac{\partial \xi}{\partial p_i} \right\rangle_{\mathrm{Q}} = \frac{3c^3}{2} \left\langle \frac{\partial \xi}{\partial p_i} \hat{\mathcal{D}}_i \right\rangle_{\mathrm{Q}} . \tag{4.111}$$

Equation (4.102) is just the particular case of (4.111) for $\xi = H$. The factor e^2 has vanished from this equation (written to the lowest order of approximation), transforming it into an apparently mechanical balance condition for the integrals of motion, which can be applied even to neutral particles (particles with a vanishing charge). In Boyer (1975) this limit is termed 'random mechanics'. Equation (4.111) evokes the quantization conditions of old QM, all of them apparently mechanical. Recall for example the quantum condition for the angular momentum: it has been universally interpreted as a mechanical constraint, yet it is a particular case of (4.111) under the quantum regime.

4.4.5 Schrödinger's *i*

A point of much significance concerning the Schrödinger equation is the imaginary coefficient i in the time derivative, which makes it differ essentially from any other *fundamental* equation of physics. This is true even for the fundamentally undulatory or wave equations, such as those of optics, electromagnetism or sound. It is only in derived equations where imaginary coefficients are present—which suggests that Eq. (4.110) should actually be interpreted as a derived equation, just as it appears in the present treatment.

It is well known that the solutions ψ of Eq. (4.78) represent an undulatory phenomenon. However, the usual wave equation is written in terms not of a Laplacian, but of a d'Alambertian containing a second-order time derivative, whereas Eq. (4.78) contains a first-order time derivative. This difference is crucial. The wave equation

$$\nabla^2 w - \frac{1}{c^2}\frac{\partial^2 w}{\partial t^2} = 0 \tag{4.112}$$

is hyperbolic and describes a (time-)reversible process, which represents an oscillating (sustained) wave. By contrast, the equation

$$K\nabla^2 w - \frac{\partial w}{\partial t} = 0 \tag{4.113}$$

with K real, is parabolic and describes a (time-)irreversible process, which fades out and propagates with no return. Typical examples to which this equation applies are the propagation of heat, or the diffusion of particles in a solvent (or evanescent 'waves').

The Schrödinger equation lies somewhere in between these two structures, since it has the form of Eq. (4.113) though it belongs to the family of wave equations. This can be seen by replacing the real coefficient K in Eq. (4.113) by an imaginary coefficient, or equivalently replacing t by it,

$$i\frac{\partial w}{\partial t} = -K\nabla^2 w. \tag{4.114}$$

The parabolic equation is thus transformed into a hyperbolic one: the operator $i(\partial/\partial t)$ takes the place of the second-order operator $(\partial^2/\partial t^2)$, and the solution w becomes oscillatory and reversible. In other words, on comparing Eqs. (4.113) and (4.114) we observe that the change $K \to iK$ (or rather $t \to it$) transforms the diffusion equation into a wave equation, and vice versa. We find in Schrödinger's i the (formal) root of the wave properties in the quantum description, although the structure of the Schrödinger equation differs from that of a legitimate wave equation. This transformation, which is known as a Wick rotation and converts a Lorentz metric into a Cartesian one (or vice versa), is often used in the formal treatment of second-order differential equations.[15] This also prepares us to find that the 'waves' of QM differ in several aspects from waves in a physical medium, as discussed, e.g. in Ballentine (1990, 1998).

4.5 Further Insights into the Quantum Description

Let us now apply some of the results obtained in the foregoing sections to obtain further insight into the nature of the quantum description. This also gives us an opportunity to introduce some preparatory material that will be revisited in the following chapters.

---

[15] A related subject is *stochastic quantization* (Parisi and Wu 1981; Masujima 2009), which makes use of an imaginary time τ related to the real time by $\tau = it$. This transforms the time-dependent Schrödinger equation into a diffusion-like equation, so the expansion in terms of eigenfunctions of the Hamiltonian takes the form of a partition function of statistical mechanics with τ interpreted as the inverse temperature, $\tau \to \beta = 1/k_B T$,

$$\Psi(x,t) = \sum_n \psi_n \exp(-i\mathcal{E}_n t) = \sum_n \psi_n \exp(-\tau\mathcal{E}_n) = \sum_n \psi_n \exp(-\beta E_n).$$

This procedure has proved to be of value in several applications, particularly in quantum field theory, by allowing for a treatment of quantum problems with the methods of statistical mechanics or stochastic processes. Of course, stochastic quantization is just a formal method of calculation; it is not intended to improve the interpretation of QM.

4.5.1 Fluctuations of the Momentum

As remarked in Sect. 4.4.1, the term proportional to η^2 in Eq. (4.85b) bears the footprint of the fluctuations in momentum space, given by Eq. (4.82) (with $\eta = \hbar/2$), namely

$$\left\langle p^2 \right\rangle_x - \langle p \rangle_x^2 = -\frac{\hbar^2}{4} \nabla^2 \ln \rho. \tag{4.115}$$

Upon an integration by parts one obtains for the average of this term

$$-\frac{\hbar^2}{4} \left\langle \nabla^2 \ln \rho \right\rangle = -\frac{\hbar^2}{4} \int \rho \nabla^2 \ln \rho dx = \frac{\hbar^2}{4} \left\langle \left(\frac{\nabla \rho}{\rho} \right)^2 \right\rangle, \tag{4.116}$$

whence Eq. (4.115) gives, resorting to Eqs. (4.57) and (4.90),

$$\left\langle p^2 \right\rangle - m^2 \left\langle v^2 \right\rangle = \frac{\hbar^2}{4} \left\langle \left(\frac{\nabla \rho}{\rho} \right)^2 \right\rangle. \tag{4.117}$$

This result discloses a contribution to the mean kinetic energy (additional to that due to the flux velocity v) originating in the local mean deviations of the momentum from its local mean value, $\left\langle p^2 \right\rangle_x - \langle p \rangle_x^2$. As follows from (4.115), such deviations are due to diffusion, which is always present because of the fluctuations impressed by the field on the momentum. Equation (4.117) thus suggests to introduce the concept of *diffusive* (also called *stochastic*) *velocity* u, defined as

$$u = \frac{\hbar}{2m} \nabla \ln \rho = \frac{\hbar}{2m} \frac{\nabla \rho}{\rho}. \tag{4.118}$$

Notice that this equation assigns the value $D = \hbar/2m$ to the diffusion coefficient introduced in Chap. 2, as was (correctly) *assumed* there. Considering as usual that ρ vanishes at infinity, u averages to zero,

$$\langle u \rangle = \frac{\hbar}{2m} \int \rho \nabla \ln \rho \, dx = 0. \tag{4.119}$$

In terms of the diffusive velocity Eq. (4.117) reads

$$\left\langle p^2 \right\rangle = m^2 \left\langle u^2 + v^2 \right\rangle, \tag{4.120}$$

and the dispersion of the momentum

$$\sigma_p^2 = \left\langle p^2 \right\rangle - \langle p \rangle^2 \tag{4.121}$$

becomes

$$\sigma_p^2 = m^2\sigma_u^2 + m^2\sigma_v^2. \tag{4.122}$$

According to the above results, the difference $\langle p^2 \rangle_x - \langle p \rangle_x^2$ does *not* represent the total local dispersion of the momentum (it does not average to σ_p^2), but only the contribution from $m\boldsymbol{u}$. Chapter 8 will provide an occasion to discuss more in depth the significant role played by $\boldsymbol{u}$ in the quantum description, showing that its presence represents a departure from classical physics. In particular, σ_u^2 will be seen to become essential for an understanding of the irreducible quantum fluctuations, and for endowing them with a *causal* origin, rooted in the presence of the ZPF.[16]

The above analysis can be extended to include the (radiative) corrections due to the tensor $\tilde{\Sigma}$ and the vector $\boldsymbol{g}$ defined in Eqs. (4.66a, 4.66b) and (4.67a, 4.67b). Indeed, Eq. (4.67a) with $i = j$ gives

$$\left\langle p^2 \right\rangle_x - \langle p \rangle_x^2 = -\frac{\hbar^2}{4}\nabla^2 \ln \rho + \mathrm{Tr}\tilde{\Sigma}. \tag{4.123}$$

The average of this expression leads to a formula for the total momentum dispersion (including radiative effects due to $\boldsymbol{A}_\chi$)

$$\sigma_p^2 = m^2\sigma_u^2 + \sigma_v^2 + \left\langle \mathrm{Tr}\tilde{\Sigma} \right\rangle, \tag{4.124}$$

where σ_v^2 may include corrections arising from the vector $\boldsymbol{g}$.

4.5.2 Local Velocities: 'Hidden' Information Contained in ψ

The diffusive velocity $\boldsymbol{u}$ and the flux velocity $\boldsymbol{v}$ defined in Eq. (4.57), appear on a similar footing in the present approach. As follows from Eq. (4.81) (with $\eta = \hbar/2$) and (4.118), these local average velocities can be expressed in terms of the wave function $\psi(\boldsymbol{x}, t)$ as (see also Eqs. (2.62) and (2.63) with $\lambda = 1$, $D = \hbar/2m$, $\boldsymbol{A} = 0$)

$$\boldsymbol{v} = \frac{\hbar}{m}\nabla S = \frac{i\hbar}{2m}\left(\frac{\nabla\psi^*}{\psi^*} - \frac{\nabla\psi}{\psi}\right) \tag{4.125}$$

[16] There exists a profuse quantum literature in which the term related to the momentum fluctuations enters through one door or another (for a discussion and several examples see Carroll 2010). Their contribution is rarely identified as coming from fluctuations in the momentum space, and almost never as due to the ZPF. Most frequently they are simply taken as 'quantum fluctuations', a term that conveys the idea that they are spontaneous, i.e., causeless. As a result, the momentum fluctuations term appears in the literature under several guises. In the stochastic theory of QM it is identified as produced by the velocity $\boldsymbol{u}$; see Chap. 2 and references therein. In Olavo (2000) it is interpreted as coming from a local entropy due to *spontaneous* local fluctuations in positions. In the Bohmian theory it is known as the quantum potential (see Chap. 8 and, e.g., Holland 1993), which in its turn is 'explained' as a Fisher information (Frieden 1998), Roy (1986) relates it to fluctuations of the metric, and so on.

and

$$\boldsymbol{u} = \frac{\hbar}{2m}\nabla \ln \rho = \frac{\hbar}{2m}\left(\frac{\nabla\psi^*}{\psi^*} + \frac{\nabla\psi}{\psi}\right). \tag{4.126}$$

Despite its importance, Eq. (4.125), along with its identification as a physically sensible local mean velocity, rarely appears in textbooks on QM, an (excellent) exception being Ballentine (1990, 1998). What is usually presented in the quantum literature is not $\boldsymbol{v}$, but the current density or probability flux,

$$\boldsymbol{j}(\boldsymbol{x}, t) = \rho\boldsymbol{v} = \frac{i\hbar}{2m}\left(\psi\nabla\psi^* - \psi^*\nabla\psi\right). \tag{4.127}$$

As for the velocity $\boldsymbol{u}$, its very presence and of course the physical meaning we have assigned to it, are completely foreign to the standard quantum discourse. Yet, Eq. (4.125) and (4.126) show that the wave function carries information about *both* components of the local velocity: its phase bears information on $\boldsymbol{v}$, whereas its modulus bears information about $\boldsymbol{u}$. Thus, they are concealed in the quantum formalism.

In order to exhibit the presence of the local velocities in the quantum description, we combine Eqs. (4.125) and (4.126) to find that

$$\hat{\boldsymbol{p}}\psi = -i\hbar\nabla\psi = m(\boldsymbol{v} - i\boldsymbol{u})\psi, \tag{4.128}$$

where $\hat{\boldsymbol{p}} = \hat{\boldsymbol{p}}_{\eta=\hbar/2}$ [cf. Eq. (4.89)]. We see that the application of the quantum operator $\hat{\boldsymbol{p}}$ to the wave function reproduces both velocities (or momenta). Since $\langle \boldsymbol{u}\rangle = 0$, when averaging over x only the contribution of $\boldsymbol{v}$ remains,

$$\langle\hat{\boldsymbol{p}}\rangle = \int \psi^*\hat{\boldsymbol{p}}\psi dx = \int m(\boldsymbol{v} - i\boldsymbol{u})\rho dx = m\langle\boldsymbol{v}\rangle = \langle\boldsymbol{p}\rangle, \tag{4.129}$$

but when considering the second moment $\left\langle\hat{\boldsymbol{p}}^2\right\rangle = \langle\boldsymbol{p}^2\rangle$ the role of $\boldsymbol{u}$ becomes evident, as follows from (4.120).

Equation (4.128) suggests the introduction of the complex vector $\boldsymbol{\pi}$

$$\hat{\boldsymbol{p}} \rightarrow \boldsymbol{\pi} \equiv m(\boldsymbol{v} - i\boldsymbol{u}) \tag{4.130}$$

as a means to calculate the quantum average of a function of the *operator* $\hat{\boldsymbol{p}}$ by averaging over a complex vector (a *c-number*). The second moment $\left\langle\hat{\boldsymbol{p}}^2\right\rangle$ would thus be reproduced according to the usual rule for calculating the second moments of a complex variable (see e.g. Papoulis 1991, Sect. 8.1). A direct calculation leads indeed to

$$\left\langle\boldsymbol{\pi}\cdot\boldsymbol{\pi}^*\right\rangle = m^2\left\langle\boldsymbol{v}^2 + \boldsymbol{u}^2\right\rangle = \left\langle\hat{\boldsymbol{p}}^2\right\rangle. \tag{4.131}$$

The need to resort to a complex vector to reproduce the quantum expectation values is linked to the fact that, as we have seen, $\hat{\boldsymbol{p}}$ extracts from the (complex) wave function information regarding *two* velocities, namely $\boldsymbol{v}$ and $\boldsymbol{u}$. In Chap. 8 we will come back to the correspondence $\hat{\boldsymbol{p}} \to \boldsymbol{\pi}$ in the context of bipartite systems, where it will be of value for the understanding of quantum correlations.

The introduction of a complex momentum $\boldsymbol{\pi}$ to reproduce the quantum averages of the $\hat{\boldsymbol{p}}$ operator using c-numbers can also be extended to its time-counterpart, the Hamiltonian. We observe that a direct calculation gives

$$\hat{H}\psi = i\hbar\frac{\partial}{\partial t}\psi = \left(H_S + iH_\rho\right)\psi = \mathcal{H}\psi, \tag{4.132}$$

where

$$H_S = -\hbar\frac{\partial S}{\partial t}, \quad H_\rho = \frac{\hbar}{2}\frac{\partial \ln\rho}{\partial t}. \tag{4.133}$$

A comparison with (4.130) shows that in a certain sense, H_S is to $\hat{H}$ what $m\boldsymbol{v}$ is to $\hat{\boldsymbol{p}}$, and similarly H_ρ is to $\hat{H}$ what $m\boldsymbol{u}$ is to $\hat{\boldsymbol{p}}$. In analogy with Eqs. (4.131), (4.129), and (4.122) we obtain

$$\begin{aligned}
\langle \mathcal{H}\mathcal{H}^* \rangle &= \left\langle \hat{H}^2 \right\rangle \\
\left\langle \hat{H} \right\rangle &= \langle H_S \rangle, \\
\sigma_H^2 &= \sigma_{H_S}^2 + \sigma_{H_\rho}^2.
\end{aligned}$$

The couple of equations

$$m\boldsymbol{u} = \frac{\hbar}{2}\boldsymbol{\nabla}\ln\rho, \quad H_\rho = \frac{\hbar}{2}\frac{\partial \ln\rho}{\partial t}, \tag{4.134}$$

thus suggests to interpret $(\hbar/2)\ln\rho$ as a sort of (diffusive) action that results in additional contributions to the momentum and energy fluctuations.

4.5.3 A Comment on Operator Ordering

In connection with the operator formalism of QM, it seems interesting to discuss the meaning that can be ascribed to the ordering of operators from the standpoint of the present theory. Limiting for simplicity the discussion to the one-dimensional case and to the products of the fundamental operators $\hat{x}$ and $\hat{p}$, we observe that an integration by parts gives

$$\begin{aligned}
\int \psi^*(\hat{x}\hat{p} + \hat{p}\hat{x})\psi dx &= -i\hbar\int \psi^*\left[x\frac{\partial\psi}{\partial x} + \frac{\partial}{\partial x}(x\psi)\right]dx \\
&= -i\hbar\int x\left(\psi^*\frac{\partial\psi}{\partial x} - \psi\frac{\partial\psi^*}{\partial x}\right)dx = 2m\int xv\rho dx = 2m\langle xv\rangle,
\end{aligned} \tag{4.135}$$

where we used Eq. (4.125). Hence the usual (symmetrized) quantum correlation $\langle \hat{x}\hat{p} + \hat{p}\hat{x} \rangle /2$ gives the correlation between the position coordinate and the (local mean) systematic velocity v,

$$\tfrac{1}{2} \langle \hat{x}\hat{p} + \hat{p}\hat{x} \rangle = m \langle xv \rangle . \tag{4.136}$$

A similar calculation, now using Eq. (4.126), gives for the commutator

$$\begin{aligned} \int \psi^* (\hat{x}\hat{p} - \hat{p}\hat{x})\psi dx &= -i\hbar \int \psi^* \left[x\frac{\partial \psi}{\partial x} - \frac{\partial}{\partial x}(x\psi) \right] dx \\ &= -i\hbar \int x \left(\psi^* \frac{\partial \psi}{\partial x} + \psi \frac{\partial \psi^*}{\partial x} \right) dx = -2im \int xu\rho dx = -2im \langle xu \rangle . \end{aligned} \tag{4.137}$$

Therefore the expectation value of the commutator $[\hat{x}, \hat{p}]$ is proportional to the correlation between the position coordinate and the diffusive velocity u,

$$\tfrac{1}{2} \langle \hat{x}\hat{p} - \hat{p}\hat{x} \rangle = -im \langle xu \rangle . \tag{4.138}$$

By adding and subtracting the above results one obtains

$$\begin{aligned} \langle \hat{x}\hat{p} \rangle &= m \langle x(v - iu) \rangle , \\ \langle \hat{p}\hat{x} \rangle &= m \langle x(v + iu) \rangle . \end{aligned} \tag{4.139}$$

These expressions show that the order of the operators in the calculation of the expectation value of their product has nothing to do with a temporal order of observations or measurements, but rather with the sign of the imaginary contribution. As was briefly mentioned following Eq. (4.122), the effects of the presence of the velocity u have no classical analogue. Though this will be clarified in more detail in Chap. 8, Eq. (4.138) already highlights the fact that it is precisely the correlation between x and u what prevents the commutator $[\hat{x}, \hat{p}]$ from vanishing.

4.5.4 Trapped Motions

An interesting result can be extracted from Eq. (4.115) by applying a variational principle (de la Peña and Cetto 1977a, b). For a stationary state with $\boldsymbol{v} = 0$ this equation (in the one-dimensional case) reduces to

$$\langle p^2 \rangle_x = -\frac{\hbar^2}{4} \frac{\partial^2}{\partial x^2} \ln \rho . \tag{4.140}$$

The mean energy of the system is then

$$\langle H \rangle = \int \left[-\frac{\hbar^2}{8m} \frac{d^2}{dx^2} \ln \rho + V(x) \right] \rho(x) dx. \tag{4.141}$$

Let us now demand that $\langle H \rangle$ acquires an extremum value, subject to the condition that

$$\int \rho(x) dx = 1. \tag{4.142}$$

A couple of integrations by parts gives for $\varphi(x) \equiv \sqrt{\rho(x)}$

$$\delta \int \varphi^2 \left[-\frac{\hbar^2}{8m} \frac{d^2}{dx^2} \ln \varphi^2 + V(x) \right] dx = \int \left(-\frac{\hbar^2}{m} \frac{d^2 \varphi}{dx^2} + 2V(x)\varphi \right) \delta\varphi dx.$$

To this we add the constraint (4.142), in the form $-2 \langle H \rangle \int \varphi \delta\varphi dx = 0$ ($\langle H \rangle$ is here a Lagrange multiplier). The solution of this variational problem is the Euler-Lagrange equation (Morse and Feshbach 1953; Hassani 1999)

$$-\frac{\hbar^2}{2m} \frac{d^2 \varphi}{dx^2} + V\varphi = \langle H \rangle \varphi. \tag{4.143}$$

This result emphasizes the remarkable role played by the local dispersion of the momentum, Eq. (4.140): it leads directly to the (stationary) Schrödinger equation for $\psi = \varphi e^{-i\langle H \rangle t/\hbar}$ and guarantees that the stationary distribution of particles corresponds to an extremum (normally a minimum) of the mean energy of the system. These extrema of the mean energy correspond to the quantized solutions: here we witness a complementary side of the demand of stationarity.

The observation that the energy of the stationary states is a local minimum is very suggestive. Of course each particle finds eventually its own specific trajectory, with a certain mean energy (averaged over the trajectory) and more or less stable. The ensemble of such trajectories acquires eventually a minimum mean energy, which corresponds to the set of most robust motions and generates a certain spatial probability distribution of particles. Usual QM does not account for such trajectories, and their (approximate) knowledge requires a simulation of the random field. A first image that comes to mind is that the orbits are 'trapped' once they are close enough to the stationarity condition. In Chap. 5 this picture will become more transparent with the demonstration that the stationary states correspond to situations that satisfy an ergodic condition. The trapping of the orbits occurs in the long run, because the stochastic field forces the particle to explore the neighboring phase-space regions; as long as the particle is not trapped, it will continue probing the phase-space. This means that the stationary orbits are qualitatively analogous to limit cycles, although here the attractive basin is formed by the lowest average energy. Similar possibilities have been suggested in the past by several authors (see Surdin 1970; de la Peña and Cetto 1995; 't Hooft 2002).

4.5.5 'Schrödinger' Equation for a Classical System?

One reason for the difficulties found when interpreting QM is rooted in the fact that the description of the quantum particles is made in field-theoretic terms. One can perceive the underlying meaning of this assertion by using the simple trick of inverting the argument, and attempting to describe a problem of classical mechanics in field-theoretic terms similar to those employed for the quantum description. For this purpose consider an ensemble of classical particles with density in configuration space $\rho(\boldsymbol{x}, t)$ and principal function $S(\boldsymbol{x}, t)$; the Lagrangian density can then be written in the form

$$\mathcal{L} = -\rho \left[\frac{\partial S}{\partial t} + \frac{1}{2m} (\nabla S)^2 + V \right]. \tag{4.144}$$

For this to be correct the variables ρ and S must be treated as canonically conjugate fields (ρ playing the role of position coordinate and S that of the corresponding momentum). Under the demand that the action $\int_{t_1}^{t_2} \mathcal{L} dx dt$ be stationary under infinitesimal arbitrary variations (with fixed end points), first of ρ and then of S, the ensuing equations of motion are

$$\frac{\partial S}{\partial t} + \frac{1}{2m} (\nabla S)^2 + V = 0, \tag{4.145}$$

$$\frac{\partial \rho}{\partial t} + \nabla \cdot \left(\rho \frac{\nabla S}{m} \right) = 0. \tag{4.146}$$

These are the correct dynamical equations for a congruency of classical particles, the first one being the Hamilton-Jacobi equation for particles subject to the external potential V, and the second one the corresponding continuity equation for an ensemble of such (conserved) particles.

Once in possession of the field-theoretic description, one can take the quantum route (Schiller 1962; Rosen 1964, 1986) by introducing a *wave function* via the definitions

$$\psi = R\, e^{iS/a}, \quad R = \sqrt{\rho}, \tag{4.147}$$

with a a constant having dimensions of action. This seems to be an acceptable transformation from (ρ, S) to the fields (ψ, ψ^*). The Lagrangian density becomes

$$\begin{aligned} \mathcal{L} = i\frac{a}{2} \left(\psi^* \frac{\partial \psi}{\partial t} - \psi \frac{\partial \psi^*}{\partial t} \right) \\ + \frac{a^2}{8m} \left(\psi^* \nabla \psi - \psi \nabla \psi^* \right) - V \psi^* \psi. \end{aligned} \tag{4.148}$$

The action is now to be varied with respect to the fields ψ and ψ^*. Indeed from the last expression for $\mathcal{L}$ it follows that ψ and ψ^* are canonically conjugate variables, $i\psi^*$ taking the place (up to a constant) of the momentum coordinate, since

$$\frac{\partial \mathcal{L}}{\partial \dot{\psi}} = i\frac{a}{2}\psi^*. \tag{4.149}$$

By varying the Lagrangian with respect to the fields ψ and ψ^* one gets

$$ia\frac{\partial \psi}{\partial t} = -\frac{a^2}{2m}\nabla^2\psi + V\psi + \frac{a^2}{2m}\frac{\nabla^2 R}{R}\psi \tag{4.150}$$

and its complex conjugate. Apparently the Schrödinger equation has been recovered for the *classical* ensemble of particles,[17] but now these are subject to the 'potential'

$$V' = V + \frac{a^2}{2m}\frac{\nabla^2 R}{R}. \tag{4.151}$$

Assume for a moment that $a = \hbar$; then the last term becomes the quantum potential V_Q with the sign reversed [see Eq. (4.86) with $\eta = \hbar/2$]. It pops up here, because it is absent from the classical Hamilton-Jacobi equation (4.145). This turns Eq. (4.150) into a highly nonlinear equation. Furthermore, in the present treatment the constant a is totally arbitrary; there is no reason to conclude that it has a universal value nor to identify it with Planck's constant, so the system described by Eq. (4.150) can have any (macroscopic) scale. Many attempted derivations of the Schrödinger equation from not explicitly quantum principles make this kind of gratuitous identification $a \sim \hbar$ (see related discussions in Chap. 2, Sect. 2.4). Further, the above result confirms that the crucial term responsible for the divide between the quantum and the classical world is the quantum potential (see Chap. 8 for further discussions related to V_Q).

Notice moreover that in the present instance there are no fluctuations; the dispersions, if any, are due to initial conditions on the functions ρ and S [cf. discussion following Eq. (2.81)]. Further, the trajectories are strictly classical, so the 'wave' description is not guaranteed. However, in the quantum case such description does make sense because the term in the quantum Hamilton-Jacobi equation cancels out the one in Eq. (4.150), leading thus to a linear equation for ψ. This marvelous linearization makes the Schrödinger equation extremely valuable and mathematically appropriate for the description of quantum systems. Extending the procedure to describe classical corpuscles by means of fields (or waves) is generally misleading and fanciful.

[17] Though this result may be quite surprising, in fact it is not, since the continuity equation along with the factorization $\rho = \psi^*\psi$ readily leads to an equation having the general form of the Schrödinger equation, as has been known for a long time (see e.g. de la Peña 1967; Jammer 1974; Kracklauer 1992). This shows that the structure of the Schrödinger equation, rather than being specific to QM, is a kind of generic framework in the presence of fluctuations (see e.g. Carroll 2010). Of course, the term proportional to $\psi\left(\nabla^2 R\right)/R$ in Eq. (4.150) disappears if it is subtracted from the potential in Eq. (4.145).

4.6 Phase-Space Distribution and the Wigner Function

The starting point for the derivation of the quantum description in configuration space was the GFPE (4.13) in the phase space of the particle. From this equation one can go either to the momentum space or to the configuration space, which was the case we analyzed, obtaining eventually the Schrödinger equation in the time-asymptotic regime. One may therefore reasonably ask: is it possible to proceed in the opposite sense, starting from the usual quantum-mechanical description provided separately in configuration or momentum space, and recover a *unique* full phase-space description? That this question cannot be answered in the positive is a well-established fact.[18] Let us briefly look into the matter from the present perspective, and disclose the reason for this difficulty.

Equation (4.46) for $\tilde{Q}(\boldsymbol{x}, z, t)$ can be inverted and combined with Eqs. (4.59) and (4.62) to obtain (with $\eta = \hbar/2$)

$$\begin{aligned} Q(\boldsymbol{x}, \boldsymbol{p}, t) &= \frac{1}{2\pi} \int \tilde{Q}(\boldsymbol{x}, z, t) e^{-i\boldsymbol{p}\cdot z} dz \\ &= \frac{1}{2\pi} \int q(\boldsymbol{x} + \tfrac{\hbar}{2}z, t) q^*(\boldsymbol{x} - \tfrac{\hbar}{2}z, t) \chi(\boldsymbol{x} + \tfrac{\hbar}{2}z, \boldsymbol{x} - \tfrac{\hbar}{2}z, t)\, e^{-i\boldsymbol{p}\cdot z} dz. \end{aligned} \tag{4.152}$$

By construction, $Q(\boldsymbol{x}, \boldsymbol{p}, t)$ furnishes a true (Kolmogorovian) probability density in phase space. This means that *if* the exact solutions for q and χ were known for all values of z and at all times, one would have a full phase-space description for the particle. However, in the radiationless approximation and in the time-asymptotic limit—a regime that leads to the more restricted Schrödinger equation—all we have at hand, at most, are the terms of the Taylor series expansion of $\chi(\boldsymbol{x} + \frac{\hbar}{2}z, \boldsymbol{x} - \frac{\hbar}{2}z, t)$ up to second order in z (as shown in Sect. 4.4.1). Hence the integral in (4.152), which extends over the entire z-space, cannot be evaluated correctly in general. One can, instead, construct an *approximate* form $W(\boldsymbol{x}, \boldsymbol{p}, t)$ obtained by fixing $\chi = 1$ while allowing z to remain as a Fourier variable (putting $q = \psi$ as in [4.80)],

$$W(\boldsymbol{x}, \boldsymbol{p}, t) = \frac{1}{\pi\hbar} \int \psi(\boldsymbol{x} + \boldsymbol{y}, t) \psi^*(\boldsymbol{x} - \boldsymbol{y}, t)\, e^{-i2\boldsymbol{p}\cdot\boldsymbol{y}/h} d\boldsymbol{y}, \tag{4.153}$$

with $\boldsymbol{y} = \hbar z/2$. This is the well-known Wigner phase-space function (Wigner 1932, Moyal 1949).[19] As a result of the approximations made in passing from Eqs. (4.152)

---

[18] Entire books have been dedicated to analyze and discuss the problem of a quantum-mechanical phase-space distribution. See, e.g., Zachos et al. (2005) .

[19] The first phase-space description of a quantum system was made in Weyl (1927); the Wigner function was introduced and studied firstly in Dirac (1930). It was later proposed independently in Heisenberg (1931) and Wigner (1932). The theory of the latter was substantially developed in Moyal (1949). A discussion of the fundamentals of the distribution functions is given in Hillery et al. (1984) (see also Tatarskii 1983); for an introductory account of the Wigner function see Case (2008).

to (4.153), there is no guarantee that W is a true Kolmogorovian probability. And indeed, despite its recognized value, it is not, since as is well known it can take on negative values (hence the name pseudo-probability distribution) in some regions of phase space for almost all states and systems—the exception being the Gaussian states, which correspond to the separable case $\chi(\boldsymbol{x}+\hbar\boldsymbol{z}/2, \boldsymbol{x}-\hbar\boldsymbol{z}/2, t) = 1$ (Urbanik 1967; Hudson 1974; Piquet 1974; Soto and Claverie 1983, 1983). The right solution to this long-standing problem is of course to recognize the intrinsic limitations of W that ensue from its approximate nature, and to revert to the full distribution $Q(\boldsymbol{x}, \boldsymbol{p}, t)$ and learn how to distil the appropriate solutions.

The point referring to the function χ introduced through (4.60) as a factor of $\tilde{Q}(\boldsymbol{x}, \boldsymbol{z}, t)$ is of importance for an understanding of the difference between the phase-space distributions in Eqs. (4.152) and (4.153). The derivations in Sect. 4.4.1 show that in the general case, both Σ and χ may differ from the constant values normally assigned to them. This has consequences not only for the construction of a true phase-space quantum distribution, but more generally for the description of the quantum behavior of particles, at least in some regions of the configuration space and for certain problems, since any value of $\chi \neq 1$ or $\Sigma \neq 0$ implies a deviation from the quantum-mechanical description.

As was shown in Sect. 4.4.1, the selection $\Sigma = 0$ decouples the first two equations of the hierarchy from the rest. This leads to a substantial simplification of the problem at hand, since instead of an infinity of equations, just two—the simplest of them—suffice for the quantum description in configuration space. But this comes with a high price: the loss of a true phase-space description.

We thus conclude that in point of fact a true phase-space probability density for the problem should exist, but that this density function is more complicated than the usual Wigner function or any other of the many already proposed. A most important problem is thus the investigation of the possibilities offered by the full phase-space probability (4.152), without making appeal to a power expansion of the diffusion term, and even less to the Markovian approximation. This seems to constitute a fruitful demanding research programme, but one that is beyond the scope of the present volume.

4.7 What We Have Learned So Far About Quantum Mechanics

The thermostatistical analysis of the radiation field in equilibrium, made in Chap. 3, showed that in the presence of its zero-point component the energy of the field becomes quantized. The statistical analysis of the dynamics of matter in the presence of the zero-point field, made in this chapter, shows that also the material system becomes quantized. In summary, the quantum properties of both matter and field emerge from the consideration of the existence of a real, ubiquitous random zero-

(Footnote 19 continued)
General formulas for quantum phase-space distributions (which apply to the Wigner function as a special case) are given in Cohen (1976) and Cohen and Zaparovanny (1980). A general overview with selected papers is Zachos et al. (2005).

point radiation field. These results substantiate the success of a whole series of previous works carried out within SED.[20]

The description provided by the Schrödinger equation has been obtained from a generalized phase-space Fokker-Planck equation for the particle, in the time-asymptotic limit when a balance is eventually reached between the diffusive and dissipative terms contributing to the mean energy. The onset of the quantum regime is thus conditioned by energy balance. This is the secret of QM: the energy-balance condition confirms the validity of the structure of the Schrödinger equation, univocally fixing the spectral density of the background field that is needed to sustain this balance, as well as the value of $\eta = \hbar/2$. Yet the approximations and simplifications made along the derivations have had the effect of deleting from the final description every explicit reference to the ZPF—the very cause of the quantum behavior! Consequently, the reason for the stochasticity becomes concealed and the fluctuations appear as causeless. By exhibiting them as real, objective fluctuations resulting from the permanent interaction of the atomic system (or whatever quantum structure is under study) with the ZPF, the theory puts the so-called *quantum fluctuations* on a mundane perspective. Further, it defines the role played by Planck's constant in determining the characteristic size of the quantum fluctuations.

Of crucial importance is the fact that the source of the noise is a radiation field; the spatial and temporal coherence of the modes of the field is central to support the stationary states for relatively long times. This introduces an essential difference between the action of the ZPF and that of a purely noisy background (as the Brownian-motion one), and leads to the existence of, for instance, stable orbital motions in the atomic case.

It is clear that from the present perspective, the description afforded by the Schrödinger equation refers to a statistical ensemble, not to an individual particle. The particle remains always a particle, the (physical) wave (the background field) remaining literally in the background. In the configuration-space description, mean trajectories that belong to subensembles are characterized by the local mean velocities $\boldsymbol{v}(\boldsymbol{x})$ and $\boldsymbol{u}(\boldsymbol{x})$; due to its intrinsically statistical nature the theory cannot be applied in general to singular events, so individual trajectories appear as unknown. These conclusions appear as inescapable, and mark a clear departure from the usual (Copenhagen) interpretation of QM in favour of the less popular ensemble (or statistical) interpretation. Briefly, the theory recovers for physics causality, determinism, realism and locality.[21]

---

[20] Some recent independent investigations are of particular relevance to the present theory. We recall the important numerical simulations in Cole (2006) and Cole and Zou 2003–2004 leading to a correct statistical prediction of the ground state orbit for the H-atom (and some results for its excitations in Cole and Zou 2009). In Huang and Batelaan (2012a, b), the modes of a classical one-dimensional harmonic oscillator immersed in the ZPF and excited by an electromagnetic pulse are studied by numerical simulation, with results that are in excellent agreement with the quantum predictions. The authors are indebted to Khaled Dechoum and Emilio Santos for having drawn their attention to this work at an early stage.

[21] It could be argued that the quantum description is indeterministic. This is obviously true, and is in consonance with any statistical description. The point is that the starting equation of motion,

A frequently posed question relates to the application of the present theory to neutral particles. The general answer is that all known particles have electromagnetic interactions; even a neutral particle possesses at least a multipole moment that couples to the ZPF.[22] Coupling of the particle through higher moments is expected to lead to the same results obtained here, although stationarity would be reached perhaps somewhat more slowly. It has also been argued that other vacuum fields, and even fluctuations of the space-time metric, could be important for the understanding of the quantum properties of matter. The present results indicate that due consideration of the ZPF suffices to get an understanding of the quantum behavior of both matter and field, in the nonrelativistic approximation. Of course, for a relativistic description one should expect other fields to play a role, in particular the electron-positron vacuum. Regarding metric fluctuations, a comment is included in Chap. 10.

Equation (4.79) defines the meaning of ψ: it is an abstract mathematical object that plays the role of an amplitude (or density) of probability (in configuration space, in the present description), rather than representing directly a physical wave.[23] This is an important result: the Born rule $\rho(\boldsymbol{x}, t) = \psi(\boldsymbol{x}, t)\psi^*(\boldsymbol{x}, t)$ is not a *postulate* of the theory; it is a natural consequence of the statistical description.[24] Theories in which the probability density ρ is conveniently expressed in terms of a product of two complex amplitudes, have been criticized in the past (see Takabayasi 1952; Wallstrom 1989, 1994) on the argument that the sets of solutions of the stochastic equations and of the Schrödinger equation are different.[25] The critique certainly does

(Footnote 21 continued)
valid for an individual member of the ensemble, is deterministic. Indeterminism enters because the specific realization of the field in the individual case is unknown. Thus quantum indeterminism should not be understood as intrinsic to matter at the microscopic level, but rather of a nature similar to that of statistical physics. Something similar can be said about nonlocality: the initial theory is local; the final statistical and partial description is the one that acquires nonlocal properties. For related discussions on causality, determinism, realism and locality see Chap. 1.

[22] Recently Huang and Batelaan (2012a, b) have proposed another form of visualizing the problem, by considering that the random motion gives rise to instantaneous multipolar moments that couple to the corresponding modes of the radiation field. Strictly speaking, this can be applied also to neutral (structured) particles.

[23] The most obvious instance in which this is manifest is a system of N particles in three dimensions; then ψ lives in an abstract $3N$-dimensional space, while physical waves (fields) live in three dimensions. This point, raised for the first time by Pauli, was a subject of much discussion during the early phase of QM; a detailed account can be seen in Bacciagaluppi and Valentini (1927). However, in the case of particles (fermions) the problem disappears using the number representation of the state vector. Then the $3N$ coordinates (if introduced at all) represent merely N points in three-dimensional space.

[24] The fact that in QM the Born rule is introduced as a postulate, is not a minor point. Indeed it is so important that serious efforts have been made for many years to demonstrate it from within QM. Probably the most far-reaching result of such attempts is due to Graham (1973), who proves that the probability of state n in the superposition of states $\sum_1^N c_n\psi_n$ tends to $|c_n|^2$ for large values of N. See also Hartle (1968).

[25] The factorization $\rho = \psi^*\psi$ taken as a minor mathematical liberty was apparently first introduced in Collins (1977) and has been repeatedly used by several authors. But in fact it is not inconsequential, since it opens the possibility to introduce a phase function, not present in the absence of such

not apply to the present case, in which the quantization is legitimated by the demand of detailed energy balance, and Born's rule is an integral and derived part of the theory.

Since from the perspective of the results presented here, present-day QM furnishes an approximate, time-asymptotic, partially averaged description of the physical phenomenon, there exists plenty of room for further and deeper investigations. For instance, it is clear that the transition from the Fokker-Planck equation to the Schrödinger equation is an irreversible procedure; one cannot reconstruct from the latter the probability density in phase space by purely logical steps backward. It therefore remains to explore the consequences of making a more complete use of the density $Q(\boldsymbol{x}, \boldsymbol{p}, t)$, or to investigate the behavior of the system before it reaches the state of energy balance (the quantum regime), when the approximations still do not apply. Which is the correct description at such short times? One should expect an entirely unknown behavior of matter in this initial nonequilibrium regime, which can neither be classical because the $\hbar$ due to the interaction with the field is part of the picture, nor quantum-mechanical because the conditions to apply such description have not yet been reached. Undoubtedly an exploration into this realm would represent a new adventure in physics, with interesting outcomes.

Appendix A: Derivation of the Generalized Fokker-Planck Equation

In this appendix the generalized Fokker-Planck equation associated with the stochastic Eq. (4.3) is derived, borrowing from de la Peña and Cetto (1977a). For the sake of simplicity the derivation is presented in one dimension. The generalization to three dimensions is made at the end of the appendix.

As discussed in the text, for any given realization of the field the density $R(x, p, t)$ of points in the phase space of the particle satisfies the continuity Eq. (4.12),

$$\frac{\partial R}{\partial t} + \frac{1}{m}\frac{\partial}{\partial x} pR + \frac{\partial}{\partial p}\left(f + m\tau\dddot{x} + eE\right) R = 0. \tag{A.1}$$

The differential equation for Q, the mean value of R over the field realizations $\{(i)\}$, can be constructed by means of the smoothing method (see e.g. Frisch 1968), as follows. A smoothing operator $\hat{P}$ is introduced, which acts on any phase function $A(x, p, t)$ by giving its local (in the particle phase space) average,

$$\hat{P}A = \overline{A}^{(i)}, \quad \text{so} \quad A = \overline{A}^{(i)} + (1 - \hat{P})A. \tag{A.2}$$

(Footnote 25 continued)
factorization. Only if a physical meaning and well-defined mathematical properties can be attributed to such function, so that it belongs naturally to the theory, the procedure can be considered acceptable. For other comments on this important matter see Sect. 2.6.

Clearly $\delta A = (1 - \hat{P})A$ is the random component of A, which means that the second Equation in (A.2) is a decomposition of A into its average $\overline{A}$ plus its fluctuating part. Further, $\hat{P} = \hat{P}^2$, so $\hat{P}$ is a projection (idempotent) operator. The application of this smoothing operator to the density R separates it into its average and its random parts Q and δQ, respectively,

$$R = Q + \delta Q, \quad Q = \hat{P}R, \quad \delta Q = (1 - \hat{P})R. \tag{A.3}$$

We are interested in constructing the differential equation for Q. For this purpose we rewrite Eq. (A.1) in the form

$$\frac{\partial}{\partial t}(Q + \delta Q) + \hat{L}(Q + \delta Q) = -e\frac{\partial}{\partial p}E(Q + \delta Q), \tag{A.4}$$

where $\hat{L}$ stands for the (nonrandom) Liouville operator for the particle, including the radiation-reaction force $m\tau\dddot{x}$—strictly speaking, the operator $\hat{L}$ differs from a true Liouville operator due to the (small) radiation reaction term—,

$$\hat{L} = \frac{1}{m}\frac{\partial}{\partial x}p + \frac{\partial}{\partial p}(f + m\tau\dddot{x}). \tag{A.5}$$

Equation (A.4) becomes separated into its nonstochastic and fluctuating parts by applying to it the projection operators $\hat{P}$ and $1 - \hat{P}$ in succession. Using that $\hat{P}E = 0$ [see Eq. (4.4)], one thus obtains the couple of equations

$$\left(\frac{\partial}{\partial t} + \hat{L}\right) Q = -e\frac{\partial}{\partial p}\hat{P}E\delta Q, \tag{A.6}$$

$$\left(\frac{\partial}{\partial t} + \hat{L}\right) \delta Q = -e\frac{\partial}{\partial p}EQ - e\frac{\partial}{\partial p}\left(1 - \hat{P}\right) E\delta Q. \tag{A.7}$$

The next step is to eliminate δQ from these equations. This can be achieved by introducing the operator $\hat{G} = (\partial/\partial t + \hat{L})^{-1}$, which corresponds to the Green function of the differential operator $\partial/\partial t + \hat{L}$, so for any phase function $A(x, p, t)$ one has

$$\hat{G}A(x, p, t) = \int_{-\infty}^{t} e^{-\hat{L}(t-t')}A(x, p, t')dt'. \tag{A.8}$$

The operator $\hat{G}$ is now used to invert Eq. (A.7),

$$\delta Q = -e\hat{G}\frac{\partial}{\partial p}EQ - e\hat{G}\frac{\partial}{\partial p}\left(1 - \hat{P}\right) E\delta Q, \tag{A.9}$$

or, even better,

$$\left[1+e\hat{G}\frac{\partial}{\partial p}\left(1-\hat{P}\right)E\right]\delta Q=-e\hat{G}\frac{\partial}{\partial p}EQ. \tag{A.10}$$

Applying from the left the inverse of the operator in square brackets gives an expression for δQ, which combined with Eq. (A.6) gives a complicated integro-differential equation for Q,

$$\left(\frac{\partial}{\partial t}+\hat{L}\right)Q=e^2\frac{\partial}{\partial p}\hat{P}E\left[1+e\hat{G}\frac{\partial}{\partial p}\left(1-\hat{P}\right)E\right]^{-1}\hat{G}\frac{\partial}{\partial p}EQ. \tag{A.11}$$

This is the GFPE for the problem. However, this is a formal expression in which the random field and the operator $\hat{G}$ appear in a form which makes it quite impractical to use. A more manageable form is obtained by formally expanding the expression within square brackets into a power series,

$$\begin{aligned}\left(\frac{\partial}{\partial t}+\hat{L}\right)Q&=e^2\frac{\partial}{\partial p}\hat{P}E\sum_{k=0}^{\infty}\left[-e\hat{G}\frac{\partial}{\partial p}\left(1-\hat{P}\right)E\right]^k\hat{G}\frac{\partial}{\partial p}EQ\\&=-e\frac{\partial}{\partial p}\hat{P}E\sum_{k=0}^{\infty}\left[-e\hat{G}\frac{\partial}{\partial p}\left(1-\hat{P}\right)E\right]^{k+1}Q\\&=-e\frac{\partial}{\partial p}\hat{P}E\sum_{k=1}^{\infty}\left[-e\hat{G}\frac{\partial}{\partial p}\left(1-\hat{P}\right)E\right]^{k}Q.\end{aligned} \tag{A.12}$$

To write the second equality we used the fact that EQ can also be written as $(1-\hat{P})EQ$.

To somewhat simplify Eq. (A.12) we take into consideration some statistical properties of the random field. First of all, for the ZPF the distribution is symmetric and centered around zero, so that the average of products of an odd number of factors vanishes,

$$\hat{P}E(t_1)E(t_2)\ldots E(t_{2n+1})\hat{P}A=0. \tag{A.13a}$$

Assuming the distribution to be Gaussian, the average of products with an even number of factors take the form (Gardiner 1983, Sect. 2.8; Wang and Uhlenbeck in Wax 1954/1985, Sect. 9a)[26]

$$\hat{P}E(t_1)E(t_2)\ldots E(t_{2n})\hat{P}A=\sum\overline{\overline{E(t_i)E(t_j)}\ldots\overline{E(t_r)E(t_s)}}^{(i)}\,\overline{A}^{(i)}, \tag{A.13b}$$

[26] As shown in these references, it is correct to assume that the distribution is Gaussian in the case of the free field. Strictly speaking $\boldsymbol{E}(t)$ cannot be taken as a free field, since it is somehow modified by its interaction with the mechanical subsystem. However, the modifications affect only a very reduced set (which can be considered of measure zero) of the (averaged) modes of the field in the vicinity of the particle, by introducing correlations among the phases of some of these modes. For the rest, the field remains essentially unchanged. For details see Peña et al. (2009), or Chap. 5.

where the sum is to be effected over all possible different pairs of factors.

Now from Eq. (A.13a) we note that all terms on the right-hand side of (A.12) with even k vanish, so the equation simplifies into

$$\left(\frac{\partial}{\partial t}+\hat{L}\right)Q=e\frac{\partial}{\partial p}\hat{P}E\sum_{k=0}^{\infty}\left[e\hat{G}\frac{\partial}{\partial p}\left(1-\hat{P}\right)E\right]^{2k+1}Q, \tag{A.14}$$

which is equivalent to

$$\left(\frac{\partial}{\partial t}+\hat{L}\right)Q=e^{2}\frac{\partial}{\partial p}\hat{P}E\hat{G}\frac{\partial}{\partial p}E\sum_{k=0}^{\infty}\left[e\hat{G}\frac{\partial}{\partial p}\left(1-\hat{P}\right)E\right]^{2k}Q. \tag{A.15}$$

This is the generalized Fokker-Planck equation (GFPE) that we use here. The generalization to three dimensions is straightforward and gives (summation over repeated indices is to be understood)

$$\left(\frac{\partial}{\partial t}+\hat{L}\right)Q=e^{2}\frac{\partial}{\partial p_i}\hat{P}E_i\hat{G}\frac{\partial}{\partial p_j}E_j\sum_{k=0}^{\infty}\left[e\hat{G}\frac{\partial}{\partial p_l}\left(1-\hat{P}\right)E_l\right]^{2k}Q, \tag{A.16}$$

where

$$\hat{L}=\frac{1}{m}\frac{\partial}{\partial x_i}p_i+\frac{\partial}{\partial p_i}(f_i+m\tau\dddot{x}_i) \tag{A.17}$$

is the Liouville operator with the radiation reaction force added as an 'external' force. Substitution of this expression into (A.16) gives Eq. (4.13).

The above derivations have been made in terms of operators; a more common form involves the Green function, as follows. The differential equation for the Green function $\mathcal{G}$ of the Liouville equation is

$$\left(\frac{\partial}{\partial t}+L\right)\mathcal{G}=\delta(\boldsymbol{x},\boldsymbol{x}')\delta(\boldsymbol{p},\boldsymbol{p}'), \tag{A.18}$$

in terms of which the evolution of a dynamical variable $A(\boldsymbol{x},\boldsymbol{p},t)$, described above with the aid of the operator $e^{-\hat{L}(t-t')}$, is given by the expression

$$e^{-\hat{L}(t-t')}A(\boldsymbol{x},\boldsymbol{p},t')=\int dx'dp'\mathcal{G}(\boldsymbol{x},\boldsymbol{p};\boldsymbol{x}',\boldsymbol{p}';t-t')A(\boldsymbol{x}',\boldsymbol{p}',t'), \tag{A.19}$$

where the prime refers to the values of the dynamical variables at $t'<t$, subject to the condition $A(\boldsymbol{x}',\boldsymbol{p}',t')\big|_t=A(\boldsymbol{x},\boldsymbol{p},t)$. The evolution from $(\boldsymbol{x}',\boldsymbol{p}')$ to $(\boldsymbol{x},\boldsymbol{p})$ is deterministic, as follows from the (modified) Liouville operator (A.17).

Appendix B: Diffusion Coefficients in the Markovian Approximation

We now proceed to the derivation of a simpler version of Eq. (4.13), by considering the approximation to second order of the GFPE, known as Markovian approximation. The results obtained are suitable to develop the SED theory further, once the system has reached a reversible condition, which is called quantum regime for the reasons mentioned in Sect. (4.4.4). It is important to remark that between the GFPE with an infinity of terms and the FPE written to order $n = 2$ in the derivatives, there is no intermediate approximation for a positive probability density. In fact, any truncation of the GFPE above $n = 2$ will automatically revert the expansion to second order for a nonnegative Q (all higher-order coefficients vanish). Thus, there are only three nontrivial possibilities: truncation at $n = 1$, which corresponds to deterministic (Newtonian) processes (described by the Liouville equation); truncation at $n = 2$, which corresponds to diffusions (Markovian) processes (described by a true FPE, which may be only approximate); and, finally, no truncation at all, which corresponds to the GFPE of infinite order (A.16). This is the essential content of Pawula's theorem, a detailed discussion of which can be seen in Risken (1984).

To get the (true, but approximate) FPE (4.19),

$$\begin{aligned}\frac{\partial Q}{\partial t} &+ \frac{1}{m}\frac{\partial}{\partial x_i} p_i Q + \frac{\partial}{\partial p_i}\left(f_i + m\tau \dddot{x}_i\right) Q \\ &= \frac{\partial}{\partial p_i} D_{ij}^{pp} \frac{\partial Q}{\partial p_j} + \frac{\partial}{\partial p_i} D_{ij}^{px} \frac{\partial Q}{\partial x_j},\end{aligned} \tag{B.1}$$

we start from the GFPE (4.13). The first step is to write Eq. (4.14) to first order in e^2 (i.e., to take only the term of the sum with $k = 0$), which is

$$\hat{\mathcal{D}}_i Q\Big|_{k=0} = \hat{P} E_i \hat{G} \frac{\partial}{\partial p_j} E_j Q = \hat{P} E_i(t) \int_{-\infty}^{t} dt' e^{-\hat{L}(t-t')} \frac{\partial}{\partial p_j} (E_j Q)(t'), \tag{B.2}$$

where Eq. (4.15) has been used. Since E_j depends on time only, this can be written in the form

$$\hat{\mathcal{D}}_i Q\Big|_{k=0} = \int_{-\infty}^{t} dt' \overline{E_i(t) E_j(t')}^{(i)} e^{-\hat{L}(t-t')} \frac{\partial}{\partial p_j} Q(t'). \tag{B.3}$$

Now, the evolution law

$$Q(t) = e^{-\hat{L}(t-t')} Q(t') \tag{B.4}$$

allows us to write, inserting the identity operator $e^{\hat{L}(t-t')} e^{-\hat{L}(t-t')}$,

$$e^{-\hat{L}(t-t')} \frac{\partial}{\partial p_j} Q(t') = \left(e^{-\hat{L}(t-t')} \frac{\partial}{\partial p_j} e^{\hat{L}(t-t')}\right) \left(e^{-\hat{L}(t-t')} Q(t')\right)$$

$$= \frac{\partial}{\partial p'_j} Q(t), \tag{B.5}$$

where, as explained in Appendix A, the prime refers to the values of the dynamical variables at $t' \leq t$, from where they follow a deterministic evolution to their end values at time t. Substitution into Eq. (B.3) using Eq. (4.9) gives

$$\hat{\mathcal{D}}_i Q\Big|_{k=0} = \int_{-\infty}^{t} dt' \varphi(t-t') \delta_{ij} \frac{\partial}{\partial p'_j} Q(x, p, t). \tag{B.6}$$

From the chain rule for the derivation,

$$\frac{\partial Q}{\partial p'_j} = \frac{\partial p_k}{\partial p'_j} \frac{\partial Q}{\partial p_k} + \frac{\partial x_k}{\partial p'_j} \frac{\partial Q}{\partial x_k}, \tag{B.7}$$

it follows that

$$\hat{\mathcal{D}}_i Q\Big|_{k=0} = \left(\int_{-\infty}^{t} dt' \varphi(t-t') \frac{\partial p_j}{\partial p'_i} \right) \frac{\partial Q}{\partial p_j} + \left(\int_{-\infty}^{t} dt' \varphi(t-t') \frac{\partial x_j}{\partial p'_i} \right) \frac{\partial Q}{\partial x_j}. \tag{B.8}$$

Direct substitution into (4.13) gives the (approximate) equation

$$\frac{\partial Q}{\partial t} + \frac{1}{m} \frac{\partial}{\partial x_i} p_i Q + \frac{\partial}{\partial p_i} f_i Q + m\tau \frac{\partial}{\partial p_i} \dddot{x}_i Q = \frac{\partial}{\partial p_i} \left(D_{ij}^{pp} \frac{\partial Q}{\partial p_j} + D_{ij}^{px} \frac{\partial Q}{\partial x_j} \right), \tag{B.9}$$

with the diffusion coefficients in the Markovian approximation given by

$$D_{ij}^{pp} = e^2 \int_{-\infty}^{t} dt' \varphi(t-t') \frac{\partial p_j}{\partial p'_i}, \tag{B.10a}$$

$$D_{ij}^{px} = e^2 \int_{-\infty}^{t} dt' \varphi(t-t') \frac{\partial x_j}{\partial p'_i}. \tag{B.10b}$$

Now, the dominant contribution to these time integrals comes from times t' close to t, when the system is already in the regime in which Eq. (4.78) controls the dynamics. This means that the dynamical variables are now represented by operators (see Sect. 4.4.2), and the factors $\partial p_j/\partial p'_i$, $\partial x_j/\partial p'_i$ should be properly expressed in terms of them. By noticing that for the corresponding classical variables the equalities

$$\frac{\partial p_j}{\partial p'_i} = \left[x'_i, p_j\right]_{\text{PB}}, \tag{B.11a}$$

$$\frac{\partial x_j}{\partial p'_i} = \left[x'_i, x_j\right]_{\text{PB}} \tag{B.11b}$$

apply, in terms of operators the following substitutions must be made,

$$\frac{\partial p_j}{\partial p'_i} \Rightarrow \frac{1}{2i\eta}\left[\hat{x}'_i, \hat{p}_j\right], \tag{B.12a}$$

$$\frac{\partial x_j}{\partial p'_i} \Rightarrow \frac{1}{2i\eta}\left[\hat{x}'_i, \hat{x}_j\right]. \tag{B.12b}$$

These can be obtained from the general rules

$$\frac{\partial}{\partial p_j} \Rightarrow \frac{1}{2i\eta}\left[\hat{x}_j, \ \right], \quad \frac{\partial}{\partial x_j} \Rightarrow \frac{1}{2i\eta}\left[\ , \hat{p}_j\right]. \tag{B.13}$$

With (B.12a, B.12b), Eqs. (B.10a), (B.10b) take the form

$$D^{pp}_{ij} = \frac{e^2}{2i\eta}\int_{-\infty}^{t} dt'\varphi(t-t')\left[\hat{x}'_i, \hat{p}_j\right], \tag{B.14a}$$

$$D^{px}_{ij} = \frac{e^2}{2i\eta}\int_{-\infty}^{t} dt'\varphi(t-t')\left[\hat{x}'_i, \hat{x}_j\right]. \tag{B.14b}$$

It is now straightforward to show that the diffusion coefficients [Eqs. (B.10a), (B.10b)] comply with the relation

$$\frac{\partial D^{pp}_{ij}}{\partial p_j} + \frac{\partial D^{px}_{ij}}{\partial x_j} = 0. \tag{B.15}$$

For this purpose we apply Eqs. (B.13) to (Eqs. B.14a), (B.14b), thus obtaining

$$\frac{\partial D^{pp}_{ij}}{\partial p_j} + \frac{\partial D^{px}_{ij}}{\partial x_j} = \frac{e^2}{(2i\eta)^2}\int_{-\infty}^{t} dt'\varphi(t-t')\left\{\left[\hat{x}_j, \left[\hat{x}'_i, \hat{p}_j\right]\right] + \left[\left[\hat{x}'_i, \hat{x}_j\right], \hat{p}_j\right]\right\}. \tag{B.16}$$

Resorting to the Jacobi identity (which is valid both for Poisson brackets and for commutators)

$$\left[x_j, \left[x'_i, p_j\right]\right] + \left[x'_i, \left[p_j, x_j\right]\right] + \left[p_j, \left[x_j, x'_i\right]\right] = 0, \tag{B.17}$$

and noticing that the second term vanishes, we find

$$\left[x_j, \left[x'_i, p_j\right]\right] = -\left[p_j, \left[x_j, x'_i\right]\right] = -\left[\left[x'_i, x_j\right], p_j\right], \tag{B.18}$$

and therefore the right-hand side of (B.16) is null, which proves Eq. (B.15). A direct consequence of this result is that Eq. (4.44) reduces to (4.45).

Appendix C: Detailed Derivation of the 'Generalized' Schrödinger Equation

The starting point for the present derivation is the couple of Eq. (4.68)

$$\frac{\partial \rho}{\partial t} + \frac{\partial}{\partial x_j}(v_j \rho) = 0, \tag{C.1a}$$

$$m\frac{\partial}{\partial t}(v_i\rho) + m\frac{\partial}{\partial x_j}(v_i v_j \rho) - \frac{\eta^2}{m}\frac{\partial}{\partial x_j}\left(\rho \frac{\partial^2}{\partial x_i \partial x_j}\ln \rho\right) + \frac{1}{m}\frac{\partial}{\partial x_j}\Sigma_{ij}\rho - f_i\rho \tag{C.1b}$$
$$= \tau v_j \frac{\partial f_i}{\partial x_j}\rho - e^2 \left.\widetilde{(\hat{\mathcal{D}}Q)_i}\right|_{z=0}.$$

Inserting the continuity equation into (C.1b) and multiplying by ρ^{-1} results in

$$m\frac{\partial \boldsymbol{v}}{\partial t} + \frac{m}{2}\nabla \boldsymbol{v}^2 - m\boldsymbol{v} \times (\nabla \times \boldsymbol{v}) + \nabla\left(-\frac{2\eta^2}{m}\frac{\nabla^2\sqrt{\rho}}{\sqrt{\rho}} + V\right) = \boldsymbol{F}_{\text{rad}} + \boldsymbol{F}_{\Sigma}, \tag{C.2}$$

since $(\boldsymbol{v}\cdot\nabla)\,\boldsymbol{v} = \frac{1}{2}\nabla \boldsymbol{v}^2 - \boldsymbol{v}\times(\nabla\times\boldsymbol{v})$. The i components of the vectors $\boldsymbol{F}_{\text{rad}}$ and $\boldsymbol{F}_{\Sigma}$ are, respectively,

$$F_{i\text{rad}} = \tau v_j \frac{\partial f_i}{\partial x_j} - \frac{e^2}{\rho}\left.\widetilde{(\hat{\mathcal{D}}Q)_i}\right|_{z=0}, \tag{C.3}$$

$$F_{i\Sigma} = -\frac{1}{m\rho}\frac{\partial}{\partial x_j}\Sigma_{ji}\rho. \tag{C.4}$$

Now, according to Eq. (4.66a) $m\boldsymbol{v}$ decomposes as

$$m\boldsymbol{v} = -i\eta\nabla \ln \frac{q(\boldsymbol{x},t)}{q^*(\boldsymbol{x},t)} + \boldsymbol{g} = 2\eta\nabla S + \boldsymbol{g}, \tag{C.5}$$

where the last equality follows from writing $q(\boldsymbol{x},t)$ in its polar form

$$q(\boldsymbol{x},t) = \sqrt{\rho}e^{iS(\boldsymbol{x},t)}. \tag{C.6}$$

Substitution of Eqs. (C.5) into (C.2) gives (using $\nabla \times m\boldsymbol{v} = \nabla \times \boldsymbol{g}$)

$$\nabla M = \boldsymbol{F}_{\text{rad}} + \boldsymbol{F}_{\Sigma} - \frac{\partial \boldsymbol{g}}{\partial t} + \boldsymbol{v}\times(\nabla\times\boldsymbol{g}), \tag{C.7}$$

with

$$M = 2\eta \frac{\partial S}{\partial t} + \frac{1}{2} m \boldsymbol{v}^2 - \frac{2\eta^2}{m} \frac{\nabla^2 \sqrt{\rho}}{\sqrt{\rho}} + V. \tag{C.8}$$

This is basically the Hamilton-Jacobi-type equation of Bohm's theory [see Sect. (4.4.1)], when $M = 0$ (and $\eta = \hbar/2$), which means that it should be possible to arrive at the Schrödinger equation from the above expressions. For this purpose we proceed as follows.

From Eq. (C.6) it follows $\ln q = (1/2) \ln \rho + iS$, which combined with the continuity Eq. (C.1a) gives

$$\frac{\partial S}{\partial t} = -\frac{i}{q} \frac{\partial q}{\partial t} - \frac{i}{2} \nabla \cdot \boldsymbol{v} - i \boldsymbol{v} \cdot \frac{\nabla q}{q} - \boldsymbol{v} \cdot \nabla S. \tag{C.9}$$

Using here Eq. (C.5) leads to

$$\begin{aligned} \frac{\partial S}{\partial t} = & -\frac{i}{q} \frac{\partial q}{\partial t} - \frac{i\eta}{m} \nabla^2 S - \frac{i}{2m} \nabla \cdot \boldsymbol{g} - \frac{2i\eta}{m} \nabla S \cdot \frac{\nabla q}{q} - \\ & - \frac{i}{m} \boldsymbol{g} \cdot \frac{\nabla q}{q} - \frac{2\eta}{m} (\nabla S)^2 - \frac{1}{m} \boldsymbol{g} \cdot \nabla S. \end{aligned} \tag{C.10}$$

This expression, together with $m\boldsymbol{v}^2 = (1/m)(2\eta \nabla S + \boldsymbol{g})^2$, allows to recast Eq. (C.8) in the form

$$\begin{aligned} M = & \frac{1}{q} \left[-2i\eta \frac{\partial q}{\partial t} + \frac{1}{2m} (-2i\eta \nabla + \boldsymbol{g})^2 q + Vq \right] - \\ & - \frac{2\eta^2}{m} \left[i \nabla^2 S + \frac{\nabla^2 \sqrt{\rho}}{\sqrt{\rho}} + (\nabla S)^2 + 2i \nabla S \cdot \frac{\nabla q}{q} - \frac{\nabla^2 q}{q} \right]. \end{aligned} \tag{C.11}$$

Now we notice that as a consequence of Eq. (C.6),

$$\frac{\nabla q}{q} = \frac{1}{2} \frac{\nabla \rho}{\rho} + i \nabla S, \tag{C.12}$$

whence

$$\frac{\nabla^2 q}{q} = \frac{\nabla^2 \sqrt{\rho}}{\sqrt{\rho}} + i \nabla^2 S + i \nabla S \cdot \frac{\nabla \rho}{\rho} - (\nabla S)^2. \tag{C.13}$$

From here it follows that the term in the second line of Eq. (C.11) vanishes. Consequently M reduces to

$$M = \frac{1}{q}\left[-2i\eta\frac{\partial q}{\partial t} + \frac{1}{2m}\left(-2i\eta\nabla + g\right)^2 q + Vq\right] = \frac{1}{q}\hat{M}q, \tag{C.14}$$

with $\hat{M}$ the linear operator

$$\hat{M} = -2i\eta\frac{\partial}{\partial t} + \frac{1}{2m}\left(-2i\eta\nabla + g\right)^2 + V. \tag{C.15}$$

Finally, inserting Eq. (C.14) into (C.7) we arrive at

$$\nabla\left(\frac{1}{q}\hat{M}q\right) = \boldsymbol{F}_{\text{rad}} + \boldsymbol{F}_{\Sigma} - \frac{\partial \boldsymbol{g}}{\partial t} + \boldsymbol{v}\times\left(\nabla\times\boldsymbol{g}\right). \tag{C.16}$$

We call this the *generalized Schrödinger equation*, since as is shown in Sect. 4.4.1, it reduces to the Schrödinger equation in the radiationless approximation. Equation (C.16) and its adjoint are equivalent to the first two equations of the hierarchy, (C.1a, C.1b).

Appendix D: Diffusive Contribution to the Energy Balance

This appendix is devoted to the calculation of the right-hand side of the energy-balance condition (4.37). The particle is considered in equilibrium with the ZPF, which means that it must be in its ground state, represented by the solution $\psi_0(x)$ of the Schrödinger equation (4.78) (still in terms of η). We recall that in the time-asymptotic limit, the Markovian approximation holds, described by the FPE (B.1). This means that one may use the simpler Eq. (4.103), which in one dimension reads

$$\tau\left\langle \dddot{x}\, p\right\rangle_0 = -\frac{1}{m}\left\langle D^{pp}\right\rangle_0. \tag{D.1}$$

Introducing the diffusion coefficient D^{pp} given by Eq. (B.14a), one obtains

$$-\frac{1}{m}\left\langle D^{pp}\right\rangle_0 = \frac{ie^2}{2\eta m}\int_{-\infty}^{t} dt'\varphi(t-t')\left\langle\left[\hat{x}'_i, \hat{p}_j\right]\right\rangle_0, \tag{D.2}$$

With $\varphi(t-t')$ given by (4.10) and $\rho = \rho_0$ given by (4.101), this becomes (recall that $\tau = 2e^2/3mc^3$)

$$-\frac{1}{m}\left\langle D^{pp}\right\rangle_0 = \frac{i\hbar\tau}{2\eta\pi}\int_0^{\infty} d\omega\,\omega^3 \int_{-\infty}^{t} dt' \cos\omega(t-t')\,\left\langle\left[\hat{x}_i', \hat{p}_j\right]\right\rangle_0, \tag{D.3}$$

with

$$\left\langle\left[\hat{x}', \hat{p}\right]\right\rangle_0 = \sum_k \left(\hat{x}_{0k}'\hat{p}_{k0} - \hat{p}_{0k}\hat{x}_{k0}'\right), \tag{D.4}$$

Equations (4.104) and (4.105) can be used to write $x_{kn}(t) = e^{i\omega_{kn}t}x_{kn}$ and $p_{nm}(t) = im\omega_{nm}e^{i\omega_{nm}t}x_{nm}$, which gives

$$\left\langle\left[\hat{x}', \hat{p}\right]\right\rangle_0 = 2im\sum_k \omega_{k0}\left|x_{0k}\right|^2 \cos\omega_{k0}(t-t'). \tag{D.5}$$

Introducing this into Eq. (D.3) leads to

$$\begin{aligned}-\frac{1}{m}\left\langle D^{pp}\right\rangle_0 = &-\frac{\hbar m\tau}{\pi\eta}\sum_k \omega_{k0}\left|x_{0k}\right|^2 \\ &\times\int_0^{\infty} d\omega\,\omega^3 \int_{-\infty}^{t} dt' \cos\omega(t-t')\cos\omega_{k0}(t-t').\end{aligned} \tag{D.6}$$

The integral over time can be calculated with the formula

$$\int_{-\infty}^{\infty} dk e^{ikx} = \int_{-\infty}^{\infty} dk\cos kx = 2\pi\delta(x). \tag{D.7}$$

This gives

$$\int_{-\infty}^{t} dt' \cos\omega(t-t')\cos\omega_{k0}(t-t') = \frac{\pi}{2}\left[\delta(\omega-\omega_{k0}) + \delta(\omega+\omega_{k0})\right]. \tag{D.8}$$

For the ground state there are no negative frequencies, i.e. all $\omega_{k0} > 0$, whence the second term in the right-hand side does not contribute to (D.6), and therefore

$$-\frac{1}{m}\left\langle D^{pp}\right\rangle_0 = -\frac{\hbar m\tau}{2\eta}\sum_k \omega_{k0}^4\left|x_{0k}\right|^2. \tag{D.9}$$

References

Alcubierre, M., Lozano N.: *Tratamiento de Sistemas Multiperiódicos en la Electrodinámica Estocástica*, Professional thesis (Universidad Nacional Autónoma de México, México) (1988)

Bacciagaluppi, G., Valentini, A.: Quantum Theory at the Crossroads. Reconsidering the: Solvay Conference. Cambridge University Press, Cambridge (1927/2009)

Balescu, R.: Equilibrium and Nonequilibrium Statistical Mechanics. Wiley, New York (1975)

Ballentine, L.E.: Quantum Mechanics. Prentice Hall, New Jersey (1990)

Ballentine, L.E.: Quantum Mechanics. A Modern Development. World Scientific, Singapore (1998)

Boyer, T.H.: Random electrodynamics: The theory of classical electrodynamics with classical electromagnetic zero-point radiation. Phys. Rev. D **11, 790 (1975)

*Boyer, T.H.: Equilibrium of random classical electromagnetic radiation in the presence of a non-relativistic nonlinear electric dipole oscillator. Phys. Rev. D **13**, 2832 (1976)

Boyer, T.H.: A brief survey of stochastic electrodynamics. In: Barut, A.O. (ed.) Foundations of Radiation Theory and Quantum Electrodynamics. Plenum, New York (1980)

Carroll, R.: On the Emergence Theme of Physics. World Scientific, Singapore (2010)

Case, W.B.: Wigner functions and Weyl transforms for pedestrians. Am. J. Phys. **76**, 937 (2008)

Cetto, A.M., de la Peña, L., Velasco, R.M.: Generalized Fokker-Planck equations for coloured, multiplicative Gaussian noise. Rev. Mex. Fís. **31**, 83 (1984)

*Cetto, A.M., de la Peña, L., Valdés-Hernandez, A.: Quantization as an emergent phenomenon due to matter-zeropoint field interaction. J. Phys. JPCS **361**, 012013 (2012)

Claverie, P., Diner, S.: Stochastic electrodynamics and quantum theory. Int. J. Quantum Chem. **12**, S1, 41 (1977)

Cohen, L.: Quantization problem and variational principle in the phase-space formulation of quantum mechanics. J. Math. Phys. **17**, 1863 (1976)

Cohen, L., Zaparovanny, Y.I.: Positive quantum joint distributions. J. Math. Phys. **21**, 794 (1980)

Cohen-Tannoudji, C., Dupont-Roc, J., Grynberg, G.: Photons and Atoms. Introduction to Quantum Electrodynamics. Wiley, New York (1989)

Cole, D.C.: Simulation results related to stochastic electrodynamics. In: Adenier, G., Khrennikov, Y., Nieuwenhuizen, T.M. (eds.) Quantum Theory. Reconsideration of Foundations-3, AIP Conference Proceedings 810, New York, vol. 99 (2006)

Cole, D.C., Zou, Y.: Quantum mechanical ground state of hydrogen obtained from classical electrodynamics. Phys. Lett. A **317**, 14 (2003)

Cole, D.C., Zou, Y.: Analysis of orbital decay time for the classical hydrogen atom interacting with circularly polarized electromagnetic radiation. Phys. Rev. E **69**, 016601 (2004a)

Cole, D.C., Zou, Y.: Simulation study of aspects of the classical hydrogen atom interacting with electromagnetic radiation: circular orbits. J. Sci. Comput. **20**, 43 (2004b)

Cole, D.C., Zou, Y.: Simulation study of aspects of the classical hydrogen atom interacting with electromagnetic radiation: elliptical orbits. J. Sci. Comput. **20**, 379 (2004c)

Cole, D.C., Zou, Y.: Perturbation analysis and simulation study of the effects of phase on the classical hydrogen atom interacting with circularly polarized electromagnetic radiation. J. Sci. Comput. **21**, 145 (2004d)

Cole, D.C., Zou, Y.: Subharmonic resonance behavior for the classical hydrogen atomic system. J. Sci. Comput. **39**(1) (2009)

Collins, R.E.: Quantum theory: a Hilbert space formalism for probability theory. Found. Phys. **7**, 475 (1977)

de la Peña, L.: A simple derivation of the Schrödinger equation from the theory of Markov processes. Phys. Lett. A **24**, 603 (1967)

**de la Peña, L.: Stochastic electrodynamics: its development, present situation and perspectives. In: Gmez, B., et al. (eds.) Stochastic Processes Applied to Physics and Other Related Fields. World Scientific, Singapore (1983)

de la Peña, L., Cetto, A.M.: Stronger form for the position-momentum uncertainty relation. Phys. Lett. A **39**, 65 (1972) de la Peña, L., Cetto, A.M.: Derivation of quantum mechanics from stochastic electrodynamics. J. Math. Phys. **18**, 1612 (1977a)

de la Peña, L., Cetto, A.M.: Why Schrödinger's equation? Int. J. Quantum Chem. XII Supl. **1**, 23 (1977b)

de la Peña, L., Cetto, A.M.: Is quantum mechanics a limit cycle theory? In: Ferrero, M., van der Merwe, A. (eds.) Fundamental Problems in Quantum Physics, p. 47. Kluwer Academic, Dordrecht (1995)

*de la Peña, L., Cetto, A.M.: The Quantum Dice An Introduction to Stochastic Electrodynamics. Kluwer Academic, Dordrecht (1996) (Referred to in the book as *The Dice*)

*de la Peña, L., Cetto, A.M.: Contribution from stochastic electrodynamics to the understanding of quantum mechanics. http://arxiv.org/abs/quant-ph/0501011arXiv:quant-ph/0501011v2 (2005)

*de la Peña, L., Cetto, A.M.: Recent developments in linear stochastic electrodynamics. In: Adenier, G., Khrennikov, A.Y., Nieuwenhuizen, T. M. (eds.) Quantum Theory: Reconsideration of Foundations-3, AIP Conference Proceedings no. 810. AIP, New York. Extended version in arXivquant-ph0501011 (2006)

*de la Peña, L., Cetto, A.M.: On the ergodic behaviour of atomic systems under the action of the zero-point radiation field. In: Nieuwenhuizen, T.M., et al. (eds.) Beyond the Quantum. World Scientific, Singapore (2007)

*de la Peña, L., Valdés-Hernández, A., Cetto, A.M.: Quantum mechanics as an emergent property of ergodic systems embedded in the zero-point radiation field. Found. Phys. **39**, 1240 (2009)

*de la Peña, L., Cetto, A.M., Valdés-Hernández, A.: Quantum behavior derived as an essentially stochastic phenomenon. Phys. Scr. **T151**, 014008 (2012a)

*de la Peña, L., Cetto A.M, Valdés-Hernández, A.: The emerging quantum. An invitation, Advanced School on Quantum Foundations and Open Quantum Systems, http://www.fisica.ufpb.br/asqf2012/index.php/ invited-lectures (2012b)

Dirac, P.A.M.: Note on exchange phenomena in the Thomas atom. Proc. Camb. Phil. Soc. **26**, 376 (1930)

Dirac, P.A.M.: In: Proceedings of the second Canadian mathematical congress (University of Toronto), 10 (1951)

Edwards, S.F., McComb, W.D.: Statistical mechanics far from equilibrium. J. Phys. A **2**, 157 (1969)

Einstein, A.: Autobiographical notes and Einstein's reply, In: Schilpp, P.A. (ed.) Albert Einstein: Philosopher-Scientist. Harper and Row, New York (1949)

Einstein, A., Hopf, L.: Statistische Untersuchung der Bewegung eines Resonators in einem Strahlungsfeld. Ann. der Physik **33**, 1096 (1910)

Frieden, B.R.: Physics from Fisher Information. Cambridge University Press, Cambridge (1998)

Frisch, U.: Wave propagation in random media. In: Bharucha-Reid, A.T. (ed.) Probabilistic methods in applied mathematics, vol. I. Academic, New York (1968)

Fujita, S., Godoy, S.V.: Mathematical Physics. Wiley, Weinheim (2010)

Gardiner, C.W.: Handbook of Stochastic Methods. Springer, Berlin (1983)

Graham, N.: The measurement of relative frequency. In: DeWitt, B.S., Graham, N. (eds.) The Many Worlds Interpretation of Quantum Mechanics. Princeton University Press, Princeton (1973)

Hartle, J.B.: Quantum mechanics of individual systems. Am. J. Phys. **36**, 704 (1968)

Hassani, S.: Mathematical Physics. Springer, New York (1999)

Heisenberg, W.: Über die inkohärente Streuung von Röntgenstrahlen. Zeit. Phys. **32**, 737 (1931)

Hillery, M., O'Connell, R.F., Scully, M.O., Wigner, E.P.: Distribution functions in physics fundamentals. Phys. Rep. **106**, 121 (1984)

Holland, P.R.: The Quantum Theory of Motion. Cambridge University Press, Cambridge (1993)

Huang, W.C.W., Batelaan, H.: Dynamics underlying the Gaussian distribution of the classical harmonic oscillator in zero-point radiation. arXiv/quant-ph:1206.5323v1 (2012a)

Huang W.C.W., Batelaan H.: Quantized Excitation Spectrum of the Classical Harmonic Oscillator in Zero-Point Radiation. arXiv/quant-ph:1206.6891v1 (2012b)

Hudson, R.L.: When is the Wigner quasi-probability density non-negative? Rep. Math. Phys. (Torun) **6**, 249 (1974)

Jammer, M.: The Philosophy of Quantum Mechanics. The interpretation of quantum mechanics in historical perspective. Wiley, New York (1974)
Kracklauer, A.F.: An intuitive paradigm for quantum mechanics. Phys. Essays **5**, 226 (1992)
Landau, L., Lifshitz, E.: The Classical Theory of Fields. Addison-Wesley, Cambridge (1951)
Madelung, E.: Quantentheorie in hydrodynamischer Form. Zeit. Phys. **40**, 322 (1926)
Marshall, T.W., Claverie, P.: Stochastic electrodynamics of nonlinear systems. I. Particle in a central field of force. J. Math. Phys. **21**, 1819 (1980)
Masujima, M.: Path Integral Quantization and Stochastic Quantization. Springer, Berlin (2009)
Morse, P., Feshbach, H.: Methods of Theoretical Physics. McGraw-Hill, New York (1953)
Moyal, J.E.: Quantum mechanics as a statistical theory. Proc. Camb. Phil. Soc. **45**, 99 (1949)
Nernst, W.: Über einen Versuch, von quantentheoretischen Betrachtungen zur Annahme stetiger Energieänderungen zurückzukehren. Verh. Deutsch. Phys. Ges. **18**, 83 (1916)
Olavo, L.S.F.: Foundations of quantum mechanics: connection with stochastic processes. Phys. Rev. A **61**, 052109 (2000)
Papoulis, A.: Probability, Random Variables, and Stochastic Processes, Chap. 6. McGraw-Hill, Boston (1991)
Parisi, G., Wu, Y.S.: Perturbation theory without Gauge fixing. Sci. Sin. **24**, 483 (1981)
Pauli, W.: Über Gasentartung und Paramagnetismus. Zeit. Phys. **41**, 81 (1927)
Piquet, C.: Fonctions de type positif associées a deux opérateurs hermitiens. C. R. Acad. Sci. Paris A **279**, 107 (1974)
Risken, H.: The Fokker-Planck Equation. Methods of Solution and Applications. Springer, Berlin (1984)
Rohrlich, F.: Classical Charged Particles. Foundations of Their Theory. Addison-Wesley, Reading (1965)
Rosen, N.: The relation between classical and quantum mechanics. Am. J. Phys. **32**, 597 (1964)
Rosen, N.: Quantum particles and classical particles. Found. Phys. **16**, 687 (1986)
Roy, S.: Stochastic geometry and origin of quantum potential. Phys. Lett. A **115**, 256 (1986)
Schiller, R.: Quasi-classical theory of the nonspinning electron. Phys. Rev. **125**, 1100 (1962a)
Schiller, R.: Quasi-classical transformation theory. Phys. Rev. **125**, 1109 (1962b)
Soto-Eguibar, F., Claverie, P.: When is the Wigner function of multi-dimensional systems nonnegative? J. Math. Phys. **24**, 97 (1983a)
Soto-Eguibar, F., Claverie, P.: Time evolution of the Wigner function. J. Math. Phys. **24**, 1104 (1983b)
Stratonovich, R.L.: Topics in the Theory of Random Noise, Vol. I: General Theory of Random Processes. Gordon and Breach, New York (1963)
Surdin, M.: L'état fondamental de l'oscillateur harmonique est-il un cycle limite? Ann. Inst. Henri Poincaré **13**, 363 (1970)
Takabayasi, T.: On the formulation of quantum mechanics associated with classical pictures. Progr. Theor. Phys. **8**, 143 (1952)
Tatarskii, V.I.: The Wigner representation of quantum mechanics. Sov. Phys. Usp. **26**, 311 (1983)
't Hooft, G.: Determinism beneath Quantum Mechanics. http://arxiv.org/abs/quant-ph/0212095 v1arXiv:quant-ph/0212095v1 (2002)
Urbanik, K.: Joint probability distributions of observables in quantum mechanics. Studia Math. **21**, 117 (1967)
van Kampen, N.G.: Stochastic differential equations. Phys. Rep. **24**, 171 (1976)
van Vleck, J.H.: The absorption of radiation by multiply periodic orbits, and its relation to the correspondence principle and the Rayleigh-Jeans law: Part II. Calculation of absorption by multiply periodic orbits. Phys. Rev. **24**, 347 (1924)
van Vleck, J.H., Huber, D.L.: Absorption, emission, and linebreadths: a semihistorical perspective. Rev. Mod. Phys. **49**, 939 (1977)
Wallstrom, T.C.: On the derivation of the Schrödinger equation from stochastic mechanics. Found. Phys. Lett. **2**, 113 (1989)

Wallstrom, T.C.: Inequivalence between the Schrödinger equation and the Madelung hydrodynamic equations. Phys. Rev. A **49**, 1613 (1994)
Wax, N. (ed.): Selected Papers on Noise and Stochastic Processes. Dover, New York (1954/1985)
Weinberg, S.: The cosmological constant problem. Rev. Mod. Phys. **61**, 1 (1989)
Wesson, P.S.: Cosmological constraints on the zero-point electromagnetic field. Astrophys. J. **387**, 466 (1991)
Weyl, H.: Quantenmechanik und Gruppentheorie. Zeit. Phys. **46**, 1 (1927)
Wigner, E.: On the quantum correction for thermodynamic equilibrium. Phys. Rev. **40**, 749 (1932)
Zachos, C.K., Fairlie, D.B., Curtright, T.L. (eds.) Quantum Mechanics in Phase Space. World Scientific, Singapore (2005)

Chapter 5
The Road to Heisenberg Quantum Mechanics

> *... quantum phenomena do not occur in a Hilbert space, they occur in a laboratory.*
>
> A. Peres (1995, p. XI)

Once the Schrödinger equation has been obtained from fundamental principles, one might consider that the task of arriving at quantum mechanics is over, and that all that is left is to develop and apply it. However, a great deal still remains to be understood concerning the underlying meaning of the quantum formalism.

This chapter is devoted to an alternative approach leading to the quantization of matter within the framework of SED. The path followed here is perhaps less intuitive, but on the other hand it is highly revealing and illustrative of some of the intricacies of QM. Although based on the same basic principles used so far, this approach takes us to the quantum formalism in its matrix form. Heisenberg's theory appears thus not only as a mere mathematical transformation of Schrödinger's theory—as Dirac, Schrödinger himself and others succeeded to show—but as a description that exhibits complementary aspects of the quantum world.

5.1 The Same System: A Fresh Approach

The main character in this chapter is again a charged particle—typically an atomic electron—immersed in the background radiation field and subject to an external force. We recall that the radiation field may or may not contain excitations, but it *always* includes the random ZPF of mean energy $\hbar\omega/2$ per normal mode. A central conclusion drawn from Chap. 4 is that once an energy balance has been attained in the mean between the diffusive and dissipative radiative terms, the statistical behavior of the particle is correctly described by the Schrödinger equation, in the radiationless approximation. This means that, for a particle subject to a conservative, binding force, the description can be made in terms of stationary states, which are energy-conserving states, i.e., eigenstates of the Hamiltonian.

L. de la Peña et al., *The Emerging Quantum*,
DOI: 10.1007/978-3-319-07893-9_5,

It seems now opportune to take a more straightforward approach for the analysis of these stationary states in the quantum regime, based again on the (stochastic) Abraham-Lorentz equation. We shall therefore use this equation as the starting point, as in Chap. 4; yet very soon our train of thoughts will take a different path, one that brings us closer to the formalism developed in the famous paper by Born, Heisenberg and Jordan in 1926. The purpose of this new approach is to throw additional light on what it means for the system to reach stationarity and become quantized. What are the specificities of the matter-field interaction in the quantum regime? How do these specificities manifest themselves in the quantum formalism? What is the physical meaning of the quantities involved in the Heisenberg description?

As we proceed with the analysis, several interesting features of the underlying physics will be disclosed. A central one is the linear, resonant response of the mechanical system in the quantum regime to a certain, well-defined set of modes of the radiation field. Not surprisingly, the corresponding resonance frequencies are eventually identified with the usual (atomic) transition frequencies. Furthermore, the ergodic properties of the system turn out to play a major role in quantization. Along the process, we are led in a natural way to matrix mechanics as a suitable alternative tool for the description of the dynamics of the particle in the quantum regime. The results obtained endow the Heisenberg formalism with a revealing physical meaning, as is discussed along the text and more in detail in the final part of the chapter.[1]

5.1.1 Description of the Mechanical Subsystem

As in Chap. 4, we consider a particle with mass m and charge e subject to an external (binding, conservative) force $\boldsymbol{f}(\boldsymbol{x})$, in addition to the radiation reaction force $m\tau\dddot{\boldsymbol{x}}$ and the electric component of the Lorentz force (in line with the nonrelativistic approximation) due to the random background field. The motion of the particle is thus governed by the Abraham-Lorentz equation

$$m\ddot{\boldsymbol{x}} = \boldsymbol{f}(\boldsymbol{x}) + m\tau\dddot{\boldsymbol{x}} + e\boldsymbol{E}(\boldsymbol{x},t), \quad \tau = \frac{2e^2}{3mc^3}. \tag{5.1}$$

Our focus will be the *stationary* solutions of Eq. (5.1), whenever they exist. For a study of these, a more detailed description of the random electric field $\boldsymbol{E}(\boldsymbol{x},t)$ is required. For this purpose we expand it in terms of plane waves of frequency $\omega_k = c\,|\boldsymbol{k}|$, wave vector $\boldsymbol{k}$, polarization λ and random amplitudes $a_{\boldsymbol{k}}^{\lambda}$,[2]

[1] The present exposition draws largely from de la Peña and Cetto (2006a, b) (2007), de la Peña et al (2009), Valdés-Hernández (2010), Cetto et al. (2012), and references therein.

[2] This kind of expansion in terms of plane waves is usual, both in dealing with stochastic forces (see e.g. Rice 1954) and in QED [see e.g. Milonni (1994)]. The difference between a classical field and a quantum field lies in the factor $a_{\boldsymbol{k}}^{\lambda}$, which takes the value 1 in the classical instance, and represents an operator in the quantum situation. Here, $a_{\boldsymbol{k}}^{\lambda}$ stands for a random variable.

$$E(x,t) = \sum_{k,\lambda} \tilde{E}(\omega_k)\hat{\epsilon}_k^\lambda a_k^\lambda e^{-ik\cdot x} e^{i\omega_k t} + \text{c.c.} \tag{5.2}$$

The factor $\tilde{E}(\omega_k)$ is determined by the mean energy of the corresponding modes—which in the case of the ZPF is given by $\hbar\omega/2$—and $\hat{\epsilon}_k^\lambda$ is the polarization unit vector. By performing an angular average over all field modes having the same frequency ω_k and summing (which, up to a factor 2, is equivalent to averaging) over the polarizations,[3] Eq. (5.2) can be recast in the form

$$\begin{aligned} E(x,t) &= \sum_{\omega_k} \tilde{E}\,(\omega_k)\left[\int_{\Omega_k} d\Omega_k \left(\sum_\lambda \hat{\epsilon}_k^\lambda a_k^\lambda\right) e^{-ik\cdot x}\right] e^{i\omega_k t} + \text{c.c.} \\ &= \sum_{\omega_k} \tilde{E}(\omega_k) a_k(x) e^{i\omega_k t} + \text{c.c.}, \end{aligned} \tag{5.3}$$

with $a_k(x)$ defined as

$$a_k(x) \equiv a(\omega_k, x) = \int_{\Omega_k} d\Omega_k \left(\sum_\lambda \hat{\epsilon}_k^\lambda a_k^\lambda\right) e^{-ik\cdot x}. \tag{5.4}$$

The expression $a(\omega_k, x)$ does not imply a functional dependence of a on ω_k; rather, the notation stresses that to each frequency ω_k there corresponds a stochastic variable $a_k(x)$. For a maximally disordered field, which the ZPF is normally assumed to be in the absence of matter [as done already in Einstein and Stern (1913) and in Chap. 3], the original a_k^λ are statistically independent random variables with a normal distribution. However, the angular integration smoothes out the fluctuations, so that the ensuing $a_k(x)$ represent a sort of coarse-grained amplitudes. We therefore make the assumption that the integrated $a_k(x)$ have roughly fixed amplitudes and random phases.[4]

Further and as before, the amplitude of the relevant modes of the field is assumed not to vary appreciably in space within the characteristic deviations of the particle from its mean position x^0, so that $E(x,t)$ can be replaced by $E(x^0,t)$ in Eqs. (5.1)–(5.4); this is the long-wavelength approximation, the validity of which will

[3] The summation over λ can be performed whenever the polarization of the radiation field is irrelevant. However, this is not always the case, as will become evident in Chap. 6, in connection with the spin of the electron; in this case one must keep the distinction between the two different polarizations.

[4] Several authors in SED, in particular Boyer, use this representation systematically. Still, the fixed amplitudes can be taken as normally distributed, but with a negligible dispersion. See also the discussion at the end of Appendix A.

be confirmed in due time.[5] In the one-dimensional case, for simplicity (with the direction of motion along $\hat{\boldsymbol{x}}$ and with the origin fixed at some $\boldsymbol{x}^0$), (5.3) reduces to

$$E(t) = \boldsymbol{E}(t) \cdot \hat{\boldsymbol{x}} = \sum_{\omega_k} \tilde{E}(\omega_k) a(\omega_k) e^{i\omega_k t} + \text{c.c.} = \sum_k \tilde{E}_k a_k e^{i\omega_k t} + \text{c.c}, \tag{5.5}$$

were $a(\omega_k)$ is the dimensionless stochastic variable $a(\omega_k)= \boldsymbol{a}_k(0) \cdot \hat{\boldsymbol{x}}$ and we introduced the succinct notation

$$\tilde{E}_k a_k = \tilde{E}(\omega_k) a(\omega_k), \tag{5.6}$$

with

$$\tilde{E}(\omega_k) a(\omega_k) = \tilde{E}^*(-\omega_k) a^*(-\omega_k). \tag{5.7}$$

In addition we have

$$a(\omega_k) = e^{i\varphi_k}, \tag{5.8}$$

with φ_k a random phase in $[0, 2\pi]$; we have taken into account the fixed amplitude of $a(\omega_k)$, $|a(\omega_k)| = 1$. The size of the coefficients $\tilde{E}(\omega_k)$ will therefore be determined by the intensity (or the mean energy) of the corresponding modes. The coefficients $\tilde{E}(\omega_k)$ in the above expressions differ from some of those appearing in previous chapters, for example, in Eqs. (4.7) and (4.8). In that case the stochastic amplitudes $a(\omega_k)$ are contained in the coefficients, whereas here they are written separately.

With the above approximations the equation of motion becomes

$$m\ddot{x} = f(x) + m\tau\dddot{x} + eE(t). \tag{5.9}$$

5.1.2 Resonant Solutions in the Stationary Regime

As stated above, we are interested in the stationary solutions of Eq. (5.9). Each of these can be decomposed into a time-independent contribution, which coincides with the time average defined as

$$\overline{g(t)}^t = \lim_{T\to\infty} \frac{1}{2T} \int_{-T}^{T} g(t) dt, \tag{5.10}$$

[5] The fact that the long-wavelength approximation does not hold for the field components $\tilde{E}(\omega_k)\boldsymbol{a}_k(\boldsymbol{x})$ of high frequency is irrelevant here, as their contribution amounts basically to an inconsequential noise. As explained in Chaps. 4 and 9, for light atoms the relevant wavelengths are $\lambda_k \gg a_B$ (a_B is the Bohr radius). For heavier atoms, $\lambda_{\text{Ryd}} \geq 4\pi\hbar^3 c/(me^4) \sim a_B/\alpha$.

and an oscillatory (and fluctuating) contribution that averages to zero. Thus we look for solutions of the form

$$x^{s}(t) = \sum_{k} \tilde{x}_k a_k e^{i\omega_k t} + \text{c.c}, \tag{5.11}$$

with $\tilde{x}_k$ the coefficient associated with the frequency ω_k and

$$\overline{x^{s}(t)}^{t} = \tilde{x}_0 a_0 + \tilde{x}_0^* a_0^*, \tag{5.12}$$

where the subindex 0 corresponds to the null frequency $\omega_{k=0} = 0$. Notice that since $x^{s}(t)$ is a real quantity, the coefficients $\tilde{x}_k$ satisfy (cf. Eq. 5.7)

$$\tilde{x}_k = \tilde{x}(\omega_k) = \tilde{x}^*(-\omega_k). \tag{5.13}$$

Assuming that the external force $f(x)$ can be expanded as a power series of x (and excluding the case of a constant force, for simplicity),

$$f(x(t)) = c_1 x(t) + c_2 x^2(t) + c_3 x^3(t) + \cdots, \tag{5.14}$$

we decompose it in the stationary state in a form analogous to (5.11),

$$f^{s}(t) = \sum_{k} \tilde{f}_k a_k e^{i\omega_k t} + \text{c.c.} \tag{5.15}$$

It is important to stress that here as in Eq. (5.11), each component of frequency ω_k is conveniently written in the form $\tilde{f}_k a_k$, although the quantities $\tilde{f}_k$ themselves (in particular $\tilde{x}_k$) depend in general (for nonlinear forces) on the variables $\{a_k\}$, but not on time. In other words, neither Eq. (5.11) nor Eq. (5.15) are *explicit* expansions in the field stochastic variables a_k; yet the time dependence in both equations is fully expressed through the periodic factors $e^{i\omega_k t}$.

Introducing Eqs. (5.5), (5.11) and (5.15) into (5.9) leads, after separating the terms that oscillate with frequency ω_k, to

$$-m\omega_k^2 \tilde{x}_k = \tilde{f}_k - im\tau\omega_k^3 \tilde{x}_k + e\tilde{E}_k, \tag{5.16}$$

or

$$\tilde{x}_k = -\frac{e}{m}\frac{\tilde{E}_k}{\Delta_k}, \tag{5.17a}$$

$$\text{with} \quad \Delta_k \equiv \omega_k^2 - i\tau\omega_k^3 + \frac{\tilde{f}_k}{m\tilde{x}_k}. \tag{5.17b}$$

The dominant contributions to the solution $x(t)$ come from the poles of $\tilde{x}_k$, i.e., for $\Delta_k \simeq 0$. They correspond to those frequencies that satisfy the (approximate) equation

$$\omega_k^2 \approx -\frac{\tilde{f}_k}{m\tilde{x}_k}. \tag{5.18}$$

The resonances at these frequencies are extremely sharp due to the small value of τ (recall that $\tau \sim 10^{-23}$ s for electrons). Indeed, for atomic frequencies of interest $\tau\omega \lesssim \tau\omega_{\mathrm{Ryd}} \sim \alpha^3$, where $\omega_{\mathrm{Ryd}} = 2\pi c/\lambda_{\mathrm{Ryd}}$ and $\alpha = e^2/\hbar c$ is the fine-structure constant. We denote the set of solutions of Eq. (5.18) as $\{\omega_k\}_{\mathrm{res}}$ and refer to its elements as *resonance frequencies*.

One should assume that there may be in principle more than one stationary solution. Since we are dealing with conservative forces, the different stationary solutions would correspond to definite values of the mechanical energy $\mathcal{E}$. A particular solution will be therefore labeled with an index α that is in direct correspondence with the energy attained, denoted by $\mathcal{E}_\alpha$.[6] It is clear that for every stationary solution, some frequencies (to be determined by the theory itself) play an important role in the dynamics, while others not. Thus, every state α has associated with it a set of frequencies, which will be referred to as *relevant frequencies*. We denote the elements of this set as $\omega_{\alpha\beta}$, where the index β labels the different frequencies in such a way that for $\beta \neq \beta'$, $\omega_{\alpha\beta} \neq \omega_{\alpha\beta'}$, and write (5.11), (5.5) and (5.15) for a state α as

$$x_\alpha^{\mathrm{s}}(t) = \sum_\beta \tilde{x}_{\alpha\beta} a_{\alpha\beta} e^{i\omega_{\alpha\beta}t} + \mathrm{c.c.}, \tag{5.19a}$$

$$E_\alpha(t) = \sum_\beta \tilde{E}_{\alpha\beta} a_{\alpha\beta} e^{i\omega_{\alpha\beta}t} + \mathrm{c.c.}, \tag{5.19b}$$

$$f_\alpha^{\mathrm{s}}(t) = \sum_\beta \tilde{f}_{\alpha\beta} a_{\alpha\beta} e^{i\omega_{\alpha\beta}t} + \mathrm{c.c.}, \tag{5.19c}$$

in terms of the shorthand notation

$$\tilde{A}_{\alpha\beta} = \tilde{A}(\omega_{\alpha\beta}), \quad a_{\alpha\beta} = a(\omega_{\alpha\beta}), \tag{5.20}$$

which does not imply a functional dependence on the frequency $\omega_{\alpha\beta}$, but simply a correspondence between the respective quantity and the frequency $\omega_{\alpha\beta}$. Further, in line with Eq. (5.8), the stochastic variable $a_{\alpha\beta}$ can be expressed as

$$a_{\alpha\beta} = e^{i\varphi_{\alpha\beta}}, \tag{5.21}$$

[6] As will be seen below, the stationary states labeled with α will turn out to be the stationary states predicted by the Schrödinger equation and derived in Chap. 4. However, for the present approach to be self-contained, in this chapter we are developing the argument with independence from our previous results.

with $\varphi_{\alpha\beta}$ a stochastic phase in $[0, 2\pi]$.

In analogy with Eq. (5.12), the time average of $x_{\alpha}^{s}(t)$ is given by

$$\overline{x_{\alpha}^{s}(t)}^{t} = \tilde{x}_{\alpha\beta_0} a_{\alpha\beta_0} + \tilde{x}_{\alpha\beta_0}^{*} a_{\alpha\beta_0}^{*}, \tag{5.22}$$

where β_0 is defined so that

$$\omega_{\alpha\beta_0} = 0. \tag{5.23}$$

An entirely analogous expression follows of course for $\overline{f_{\alpha}^{s}(t)}^{t}$. Again, the coefficients $\tilde{x}_{\alpha\beta}$, $\tilde{f}_{\alpha\beta}$ depend in principle on the field variables $a_{\alpha\beta}$, and hence the expressions (5.19a) and (5.19c) should not be understood as explicit expansions on such variables. It should be noticed that in (5.19b), E_{α} does not represent the expansion of the free field, but rather the expansion of the field which contains only those modes of frequencies that are relevant for the dynamics of the mechanical system when the latter has attained the stationary state α. The rest of components generate a background noise that is being neglected, on account of the strong response to the relevant frequencies.

One of the main goals of the present theory is to determine the set $\{\omega_{\alpha\beta}\}$ for a given stationary state α; consequently we must in principle allow $\omega_{\alpha\beta}$ to acquire both positive and negative values. Therefore the expansions (5.19a, 5.19b, 5.19c), at variance with those of the form (5.15), do not necessarily refer to explicit expansions in terms of positive and negative frequencies. The transition $\sum_k \to \sum_\beta$ may thus require a reordering of terms in the sums.

In order to determine the equation for the resonance frequencies for a given state α, we introduce the expansions (5.19a, 5.19b, 5.19c) into Eq. (5.9), which thus becomes

$$\begin{aligned} &-\sum_{\beta} m\omega_{\alpha\beta}^{2} \tilde{x}_{\alpha\beta} a_{\alpha\beta} e^{i\omega_{\alpha\beta}t} + \text{c.c.} \\ &= \sum_{\beta} (\tilde{f}_{\alpha\beta} - im\tau\omega_{\alpha\beta}^{3}\tilde{x}_{\alpha\beta} + e\tilde{E}_{\alpha\beta}) a_{\alpha\beta} e^{i\omega_{\alpha\beta}t} + \text{c.c.} \end{aligned} \tag{5.24}$$

Thus for each relevant frequency $\omega_{\alpha\beta}$,

$$-m\omega_{\alpha\beta}^{2}\tilde{x}_{\alpha\beta}e^{i\omega_{\alpha\beta}t} = \tilde{f}_{\alpha\beta}e^{i\omega_{\alpha\beta}t} - im\tau\omega_{\alpha\beta}^{3}\tilde{x}_{\alpha\beta}e^{i\omega_{\alpha\beta}t} + e\tilde{E}_{\alpha\beta}e^{i\omega_{\alpha\beta}t}. \tag{5.25}$$

This equation can be rewritten as

$$m\frac{d^{2}\tilde{x}_{\alpha\beta}(t)}{dt^{2}} = \tilde{f}_{\alpha\beta}(t) + m\tau\frac{d^{3}\tilde{x}_{\alpha\beta}(t)}{dt^{3}} + e\tilde{E}_{\alpha\beta}(t), \tag{5.26}$$

with

$$\tilde{A}_{\alpha\beta}(t) \equiv \tilde{A}_{\alpha\beta} e^{i\omega_{\alpha\beta} t}. \tag{5.27}$$

From Eq. (5.25) one obtains, in analogy with (5.17a, 5.17b),

$$\tilde{x}_{\alpha\beta} = -\frac{e}{m} \frac{\tilde{E}_{\alpha\beta}}{\Delta_{\alpha\beta}}, \tag{5.28a}$$

$$\Delta_{\alpha\beta} = \omega_{\alpha\beta}^2 - i\tau\omega_{\alpha\beta}^3 + \frac{\tilde{f}_{\alpha\beta}}{m\tilde{x}_{\alpha\beta}}. \tag{5.28b}$$

This means that the mechanical system in state α responds resonantly to those frequencies that solve, in analogy with Eq. (5.18), the system of equations

$$\omega_{\alpha\beta}^2 \approx -\frac{\tilde{f}_{\alpha\beta}}{m\tilde{x}_{\alpha\beta}}. \tag{5.29}$$

Equation (5.29) represents in general a system of coupled nonlinear equations for the $\tilde{x}_{\alpha\beta}$'s and $\omega_{\alpha\beta}$'s. Each one of the quantities $\omega_{\alpha\beta}$, $\tilde{x}_{\alpha\beta}$ and $\tilde{f}_{\alpha\beta}$ depends (for nonlinear systems) on other members of the same set. In the present condensed writing all the nonlinearities of the problem are concealed.

Those relevant frequencies $\omega_{\alpha\beta}$ that satisfy the above equation are precisely the resonance frequencies corresponding to the state α—the ones that determine the main contributions to $x_\alpha^s(t)$—which of course constitute a subset of the solutions of Eq. (5.18). As seen from Eq. (5.19c), the remaining relevant frequencies are just the oscillation frequencies that contribute dominantly to the expansion of an arbitrary power series of x (or p, as will be clear from Appendix B); therefore, the entire set of frequencies that are of relevance for a particular problem is in general larger than the set of resonance frequencies. As we go ahead, the meaning of the relevant frequencies will be further clarified. In particular, in Appendix B a specific example is studied and the problem of determining the relevant frequencies for higher powers of x is carried out in detail. At the root of the resonant behavior is the fact that the field contains modes of all frequencies, each with a relatively large spatial and temporal coherence and thus able to act coherently on the particle for times longer than the relaxation time of the system.

The fact that the equation of motion (5.24) decomposes into Eq. (5.26) and its complex conjugate, allows us to restrict the study to the solutions of (5.26) only. We thus observe that this latter is the detailed form, satisfied term by term, of the equation

$$m\ddot{x}_\alpha = f_\alpha + m\tau\dddot{x}_\alpha + eE_\alpha, \tag{5.30}$$

with

$$x_\alpha(t) = \sum_\beta \tilde{x}_{\alpha\beta} a_{\alpha\beta} e^{i\omega_{\alpha\beta} t}, \tag{5.31a}$$

$$f_\alpha(t) = \sum_\beta \tilde{f}_{\alpha\beta} a_{\alpha\beta} e^{i\omega_{\alpha\beta} t}, \tag{5.31b}$$

$$E_\alpha(t) = \sum_\beta \tilde{E}_{\alpha\beta} a_{\alpha\beta} e^{i\omega_{\alpha\beta} t}. \tag{5.31c}$$

From this point on we will work with Eq. (5.30)—which is in direct correspondence with the original equation of motion—and with expansions of the form (5.31a, 5.31b, 5.31c), where the coefficients are given by the solutions of Eq. (5.25).

5.2 The Principle of Ergodicity

Due to the stochasticity of the field, the solutions (5.11) refer to a given realization (i) of the random ZPF. This means that instead of $x^{\mathrm{s}}(t)$, for example, one ought to write $x^{\mathrm{s}(i)}(t)$, where the index (i) (omitted for simplicity in the previous expansions) stresses the dependence of $x^{\mathrm{s}}(t)$ on the specific field realization through the random coefficients $a \equiv a(i)$.

The set $\{i\}$ of all the realizations of the field determines the ensemble of all possible realizations for an individual system. This statistical set can therefore be reproduced by considering not one, but an ensemble of particles, each of which is subject to a different realization of the field. Clearly such ensemble includes particles that, once stationarity has been attained, have reached one of the possible stationary states. Thus, when focusing on a specific state α, we are working only with a subensemble of particles, or equivalently, a subensemble $\{i\}_\alpha$ of the field realizations, such that $\{i\} = \bigcup_\alpha \{i\}_\alpha$.[7] Hence, for example, $x_\alpha(t)$ should be written in an explicitly realization-dependent form as $x_\alpha^{(i)}(t)$, with $(i) \in \{i\}_\alpha$.

According to the above, in what follows, when referring to a given stationary state α, the averages over the ensemble of realizations of the field (denoted as $\overline{(\cdot)}^{(i)}$) are considered to be taken over $(i) \in \{i\}_\alpha$; alternatively, they can be determined by

[7] The proposed decomposition is similar to the one that occurs in the harmonic oscillator case studied in Chap. 3. Thus, for a system of oscillators in equilibrium with the background field, one can resort to Eq. (3.84) applied to $f(\mathcal{E}) = \mathcal{E}$ to write the mean energy of the ensemble as $\langle \mathcal{E} \rangle = \sum_n w_n \mathcal{E}_n = \sum_i P_i \mathcal{E}^{(i)} = \overline{\mathcal{E}^{(i)}}^{(i)}$, where $\mathcal{E}_n$ and w_n are given by Eqs (3.80) and (3.85), respectively, and P_i is the weight function (with respect to the whole event space $\{i\}$) associated with the specific realization (i). According to the discussion following Eq. (3.85), the index n distinguishes among the different stationary states accessible to the mechanical subsystem (thus n here plays the role of α). By contrast, consider an ensemble of classical (Brownian) harmonic oscillators with $\mathcal{E}_0 = 0$. In this case the decomposition $\langle \mathcal{E} \rangle = \sum_{n=0}^{\infty} w_n \mathcal{E}_n$ becomes trivial, since the only stationary state corresponds to $\mathcal{E}_0 = 0$.

averaging over the subensemble of particles (denoted as $\langle\cdot\rangle$) that have reached such state.

We now introduce a hypothesis that will turn out to be central for the development of the present theory, namely that once in a stationary state, the system has acquired ergodic properties. This is a natural assumption for a particle that performs a random but stationary, hence recurring motion when subject to a binding potential. It means that the system becomes recurrent in statistical terms, so that it eventually fills the available subspace surface. The ergodic principle implies that the time averages of the dynamical variables coincide with their corresponding field-realization (or ensemble) averages, that is,

$$\overline{g(i,t;\alpha)}^{t} = \overline{g(i,t;\alpha)}^{(i)} = \langle g(i,t;\alpha)\rangle \tag{5.32}$$

for every function g that depends on time and on the field realization, for a given stationary state α. This means that

$$\overline{g(i,t;\alpha)}^{t} \text{ is independent of } (i). \tag{5.33}$$

In the following we explore the consequences of introducing such condition. Throughout this section we make the i-dependence explicit in every quantity that depends on (i) when convenient, yet for simplicity in the writing we shall omit it in general in the rest of the sections.

Let us focus on an arbitrary dynamical variable A that in state α has the form

$$A_{\alpha}(t) = \sum_{\beta} \tilde{A}_{\alpha\beta} a_{\alpha\beta} e^{i\omega_{\alpha\beta}t}. \tag{5.34}$$

By decomposing (5.34) into its time-independent contribution plus oscillating terms that average to zero, one gets (recall from Eq. (5.23) that $\omega_{\alpha\beta_0} = 0$)

$$A_{\alpha}(t) = \tilde{A}_{\alpha\beta_0} a_{\alpha\beta_0} + \sum_{\beta(\neq\beta_0)} \tilde{A}_{\alpha\beta} a_{\alpha\beta} e^{i\omega_{\alpha\beta}t}, \tag{5.35}$$

whence

$$\overline{A_{\alpha}^{(i)}(t)}^{t} = \tilde{A}^{(i)}_{\alpha\beta_0} a^{(i)}_{\alpha\beta_0}. \tag{5.36}$$

The ergodic condition in its form (5.33) implies that the right hand side of (5.36) is i-independent, whence

$$\tilde{A}^{(i)}_{\alpha\beta_0} a^{(i)}_{\alpha\beta_0} = \tilde{A}_{\alpha\beta_0} a_{\alpha\beta_0}. \tag{5.37}$$

We will come back to this result below.

The second term in (5.35) corresponds to the deviations of $A_\alpha(t)$ from its mean value, and its modulus allows us to calculate the variance $\sigma^2_{A_\alpha}$ defined as

$$\sigma^2_{A_\alpha} = \overline{\left| A_\alpha - \overline{A_\alpha(t)}^t \right|^2}^t, \tag{5.38}$$

where

$$\begin{aligned} \left| A_\alpha - \overline{A_\alpha(t)}^t \right|^2 &= \sum_{\beta'(\neq\beta_0),\beta''(\neq\beta_0)} \tilde{A}_{\alpha\beta'} \tilde{A}^*_{\alpha\beta''} a_{\alpha\beta'} a^*_{\alpha\beta''} e^{i(\omega_{\alpha\beta'}-\omega_{\alpha\beta''})t} \\ &= \sum_{\beta'(\neq\beta_0)} \left| \tilde{A}_{\alpha\beta'} a_{\alpha\beta'} \right|^2 \\ &+ \sum_{\beta'(\neq\beta_0)\neq\beta''(\neq\beta_0)} \tilde{A}_{\alpha\beta'} \tilde{A}^*_{\alpha\beta''} a_{\alpha\beta'} a^*_{\alpha\beta''} e^{i(\omega_{\alpha\beta'}-\omega_{\alpha\beta''})t}. \end{aligned} \tag{5.39}$$

The first term in the second line is time-independent, whilst the remaining terms oscillate with frequency $\omega_{\alpha\beta'} - \omega_{\alpha\beta''} \neq 0$ for $\beta' \neq \beta''$. Thus, by identifying the non oscillating contribution in Eq. (5.39) with the time average of $\left| A_\alpha - \overline{A_\alpha(t)}^t \right|^2$ we obtain the following expression for the variance of A_α,

$$\sigma^{2(i)}_{A_\alpha} = \overline{\left| A^{(i)}_\alpha - \overline{A^{(i)}_\alpha(t)}^t \right|^2}^t = \sum_{\beta(\neq\beta_0)} \left| \tilde{A}^{(i)}_{\alpha\beta} a^{(i)}_{\alpha\beta} \right|^2. \tag{5.40}$$

As before, it follows from condition (5.33) that the right-hand side of Eq. (5.40) must be independent of the realization. Further, since according to Eq. (5.8) the modulus of $a^{(i)}_{\alpha\beta}$ is 1, we get

$$\sigma^2_{A_\alpha} = \sum_{\beta(\neq\beta_0)} \left| \tilde{A}^{(i)}_{\alpha\beta} \right|^2 \text{ independent of } (i). \tag{5.41}$$

This equation must be satisfied for every α and for any variable A_α that decomposes as (5.34). Moreover, the number of (statistically independent) terms that contribute to the sum depends on the particular system and on the specific variable A, hence the sum may be either finite or infinite. The solution to Eq. (5.41) is that *each* term of the respective sum is independent of the realization.[8] As to the term with $\beta = \beta_0$,

[8] For a finite number of terms the proposed solution is the general one. For an infinite sum of statistically independent terms, according to the central-limit theorem [see e.g. Papoulis (1991)] the variable $\sigma^2_{A_\alpha}$ follows a normal distribution. Yet the condition (5.41) implies that $\sigma^2_{A_\alpha}$ must be a sure, nonstochastic variable, so that the normal distribution must have zero width. This implies that each term in the sum has zero variance, whence it is a sure quantity.

we can resort to Eqs. (5.8) and (5.37) to conclude that $\left|\tilde{A}^{(i)}_{\alpha\beta_0}\right|^2$ is also independent of i. This allows us to write

$$\tilde{A}^{(i)}_{\alpha\beta} = \tilde{A}_{\alpha\beta} e^{i\theta^{(i)}_{\alpha\beta}(A)} \tag{5.42}$$

for all β, with $\tilde{A}_{\alpha\beta}$ a nonstochastic (sure) complex number, and $\theta^{(i)}_{\alpha\beta}(A)$ a stochastic phase that, in principle, depends on the variable A. In what follows we draw important conclusions regarding such phase.

Consider two variables A_1 and A_2 that in the stationary state α are given by expansions of the form (5.34). Clearly the sum $A_{3\alpha} = A_{1\alpha} + A_{2\alpha}$ can also be expanded as in (5.34), so that the coefficient $\tilde{A}^{(i)}_{3\alpha\beta}$ corresponding to the frequency $\omega_{\alpha\beta}$ is given by

$$\tilde{A}^{(i)}_{3\alpha\beta} = \tilde{A}^{(i)}_{1\alpha\beta} + \tilde{A}^{(i)}_{2\alpha\beta}. \tag{5.43}$$

Multiplying this equation by its complex conjugate and using Eq. (5.42) for each $\tilde{A}^{(i)}_{n\alpha\beta}$ $(n = 1, 2, 3)$, one obtains

$$\left|\tilde{A}_{3\alpha\beta}\right|^2 = \left|\tilde{A}_{1\alpha\beta}\right|^2 + \left|\tilde{A}_{2\alpha\beta}\right|^2 + 2\mathrm{Re}\tilde{A}_{1\alpha\beta}\tilde{A}^*_{2\alpha\beta} e^{i\Theta^{(i)}_{\alpha\beta}}, \tag{5.44}$$

with

$$\Theta^{(i)}_{\alpha\beta} = \theta^{(i)}_{\alpha\beta}(A_1) - \theta^{(i)}_{\alpha\beta}(A_2). \tag{5.45}$$

Since the left-hand side of (5.44) is a sure quantity, and $\Theta^{(i)}_{\alpha\beta}$ is the only realization-dependent term on the right-hand side, the equation requires $\Theta^{(i)}_{\alpha\beta}$ to be a sure, nonstochastic phase. This means that for all α, β, and irrespective of A_1 and A_2, the stochasticity of $\theta^{(i)}_{\alpha\beta}(A_1)$ is the same as that of $\theta^{(i)}_{\alpha\beta}(A_2)$, hence $\theta^{(i)}_{\alpha\beta}$ in Eq. (5.42) is the same for all A. This result implies that the (common) stochasticity of $\tilde{A}^{(i)}_{\alpha\beta}$ is physically irrelevant when considering equations involving coefficients of the form $\tilde{A}^{(i)}_{\alpha\beta}$, whence we may drop the phase $\theta^{(i)}_{\alpha\beta}$ and write

$$\tilde{A}^{(i)}_{\alpha\beta} = \tilde{A}_{\alpha\beta}. \tag{5.46}$$

In particular, putting A_α equal to x_α and $\dot{x}_\alpha$, we obtain, respectively,

$$\tilde{x}^{(i)}_{\alpha\beta} = \tilde{x}_{\alpha\beta}, \quad \tilde{\dot{x}}^{(i)}_{\alpha\beta} = \tilde{\dot{x}}_{\alpha\beta} = i\omega_{\alpha\beta}\tilde{x}_{\alpha\beta}. \tag{5.47}$$

With Eq. (5.46) at our disposal, we are in a position to return to Eqs. (5.37) and (5.21) and obtain

$$a^{(i)}_{\alpha\beta_0} = a_{\alpha\beta_0}, \tag{5.48}$$

$$\varphi^{(i)}_{\alpha\beta_0} = \varphi_{\alpha\beta_0}. \tag{5.49}$$

Equation (5.46) is a most notable outcome of the principle of ergodicity. The meaning and implications of it are examined below when applied, in particular, to the coefficients of expansions such as $(x^n)_\alpha$. The fact that the $\tilde{A}_{\alpha\beta}$ (in particular $\tilde{x}_{\alpha\beta}$) and the $\omega_{\alpha\beta}$ are nonrandom variables in the present approximation, leads to the conclusion that Eqs. (5.31a, 5.31b, 5.31c) are *explicit* expansions in the variables $a^{(i)}_{\alpha\beta}$, i.e., *linear* functions of the stochastic components of the field. In particular, with $\tilde{x}_{\alpha\beta}$ given by (5.28a), we have

$$x^{(i)}_\alpha(t) = -\frac{e}{m}\sum_\beta \frac{\tilde{E}_{\alpha\beta}}{\Delta_{\alpha\beta}} a^{(i)}_{\alpha\beta} e^{i\omega_{\alpha\beta}t}. \tag{5.50}$$

For this reason, the theory that ensues as a result of the condition of ergodicity is called *Linear Stochastic Electrodynamics* (LSED).[9]

5.2.1 The Chain Rule

The ergodic properties just discussed have far-reaching consequences, both for the physical behavior of the system and for the mathematical formalism used to describe it. We focus first on the latter and leave the former for the following sections. Since the derivations are quite lengthy they are presented in Appendix A; the following is a summary of the main results.

A central aspect relates to the way in which the expansion of the force, or more generally of a nonlinear function of x, must be constructed when the mechanical system is in the stationary state α, so as to comply with the demand of ergodicity. Take, for example, the expression for the variable x^2,

$$(x^2)_\alpha = \sum_\beta \widetilde{(x^2)}_{\alpha\beta} a_{\alpha\beta} e^{i\omega_{\alpha\beta}t}. \tag{5.51}$$

As shown in Appendix A [see Eq. (A.11)], for Eq. (5.46) to be satisfied the coefficients in this expansion must be given by

[9] The foundations of LSED can be traced to the early papers by de la Peña and Cetto (1991–1995) and Cetto and de la Peña (1991). A detailed account of this initial stage of the theory can be found in *The Dice*, Chap. 10. As mentioned there, this line of research was motivated by the need to solve some of the critical challenges faced by SED in the 1980-90s. More recent work, as of de la Peña and Cetto (1999, 2001), and especially the references cited in footnote 1, deal with a more developed form of the theory that in some aspects differs noticeably from the original one.

$$(\widetilde{x^2})_{\alpha\beta} = \widetilde{x^2}(\omega_{\alpha\beta}) = \sum_{\beta'} \tilde{x}(\omega_{\alpha\beta'})\tilde{x}(\omega_{\alpha\beta} - \omega_{\alpha\beta'}), \tag{5.52}$$

and the stochastic variables must fulfill Eq. (A.10), namely (in this and the following expressions no summation over repeated indices is assumed)

$$a(\omega_{\alpha\beta'})a(\omega_{\alpha\beta} - \omega_{\alpha\beta'}) = a(\omega_{\alpha\beta}). \tag{5.53}$$

Equation (5.52) shows that the dominant (nonnoisy) contributions to $(x^2)_\alpha$ come from those frequencies $\omega_{\alpha\beta}$ such that $\omega_{\alpha\beta} - \omega_{\alpha\beta'}$ and $\omega_{\alpha\beta'}$ are resonance frequencies (i.e., frequencies corresponding to dominant contributions to x_α). The condition that both $\omega_{\alpha\beta} - \omega_{\alpha\beta'}$ and $\omega_{\alpha\beta'}$ satisfy an equation of the form (5.29) is the one that selects, from the complete set of relevant frequencies $\{\omega_{\alpha\beta}\}$, those that are important for the variable $(x^2)_\alpha$. For other variables this condition will change, and other relevant frequencies will come into play, as shown in the example presented in Appendix B.

It is further argued in Appendix A that the summation indices (β) in the above expansions are much more than mere labels introduced to distinguish between the different relevant frequencies $\omega_{\alpha\beta}$. Instead they are endowed with a physical meaning by denoting additional stationary states accesible to the mechanical system, α being only one of them (hence $\alpha \in \{\beta\}$). Consequently, new frequencies $\omega_{\beta'\beta''}$ (with β', $\beta'' \in \{\beta\}$) appear in the description, satisfying the relations [Eqs. (A.28) and (A.29)]

$$\omega_{\beta\alpha} = -\omega_{\alpha\beta}, \tag{5.54a}$$

$$\omega_{\alpha\beta'} + \omega_{\beta'\beta''} + \cdots + \omega_{\beta^{(n-1)}\beta} = \omega_{\alpha\beta}. \tag{5.54b}$$

The antisymmetry of $\omega_{\alpha\beta}$ suggests writing it in the form (see A.27)

$$\omega_{\alpha\beta} = \Omega_\alpha - \Omega_\beta, \tag{5.55}$$

with the physical meaning of the parameters $\Omega_{\alpha,\beta}$ to be determined in Sect. 5.4.1. Equation (5.54b) involves an arbitrary number of relevant frequencies $\omega_{\beta^{(n)}\beta^{(m)}}$ that do not necessarily represent resonance frequencies, indicating that appropiate linear combinations of relevant frequencies result in another relevant frequency (see Appendix B).

From Eqs. (5.54a, 5.54b) one obtains, in particular,

$$\omega_{\alpha\beta} + \omega_{\beta\alpha} = \omega_{\alpha\alpha} = 0, \tag{5.56}$$

whence the index β_0 introduced in Eq. (5.23) is just $\beta_0 = \alpha$. It also follows from (5.54a, 5.54b) that $\omega_{\alpha\beta} - \omega_{\alpha\beta'} = \omega_{\beta'\beta}$, whence Eq. (5.52) reduces to [see Eq. (A.37)]

$$(\widetilde{x^2})_{\alpha\beta} = \sum_{\beta'} \tilde{x}(\omega_{\alpha\beta'})\tilde{x}(\omega_{\beta'\beta}) = \sum_{\beta'} \tilde{x}_{\alpha\beta'}\tilde{x}_{\beta'\beta}, \tag{5.57}$$

and (5.53) becomes

$$a_{\alpha\beta'}a_{\beta'\beta} = a_{\alpha\beta}. \tag{5.58}$$

This last result is readily generalized to an arbitrary number of factors by a successive (chained) application of it, [see Eq. (A.31)]

$$a_{\alpha\beta'}a_{\beta'\beta''}a_{\beta''\beta'''}\cdots a_{\beta^{(n-1)}\beta} = a_{\alpha\beta}. \tag{5.59}$$

As shown in the appendix (Eq. A.34) it follows that $a_{\alpha\beta}$ is of the form

$$a_{\alpha\beta} = e^{i\varphi_{\alpha\beta}} = e^{i(\phi_\alpha - \phi_\beta)}, \tag{5.60}$$

with ϕ_λ a random phase. Thus,

$$a_{\alpha\beta} = a^*_{\beta\alpha}, \tag{5.61}$$

which combined with (5.59) gives

$$a_{\alpha\beta}a_{\beta\alpha} = \left|a_{\alpha\beta}\right|^2 = a_{\alpha\alpha} = 1. \tag{5.62}$$

Equations (5.54b) and (5.59) are the *chain rules* established for the frequencies and the corresponding stochastic variables, respectively. It is important to note that they ultimately ensue from the ergodic demand, specifically as a consequence of imposing condition (5.46), and turn out to be decisive on the mathematical structure of the description, as will be seen in the next section. Equation (5.59) tells us that the amplitudes of the field pertaining to relevant modes become partially correlated, indicating that not only the material part, but also the near background field is affected during the evolution of the complete system towards equilibrium.

5.2.2 Matrix Algebra

We now come back to Eq. (5.57), namely

$$(\widetilde{x^2})_{\alpha\beta} = \sum_{\beta'} \tilde{x}_{\alpha\beta'}\tilde{x}_{\beta'\beta}, \tag{5.63}$$

and observe that the right-hand side embodies the rule for matrix multiplication, with the (square) matrix $\hat{x}$ having as elements the coefficients $\tilde{x}_{\beta^{(n)}\beta^{(m)}}$ $(\beta^{(n)}, \beta^{(m)} \in \{\beta\})$. The matrix $\hat{x}$ thus defined is Hermitian, as follows from considering the expressions for $x_\alpha(t)$ and $x^*_\alpha(t)$,

$$x_\alpha(t) = \sum_\beta \tilde{x}_{\alpha\beta} a_{\alpha\beta} e^{i\omega_{\alpha\beta} t}, \tag{5.64a}$$

$$x^*_\alpha(t) = \sum_\beta \tilde{x}^*_{\alpha\beta} a_{\beta\alpha} e^{i\omega_{\beta\alpha} t}, \tag{5.64b}$$

where the second line follows from (5.54a) and (5.61). Indeed, these equations show that with respect to $a_{\alpha\beta}$ and $\omega_{\alpha\beta}$, the conjugation amounts to an exchange of the indices α and β, whence

$$\tilde{x}^*_{\alpha\beta} = \tilde{x}_{\beta\alpha}. \tag{5.65}$$

The antisymmetry of $\omega_{\alpha\beta}$ indicates that this result is consistent with the property $\tilde{x}^*(\omega_{\alpha\beta}) = \tilde{x}(-\omega_{\alpha\beta})$, satisfied by the coefficients $\tilde{x}_{\alpha\beta} = \tilde{x}(\omega_{\alpha\beta})$ of the expansion (5.19a) for the real variable x^s_α [see also Eq. (5.13)].

In line with Eq. (A.39), the expression (5.57) can be generalized to higher powers of x, thus obtaining

$$(\widetilde{x^n})_{\alpha\beta} = \sum_{\beta' \dots \beta^{(n-1)}} \tilde{x}_{\alpha\beta'} \tilde{x}_{\beta'\beta''} \dots \tilde{x}_{\beta^{(n-1)}\beta}, \tag{5.66}$$

whence for the n-th power of x in the state α we have

$$(x^n)_\alpha = \sum_\beta (\widetilde{x^n})_{\alpha\beta} a_{\alpha\beta} e^{i\omega_{\alpha\beta} t}, \tag{5.67}$$

with $(\widetilde{x^n})_{\alpha\beta}$ given by the element $\alpha\beta$ of the corresponding matrix product, $(\widetilde{x^n})_{\alpha\beta} = (\hat{x}^n)_{\alpha\beta}$.

With the chain rule applied to the frequencies, Eq. (5.54b), one may transfer the time dependence to every single factor $\tilde{x}$ in Eq. (5.67) and define an *evolving* matrix $\hat{x}(t)$ as follows,

$$\begin{aligned}
(x^n)_\alpha &= \sum_\beta \left(\sum_{\beta', \beta'' \dots \beta^{(n-1)}} \tilde{x}_{\alpha\beta'} \tilde{x}_{\beta'\beta''} \dots \tilde{x}_{\beta^{(n-1)}\beta} \right) a_{\alpha\beta} e^{i\omega_{\alpha\beta} t} \\
&= \sum_\beta \left[\sum_{\beta', \dots, \beta^{(n-1)}} (\tilde{x}_{\alpha\beta'} e^{i\omega_{\alpha\beta'} t})(\tilde{x}_{\beta'\beta''} e^{i\omega_{\beta'\beta''} t}) \dots (\tilde{x}_{\beta^{(n-1)}\beta} e^{i\omega_{\beta^{(n-1)}\beta} t}) \right] a_{\alpha\beta} \\
&= \sum_\beta \left[\hat{x}^n(t) \right]_{\alpha\beta} a_{\alpha\beta},
\end{aligned} \tag{5.68}$$

with $\hat{x}_{\alpha\beta}(t) = \tilde{x}_{\alpha\beta} e^{i\omega_{\alpha\beta} t}$.

The above results hold also for arbitrary powers of $\dot{x}(t)$, so that

$$\left(\dot{x}^n\right)_\alpha = \sum_\beta (\widehat{\dot{x}^n})_{\alpha\beta} a_{\alpha\beta} e^{i\omega_{\alpha\beta}t} \tag{5.69}$$

with $\widehat{\dot{x}}_{\alpha\beta} = i\omega_{\alpha\beta}\tilde{x}_{\alpha\beta}$ the $\alpha\beta$ element of the Hermitian matrix $\widehat{\dot{x}}$. Thus, every dynamical variable A that can be expressed as a power series of x or $\dot{x}$—or, more generally, as a power series of the form $h(x) + g(\dot{x})$—has a square Hermitian matrix $\hat{A}$ associated with it, with elements $\alpha\beta$ given by the coefficients $\tilde{A}_{\alpha\beta}$. Functions that are expressed in terms of products of the form $\Sigma_{nm} c_{nm} x^n p^m$ are left aside for the time being, since a rule for the ordering of the corresponding matrices would be required. In the following the matrix elements $\tilde{A}_{\alpha\beta}$ in expansions such as (5.19a) and (5.31a, 5.31b, 5.31c) will be simply denoted as $A_{\alpha\beta}$. In state α the expansion for A reads therefore

$$A_\alpha(t) = \sum_\beta A_{\alpha\beta} a_{\alpha\beta} e^{i\omega_{\alpha\beta}t}. \tag{5.70}$$

Moreover, the matrix $\hat{A}$ can absorb the time dependence, so that

$$A_{\alpha\beta}(t) = A_{\alpha\beta} e^{i\omega_{\alpha\beta}t}. \tag{5.71}$$

Extracting from the sum (5.70) the element corresponding to $\beta = \alpha$ and using Eqs. (5.56) and (5.62), one gets

$$A_\alpha(t) = A_{\alpha\alpha} + \sum_{\beta\neq\alpha} A_{\alpha\beta} a_{\alpha\beta} e^{i\omega_{\alpha\beta}t}, \tag{5.72}$$

which shows that the diagonal element $A_{\alpha\alpha}$ coincides with the mean value

$$A_{\alpha\alpha} = \overline{A_\alpha(t)}^t = \langle A_\alpha \rangle . \tag{5.73}$$

With the above results, the well-known (formal) correspondence between physical variables and (Hermitian) operators is established. Contrary to what happens in the usual formalism, however, these results point to a direct and well-defined connection between a matrix $\hat{A}$ and the associated variable A. Indeed, the expansion (5.70) expresses the relation between the variable $A(t)$ in the state α and the α-th row of the matrix $\hat{A}(t)$. Thus, it is the set $\{A_\alpha(t)\}$ (including *all* states α) what is in correspondence with the matrix $\hat{A}(t)$. In particular, since the set of stationary solutions of Eq. (5.9) contains all the possible stationary states (recall the discussion at the beginning of Sect. 5.2), we conclude that the matrix $\hat{x}$ is in direct correspondence with the (family of) solutions of the original equation of motion (5.9).

It should be borne in mind that the $A_\alpha(t)$ given by (5.70) still contains the random amplitudes $a_{\alpha\beta}$, which depend on the realization (i). Therefore the present description refers to an element of a (sub)ensemble of particles, and the information contained in it must be considered only as representative of this ensemble. Consequently the quantum operator $\hat{A}$ does not correspond, as is usually assumed, to the variable $A(t)$ of a *single* system; rather, it encapsulates the dynamical information that is common to all members of the ensemble.

While according to (5.73) the diagonal elements of the matrix $\hat{A}$ carry information about the mean values of the corresponding variable A in all possible stationary states, the nondiagonal elements bear information about the deviations from those mean values. This can be seen by resorting to Eq. (5.41) (with $\tilde{A}^{(i)}_{\alpha\beta} = \tilde{A}_{\alpha\beta} = A_{\alpha\beta}$ and $\beta_0 = \alpha$, as explained above) for the variance of A in state α,

$$\sigma^2_{A_\alpha} = \sum_{\beta(\neq\alpha)} \left|A_{\alpha\beta}\right|^2 . \tag{5.74}$$

Since in Eq. (5.70) these deviations are accompanied by the stochastic field amplitudes, we conclude that the nondiagonal elements of $\hat{A}$ are nontrivial as a consequence of the presence of the ZPF. We refer the reader to Sect. 5.4.4 for a discussion on the relation between fluctuations and the off-diagonal elements of $\hat{A}$.

5.3 Physical Consequences of the Ergodic Principle

5.3.1 Establishing Contact with Quantum Theory

In the preceding sections we have seen that the chain rule deriving from the ergodic principle entails a matrix algebra for the generic amplitudes $A_{\alpha\beta}$, which are the coefficients in the expansion of $A_\alpha(t)$. With these results Eq. (5.26) takes the form

$$m\frac{d^2\hat{x}(t)}{dt^2} = \hat{f}(t) + m\tau\frac{d^3\hat{x}(t)}{dt^3} + e\hat{E}(t). \tag{5.75}$$

It should be stressed that (5.75) is much more than a new form of writing Eq. (5.26). As a result of the ergodic demand, neither $x_{\alpha\beta}$ ($\hat{x}$) nor $f_{\alpha\beta}$ ($\hat{f}$) depend on the $a_{\alpha\beta}$'s, and hence the random variables have been cancelled out. That is, the stochastic equation of motion (5.30), written in terms of c-(random) numbers, has been transformed into a nonstochastic matrix equation (q-numbers). Yet in contrast with (5.30), which refers to a specific stationary state α, Eq. (5.75) involves matrices that contain information about all possible stationary states.

Equation (5.75) goes beyond the quantum-mechanical description; it is the equation of motion proper of nonrelativistic QED. As will be discussed in Chap. 6, the radiative terms in (5.75) prevent any excited state from having an infinite lifetime.[10] This confirms our conclusion that the transition to QM, and hence to a description that admits (formally) truly stationary states, entails neglecting the radiative terms in Eq. (5.75). The radiationless approximation must be made (and will be made in the following section), in order to obtain the quantum-mechanical description, made in terms of stationary states that satisfy the ergodic (recurrent) condition. This should be compared with the results of Chap. 4, where stationarity (a recurrent behavior for bounded systems) is reached under the condition of energy balance. Both conditions can be captured under the broader term *quantum regime*, already introduced in Chap. 4.

The fact that we have arrived at Eq. (5.75), which bears information regarding the dynamics of the mechanical system in the ergodic regime, indicates the convergence of the present theory and (nonrelativistic, spinless) QED. We recall that the results of Chap. 3 already pointed to the conclusion that, when the ZPF is duly taken into account, the field in equilibrium with matter [which is the field entering in Eq. (5.75)] appears quantized. It should be clear by now, however, that the equivalence between the descriptions afforded by QED and LSED refers to their formal features; they are conceptually distinct theories, with important differences in their physical outlook.[11]

5.3.2 The Radiationless Approximation

Since in the quantum regime the contribution of the radiative terms in the equation of motion (5.75) reduces to radiative corrections, these terms can be neglected in a first approximation, whence Eq. (5.75) becomes

$$m\frac{d^2\hat{x}(t)}{dt^2} = \hat{f}(t). \tag{5.76}$$

It must be noticed that even though this equation is the result of a legitimate approximation, it differs crucially from Eq. (5.75) in the sense that it lacks any reference to the field that has driven the system to the quantum regime. With the disappearance of the cause of the quantum behavior, the description becomes acausal, and with the disappearance of any reference to the background field it becomes an abstract expression; quantum mechanics is then (futilely) doomed to explain itself by its own bootstraps.

---

[10] In QM it is customary to call *stationary* the eigenstates of the time-independent (stationary) Schrödinger equation. Strictly speaking, all excited atomic eigenstates have finite lifetimes. Since the atomic lifetimes are $\sim 10^6$ times a typical atomic period, such states can be appropriately called quasi-stationary.

[11] Recall that the present theory has focused on the description of the states attained by the mechanical system only, leaving aside the description of the evolution of the field. Therefore the stated equivalence between the present theory and QED refers basically to the mechanical subsystem.

In the radiationless approximation the Hamiltonian matrix for the particle reduces to

$$\hat{H} = \frac{\hat{p}^2}{2m} + V(\hat{x}), \tag{5.77a}$$

and $\hat{x}$ and $\hat{p}$ satisfy the equations

$$\hat{p} = m\frac{d\hat{x}}{dt}, \quad \frac{d\hat{p}}{dt} = \hat{f}. \tag{5.78}$$

Recalling that the elements of $\hat{A}(t)$ are $A_{\alpha\beta}(t) = A_{\alpha\beta}e^{i\omega_{\alpha\beta}t}$ [see Eq. (5.71)], this gives

$$\begin{aligned} p_{\alpha\beta} &= im\omega_{\alpha\beta}x_{\alpha\beta}, \\ f_{\alpha\beta} &= i\omega_{\alpha\beta}p_{\alpha\beta}. \end{aligned} \tag{5.79}$$

With these results one may proceed to construct the law of evolution for the matrix $\hat{A}(t)$ associated with a dynamical variable A. On one hand, by taking the time derivative of $A_\alpha(t)$ and using Eqs. (5.71), (5.55), one arrives at

$$\begin{aligned} \frac{dA_\alpha(t)}{dt} &= \sum_\beta i\left[\Omega_\alpha A_{\alpha\beta}(t) - \Omega_\beta A_{\alpha\beta}(t)\right]a_{\alpha\beta} \\ &= \sum_\beta i\left[\hat{\Omega}, \hat{A}(t)\right]_{\alpha\beta} a_{\alpha\beta}, \end{aligned} \tag{5.80}$$

where $\hat{\Omega}$ is the diagonal matrix with elements

$$\Omega_{\alpha\beta} = \Omega_\alpha\delta_{\alpha\beta}. \tag{5.81}$$

On the other hand, one can write

$$\frac{dA_\alpha(t)}{dt} = \left(\frac{dA(t)}{dt}\right)_\alpha = \sum_\beta \dot{A}_{\alpha\beta}(t)a_{\alpha\beta}. \tag{5.82}$$

Equations (5.80) and (5.82) taken together give the evolution law $\dot{A}_{\alpha\beta} = i\left[\hat{\Omega}, \hat{A}(t)\right]_{\alpha\beta}$, or in closed matrix notation,

$$i\frac{d\hat{A}(t)}{dt} = \left[\hat{A}(t), \hat{\Omega}\right]. \tag{5.83}$$

This result is a direct consequence of the structure of the expansion of $A_\alpha(t)$ and the antisymmetry of the frequencies $\omega_{\alpha\beta}$, a property that is therefore at the root of an evolution law involving an algebra of commutators.

Equation (5.83) shows that the matrix $\hat{\Omega}$ plays a central role in determining the evolution of the mechanical subsystem. Since this latter is controlled by the Hamiltonian, it follows that the matrix $\hat{\Omega}$ must be related to $\hat{H}$. The relation between $\hat{\Omega}$ and $\hat{H}$, to be established below, acquires importance not only because it allows to express (5.83) in terms of $\hat{H}$, but also because it relates the relevant frequencies $\omega_{\alpha\beta} = \Omega_\alpha - \Omega_\beta$ with the elements of the Hamiltonian matrix.

As a first step for establishing the connection between $\hat{\Omega}$ and $\hat{H}$ we expand the (mechanical) Hamiltonian function in the state α in the form

$$H_\alpha = \mathcal{E}_\alpha + \sum_{\beta \neq \alpha} H_{\alpha\beta} a_{\alpha\beta} e^{i\omega_{\alpha\beta} t}, \tag{5.84}$$

since $H_{\alpha\alpha} = \overline{H_\alpha}^t = \mathcal{E}_\alpha$ [see Eq. (5.73)]. In the stationary radiationless regime H_α does not evolve in time and is a sure, nonstochastic quantity. Therefore,

$$\frac{dH_\alpha}{dt} = \sum_{\beta \neq \alpha} i\omega_{\alpha\beta} H_{\alpha\beta} a_{\alpha\beta} e^{i\omega_{\alpha\beta} t} = 0. \tag{5.85}$$

This equation holds only if $\omega_{\alpha\beta} H_{\alpha\beta} = 0$ for $\alpha \neq \beta$, whence

$$H_{\alpha\beta} = \mathcal{E}_\alpha \delta_{\alpha\beta}. \tag{5.86}$$

Since in the present description both $\hat{\Omega}$ and $\hat{H}$ are diagonal matrices, they relate to each other via a third diagonal matrix $\hat{D}$, so that,

$$\hat{\Omega} = \hat{D}\hat{H} = \hat{H}\hat{D}, \tag{5.87}$$

where the elements of $\hat{D}$ are functions, still to be determined, of the elements of $\hat{H}$, i.e.,

$$\hat{D}_{\alpha\beta} = \zeta_\alpha(\hat{H})\delta_{\alpha\beta}. \tag{5.88}$$

Equation (5.87) thus gives

$$\Omega_\alpha = \zeta_\alpha(\hat{H})\mathcal{E}_\alpha, \tag{5.89}$$

which combined with (5.55) results in

$$\omega_{\alpha\beta} = \zeta_\alpha \mathcal{E}_\alpha - \zeta_\beta \mathcal{E}_\beta. \tag{5.90}$$

We now introduce (5.87) into Eq. (5.83) applied to $\hat{A} = \hat{x}$, resort to (5.77a) and the first equation in (5.78), and arrive thus at

$$\frac{i}{m}\hat{p} = \hat{H}\left[\hat{x}, \hat{D}\right] + \frac{1}{2m}\left(\hat{C}\hat{p} + \hat{p}\hat{C}\right)\hat{D}, \tag{5.91}$$

with $\hat{C}$ the canonical commutator,

$$\hat{C} = \left[\hat{x}, \hat{p}\right]. \tag{5.92}$$

The first expression in (5.79), together with (5.86) and (5.88), shows that equation (5.91) establishes an algebraic relation between the relevant frequencies and the elements of the matrices $\hat{D}$, $\hat{H}$ and $\hat{C}$. Thus, by determining the elements of the matrix $\left[\hat{x}, \hat{p}\right]$ we will be able, after comparison of Eqs. (5.90) and (5.91), to disclose the relation between $\hat{\Omega}$ and $\hat{H}$. In the following we tackle the problem of calculating the commutator (5.92), to then come back and give final form to the evolution Eq. (5.83).

5.3.3 *The Canonical Commutator* $\left[\hat{x}, \hat{p}\right]$

To calculate the commutator $\hat{C} = \left[\hat{x}, \hat{p}\right]$ we start by writing its element $\alpha\beta$ explicitly, using the chain rule (5.54b) in the form $\omega_{\alpha\beta} = \omega_{\alpha\gamma} + \omega_{\gamma\beta}$. This gives

$$\begin{aligned} \omega_{\alpha\beta}C_{\alpha\beta} &= \sum_\gamma \left(x_{\alpha\gamma}p_{\gamma\beta} - p_{\alpha\gamma}x_{\gamma\beta}\right)\left(\omega_{\alpha\gamma} + \omega_{\gamma\beta}\right) \\ &= \sum_\gamma \left(x_{\alpha\gamma}p_{\gamma\beta}\omega_{\alpha\gamma} + x_{\alpha\gamma}p_{\gamma\beta}\omega_{\gamma\beta} - p_{\alpha\gamma}\omega_{\alpha\gamma}x_{\gamma\beta} - p_{\alpha\gamma}x_{\gamma\beta}\omega_{\gamma\beta}\right). \end{aligned} \tag{5.93}$$

The first and last terms in the second row cancel each other by virtue of the equations of motion (5.79), and we are left with

$$\omega_{\alpha\beta}C_{\alpha\beta} = -i\left[\hat{x}, \hat{f}(x)\right]_{\alpha\beta} = 0. \tag{5.94}$$

Since this must be satisfied for any pair $\alpha\beta$, and in general $\omega_{\alpha\beta} \neq 0$ (except for $\alpha = \beta$), $\hat{C}$ must be diagonal,

$$C_{\alpha\beta} = C_\alpha\delta_{\alpha\beta}. \tag{5.95}$$

We can now introduce this expression into Eq. (5.91) written in terms of the matrix elements, using also Eqs. (5.86) and (5.88). The result reads

$$\frac{i}{m}p_{\alpha\beta} = \mathcal{E}_\alpha x_{\alpha\beta}\left(\zeta_\beta - \zeta_\alpha\right) + \frac{1}{2m}p_{\alpha\beta}\zeta_\beta\left(C_\alpha + C_\beta\right). \tag{5.96}$$

With the aid of the first equation in (5.79) and (5.90) we obtain, after some rearrangements (which include dividing by the factor $x_{\alpha\beta}\zeta_\beta$, assumed to be nonzero)[12]

$$\omega_{\alpha\beta}\left(C_\alpha + C_\beta\right) = 2i\left(\mathcal{E}_\alpha - \mathcal{E}_\beta\right). \tag{5.97}$$

At this point we apply the chain rule (5.54b) to the frequencies in (5.97), i.e.,

$$\omega_{\alpha\gamma} + \omega_{\gamma\beta} = 2i\left(\frac{\mathcal{E}_\alpha - \mathcal{E}_\gamma}{C_\alpha + C_\gamma} + \frac{\mathcal{E}_\gamma - \mathcal{E}_\beta}{C_\beta + C_\gamma}\right) = \omega_{\alpha\beta} = 2i\frac{\mathcal{E}_\alpha - \mathcal{E}_\beta}{C_\alpha + C_\beta}. \tag{5.98}$$

For this equation to hold, the coefficient of $\mathcal{E}_\gamma$ in the middle term must vanish, which implies that

$$C_\alpha = C_\beta = C. \tag{5.99}$$

This is a most important result: it means that the commutator $\left[\hat{x}, \hat{p}\right]$ has a *universal* value, independent of the state, hence of the particular system,

$$\left[\hat{x}, \hat{p}\right] = C\mathbb{I}. \tag{5.100}$$

We now proceed to determine the value of the constant C. Due to its universality, for simplicity in the calculation we consider an ensemble of harmonic oscillators of frequency ω_0 embedded in the background field. With the help of Eqs. (5.28a) and (5.79), the nonzero (diagonal) elements of $\left[\hat{x}, \hat{p}\right]$ can be written as

$$C_\alpha = \left[\hat{x}, \hat{p}\right]_{\alpha\alpha} = i\frac{2e^2}{m}\sum_\gamma \frac{\omega_{\gamma\alpha}}{\left|\Delta_{\gamma\alpha}\right|^2}\tilde{E}^*_{\gamma\alpha}\tilde{E}_{\gamma\alpha}. \tag{5.101}$$

Since C_α does not depend on α, one can calculate the right-hand side of this equation assuming that α represents the oscillator's ground state, $\alpha = 0$. Now, for the oscillator to reach stationarity and remain in its ground state, also the background field must be in the ground state, which means that only the vacuum field, with spectral energy density

$$\rho_0(\omega) = \frac{\hbar\omega^3}{2\pi^2 c^3}, \tag{5.102}$$

is present (see discussion in Sect. 4.4.4).

Passing to the continuum with $\omega_{\gamma 0} \to \omega$ and[13]

---

[12] Notice that the assumption $x_{\alpha\beta} \neq 0$ needed to arrive at this expression implies that $\omega_{\alpha\beta}$ in (5.97) is a resonance frequency. Otherwise the coefficient $x_{\alpha\beta} = x_{\alpha\beta}(\omega_{\alpha\beta})$ would be negligible (and would not contribute significantly to the expansion of x_α).

[13] The correspondence $\tilde{E}^*(\omega_{\gamma\alpha})\tilde{E}(\omega_{\gamma\alpha}) \to (4\pi/3)\rho(\omega)$ follows by noticing that $\tilde{E}^*(\omega_{\gamma\alpha})\tilde{E}(\omega_{\gamma\alpha})$ gives the contribution of one Cartesian component of the field to the spectral energy density in a discrete representation, which corresponds just to $(4\pi/3)\rho(\omega)$ in the continuum description.

$$\tilde{E}^*_{\gamma 0}\tilde{E}_{\gamma 0} \to \frac{4\pi}{3}\rho_0(\omega), \tag{5.103}$$

one obtains

$$C = i\hbar\frac{2\tau}{\pi}\int_0^\infty \frac{\omega^4}{|\Delta(\omega)|^2}d\omega, \tag{5.104}$$

with

$$\Delta(\omega) = \omega^2 - i\tau\omega^3 + \tilde{g}(\omega), \quad \tilde{g}(\omega) = \frac{\tilde{f}(\omega)}{m\tilde{x}(\omega)}. \tag{5.105}$$

For the harmonic oscillator $f = -m\omega_0^2 x$, so that

$$\tilde{g}(\omega) = -\omega_0^2. \tag{5.106}$$

Inspection of (5.105) thus shows that the dominant contribution to the integral comes from the resonance at $\omega = \omega_0$. One thus obtains, in terms of a new variable $z = \tau\omega$,

$$C = i\hbar\frac{2}{\pi}\int_0^\infty \frac{z^4}{z^6 + (z+z_0)^2(z-z_0)^2}dz. \tag{5.107}$$

Further, due to the sharpness of the resonance one can replace z by z_0 everywhere in the integrand except in the difference $u = z - z_0$ and extend the integral from $-\infty$ to $+\infty$, thus obtaining

$$C = i\hbar\frac{z_0^2}{2\pi}\int_{-\infty}^\infty \frac{du}{u^2 + (z_0^4/4)} = i\hbar, \tag{5.108}$$

whence

$$\left[\hat{x}, \hat{p}\right] = i\hbar\mathbb{I}. \tag{5.109}$$

This calculation serves to emphasize the role played by both the ZPF and the resonant response of the particle in fixing the value of the commutator.[14] Moreover, since the $\tilde{E}_{\alpha\beta}$ bears information about the mean square fluctuations of the field, according to Eqs. (5.102) and (5.103), the derivation shows that the value of $\left[\hat{x}, \hat{p}\right]$ is determined *exclusively* by the fluctuations of the ZPF, and that it can therefore be considered a direct measure of the intensity of the fluctuations impressed by this field upon the particle.

[14] Milonni (1981) presents a similar derivation of the commutator for the free particle within QED.

On the other hand, the variances $\sigma^2_{x_\alpha}$ and $\sigma^2_{p_\alpha}$ satisfy the Robertson-Schrödinger inequality,

$$\sigma^2_{x_\alpha}\sigma^2_{p_\alpha} \geq \tfrac{1}{4}\left|\langle[\hat{x},\hat{p}]\rangle\right|^2 + \left|\langle\tfrac{1}{2}\{\hat{x},\hat{p}\} - \langle\hat{x}\rangle\langle\hat{p}\rangle\rangle\right|^2, \tag{5.110}$$

with

$$\{\hat{x},\hat{p}\} = \hat{x}\hat{p} + \hat{p}\hat{x}. \tag{5.111}$$

We assumed here that the variance of A_α is equal to the variance of the operator $\hat{A}$ when the system is in state α; this equivalence is demonstrated below [see Eq. (5.124)]. From this strictly mathematical relation it follows that

$$\sigma^2_{x_\alpha}\sigma^2_{p_\alpha} \geq \tfrac{1}{4}\left|\langle[\hat{x},\hat{p}]\rangle\right|^2 = \frac{\hbar^2}{4}, \tag{5.112}$$

hence,

$$\left(\sigma^2_{x_\alpha}\sigma^2_{p_\alpha}\right)_{\min} = \tfrac{1}{4}\left|C\right|^2 = \frac{\hbar^2}{4}. \tag{5.113}$$

With this result we generalize Eq. (3.102) (with x instead of q) derived for an ensemble of harmonic oscillators, to an arbitrary system. The result endows the Heisenberg inequalities with a deep meaning (see the related discussion in Sect. 5.4.2).

5.3.3.1 The Commutator as an Imprint of the Zero-Point Field

The above derivation of Eq. (5.109) shows that the presence of the ZPF is at the core of one of the most fundamental quantum formulas, $\left[\hat{x},\hat{p}\right] = i\hbar\mathbb{I}$. This observation is of importance because it points to the *physical* origin of the commutator, which in the radiationless approximation becomes concealed—and in usual QM is inexistent.

Since equation $\left[\hat{x},\hat{p}\right] = i\hbar\mathbb{I}$ is a physical law that applies once the quantum regime has been reached, prior to that point its absolute validity is not warranted.[15] Notice, by contrast, that the classical equation for the Poisson bracket $[x,p]_{\mathrm{PB}} = 1$ is an *identity*, an inviolable relation. Therefore, the correspondence $[x,p]_{\mathrm{PB}} = 1 \rightarrow (i\hbar)^{-1}\left[\hat{x},\hat{p}\right]$ is legitimate provided the quantum regime has been reached (as we assume when applying it, particularly in Chaps. 4 and 6).

The basic commutator serves to define a scale for the matrix elements $x_{\alpha\beta}$. We recall that in the original equation of motion (5.30), the Fourier components $\tilde{x}_{\alpha\beta}$ of the

[15] One may conceive of a dispersionless (for the particle) initial condition, when particle and field start interacting. The evolution of the system towards the quantum regime will repair the initial violation of the law $\left[\hat{x},\hat{p}\right] = i\hbar\mathbb{I}$. Intermediate situations should be possible, and their observation or not could eventually help to prove or disprove the present theory.

solutions x_α are proportional to the coefficients $\tilde{E}_{\alpha\beta}$ [see Eq. (5.28a)], which in their turn, as has been said above, bear information about the mean square fluctuations of the field. As mentioned earlier, such information gets lost in the passage from Eq. (5.75) to Eq. (5.76); this loss is repaired by resorting to the commutator (5.109). Indeed, from Eqs. (5.79) one obtains

$$2m \sum_\beta \omega_{\beta\alpha} \left| x_{\alpha\beta}(t) \right|^2 = \hbar, \tag{5.114}$$

an expression that fixes the scale of the $x_{\alpha\beta}$—which is equivalent to fixing the variance of the variables x and p by means of equations such as (5.124) below.

5.4 The Heisenberg Description

5.4.1 Heisenberg Equation, Representations, and Quantum Transitions

As anticipated at the end of Sect. 5.3.2, we are now in a position to write down the equation of motion (5.83) in terms of $\hat{H}$. This is achieved by introducing $C_\alpha = C_\beta = i\hbar$ in Eq. (5.97)—which ensued from Eq. (5.91)—to obtain

$$\omega_{\alpha\beta} = \hbar^{-1} \left(\mathcal{E}_\alpha - \mathcal{E}_\beta \right). \tag{5.115}$$

Comparison with (5.90) gives $\zeta_\alpha = \zeta_\beta = \hbar^{-1}$, whence (5.87) becomes

$$\hat{\Omega} = \hbar^{-1} \hat{H} \tag{5.116}$$

and the final form of Eq. (5.83) is therefore

$$i\hbar \frac{d\hat{A}(t)}{dt} = \left[\hat{A}(t), \hat{H} \right], \tag{5.117}$$

i.e., the Heisenberg equation for the operator $\hat{A}(t) = \hat{A}(\hat{x}(t), \hat{p}(t))$. Recall that we have limited the study to variables of the form $A(t) = A(x(t), p(t))$, with no explicit time dependence; this is consistent with the expansion (5.70) for $A_\alpha(t)$, and explains the absence of the term $i\hbar\partial\hat{A}/\partial t$ on the right-hand side of (5.117).

The present derivation shows the intimate connection between the commutator $\left[\hat{x}, \hat{p}\right]$ and the Heisenberg equation (5.117) governing the evolution of the dynamical variables. It also shows that it is precisely through the commutator that the signature of the ZPF is recovered, through the appearance of Planck's constant. The law of evolution (5.117) bears thus a hallmark of this pervading field.

Note that even though (5.117) has been derived resorting to the so-called energy representation, i.e. to a description in terms of stationary, energy-conserving states, in which $\hat{H}$ is diagonal, this law continues in force when a change of basis is performed via a unitary transformation U that turns $\hat{H}$ into a nondiagonal matrix $\hat{H}' = U^{\dagger}\hat{H}U$. Hence (5.117) is the general evolution law in the quantum regime, valid in any representation. This observation acquires relevance in those cases where the states of the system under study are characterized by well-defined values of some other dynamical variable G, whose matrix is not diagonal in the energy representation but becomes diagonal under a transformation U, so that $G_{\gamma\gamma'} = \mathcal{G}_{\gamma}\delta_{\gamma\gamma'}$. According to the discussion in Sect. 5.2, in this case the complete ensemble must be divided into a set of subensembles $\{i\}_{\gamma}$, each containing the particles in specific states γ, such that $\{i\} = \bigcup_{\gamma}\{i\}_{\gamma}$; this subdivision is different from $\{i\}_{\alpha}$ and incompatible with it (unless $\hat{G}$ commutes with $\hat{H}$, in which case the transformation U is not necessary). In other words, every representation entails a specific division into subensembles. This explains why it is *not possible* to characterize the state of a system simultaneously in terms of two or more dynamical variables represented by noncommuting matrices. This incompatibility at the level of the statistical description does of course not preclude the dynamical variables themselves from continuing to exist simultaneously for every individual particle.

Let us now turn to Eq. (5.115), which identifies the relation $\omega_{\alpha\beta} = \Omega_{\alpha} - \Omega_{\beta}$ (Eq. 5.55) with Bohr's transition rule,

$$\hbar\omega_{\alpha\beta} = \mathcal{E}_{\alpha} - \mathcal{E}_{\beta}, \tag{5.118}$$

thus demonstrating that the transition frequencies are just the frequencies of resonance (see footnote 12). This is a highly nontrivial result, but a perfectly natural one, since it means that a (quantum) transition occurs when the mechanical system responds resonantly to a mode of the radiation field.

This explains how it is that an atomic electron 'knows' in advance the energy of the state where it will land when realizing a transition, since the energy difference is determined with precision because of the sharpness of the resonance. The resonant response of the particle to a selected set of frequencies of the random ZPF constitutes the physical mechanism responsible for the 'indeterminate' transition of the particle to one among the collection of accessible stationary states. The same applies to every allowed transition from a given state α. Which is the transition that effectively takes place in a given case, and at what instant does it take place, depends on the precise conditions of the atom and the mode of the field responsible for the transition; it cannot be inferred from the present (statistical) description. Moreover, the identification of the linear resonant response as the mechanism behind the quantum (electric-dipole) transitions explains the fact that the corresponding selection rules are expressed in terms of the matrix elements of $\hat{x}$ (i.e., resonances) rather than of any other (nonlinear) dynamical variable.

This resonant phenomenon, along with the fact that the quantities featuring in Eq. (5.118) become fixed (i.e. nonstochastic) once the ergodic condition is in force, can be ultimately considered as a quantization principle. During the evolution towards the quantum regime, those orbits that happen to be robust enough against the fluctuations of the field become selected by their ergodicity, and eventually determine the stationary quantum states. The excited (metastable) states are included alongside the ground (stable) state as if they were *truly* stationary states, only when the radiative corrections are neglected, just as in the usual quantum-mechanical description.

5.4.2 The Hilbert-Space Description and State Vectors

For a more comprehensive description of the quantum system it is convenient to introduce the state vectors of an appropriate Hilbert space. The ensuing formalism, which is of course known to be an extremely useful tool in QM, will be useful for the study of the statistical nature of the quantum description to be made below. It also will prove very valuable for the analysis of entanglement to be developed in Chap. 7.

We start by observing that the matrix $\hat{A}(t)$, represented by the set of its elementary oscillators $A_{\alpha\beta}(t) = A_{\alpha\beta}e^{i\omega_{\alpha\beta}t}$, can be expanded as follows,

$$\hat{A}(t) = \sum_{\alpha,\beta} A_{\alpha\beta}e^{i\omega_{\alpha\beta}t} |e_\alpha\rangle\langle e_\beta|, \tag{5.119}$$

with $\{|e_\alpha\rangle\langle e_\beta|\}$ a basis constructed from the vectors of a complete orthonormal basis $\{|e_\alpha\rangle\}$, that spans the Hilbert space of states of the system.

By virtue of Eq. (5.118), the above expression gives

$$\begin{aligned}\langle e_\alpha| \hat{A}(t) |e_\beta\rangle &= \langle e_\alpha| \hat{A}(0) |e_\beta\rangle e^{i\omega_{\alpha\beta}t} \\ &= \langle e_\alpha| e^{i\mathcal{E}_\alpha t/\hbar}\hat{A}(0)e^{-i\mathcal{E}_\beta t/\hbar} |e_\beta\rangle \\ &= \langle \alpha(t)| \hat{A}(0) |\beta(t)\rangle,\end{aligned} \tag{5.120}$$

where $|\alpha(t)\rangle$ is obtained from $|e_\alpha\rangle$ by means of the unitary transformation

$$|e_\alpha\rangle \to |\alpha(t)\rangle = e^{-i\mathcal{E}_\alpha t/\hbar} |e_\alpha\rangle. \tag{5.121}$$

Equation (5.120) allows to transfer the time dependence from the matrix $\hat{A}(t)$—which lies at the core of the Heisenberg representation of QM—to the vectors of a new basis $\{|\alpha(t)\rangle\}$, which now contains all the information regarding the dynamics of the system. Moreover, the $\{|\alpha(t)\rangle\}$ are directly related to the energy eigenvalues $\{\mathcal{E}_\alpha\}$ and therefore to the stationary states $\{\alpha\}$, so that $|\alpha(t)\rangle$ stands now for a *state vector* that evolves in time according to Eq. (5.121). This fact will be used below to establish contact with the Schrödinger picture of QM.

Let us now introduce a shorthand (but practical and usual) notation by writing Eq. (5.120) as

$$A_{\alpha\beta}e^{i\omega_{\alpha\beta}t} = \langle\alpha(0)|\,\hat{A}(t)\,|\beta(0)\rangle = \langle\alpha(t)|\,\hat{A}(0)\,|\beta(t)\rangle = \langle\alpha|\,\hat{A}\,|\beta\rangle\,. \tag{5.122}$$

This expression relates the Fourier coefficients of (5.70) with the elements of the Hilbert-space description found in QM. From this and Eq. (5.73) we obtain

$$\overline{A}^{t}_{\alpha} = \langle A_\alpha\rangle = A_{\alpha\alpha} = \langle\alpha|\,\hat{A}\,|\alpha\rangle\,, \tag{5.123}$$

whence the quantity $\langle\alpha|\,\hat{A}\,|\alpha\rangle$ can be legitimately called an expectation (or mean) value. Further, Eq. (5.74) rewrites as

$$\sigma^2_{A_\alpha} = \sum_{\beta(\neq\alpha)} \left|\langle\alpha|\,\hat{A}\,|\beta\rangle\right|^2 = \sum_{\beta(\neq\alpha)} \langle\alpha|\,\hat{A}\,|\beta\rangle\,\langle\beta|\,\hat{A}\,|\alpha\rangle = \langle\alpha|\,\hat{A}^2\,|\alpha\rangle - \langle\alpha|\,\hat{A}\,|\alpha\rangle^2\,, \tag{5.124}$$

where the Hermiticity of $\hat{A}$ and the completeness of the basis $\{|\beta\rangle\}$ were used. Since the left-hand side of (5.124) stands for the variance (obtained by calculating averages over time or over the realizations of the field), the result confirms that the variance as is calculated within the standard quantum formalism should be interpreted just as a *genuine statistical variance*. In addition, Eq. (5.124) indicates that the quantities $\langle\alpha|\,\hat{A}\,|\beta\rangle$ possess a statistical connotation; thus, although no explicit sign of stochasticity remains in the quantities (5.122), their statistical nature has not been lost.

Further, from Eq. (5.112) we find

$$\sigma^2_{x_\alpha}\sigma^2_{p_\alpha} \geq \tfrac{1}{4}\hbar^2. \tag{5.125}$$

That this result, here derived only for stationary states α, holds for any state, can be seen by resorting once more to an application of the Robertson-Schrödinger inequality. Therefore, the Heisenberg inequality actually involves statistical variances, and is again a consequence of the persistent action of the ZPF, so that no reference to observations or measurements is required for its interpretation. Before the quantum regime is established and the ergodic properties are in place, this inequality may of course be violated, as mentioned earlier.

5.4.3 Transition to the Schrödinger Equation

In the previous section we introduced the Schrödinger picture of QM, in which it is the basis $\{|\alpha(t)\rangle\}$ what evolves in time, whereas the operators $\hat{A}$ remain fixed, $\hat{A} = \hat{A}(0)$. In this picture we thus need a law of evolution for an arbitrary state vector, instead of Eq. (5.117) governing the dynamics of the operators. Such law of evolution is of course the Schrödinger equation, which is derived here for the sake of completeness, and in order to make contact with the results of Chap. 4. For this purpose we take the solution of Eq. (5.117),

$$\hat{A}(t) = \hat{U}^{\dagger}(t)\hat{A}(0)\hat{U}(t), \tag{5.126}$$

where $\hat{U}(t)$ is the time-evolution operator

$$\hat{U}(t) = e^{-i\hat{H}t/\hbar}. \tag{5.127}$$

This implies that in the Schrödinger picture an arbitrary initial state vector $|\psi(0)\rangle$ evolves according to

$$|\psi(t)\rangle = \hat{U}(t)\,|\psi(0)\rangle = e^{-i\hat{H}t/\hbar}\,|\psi(0)\rangle\,, \tag{5.128}$$

which immediately gives the complete Schrödinger equation

$$i\hbar\frac{d\,|\psi(t)\rangle}{dt} = \hat{H}\,|\psi(t)\rangle\,. \tag{5.129}$$

The most common form of its solutions is easily obtained by expanding $|\psi(0)\rangle$ in the canonical basis $\{|e_\alpha\rangle\}$ and resorting to the energy representation, in which $\hat{H}$ is diagonal. In this case Eq. (5.128) gives

$$|\psi(t)\rangle = \sum_\alpha c_\alpha e^{-i\hat{H}t/\hbar}\,|e_\alpha\rangle = \sum_\alpha c_\alpha e^{-i\mathcal{E}_\alpha t/\hbar}\,|e_\alpha\rangle = \sum_\alpha c_\alpha\,|\alpha(t)\rangle\,, \tag{5.130}$$

as was to be expected, since in the energy representation the privileged basis is precisely the basis $\{|\alpha(t)\rangle\}$ of stationary vectors. In particular, for a stationary state $|\psi(t)\rangle = |\alpha(t)\rangle = e^{-i\mathcal{E}_\alpha t/\hbar}\,|e_\alpha\rangle$, we arrive at the stationary Schrödinger equation

$$i\hbar\frac{d\,|\alpha(t)\rangle}{dt} = \mathcal{E}_\alpha\,|\alpha(t)\rangle\,. \tag{5.131}$$

5.4.3.1 Transition to the Configuration Space

The basis vectors $\{|\alpha(t)\rangle\}$ defined in Eq. (5.121) can be used to construct a complete set of orthonormal functions $\{\psi_\alpha(x,t)\}$ in configuration space, such that

$$\psi_\alpha(x,t) = \langle x|\,\alpha(t)\rangle = e^{-i\mathcal{E}_\alpha t/\hbar}\varphi_\alpha(x), \tag{5.132}$$

where $|x\rangle$ is the eigenvector of the position operator $\hat{x}$ with eigenvalue x (so that $\{|x\rangle\}$ constitutes a continuous basis in Hilbert space), and

$$\varphi_\alpha(x) = \psi_\alpha(x,0) = \langle x|\,\alpha(0)\rangle = \langle x|\,e_\alpha\rangle\,. \tag{5.133}$$

From (5.132), the Schrödinger equation (5.129) in configuration space follows immediately,[16]

$$i\hbar\frac{\partial\psi(x,t)}{\partial t} = \hat{H}\psi(x,t), \tag{5.134}$$

with [see Eq. (5.130)]

$$\psi(x,t) = \langle x|\ \psi(t)\rangle = \sum_{\alpha} c_\alpha e^{-i\mathcal{E}_\alpha t/\hbar}\varphi_\alpha(x). \tag{5.135}$$

It is important to observe that in the transition from the Heisenberg picture to the Schrödinger picture, the energies $\mathcal{E}_\alpha$ take on the central role played earlier by the frequencies $\omega_{\alpha\beta}$. This result represents a major shift in priority and emphasis: in the Schrödinger description the state energies are the important parameters, whereas in the Heisenberg description it is the transition frequencies that matter.

In the new representation, $A_{\alpha\beta}(t) = \langle\alpha(t)|\ \hat{A}(0)\ |\beta(t)\rangle$ becomes

$$A_{\alpha\beta}(t) = \int \psi^*_\alpha(x,t)\hat{A}_x\psi_\beta(x,t)dx, \tag{5.136}$$

where the subscript x in $\hat{A}_x$ stresses that $\hat{A}$ is in the configuration-space representation, although it can be (and generally is) omitted. The elements of the (time-independent) matrix $\hat{A}$ are thus

$$A_{\alpha\beta} = \int \varphi^*_\alpha(x)\hat{A}\varphi_\beta(x)dx. \tag{5.137}$$

For $A = A(x)$, and using Eq. (5.123), we get

$$\overline{A}^t_\alpha = \langle A\rangle_\alpha = \int A(x)\rho_\alpha(x)dx, \tag{5.138}$$

where

$$\rho_\alpha(x) = |\varphi_\alpha|^2 = |\psi_\alpha|^2. \tag{5.139}$$

From Eqs. (5.138) and (5.139) it follows that $|\psi_\alpha|^2$ is the probability density in configuration space. The wave function ψ_α is therefore the amplitude (of density) of probability in such space, which gives statistical information about the subensemble α representing the particles with mean energy $\mathcal{E}_\alpha$. This leads quite naturally to the ensemble interpretation of the Schrödinger theory, in line with the particular statistical nature of the present approach.

---

[16] The time derivative in Eq. (5.129) should be taken as a partial derivative in Eq. (5.134) because now the state function depends also on the variable x. In the Hilbert space in which Eq. (5.129) operates, time is the single variable required to describe the system.

5.4.4 The Stochastic Representation

Our derivations have led successfully to the two alternative formalisms usual in QM, namely the Schrödinger and Heisenberg pictures. As has been discussed above, one feature that is shared by these representations is the loss of any explicit trace of stochasticity. This may be considered a shortcoming, since the source of randomness that makes the quantum fluctuations causal has disappeared entirely from the description. It seems therefore interesting to construct a new representation that retains the stochastic elements inherited from the background field.

We start by observing that since the random variables $\{a_{\alpha\beta}\}$ satisfy a chain rule (Eq. 5.59), also the quantities $A_{\alpha\beta}a_{\alpha\beta}e^{i\omega_{\alpha\beta}t}$ satisfy a matrix algebra. We therefore introduce the stochastic matrix $\hat{A}^{(i)}$, which in terms of the basis vectors $\{|e_\alpha\rangle\}$ used earlier takes the form

$$\hat{A}^{(i)}(t) = \sum_{\alpha,\beta} A_{\alpha\beta}e^{i\omega_{\alpha\beta}t}a^{(i)}_{\alpha\beta}|e_\alpha\rangle\langle e_\beta|. \tag{5.140}$$

Hence, with $|e_\alpha\rangle = |\alpha(0)\rangle$,

$$\begin{aligned} \langle\alpha(0)|\,\hat{A}^{(i)}(t)\,|\beta(0)\rangle &= \langle\alpha(0)|\,\hat{A}(t)\,|\beta(0)\rangle\, a^{(i)}_{\alpha\beta} \\ &= \langle\alpha(0)|\, e^{i\phi^{(i)}_\alpha}\,\hat{A}(t)e^{-i\phi^{(i)}_\beta}\,|\beta(0)\rangle \\ &= \left\langle\alpha^{(i)}(0)\right|\hat{A}(t)\left|\beta^{(i)}(0)\right\rangle, \end{aligned} \tag{5.141}$$

where Eq. (5.60) has been used to write

$$a^{(i)}_{\alpha\beta} = e^{i\phi^{(i)}_\alpha}e^{-i\phi^{(i)}_\beta}, \tag{5.142}$$

and the *stochastic basis* $\{\left|\alpha^{(i)}(t)\right\rangle\}$ was introduced, with elements (de la Peña and Cetto 1996)

$$\left|\alpha^{(i)}(t)\right\rangle = e^{-i\phi^{(i)}_\alpha}\,|\alpha(t)\rangle = e^{-i\phi^{(i)}_\alpha}e^{-i\mathcal{E}_\alpha t/\hbar}\,|e_\alpha\rangle\,. \tag{5.143}$$

Equation (5.141) shows that the stochasticity of the matrix $\hat{A}^{(i)}(t)$ can be transferred to the basis vectors, in analogy with the transition from the Heisenberg to the Schrödinger picture, but here involving the i-dependence instead of the time dependence. Further, by resorting to the evolving vectors (5.143) we obtain

$$\begin{aligned} \langle\alpha|\,\hat{A}\,|\beta\rangle^{(i)} &\equiv \left\langle\alpha^{(i)}(0)\right|\hat{A}(t)\left|\beta^{(i)}(0)\right\rangle = \left\langle\alpha^{(i)}(t)\right|\hat{A}(0)\left|\beta^{(i)}(t)\right\rangle = \\ &= \langle\alpha(0)|\,\hat{A}^{(i)}(t)\,|\beta(0)\rangle = \langle\alpha(t)|\,\hat{A}^{(i)}(0)\,|\beta(t)\rangle\,. \end{aligned} \tag{5.144}$$

Here both lines correspond to the proper Heisenberg → Schrödinger transition, but the first one involves stochastic vectors whereas the second one involves a stochastic matrix $\hat{A}^{(i)}$. The equality

$$\langle\alpha(0)|\,\hat{A}^{(i)}(t)\,|\beta(0)\rangle = \left\langle\alpha^{(i)}(t)\right|\,\hat{A}(0)\,\left|\beta^{(i)}(t)\right\rangle \tag{5.145}$$

thus allows for the transition from the *stochastic Heisenberg picture* (stochastic, time-dependent operators) to the *stochastic Schrödinger representation* (stochastic, time-dependent vectors). The latter representation is interesting in that it includes from the outset a random phase $\phi_\alpha^{(i)}$ for each basis vector $\left|\alpha^{(i)}(t)\right\rangle$, whereas in usual QM the phase of a pure state $|\alpha(t)\rangle$ remains fully discretionary. The $\phi_\alpha^{(i)}$ enters here as a reminder of the stochastic properties of the system and has a well-defined physical meaning.

The random value of the phase factors $e^{-i\phi_\alpha^{(i)}}$ acquires a real importance when states are superposed, since in this case they give rise to random relative phases that are physically relevant. For example, for a superposition of two stationary states we have

$$\left|\psi^{(i)}\right\rangle = a\left|\alpha^{(i)}\right\rangle + b\left|\beta^{(i)}\right\rangle = e^{-i\phi_\alpha^{(i)}}\left(a\,|\alpha\rangle + e^{i(\phi_\alpha^{(i)}-\phi_\beta^{(i)})}b\,|\beta\rangle\right), \tag{5.146}$$

so that $e^{-i\phi_\alpha^{(i)}}$ is physically innocuous but $e^{(i\phi_\alpha^{(i)}-\phi_\beta^{(i)})}$ is relevant. Such random relative phases are frequently introduced by hand in the theory of measurement or of decoherence for example, to represent the random effects of the environment. Here they appear automatically as one more manifestation of the ubiquitous ZPF, which in this context plays the role of an unavoidable environment. This latter observation will prove to be of high relevance in Chap. 7, in connection with entangled states of bipartite systems.

5.5 Concluding Remarks

Chapter 4 and the present one have taken us from the same point of departure—viz equations (4.2) and (5.9)—through different itineraries to the province of quantum mechanics. In both cases a filter was introduced to select the acceptable time-asymptotic motions in the form of a strong requisite, which operates as a sort of quantum principle. The requirement in the present chapter was that of ergodicity, whereas in Chap. 4 it was the energy balance. Since these different requirements lead to equivalent results—the reduction of the description to the quantum regime—they should be closely related.

In the present case it is clear that when the ergodic, radiationless regime is eventually attained, energy balance is satisfied, so that the stationary solutions α characterized by a fixed energy correspond to surfaces of constant energy in the available phase space. Conversely, the arguments of Chap. 4 can be completed by noticing—as already argued in Sect. 4.5.4—that stationarity of an orbit, in the presence of stochasticity, means that in the long run the particle visits the entire available energy surface for the corresponding state, giving rise to the ergodic behavior. This speaks to the physical—beyond the formal—equivalence of both theories known after the names of Heisenberg and Schrödinger. They enrich and complement each other by revealing different aspects of the same quantum phenomenon. In particular, the Heisenberg formalism just derived discloses the linear and resonant response of the particle to certain modes of the background field, a property that remains hidden in Schrödinger's description, in which, as stated before, the energies $\mathcal{E}_\alpha$ and not the frequencies $\omega_{\alpha\beta}$ take on a prominent role.[17]

The ergodic principle turns out to play a crucial role in defining a matrix algebra for the description of the dynamics of the system. The original stochastic equation of motion (5.9) that describes a stationary state α, becomes transformed into the nonstochastic matrix Eq. (5.75) (that includes all accesible stationary states). The mathematical consequence of ergodicity is thus a dynamical law for *operators* that are in correspondence with the dynamical variables. Physically, in the evolution towards the quantum regime, ergodicity selects from among all the possible stationary solutions of Eq. (5.30), those that are robust with respect to the field fluctuations; these are the quantum solutions.

The transitions between states result from very sharp resonances of the particle to certain modes of the field, be it just the ZPF or otherwise. This extreme sharpness (due to the small value of τ) leads to the precisely defined value of the energy of the radiated field—the photon—in a single-step transition. As for the background field, we found that whereas the original free ZPF is maximally disordered, the relevant modes of the effective field in the neighborhood of the particle are considerably less disordered, having constant amplitudes and random phases that are partially correlated, according to the chain rule for the variables $a_{\alpha\beta}$. This change in the near field in interaction with the particle is accompanied by the emergence of its quantum properties, as already envisaged from the results of Chap. 3.

Appendix A: The Ergodic Principle and the Algebraic Description

This appendix is devoted to a detailed analysis of the implications of imposing the ergodic condition in the form (5.46), for the different elements that enter into the description.

We start by recalling from Sect. 5.1.1 that to distinguish the solutions $x(t)$ that correspond to different accessible stationary states α, the substitution

[17] Such resonant response will turn out to be crucial for the elucidation of some fundamental aspects of entanglement, as shown in Chap. 7.

$$x(t) = \sum_k \tilde{x}_k a_k e^{i\omega_k t} \rightarrow x_\alpha(t) = \sum_\beta \tilde{x}_{\alpha\beta} a_{\alpha\beta} e^{i\omega_{\alpha\beta} t} \tag{A.1}$$

has been made, where $\tilde{x}_{\alpha\beta} = \tilde{x}(\omega_{\alpha\beta})$ stands for the amplitude corresponding to the frequency $\omega_{\alpha\beta}$ and similarly for $a_{\alpha\beta}$. The purpose of the first part of the appendix is to establish the correspondence for higher powers of x, $x^n \rightarrow (x^n)_\alpha$, that is consistent with Eq. (5.46). The problem is equivalent to determining the appropriate coefficients $\tilde{A}_{\alpha\beta}$ for any dynamical variable $A_\alpha(t)$ that can be expanded as a power series of x, with A_α given by Eq. (5.34), namely

$$A_\alpha(t) = \sum_\beta \tilde{A}_{\alpha\beta} a_{\alpha\beta} e^{i\omega_{\alpha\beta} t}. \tag{A.2}$$

Let us first analyze the quadratic case, $A_\alpha = (x^2)_\alpha$. The introduction of the index α must be such as to guarantee that the expansion for $(x^2)_\alpha$ is consistent with the demands imposed by the theory. We therefore start by writing

$$x^2(t) = \sum_{k,k'} \tilde{x}_k \tilde{x}_{k'} a_k a_{k'} e^{i(\omega_k + \omega_{k'})t} \tag{A.3}$$

and define

$$\omega_K \equiv \omega_k + \omega_{k'}, \tag{A.4}$$

so that Eq. (A.3) rewrites as

$$x^2(t) = \sum_{k,K} \tilde{x}(\omega_k)\tilde{x}(\omega_K - \omega_k) a(\omega_k) a(\omega_K - \omega_k) e^{i\omega_K t}. \tag{A.5}$$

As explained in Sect. 5.1.2, passing from $x^2(t)$ to $\left(x^2\right)_\alpha(t)$ requires focusing not on the complete set $\{\omega_K\}$ but rather on the subset $\{\omega_{\alpha\beta}\}$, so that the sum over K becomes a sum over β. Similarly, we make the substitution $\omega_k \rightarrow \omega_{\alpha\beta'}$. Thus,

$$(x^2)_\alpha(t) = \sum_{\beta,\beta'} \tilde{x}(\omega_{\alpha\beta'})\tilde{x}(\omega_{\alpha\beta} - \omega_{\alpha\beta'}) a(\omega_{\alpha\beta'}) a(\omega_{\alpha\beta} - \omega_{\alpha\beta'}) e^{i\omega_{\alpha\beta} t}. \tag{A.6}$$

On the other hand, according to Eq. (A.2), $(x^2)_\alpha$ has the form

$$(x^2)_\alpha(t) = \sum_\beta \widetilde{x^2}(\omega_{\alpha\beta}) a(\omega_{\alpha\beta}) e^{i\omega_{\alpha\beta} t}. \tag{A.7}$$

Comparison of these two equations leads to

$$\widetilde{x^2}(\omega_{\alpha\beta})a^{(i)}(\omega_{\alpha\beta}) = \sum_{\beta'} \tilde{x}(\omega_{\alpha\beta'})\tilde{x}(\omega_{\alpha\beta} - \omega_{\alpha\beta'})a^{(i)}(\omega_{\alpha\beta'})a^{(i)}(\omega_{\alpha\beta} - \omega_{\alpha\beta'}), \tag{A.8}$$

where the index (i) denotes the dependence of the a's on the field realization. Since according to (5.46) $\widetilde{x^2}(\omega_{\alpha\beta})$ is a sure quantity, the sum

$$\sum_{\beta'} \tilde{x}(\omega_{\alpha\beta'})\tilde{x}(\omega_{\alpha\beta} - \omega_{\alpha\beta'})\frac{a^{(i)}(\omega_{\alpha\beta'})a^{(i)}(\omega_{\alpha\beta} - \omega_{\alpha\beta'})}{a^{(i)}(\omega_{\alpha\beta})} \tag{A.9}$$

must be a sure quantity as well, for any α, β. The only way to ensure that (A.9) is an i-independent quantity is by taking

$$a^{(i)}(\omega_{\alpha\beta'})a^{(i)}(\omega_{\alpha\beta} - \omega_{\alpha\beta'}) = a^{(i)}(\omega_{\alpha\beta}). \tag{A.10}$$

Equation (A.8) reduces thus to

$$\widetilde{x^2}(\omega_{\alpha\beta}) = \sum_{\beta'} \tilde{x}(\omega_{\alpha\beta'})\tilde{x}(\omega_{\alpha\beta} - \omega_{\alpha\beta'}). \tag{A.11}$$

Together with Eq. (A.7), this gives the expression for $\left(x^2\right)_\alpha$ that is consistent with the ergodic demand:

$$(x^2)_\alpha(t) = \sum_{\beta\beta'} \tilde{x}(\omega_{\alpha\beta'})\tilde{x}(\omega_{\alpha\beta} - \omega_{\alpha\beta'})a(\omega_{\alpha\beta})e^{i\omega_{\alpha\beta}t}. \tag{A.12}$$

We now focus on the frequency

$$\Omega^{(\alpha)}_{\beta'\beta} \equiv \omega_{\alpha\beta} - \omega_{\alpha\beta'}, \tag{A.13}$$

which is a difference between two relevant frequencies of the subensemble α. For a given α, this frequency depends on the indices β, β' only. By its definition it has the following properties:

$$\Omega^{(\alpha)}_{\beta'\beta} = -\Omega^{(\alpha)}_{\beta\beta'}, \tag{A.14a}$$

$$\Omega^{(\alpha)}_{\beta\beta'} + \Omega^{(\alpha)}_{\beta'\beta''} + \cdots + \Omega^{(\alpha)}_{\beta^{(n-1)}\beta^{(n)}} = \Omega^{(\alpha)}_{\beta\beta^{(n)}}. \tag{A.14b}$$

The second equation involves an arbitrary number of terms, with each $\beta^{(m)}$ an element of the set $\{\beta\}$ of indices introduced to enumerate the different relevant frequencies in $\{\omega_{\alpha\beta}\}$. In particular, for $\beta' = \beta_0$, $\omega_{\alpha\beta'}$ vanishes [see Eq. (5.23)], and $\Omega^{(\alpha)}_{\beta'\beta}$ reduces to

$$\Omega^{(\alpha)}_{\beta_0\beta} = \omega_{\alpha\beta}. \tag{A.15}$$

In terms of the new frequencies, and writing

$$\tilde{x}(\Omega^{(\alpha)}_{\beta\beta'}) = \tilde{x}^{(\alpha)}_{\beta\beta'}, \tag{A.16}$$

Equation (A.11) reads

$$\widetilde{x^2}(\Omega^{(\alpha)}_{\beta_0\beta}) = \sum_{\beta'} \tilde{x}^{(\alpha)}_{\beta_0\beta'}\tilde{x}^{(\alpha)}_{\beta'\beta}. \tag{A.17}$$

Now we resort to the fact that β' and β run over the same domain ($\{\beta\}$) to define a square matrix $\Omega(\alpha)$ with elements $\Omega^{(\alpha)}_{\beta'\beta}$. Thus, using the (provisional) notation

$$\{\beta\} = \{0, 1, 2, \ldots, \beta_0 - 1, \beta_0, \beta_0 + 1, \cdots\}, \tag{A.18}$$

we write

$$\Omega(\alpha) = \begin{pmatrix} \Omega^{(\alpha)}_{00} = 0 & \Omega^{(\alpha)}_{01} & \ldots & \Omega^{(\alpha)}_{0\beta_0} & \ldots \\ \Omega^{(\alpha)}_{10} & \Omega^{(\alpha)}_{11} = 0 & \ldots & & \\ \ldots & & & & \\ \omega_{\alpha 0} & \omega_{\alpha 1} & \ldots & \omega_{\alpha\beta_0} = 0 & \ldots \\ \ldots & & & & \end{pmatrix}, \tag{A.19}$$

where Eq. (A.15) was used in the β_0-th row. This shows that the set of relevant frequencies $\{\omega_{\alpha\beta}\}$ is a row in the wider set of frequencies $\Omega(\alpha)$. This, together with the fact that all the elements $\Omega^{(\alpha)}_{\beta'\beta}$ should be taken into consideration on an equal footing when calculating $\left(x^2\right)_\alpha$ [see Eq. (A.17)], leads us to extend the analysis to the entire matrix $\Omega(\alpha)$.

The order of the rows in $\Omega(\alpha)$ is physically irrelevant, since it was established via the arbitrary ordering (A.18). This means that the rows in $\Omega(\alpha)$ are physically equivalent. Thus, if the β_0-th row gives the set $\{\omega_{\alpha\beta}\}$ for the subensemble α, another row, say $\beta' = \eta \neq \beta_0$, gives the set $\{\Omega^{(\alpha)}_{\eta\beta}\}$ for the subensemble labeled with η, whose relevant frequencies are $\{\Omega^{(\alpha)}_{\eta\beta} = \omega_{\alpha\beta} - \omega_{\alpha\eta}\}$. When the system is in the corresponding stationary state η, the expansion of the dynamical variable A reads

$$A_\eta(t) \equiv \sum_\beta \tilde{A}(\Omega^{(\alpha)}_{\eta\beta})a(\Omega^{(\alpha)}_{\eta\beta})e^{i\Omega^{(\alpha)}_{\eta\beta}t}, \tag{A.20}$$

which is a generalization of

$$A_\alpha(t) = \sum_\beta \tilde{A}(\omega_{\alpha\beta})a(\omega_{\alpha\beta})e^{i\omega_{\alpha\beta}t} \tag{A.21}$$

for a variable A in the stationary state α, in the same way that $\Omega(\alpha)$ generalized the set $\{\omega_{\alpha\beta}\}$. Clearly (A.21) can be obtained from (A.20) taking $\eta = \beta_0$. A first conclusion that derives from here is that the row-index β' is in direct correspondence with a stationary state index, so that the indices in $\{\beta\}$, originally introduced as labels that distinguished each of the relevant frequencies, denote subensembles corresponding to new stationary states. It follows that the indices α and β have the same domain and the same physical interpretation. In particular, this allows us to write

$$\beta_0 = \alpha, \tag{A.22}$$

so that Eq. (A.18) rewrites as

$$\{\beta\} = \{\alpha - \alpha, \alpha - (\alpha + 1), \cdots, \alpha - 1, \alpha, \alpha + 1, \cdots\}. \tag{A.23}$$

According to the discussion preceding Eq. (A.20), the (arbitrary) state η ($\neq \alpha$) is related to α via their relevant frequencies, an observation that discloses a relation among all stationary states, and ultimately connects them all. Of course, the origin of such relation goes back to the α-dependence of $\Omega(\alpha)$ (and hence of all its rows). However, even though $\Omega(\alpha)$ was constructed taking α as a privileged ensemble, this matrix does not depend on α since the same matrix is obtained when considering the difference between two relevant frequencies of *any* subensemble η. In order to see this we construct the matrix $\Omega(\eta)$ with elements $\Omega^{(\eta)}_{\beta'\beta}$ defined, in analogy with Eq. (A.13), as the difference between two relevant frequencies of the subensemble η:

$$\Omega^{(\eta)}_{\beta'\beta} \equiv \Omega^{(\alpha)}_{\eta\beta} - \Omega^{(\alpha)}_{\eta\beta'}. \tag{A.24}$$

Resorting now to Eq. (A.14b) we obtain

$$\Omega^{(\eta)}_{\beta'\beta} = \Omega^{(\alpha)}_{\beta'\beta}, \tag{A.25}$$

hence $\Omega(\eta) = \Omega(\alpha)$. Since η is arbitrary, it follows that the matrix (A.19) is indeed α-independent so that $\Omega(\alpha) = \Omega$ and the superindex (α) in $\Omega^{(\alpha)}_{\beta'\beta}$ may be dropped. The matrix Ω can then be understood as an array (in row form) of all the relevant frequencies corresponding to all the accessible stationary states of the mechanical system. Because of equations (A.14), its elements can be written in the general form

$$\Omega_{\beta'\beta} = \Omega_{\beta'} - \Omega_{\beta}. \tag{A.26}$$

The physical meaning of the parameters Ω_β is determined in Sect. (5.4.1). Together with Eq. (A.15) (with $\beta_0 = \alpha$), (A.26) gives

$$\Omega_{\alpha\beta} = \omega_{\alpha\beta} = \Omega_\alpha - \Omega_\beta, \tag{A.27}$$

hence

$$\omega_{\beta\alpha} = -\omega_{\alpha\beta} \tag{A.28}$$

and

$$\omega_{\alpha\beta'} + \omega_{\beta'\beta''} + \ldots + \omega_{\beta^{(n-1)}\beta} = \omega_{\alpha\beta}. \tag{A.29}$$

With these results we may now go back to Eq. (A.10) and write

$$a^{(i)}(\omega_{\alpha\beta'})a^{(i)}(\omega_{\beta'\beta}) = a^{(i)}(\omega_{\alpha\beta}). \tag{A.30}$$

Using the short notation $a_{\beta'\beta} = a(\omega_{\beta'\beta})$, this relation can be easily generalized to any number of factors by a successive (chained) application of it,

$$\begin{aligned} a_{\alpha\beta'}a_{\beta'\beta''}a_{\beta''\beta'''}\ldots a_{\beta^{(n-1)}\beta} &= \left(a_{\alpha\beta'}a_{\beta'\beta''}\right) a_{\beta''\beta'''}\ldots a_{\beta^{(n-1)}\beta} \\ &= \left[\left(a_{\alpha\beta''}\right) a_{\beta''\beta'''}\right]\ldots a_{\beta^{(n-1)}\beta} \\ &= \left[a_{\alpha\beta'''}\right]\cdots a_{\beta^{(n-1)}\beta} \\ &= a_{\alpha\beta}. \end{aligned} \tag{A.31}$$

With each $a^{(i)}_{\beta^{(n)}\beta^{(m)}}$ written in polar form according to (5.21), this implies that also the stochastic phases must satisfy the relation

$$\varphi^{(i)}_{\alpha\beta'} + \varphi^{(i)}_{\beta'\beta''} + \cdots + \varphi^{(i)}_{\beta^{(n-1)}\beta} = \varphi^{(i)}_{\alpha\beta}, \tag{A.32a}$$

and can therefore be expressed as a difference of terms,

$$\varphi^{(i)}_{\alpha\beta} = \phi^{(i)}_{\alpha} - \phi^{(i)}_{\beta}, \tag{A.33}$$

where each of the ϕ_λ represents a random phase. Equation (5.21) becomes thus

$$a_{\alpha\beta} = e^{i\varphi_{\alpha\beta}} = e^{i(\phi_\alpha - \phi_\beta)}, \tag{A.34}$$

from where it follows, in particular, that

$$a_{\beta\alpha} = a^*_{\alpha\beta}. \tag{A.35}$$

The relations (A.29) and (A.31), which are fundamental for the theory, constitute what is called the *chain rule.* The frequencies that enter in the chain rule can refer to relevant frequencies of any stationary state (they are elements of the matrix Ω), and can be either resonance frequencies or linear (chained) combinations of them.

The result $|a_{\alpha\beta}| = 1$ requires a comment. In Chap. 3 it was found that the energy of the oscillators of the ZPF have an important dispersion. Here we got what seems to be a contradictory result, namely that the $a_{\alpha\beta}$'s have a fix amplitude. Consistency is recovered by considering the discussion following Eq. (5.4). Further, the strict meaning of (A.34) is that only the modes of the field that have an important role in the dynamics of the mechanical subsystem in the ergodic regime, are those that have amplitudes with a Gaussian distribution with a negligible dispersion around an average value 1. The remaining modes simply contribute to the background noise.

Finally, with the results obtained above, Eq. (A.12) becomes

$$(x^2)_\alpha(t) = \sum_{\beta\beta'} \tilde{x}_{\alpha\beta'}\tilde{x}_{\beta'\beta} a_{\alpha\beta} e^{i\omega_{\alpha\beta}t}, \tag{A.36}$$

which, when compared with (A.7), gives

$$(\widetilde{x^2})_{\alpha\beta} = \sum_{\beta'} \tilde{x}_{\alpha\beta'}\tilde{x}_{\beta'\beta}. \tag{A.37}$$

Iteration of the procedure presented above leads to the following expression for the n-th power of the variable x^n in state α

$$(x^n)_\alpha(t) = \sum_{\beta} (\widetilde{x^n})_{\alpha\beta} a_{\alpha\beta} e^{i\omega_{\alpha\beta}t}, \tag{A.38}$$

with

$$(\widetilde{x^n})_{\alpha\beta} = \sum_{\beta'\ldots\beta^{(n-1)}} \tilde{x}_{\alpha\beta'}\tilde{x}_{\beta'\beta''}\ldots\tilde{x}_{\beta^{(n-1)}\beta}. \tag{A.39}$$

Appendix B: A Simple Example: The Harmonic Oscillator

The aim of this appendix is to analyze a simple system by applying some of the methods presented in the body of the chapter, in order to clarify the meaning of the resonance and relevant frequencies, respectively. For this purpose let us consider the case of a harmonic oscillator of natural frequency ω_0. The force is $f = -m\omega_0^2 x$, so that in the state α

$$f_\alpha = -m\omega_0^2 x_\alpha. \tag{B.1}$$

Equation (5.29) becomes then

$$\omega_{\alpha\beta}^2 \approx \omega_0^2, \tag{B.2}$$

with solutions $\pm\omega_0$. The independence of the resonance frequencies from the state is characteristic of the harmonic oscillator; it does not hold in general for arbitrary forces. There exist therefore only two resonance frequencies for the state α, which will be labeled with the indices $\beta_+ = \alpha - 1$ and $\beta_- = \alpha + 1$ [see Eq. (A.23)] in such a way that

$$\omega_{\alpha,\alpha-1} = \omega_0, \quad \omega_{\alpha,\alpha+1} = -\omega_0. \tag{B.3}$$

Since these are the frequencies that contribute significantly to the expansion

$$x_\alpha(t) = \sum_\beta \tilde{x}_{\alpha\beta} a_{\alpha\beta} e^{i\omega_{\alpha\beta} t}, \tag{B.4}$$

we conclude that when the noisy terms are neglected in Eq. (B.4), the coefficients $\tilde{x}_{\alpha\beta}$ are

$$\tilde{x}_{\alpha\beta} = \tilde{x}_{\alpha,\alpha-1}\delta_{\beta,\alpha-1} + \tilde{x}_{\alpha,\alpha+1}\delta_{\beta,\alpha+1}, \tag{B.5}$$

hence

$$x_\alpha(t) = \tilde{x}_{\alpha,\alpha-1} a_{\alpha,\alpha-1} e^{i\omega_0 t} + \tilde{x}_{\alpha,\alpha+1} a_{\alpha,\alpha+1} e^{-i\omega_0 t}. \tag{B.6}$$

To determine the relevant frequencies of the system we resort to the chain rule (Eq. A.29) and to the antisymmetry $\omega_{\alpha\beta} = -\omega_{\beta\alpha}$ to combine the resonance frequencies, thus obtaining

$$\omega_{\alpha+1,\alpha} + \omega_{\alpha,\alpha-1} = \omega_{\alpha+1,\alpha-1} = 2\omega_0. \tag{B.7}$$

This defines a new (relevant) frequency $\omega_{\alpha+1,\alpha-1} = 2\omega_0$. Now, according to the discussion following Eq. (A.21), $\alpha + 1$ and $\alpha - 1$ represent new stationary states. Their resonance frequencies are again $\pm\omega_0$, since as stated above, the solutions of (B.2) are state-independent. As before, the two resonance frequencies for the state $\alpha \pm 1$ will be labeled with the indices $\beta_+ = (\alpha \pm 1) - 1$ and $\beta_- = (\alpha \pm 1) + 1$. This gives, in analogy with (B.3),

$$\begin{aligned} \omega_{\alpha+1,\alpha} &= \omega_0, \quad \omega_{\alpha+1,\alpha+2} = -\omega_0, \\ \omega_{\alpha-1,\alpha-2} &= \omega_0, \quad \omega_{\alpha-1,\alpha} = -\omega_0. \end{aligned} \tag{B.8}$$

A chained combination of these frequencies with $\omega_{\alpha+1,\alpha-1}$ in Eq. (B.7) defines two more relevant frequencies,

$$\begin{aligned} \omega_{\alpha+2,\alpha+1} + \omega_{\alpha+1,\alpha-1} &= \omega_{\alpha+2,\alpha-1} = 3\omega_0, \\ \omega_{\alpha+2,\alpha-1} + \omega_{\alpha-1,\alpha-2} &= \omega_{\alpha+2,\alpha-2} = 4\omega_0. \end{aligned} \tag{B.9}$$

It is clear that the procedure can be applied iteratively, so that the relevant frequencies for the harmonic oscillator take the form

$$\omega_{\alpha+n,\alpha+m} = \omega_{nm} = \omega_0(n-m), \tag{B.10}$$

with $n, m = 0, \pm 1, \pm 2, \ldots$

In particular, the resonance frequencies ($\pm\omega_0$) correspond to $m = n \pm 1$; the remaining relevant frequencies (those in Eq. (B.10) with $m \neq n \pm 1$) will be important in expansions of higher powers of $x(t)$ (or $p(t)$). For example, for $\widetilde{x^2}_{\alpha\beta}(t)$ one finds, using Eqs. (5.57) and (B.5), that the only terms that represent an important contribution to $\widetilde{x^2}_{\alpha\beta}(t)$ are

$$\begin{aligned} \widetilde{x^2}_{\alpha\alpha} &= \tilde{x}_{\alpha,\alpha-1}\tilde{x}_{\alpha-1,\alpha} + \tilde{x}_{\alpha,\alpha+1}\tilde{x}_{\alpha+1,\alpha}, \\ \widetilde{x^2}_{\alpha,\alpha-2}(t) &= \tilde{x}_{\alpha,\alpha-1}\tilde{x}_{\alpha-1,\alpha-2}e^{i\omega_{\alpha,\alpha-2}t} = \widetilde{x^2}_{\alpha,\alpha-2}e^{i2\omega_0 t}, \\ \widetilde{x^2}_{\alpha,\alpha+2}(t) &= \tilde{x}_{\alpha,\alpha+1}\tilde{x}_{\alpha+1,\alpha+2}e^{i\omega_{\alpha,\alpha+2}t} = \widetilde{x^2}_{\alpha,\alpha+2}e^{-i2\omega_0 t}. \end{aligned} \tag{B.11}$$

The frequencies of oscillation of $(x^2)_\alpha$ are thus 0 (for $\left\langle (x^2)_\alpha \right\rangle = \widetilde{x^2}_{\alpha\alpha}$, see Eq. (5.73)) and $\pm 2\omega_0$. The extension of this exercise to other powers allows to recover the remaining relevant frequencies (B.10).

This example serves to show that the relevant frequencies $\omega_{\alpha\beta}$ are combinations of resonance frequencies, as dictated by the chain rule (A.29). This latter is the quantum counterpart of the classical rule to construct the combination frequencies intervening in nonlinear systems, $\omega_j = \sum_i \pm n_{ji}\omega_i$, where ω_i is a fundamental frequency of oscillation and n_{ji} are integer multiples; these frequencies correspond to the different harmonic and intermodulation frequencies.

References

Born, M., Heisenberg, W., Jordan, P.: Zur Quantenmechanik II. Zeit. Phys. **34**, 858 (1926)

*Cetto, A.M., de la Peña, L.: Electromagnetic vacuum-particle interaction as the source of quantum phenomena. Found. Phys. Lett. **4**, 476 (1991)

*Cetto, A.M., de la Peña, L., Valdés-Hernández, A.: Quantization as an emergent phenomenon due to matter-zeropoint field interaction. J. Phys. JCPCS. **361**, 012013 (2012). http://iopscience.iop.org/1742-6596/361/1/ 012013 (free access)

*de la Peña, L., Cetto, A.M.: Electromagnetic vacuum-particle interaction as the source of quantum phenomena. Found. Phys. Lett. **12**, 476 (1991)

**de la Peña, L., Cetto, A.M.: The Quantum Dice. An introduction to stochastic electrodynamics. Kluwer, Dordrecht (1996) (Referred to in the book as *The Dice*)

**de la Peña, L., Cetto, A.M.: Linear stochastic electrodynamics: looking for the physics behind quantum theory, abridged version of notes XXXI ELAF (1998). In: Hacyan, S., Jáuregui, R., López-Peña, R. (eds.) New Perspectives on Quantum Mechanics, AIP Conference Proceedings 464, New York, p 151 (1999)

*de la Peña, L., Cetto, A.M.: Quantum theory and linear stochastic electrodynamics. Found. Phys. **31**, 1703 (2001). arXiv:quant-ph/0501011v2

de la Peña, L., Cetto, A.M.: Recent developments in linear stochastic electrodynamics, In: Adenier, G., Khrennikov, A.Yu., Nieuwenhuizen, T.M. (eds.) Quantum theory: reconsideration of foundations-3. AIP Conference Proceedings no. 810, AIP, NY. Extended version in arXiv:quant-ph0/501011 (2006a)

*de la Peña, L., Cetto, A.M.: The foundations of linear stochastic electrodynamics. Found. Phys. **36**, 350 (2006b)

*de la Peña, L., Cetto, A.M.: On the ergodic behaviour of atomic systems under the action of the zero-point radiation field. In: Nieuwenhuizen, T.M. et al (eds.) Beyond the Quantum. World Scientific, Singapore (2007)

*de la Peña, L., Valdés-Hernández, A., Cetto, A.M.: Quantum mechanics as an emergent property of ergodic systems embedded in the zero-point radiation field. Found. Phys. **39**, 1240 (2009)

Einstein, A., Stern, O.: Einige argumente für die annahme einer molekularen agitation beim absoluten nullpunkt. Ann. Phys. **40**, 551 (1913)

Milonni, P.W.: Radiation reaction and the nonrelativistic theory of the electron. Phys. Lett. A **82**, 225 (1981)

Milonni, P.W.: The Quantum Vacuum. An Introduction to Quantum Electrodynamics. Academic Press, San Diego (1994)

Papoulis, A.: Probability, Random Variables, and Stochastic Processes. McGraw-Hill, Boston (1991)

Peres, A.: Quantum Theory: Concepts and Methods. Kluwer Academic Publishers, Dordrecht (1995)

Rice, S.O.: In: Wax, N. (ed.) Selected Papers on Noise and Stochastic Processes. Dover, New York (1954)

*Valdés-Hernández, A.: Investigación del origen del enredamiento cuántico desde la perspectiva de la Electrodinámica Estocástica Lineal. Ph. D. Thesis, Universidad Nacional Autónoma de México, Mexico (2010)

Chapter 6
Beyond the Schrödinger Equation

The lack of a concrete picture [for the electron spin] is the most satisfactory state of affairs.
W. Pauli,quoted in van der Waerden (1960) .

... I should like to preserve this ideal of the past, to describe everything that happens in the world with distinct images. I am ready to accept other theories, on condition that one is able to re-express them in terms of clear and distinct images.
H. A. Lorentz, quoted in Bacciagaluppi and Valentini 2009.

In previous two chapters, the quantum behavior of matter has been shown to emerge as a result of the permanent interaction with the random zero-point field. Fundamental quantum results, such as the Schrödinger and the Heisenberg formalism, have emerged within this framework. In this chapter, the theory developed so far will take us beyond the realm of quantum mechanics, in two important directions.

On the one hand, arriving at the Schrödinger (or Heisenberg) description meant neglecting the radiative terms—the radiation reaction and the Lorentz force due to the fluctuating vacuum field—once they had played their main role in taking the system to the quantum regime. Due consideration of these neglected terms will now allow us to calculate the effects that they produce on the already quantized system. Specifically, the (nonrelativistic) formulas of quantum electrodynamics for the atomic lifetimes and the Lamb shift will thus be recovered. More generally, the mean evolution of any integral of motion under a breakdown of the balance equations will be analyzed.

Another most relevant quantum phenomenon that cannot be predicted by the Schrödinger theory is the spin of the electron. The second part of this chapter is therefore devoted to an inquiry about the genesis of the electron spin from the perspective of the present treatment. Just as it gives rise to position, momentum and energy fluctuations, the ZPF is seen to induce an angular momentum resulting from the instantaneous torque exerted by the Lorentz force on the particle. A close analysis based on the separation of the modes of the ZPF of given circular polarization reveals the existence of a spin angular momentum of value $\hbar/2$, as well as of a corresponding magnetic moment with a g-factor of value 2, associated with the particle. This leads

L. de la Peña et al., *The Emerging Quantum*,
DOI: 10.1007/978-3-319-07893-9_6,

us to the identification of the spin of the electron as a further emergent property, generated by the interaction of the particle with the ZPF.

6.1 Radiative Corrections. Contact with QED

It is today widely accepted that the fluctuations of the electromagnetic vacuum are responsible for important observable physical phenomena that pertain to the realm of quantum electrodynamics. Among their best known manifestations are the finite atomic lifetimes. Indeed, the vacuum fluctuations are known to contribute, along with radiation reaction, to the 'spontaneous' transitions of the excited states (see e.g. Davydov 1965; Dalibard et al. 1982; Milonni 1994). Moreover, both the atomic Lamb shift and the Casimir and van der Waals forces have been shown to be attributable to changes in the energy of the vacuum field due to the presence of matter (see e.g. Boyer 1968, 1969; Milonni 1994; Bordag et al. 2009).

Within the traditional framework of QM, the analysis of the radiative corrections implies introducing by hand the quantized electromagnetic field (including its vacuum component) and using perturbative methods for the calculation of its effects. In SED, by contrast, the field—both radiation reaction and the ZPF—is there from the very beginning. It is in fact an essential ingredient in any quantum system: both the Schrödinger and the Heisenberg description have been obtained by considering its influence to zero order in τ (or e^2). Additional effects of this field on matter show up explicitly in the equations for the averaged dynamical variables that ensue from the generalized Fokker-Planck equation (see Sect. 4.2.1). It is therefore pertinent to investigate how these neglected radiative terms can be brought back into the picture in a self-consistent approach, and to calculate the effects of such terms on the quantum-mechanical system.

Strictly speaking, in order to find the *exact* solution of the problem of the particle subject to the radiation field one should revert to the original SED Hamiltonian equations of motion, and study the evolution of the complete (particle plus field) system into the quantum regime, without making approximations along the way. This would indeed be the ideal way to proceed. However, as stated in Chap. 4, solving this problem is beyond present possibilities. The practical solution, therefore, is to use the present, more restricted approach: let the full system evolve into the quantum regime, take the corresponding (Schrödinger) zero-order solutions, and use them to calculate the effects of the radiative terms contained in the original equations. This procedure leads to closed formulas for the most important radiative corrections to the already quantized system, to lowest significative order in the fine-structure constant $\alpha = e^2/\hbar c$. The results obtained do not show any difference with respect to the (nonrelativistic) QED predictions to the same order of approximation. However, in contrast to QED where these corrections represent the *main* effects due to the vacuum fluctuations, in SED they represent *secondary* effects, since the central effect of the action of the vacuum radiation field is quantum mechanics itself.

The calculation of radiative corrections has constituted a central component of the program of SED. A number of significant results have been thus obtained in the past, although with varying degrees of success and mostly restricted to the harmonic oscillator.[1] The derivations presented in this chapter are more general and therefore applicable to the atomic case, where they lead to formulas that are directly comparable to those of QED. [2]

6.1.1 Radiative Transitions

In Chap. 4 we found that in the time-asymptotic, Markovian regime, the evolution of the mean value of the particle Hamiltonian $H(\boldsymbol{p}, \boldsymbol{x}) = (1/2m)\boldsymbol{p}^2 + V(\boldsymbol{x})$ is determined by the Eq. (4.32), namely

$$\frac{d}{dt}\langle H\rangle = \tau\langle\dddot{\boldsymbol{x}}\cdot\boldsymbol{p}\rangle - \frac{e^2}{m}\left\langle\boldsymbol{p}\cdot\hat{\boldsymbol{\mathcal{D}}}\right\rangle, \tag{6.1}$$

where the first term on the right-hand side represents the average power lost by the particle through radiation reaction, and the second one represents the mean power exchanged between the particle and the background field. Energy balance exists if and only if $(d\langle H\rangle/dt) = 0$. Under this condition, we found that the ensuing description for the mechanical system is governed by a Schrödinger-like equation (4.78), namely

$$-\frac{2\eta^2}{m}\nabla^2\psi + V\psi = 2i\eta\frac{\partial\psi}{\partial t}. \tag{6.2}$$

The value of the parameter η was determined by imposing the balance condition

$$\tau\langle\dddot{\boldsymbol{x}}\cdot\boldsymbol{p}\rangle_0 = \frac{e^2}{m}\left\langle\boldsymbol{p}\cdot\hat{\boldsymbol{\mathcal{D}}}\right\rangle_0 \tag{6.3}$$

for the particle in the ground state. The value of the right-hand side term depends, through the diffusion operator $\hat{\boldsymbol{\mathcal{D}}}$, on the spectral energy density of the field. By introducing the value corresponding to the ZPF,

$$\rho_0(\omega) = \frac{\hbar\omega^3}{2\pi^2c^3}, \tag{6.4}$$

---

[1] A representative list of related works along the years is Kalitsin (1953), Sokolov and Tumanov (1956), Braffort et al. (1965), Braffort and Taroni (1967), Surdin (1970, 1974), Boyer (1968–1980), Santos (1974), de la Peña and Cetto (1976–1979), Moore (1977, 1984), Jáuregui and de la Peña (1981), Moore and Ramírez (1982), de la Peña and Jáuregui (1982) and Cetto and de la Peña (1988a, b, c). See also Davies (1982) .

[2] Previous versions of the material presented in this section can be found in de la Peña et al. (2010, 2012), Cetto and de la Peña (2012), and Cetto et al. (2012, 2013).

the correct result, namely $\eta = \hbar/2$, was obtained. This made it clear that the Schrödinger equation contains key information about the background field with which the particle is interacting. Additionally, the result indicates that detailed energy balance for a particle in its ground state takes place only when it is in equilibrium with a field of spectral energy density equal to ρ_0.

Now, for *any* state ψ that is solution of the Schrödinger equation, this latter predicts $d\langle H\rangle_\psi/dt = 0$ (provided that H does not depend explicitly on time)[3]. Yet, for a stationary state ψ_n the equation

$$\tau\langle \dddot{x}\cdot p\rangle_n = \frac{e^2}{m}\left\langle p\cdot\hat{\mathcal{D}}\right\rangle_n \tag{6.5}$$

does not hold in general for states other than the ground one (see Sect. 6.1.4). This could appear as contradicting the balance equation (6.1) applied to the n-state. However, the apparent contradiction is solved by observing that whereas $d\langle H\rangle_n/dt = 0$ is a strictly quantum-mechanical expression, consistent with the Schrödinger equation which has been derived neglecting terms of order τ (or e^2), the right-hand-side terms in (6.1) represent the *radiative* terms —precisely those that were neglected in the Schrödinger approximation. Now we shall take them into account. As a result, when the particle or the radiation field or both are in an excited state, the energy balance is broken in general, and radiative transitions take place. The finite lifetimes of the excited states are accounted for by these radiative terms. The corresponding rate of change is, therefore, not $d\langle H\rangle_n/dt$, but a (radiative) correction of the latter, which we denote as

$$\frac{d}{dt}\langle H\rangle_{n\mathrm{r}} = \tau\langle \dddot{x}\cdot p\rangle_n - \frac{e^2}{m}\left\langle p\cdot\hat{\mathcal{D}}\right\rangle_n . \tag{6.6}$$

For actual calculations it is simpler to use instead of Eq. (6.6) its alternative form, namely

$$\frac{d}{dt}\langle H\rangle_{n\mathrm{r}} = \tau\langle \dddot{x}\cdot p\rangle_n + \frac{1}{m}\left\langle \mathrm{Tr}\boldsymbol{D}^{pp}\right\rangle_n , \tag{6.7}$$

which ensues from Eq. (4.45) with $\mathcal{G} = H$.

In Sect. 6.1.4 a similar analysis will be shown to apply to more general 'classical' integrals of motion, i.e., dynamical variables that are conserved in the absence of the radiation terms. It is understood that all the calculations that follow involve the radiative terms. Therefore, and for simplicity in the writing, the additional index r will be dropped everywhere, except where it must be kept to avoid confusion.

---

[3] That $d\langle H\rangle_\psi/dt = 0$ can be easily verified resorting to the general form of the solution

$$\psi(x,t) = \sum_n c_n e^{-i\mathcal{E}_n t/\hbar}\varphi_n(x).$$

6.1.2 Breakdown of Energy Balance

Instead of considering the mechanical system in its ground state, as was done for the calculation of the parameter η in Sect. 4.4.4, we assume now that it is in an excited state n, the background field still being in its ground state (the ZPF). Then both terms on the right-hand side of Eq. (6.7) must be recalculated. Since the quantum regime has already been attained, the calculation of such terms is performed following the same procedure as in Sect. 4.4.4. In particular, the mean value $\tau \langle \boldsymbol{p} \cdot \dddot{\boldsymbol{x}} \rangle_n$ is given, in the one-dimensional case, for simplicity, by (see Eqs. (4.106) and (4.107))

$$\tau(\widehat{\ddot{x}}\,\hat{p})_{nn} = \tau(\hat{p}\,\widehat{\ddot{x}})_{nn} = -m\tau \sum_k \omega_{nk}^4 \left|x_{nk}\right|^2, \tag{6.8}$$

with $\omega_{nk} = (\mathcal{E}_n - \mathcal{E}_k)/\hbar$.

For the second term on the right-hand side of (6.7), we proceed as in the Appendix 4D for the calculation for $\langle D^{pp} \rangle_0$, just noticing that for negative values of ω_{kn} it is the second integral on the right-hand side of equation (D.8) that contributes to the sum. One thus obtains

$$\frac{1}{m}\left\langle D^{pp}\right\rangle_n = -m\tau \sum_k \omega_{nk}^4 \left|x_{nk}\right|^2 \operatorname{sign}\omega_{nk}. \tag{6.9}$$

This expression contains a mixture of positive and negative terms, whilst in (6.8) all contributions have the same sign. As follows from equations (6.7)–(6.9), the net loss of average energy per unit time is given by (recall that we are dropping the additional index r that appears in (6.7))

$$\begin{aligned}\frac{d\langle H\rangle_n}{dt} &= -m\tau \sum_k \omega_{nk}^4 \left|x_{nk}\right|^2 (1 - \operatorname{sign}\omega_{kn}) \\ &= -2m\tau \sum_{k<n} \omega_{nk}^4 \left|x_{nk}\right|^2 .\end{aligned} \tag{6.10}$$

The upshot is that there cannot be energy balance between the ZPF and a particle in an excited state—as was to be expected, since the ZPF is the background radiation field in its ground state. Only for $n = 0$ (hence all $k > n$) Eq. (6.10) gives zero, which means that only the ground state of the particle is sustained by the ZPF. Since in (6.10) the transitions from state n to (lower-energy) states k take place without the intervention of an external radiation field, we speak of 'spontaneous' transitions (more on this in the following section). The corresponding average energy loss per unit time in each such transition, W_{nk}, is obtained by writing the total average energy loss as a sum of contributions from the various possible transitions,

$$\frac{d\langle H\rangle_n}{dt} = \sum_k W_{nk}, \tag{6.11}$$

whence

$$W_{nk} = -2m\tau\omega_{nk}^4 |x_{nk}|^2 . \tag{6.12}$$

It is clear that expressions such as (6.10) have a meaning only in a statistical sense. To understand this meaning we have to think of an ensemble of systems all prepared initially in the same excited state n, and subject to the action of the ZPF. Then according to (6.10), the members of the ensemble have a certain probability per unit time to make a transition to a lower-energy state k, which is determined by the values of ω_{nk} and $|x_{nk}|$, i.e., by the specific properties of the system. However, which transition will take place in every instance is impossible to predict with the present statistical description.

Let us now inquire whether there is *any* (excited or external) background field with which a mechanical system in an excited state n can be in equilibrium. The excited background field is defined by its spectral energy density $\rho(\omega) = \rho_0(\omega)g(\omega)$, with $g(\omega) > 1$ an even function, so that

$$\frac{d\langle H\rangle_n}{dt} = -m\tau\sum_k \omega_{nk}^4 |x_{nk}|^2 (1 - g(\omega_{kn})\mathrm{sign}\omega_{kn}). \tag{6.13}$$

This expression is the generalization of (6.10) for the general density $\rho(\omega)$. To find the answer to the above question we observe that the terms within the parentheses have different signs, depending on whether ω_{kn} refers to an upward or a downward transition (i.e., $k > n$ or $k < n$). Therefore, there is no way that detailed balance can be satisfied in general.

Nevertheless, there is a particular system that can coexist with the field in an excited state, namely the harmonic oscillator. In this case, all $|\omega_{nk}|$ that contribute to the sum in (6.13) are equal in value and coincide with the oscillator frequency ω_0. Since for the harmonic oscillator,

$$|x_{nk}|^2 = \frac{\hbar}{2m\omega_0}\left[\delta_{k,n+1}(n+1) + \delta_{k,n-1}n\right], \tag{6.14}$$

the first term on the right-hand side of Eq. (6.13) gives

$$-m\tau\sum_k \omega_{nk}^4 |x_{nk}|^2 = -\tfrac{1}{2}\hbar\tau\omega_0^3(2n+1) \tag{6.15}$$

and the second terms gives

$$m\tau\sum_k \omega_{nk}^4 |x_{nk}|^2 g(\omega_{kn})\mathrm{sign}\omega_{kn} = \tfrac{1}{2}\hbar\tau\omega_0^3 g(\omega_{n+1n}) = \tfrac{1}{2}\hbar\tau\omega_0^3 g_n(\omega_0), \tag{6.16}$$

because $g(\omega_{n+1,n}) = g_n(\omega_0)$. Therefore, detailed balance exists if $g_n(\omega_0) = 2n+1$; in other words, if the harmonic oscillator in its excited state n is embedded in a background field with spectral energy density

$$\rho(\omega) = \rho_0(\omega)(2n+1), \tag{6.17}$$

there are as many absorptions as there are emissions per unit time, all with the same frequency ω_0, so that the average energy of the oscillator does not change. This result should not come as a surprise, since this field has precisely an energy per normal mode $\hbar\omega_0(2n+1)/2$, equal to the energy of the mechanical oscillator with which it is in equilibrium.

6.1.3 Atomic Lifetimes: Einstein's A and B Coefficients

We now investigate further implications of the absence of detailed balance. This can be conveniently done by using Eq. (6.13) to calculate the average energy lost (or gained) per unit time by the atom (or a mechanical system in general) prepared in an arbitrary state n when it is subject to the action of a radiation field with an arbitrary spectral energy density $\rho(\omega) = \rho_0(\omega)g(\omega)$. It is convenient to write the adimensional function g as $g(\omega) = 1 + g_a(\omega)$, in order to separate the contribution coming from the additional background field, so that

$$\begin{aligned} \rho(\omega) = \rho_0(\omega)g(\omega) &= \rho_0(\omega) + \rho_a(\omega), \\ \text{with} \quad \rho_a(\omega) &= \rho_0(\omega)g_a(\omega). \end{aligned} \tag{6.18}$$

Equation (6.13) thus writes as

$$\begin{aligned} \frac{d\langle H\rangle_n}{dt} &= -m\tau \sum_k \omega_{nk}^4 |x_{nk}|^2 \left[1 - (1 + g_a(\omega_{nk}))\text{sign}\omega_{kn}\right] \\ &= m\tau \sum_k \omega_{nk}^4 |x_{nk}|^2 \left[(g_a)_{\omega_{kn}>0} - (2 + g_a)_{\omega_{kn}<0}\right]. \end{aligned} \tag{6.19}$$

The first term within the brackets in the second line of this equation, proportional to g_a, represents the absorptions ($k > n$) and the second one, proportional to $2 + g_a$, the emissions ($k < n$). It is clear from this expression that there can be absorptions only when the background field is excited so that there is an external component $\rho_a \neq 0$. This additional field can therefore be identified with the *photonic* radiation field. The emissions, on the other hand, may be either 'spontaneous' (in presence of just the ZPF, as in the previous section) or else stimulated by the additional field, represented by g_a.

The coefficients appearing in the various terms determine the respective rates of energy gain and energy loss; therefore, they must be directly related to Einstein's A

and B coefficientsfor the transition probabilities. We recall that the coefficient A is defined as determining the time rate for spontaneous emissions (the term independent of g_a in Eq. (6.19)),

$$\frac{d}{dt}\langle H\rangle_n^{\text{sp em}} = -\hbar\sum_{k<n}\omega_{nk}A_{nk}. \tag{6.20}$$

Thus, A_{nk} is the probability that the atom realizes a spontaneous transition from state n to a lower state k during a unit of time, with a consequent loss of energy given by $\hbar\omega_{nk}$. In its turn, the coefficient B_{nk}^{em}, associated with the rate of energy loss due to transitions induced (stimulated) by the external field, is defined through

$$\frac{d}{dt}\langle H\rangle_n^{\text{ind em}} = -\hbar\sum_{k<n}\omega_{nk}B_{nk}^{\text{em}}\rho_a(\omega_{nk}), \tag{6.21}$$

whereas the coefficient B_{kn}^{abs}, associated with the rate of energy gain due to absorptions induced by the external field, is defined according to

$$\frac{d}{dt}\langle H\rangle_n^{\text{ind abs}} = \hbar\sum_{k>n}\omega_{kn}B_{kn}^{\text{abs}}\rho_a(\omega_{nk}). \tag{6.22}$$

The B coefficients are thus transition probabilities in presence of a photonic field with spectral energy density $\rho_a(\omega_{nk})$. The total rate of energy change can therefore be rewritten in the more transparent form

$$\begin{aligned}\frac{d\langle H\rangle_n}{dt} &= \frac{d}{dt}\langle H\rangle_n^{\text{sp em}} + \frac{d}{dt}\langle H\rangle_n^{\text{ind em}} + \frac{d}{dt}\langle H\rangle_n^{\text{ind abs}}\\ &= \sum_{k>n}\hbar\omega_{kn}B_{kn}^{\text{abs}}\rho_a(\omega_{nk})\\ &\quad -\sum_{k<n}\hbar\omega_{nk}\left[A_{nk}+B_{nk}^{\text{em}}\rho_a(\omega_{nk})\right],\end{aligned} \tag{6.23}$$

which can be recast, using (6.18), as

$$\frac{d\langle H\rangle_n}{dt} = \sum_k \hbar\left|\omega_{nk}\right|\left[\left(B_{kn}^{\text{abs}}\rho_0 g_a\right)_{\omega_{kn}>0} - \left(A_{nk}+B_{nk}^{\text{em}}\rho_0 g_a\right)_{\omega_{kn}<0}\right]. \tag{6.24}$$

Comparison of this expression with Eq. (6.19) gives for the spontaneous emissions coefficient

$$A_{nk} = \frac{4e^2\omega_{nk}^3}{3\hbar c^3}\left|x_{nk}\right|^2, \quad (n>k), \tag{6.25}$$

which is just the respective QED formula (see e.g. Louisell 1973). In its turn, the coefficients B are found to be

$$B_{kn}^{\text{abs}} = B_{nk}^{\text{em}} = \frac{4\pi^2 e^2}{3\hbar^2} |x_{nk}|^2 \equiv B_{nk}. \tag{6.26}$$

These results coincide with the respective formula of QED (or QM) (Louisell 1973). It is important to note that the expressions for the coefficients A_{nk}, and B_{nk} involve each one the single frequency ω_{nk}, which means that the transition between states involves a resonance with a field component of that frequency—in line with the results of Chap. 5—and a consequent exchange of energy given by $\Delta\mathcal{E}_{nk} = \hbar\omega_{nk}$.

The ratio of the A to the B coefficients is

$$\frac{A_{nk}}{B_{nk}} = \frac{\hbar |\omega_{nk}|^3}{\pi^2 c^3} = 2\rho_0(\omega_{nk}). \tag{6.27}$$

Notice in particular the factor 2 in this equation. Given the definition of the coefficients, one could have expected this ratio to correspond exactly to the spectral density of the ZPF, which would have meant a factor of 1 instead of the factor 2. This latter seems to suggest that the ZPF has double the ability of the rest of the electromagnetic field to induce transitions. The correct explanation, however, is another: the structure of Eq. (6.19) indicates that one should actually write $2\rho_0 = \rho_0 + \rho_0$. One of these two equal contributions to spontaneous decay is due to the effect of the fluctuations impressed on the particle by the field; the second one is the expected contribution due to Larmor radiation. Not surprisingly, they turn out to be equal: it is precisely their equality what leads to the exact balance between the two contributions when the system is in its ground state, guaranteeing the stability of this state. Yet one can frequently find in the literature that all the spontaneous decay is attributed to one or the other of these two causes, more frequently to Larmor radiation. It is an important result of both the present theory and quantum electrodynamics (provided the symmetric operator ordering is used) that the two effects contribute equal shares. Interesting related discussions can be seen in Davydov (1965), Fain (1966), Fain and Khanin (1969), Dalibard et al. (1982), Milonni (1994).

The relation (6.27) and the equality of both coefficients $B_{kn}^{\text{abs}} = B_{nk}^{\text{em}}$, were predicted by Einstein on the basis of statistical considerations in his 1916 paper on the theory of radiation. It is pertinent to ask here at which point the quantization enters in Einstein's paper—a somewhat confused issue that comes to surface every now and then.[4] A current answer to this question is that quantization is introduced by assuming discrete atomic levels. However, this is wrong, as Einstein and Ehrenfest demonstrated some time after the initial paper, by redoing the calculations with a continuous distribution of atomic levels (Einstein 1917) and recovering the old results. The correct answer is that quantization enters through the assumption of a physical

[4] The derivation of the Planck distribution from the A and B coefficients goes back to Einstein's 1916 work. In Chap. 3 the same law was derived by considering the existence of the ZPF. A strong relation should therefore exist between these two different forms of arriving at Planck's law. The remark made reveals this relation: the A coefficient is proportional to the spectral density of the ZPF. Therefore, postulating the existence of $\rho_0 \neq 0$ or of $A \neq 0$ are two different but equivalent forms of treating the problem.

source that can generate 'spontaneous' transitions. This can be easily verified by redoing the Einsteinian calculation but omitting the term that allows for spontaneous emissions, which leads to absurd results, such as atomic coefficients that depend on the temperature. On the other hand, it is interesting to observe that the omission of the term associated with stimulated emissions in Eq. (6.19) (after introducing appropriate populations) leads to the approximate expression for Planck's law that was proposed by Wien (Eq. 3.63), which is a fair approximation for low temperatures, so it already contains a quantum seed.

It is easy to follow the procedure used by Einstein to arrive at the Planck distribution from the A and B coefficients by focusing on just two states n and m, with $\mathcal{E}_n - \mathcal{E}_m = \hbar\omega_{nm} > 0$ and respective populations N_n, N_m. For a system in thermal equilibrium at temperature T, Einstein used the relation (k_B stands for the Bolztmann constant)

$$N_m/N_n = \exp(\mathcal{E}_n - \mathcal{E}_m)/k_B T, \tag{6.28}$$

leaving aside possible but inconsequential degeneracies. In line with Eq. (6.19) (first with $n = m$ and then with n) the number of absorptions ($m \to n$ transitions) is proportional to $N_m g_a(\omega_{nm})$, and the number of emissions ($n \to m$ transitions) is proportional to $N_n[2 + g_a(\omega_{nm})]$. From the equilibrium condition

$$N_m g_a = N_n\,(2 + g_a) \tag{6.29}$$

and Eq. (6.28) one thus obtains indeed Planck's law (for the thermal field)

$$g_a(\omega_{nm}) = \frac{2}{e^{(\mathcal{E}_n - \mathcal{E}_m)/k_B T} - 1}. \tag{6.30}$$

Notice that the equilibrium condition (6.29) implies detailed energy balance, since the individual emissions and absorptions involve a same amount of energy, $\hbar\omega_{nk}$. Notice also that this result implies the Bohr rule $(\mathcal{E}_n - \mathcal{E}_m)/\hbar = \omega_{nm}$, which was *derived* for the first time with this procedure.

6.1.4 A More General Equation for the Balance Breakdown

The procedure just used allows to calculate the rate of change of other dynamical quantities that correspond to 'classical' conserved variables. For this purpose we go back to Eq. (4.33), which holds for any dynamical quantity $\xi(\boldsymbol{x}, \boldsymbol{p})$ that is a 'classical' integral of motion (i.e., an integral of motion when the action of the radiation field is turned off),

$$\frac{d}{dt}\langle\xi\rangle = m\tau\left\langle \dddot{x}_i \frac{\partial\xi}{\partial p_i}\right\rangle - e^2\left\langle \frac{\partial\xi}{\partial p_i}\hat{\mathcal{D}}_i\right\rangle. \tag{6.31}$$

In one-dimensional problems the only integral of motion is H, so that in order to study the balance equation for $\xi \neq H$ it is necessary to analyze the more general, multidimensional case. Since both the Hamiltonian H and ξ are constants of motion, the corresponding operators, $\hat{H}$ and $\hat{\xi}$, commute, which means that in the energy representation the matrix associated with $\hat{\xi}$ is diagonal, i.e.,

$$\xi_{nk} = \xi_n \delta_{nk}, \tag{6.32}$$

where the index $n = (n'_H, n''_\xi)$ includes the quantum numbers corresponding to eigenvalues for both $\hat{H}$ and $\hat{\xi}$. If the system is in a state n, the (radiative) time evolution of $\langle \xi \rangle_n$ is, according to Eq. (6.31),

$$\frac{d}{dt} \langle \xi \rangle_{n\mathrm{r}} = m\tau \left\langle \dddot{x}_i \frac{\partial \xi}{\partial p_i} \right\rangle_n - e^2 \left\langle \frac{\partial \xi}{\partial p_i} \hat{\mathcal{D}}_i \right\rangle_n . \tag{6.33}$$

As before, we omit the subindex r in the following. For the calculation of both terms on the right-hand side we use the relation

$$i\hbar \frac{\partial \hat{\xi}}{\partial p_i} = [\hat{x}_i, \hat{\xi}], \tag{6.34}$$

whence

$$\left(\frac{\partial \xi}{\partial p_i} \right)_{kn} \to \frac{1}{i\hbar} [\hat{x}_i, \hat{\xi}]_{kn} = \frac{1}{i\hbar} x_{ikn} (\xi_n - \xi_k). \tag{6.35}$$

This expression, together with $\dddot{x}_{ink} = -i\omega_{nk}^3 x_{ink}$, leads to

$$[\hat{\dddot{x}}_i, \frac{\partial \hat{\xi}}{\partial p_i}]_{nn} = \frac{1}{\hbar} \sum_k |x_{ink}|^2 (\xi_k - \xi_n) \left(\omega_{nk}^3 + \omega_{kn}^3 \right) = 0, \tag{6.36}$$

due to the antisymmetry of ω_{nk}. This means that

$$\left(\hat{\dddot{x}}_i \frac{\partial \hat{\xi}}{\partial p_i} \right)_{nn} = \left(\frac{\partial \hat{\xi}}{\partial p_i} \hat{\dddot{x}}_i \right)_{nn}, \tag{6.37}$$

and hence no ambiguity in the operator ordering arises when calculating the first term in Eq. (6.33), which reduces to

$$m\tau \left\langle \dddot{x}_i \frac{\partial \xi}{\partial p_i} \right\rangle_n \to m\tau \left(\hat{\dddot{x}}_i \frac{\partial \hat{\xi}}{\partial p_i} \right)_{nn} = -\frac{m\tau}{\hbar} \sum_k \omega_{nk}^3 |\boldsymbol{x}_{nk}|^2 (\xi_n - \xi_k). \tag{6.38}$$

For the second term in (6.33) we resort again to the Markovian approximation and follow the same procedure as in appendix 4D, thus arriving at

$$e^2 \left\langle \frac{\partial \xi}{\partial p_i} \hat{\mathcal{D}}_i \right\rangle_n = \frac{4\pi e^2}{3\hbar^2} \int_0^\infty d\omega \rho(\omega) \int_{-\infty}^t dt' \cos\omega(t-t') \left[\hat{x}'_i, \left[\hat{x}_i, \hat{\xi} \right] \right]_{nn}. \tag{6.39}$$

Allowing for a possible additional (external) background field so that $\rho(\omega) = \rho_0(\omega) g(\omega)$, we get

$$\begin{aligned} e^2 \left\langle \frac{\partial \xi}{\partial p_i} \hat{\mathcal{D}}_i \right\rangle_n &= \frac{2m\tau}{\pi\hbar} \sum_k |\boldsymbol{x}_{nk}|^2 (\xi_n - \xi_k) \int_0^\infty d\omega\, \omega^3 g(\omega) \times \\ &\quad \times \int_{-\infty}^t dt' \cos\omega(t-t') \cos\omega_{nk}(t-t') \\ &= \frac{m\tau}{\hbar} \sum_k |\boldsymbol{x}_{nk}|^2 (\xi_n - \xi_k) \int_0^\infty d\omega\, \omega^3 g(\omega) [\delta(\omega + \omega_{nk}) + \delta(\omega - \omega_{nk})] \\ &= \frac{m\tau}{\hbar} \sum_k |\boldsymbol{x}_{nk}|^2 (\xi_n - \xi_k) |\omega_{nk}|^3 g(\omega_{nk}). \end{aligned} \tag{6.40}$$

With (6.38) and (6.40), Eq. (6.33) becomes (recall that $g(\omega_{kn}) = g(|\omega_{kn}|)$)

$$\frac{d}{dt} \langle \xi \rangle_n = -\frac{m\tau}{\hbar} \sum_k \omega_{nk}^3 |\boldsymbol{x}_{nk}|^2 (\xi_n - \xi_k) [1 - g(\omega_{kn}) \mathrm{sign}\omega_{kn}]. \tag{6.41}$$

This means that the ZPF with spectral energy density $\rho_0(\omega)$ (i.e. $g(\omega) = 1$) guarantees not only detailed energy balance, but more generally, detailed balance of *any* ('classical') integral of motion of the form here considered, for a mechanical system in its ground state (i.e., $\mathrm{sign}\omega_{kn} = +1$ for all ω_{kn}).

In analogy with Eq. (6.19), (6.41) can be rewritten as

$$\frac{d}{dt} \langle \xi \rangle_n = \frac{m\tau}{\hbar} \sum_k \omega_{nk}^3 |\boldsymbol{x}_{nk}|^2 (\xi_n - \xi_k) [(g_a)_{\omega_{kn} > 0} - (2 + g_a)_{\omega_{kn} < 0}]. \tag{6.42}$$

In terms of the Einstein coefficients given by (6.25) and (6.26) (expressed in three-dimensional notation), this equation gives

$$\frac{d}{dt} \langle \xi \rangle_n = \sum_{\omega_{kn} > 0} (\xi_k - \xi_n) \rho_a(\omega_{nk}) B_{kn} - \sum_{\omega_{kn} < 0} (\xi_n - \xi_k) [A_{nk} + \rho_a(\omega_{nk}) B_{nk}] \tag{6.43}$$

for the net change of $\langle \xi \rangle$ per unit time due to (upward and downward) radiative transitions from state n to states k.

6.1.5 Radiative Corrections to the Energy: The Lamb Shift

The calculations presented in previous pages confirm that the radiative terms neglected in the process of deriving QM give rise to corrections to the solutions of the (unperturbed, i.e. radiationless) Schrödinger equation. The Einstein A and B coefficients for the lifetimes of atomic states pertain to this category. A further important—even if smaller—radiative correction, one that represents a major success of QED, is the shift of the atomic levels due to another residual effect of the ZPF. Indeed, the effective work realized by the fluctuating motions of the bound particle gives rise to a tiny modification of the mean kinetic energy, as is shown in the following by means of a direct approach to the subject.

To calculate the radiative energy shift let us go back to the one-dimensional version of Eq. (4.29) (where the subindex Q reminds us that the calculations are carried out according to the quantum rules),

$$\frac{d}{dt}\langle xp\rangle_{\mathrm{Q}} = \frac{1}{m}\left\langle p^2\right\rangle_{\mathrm{Q}} + \langle xf\rangle_{\mathrm{Q}} + m\tau\,\langle x\,\dddot{x}\rangle_{\mathrm{Q}} - e^2\left\langle x\hat{\mathcal{D}}\right\rangle_{\mathrm{Q}}. \tag{6.44}$$

As explained is Sect. 4.2.1, this equation is a time-dependent version of the virial theorem, with radiative corrections included and the average values taken over the ensemble instead of over time, as is customarily done. In line with the discussion in Sect. 6.1.1, the correction $\langle\delta T\rangle_{n\mathrm{r}}$ (to lowest order in $\tau\sim e^2$) to the mean kinetic energy in the quantum state n, is given by

$$\langle\delta T\rangle_{n\mathrm{r}} = -\frac{m\tau}{2}\langle x\,\dddot{x}\rangle_n + \frac{e^2}{2}\left\langle x\hat{\mathcal{D}}\right\rangle_n, \tag{6.45}$$

where the two average values on the right-hand side are calculated using the solutions of the Schrödinger equation. We shall again dispose of the additional index r, whenever this does not lead to confusion.

The first term on the right-hand side of (6.45), associated with the Larmor radiation, can be approximated to lowest order in τ, by

$$-\frac{m\tau}{2}\langle x\,\dddot{x}\rangle_n = \frac{m\tau}{2}\langle\dot{x}\,\ddot{x}\rangle_n = \frac{m\tau}{4}\frac{d}{dt}\left\langle\dot{x}^2\right\rangle_n = 0, \tag{6.46}$$

and hence it does not contribute to the energy shift in the mean. Now, it is interesting to note that when $\langle x\,\dddot{x}\rangle$ is calculated in quantum terms, there is an ambiguity in the order of the factors (since $(\hat{x}\,\widehat{\dddot{x}})_{nn}\neq(\widehat{\dddot{x}}\,\hat{x})_{nn}$), and only the symmetrized operator

$$(\hat{x}\,\widehat{\dddot{x}})^S = \tfrac{1}{2}(\hat{x}\,\widehat{\dddot{x}} + \widehat{\dddot{x}}\,\hat{x}) \tag{6.47}$$

has a real and null mean value, in accordance with (6.46), which defines the rule of correspondence in this case. The anti-Hermitian, antisymmetric combination

$$(\hat{x}\,\hat{\ddot{x}})^A = \tfrac{1}{2}(\hat{x}\,\hat{\ddot{x}} - \hat{\ddot{x}}\,\hat{x}) \tag{6.48}$$

is not devoid of physical meaning, however. To identify this, we write

$$\langle \delta T \rangle_n^A \equiv -i\frac{m\tau}{2}(\hat{x}\,\hat{\ddot{x}})_{nn}^A = \frac{1}{2}m\tau \sum_k |x_{nk}|^2\,\omega_{nk}^3. \tag{6.49}$$

In terms of the spontaneous-emission coefficient given by Eq. (6.25) we have (for $\mathcal{E}_n > \mathcal{E}_k$)

$$\langle \delta T \rangle_n^A = \frac{\hbar}{4}\sum_k A_{nk} = \frac{\hbar}{4T_n}, \tag{6.50}$$

where $T_n = (\sum_k A_{nk})^{-1}$ is the lifetime of state n against spontaneous decay. This result shows that the expectation value of $(\hat{x}\,\hat{\ddot{x}})^A$ furnishes a measure of the (inverse) lifetime of the corresponding excited state—whereas the mean value of $(\hat{x}\,\hat{\ddot{x}})^S$ represents a contribution to the energy shift (which turns out to be zero) for the same state.

From Eq. (6.45) and the previous results, it follows that the correction to the energy is due solely to the coupling of the (instantaneous) electric dipole moment of the atom $d = ex$ to the electric component of the background field, represented by the second term in Eq. (6.45),

$$\langle \delta T \rangle_n = \frac{e^2}{2}\left\langle x\hat{\mathcal{D}}\right\rangle_n = \frac{e}{2}\left\langle d\hat{\mathcal{D}}\right\rangle_n. \tag{6.51}$$

The calculation of this term is carried out in appendix A, and gives the result

$$\frac{e^2}{2}\left\langle x\hat{\mathcal{D}}\right\rangle_n = -\frac{2e^2}{3\pi c^3}\sum_k |x_{nk}|^2\,\omega_{kn}\int_0^\infty d\omega\,\frac{\omega^3}{\omega_{kn}^2 - \omega^2}. \tag{6.52}$$

The radiative correction to the mean energy is thus (writing $\langle \delta T \rangle_n = \delta\mathcal{E}_n$ and in three dimensions, for comparison purposes)

$$\delta\mathcal{E}_n = \frac{e^2}{2}\left\langle \boldsymbol{x}\cdot\hat{\boldsymbol{\mathcal{D}}}\right\rangle_n = -\frac{2e^2}{3\pi c^3}\sum_k |\boldsymbol{x}_{nk}|^2\,\omega_{kn}\int_0^\infty d\omega\,\frac{\omega^3}{\omega_{kn}^2 - \omega^2}. \tag{6.53}$$

This coincides with the formula derived by Power in 1966 for the Lamb shift on the basis of Feynman's argument of 1961. We recall that according to Feynman, the presence of the atom creates a weak perturbation on the nearby field, thereby acting as a refracting medium. The effect of this perturbation is to change the frequencies of the background field in the vicinity of the atom from ω to $\omega/n(\omega)$, n being the refractive index. The shift of the ZPF energy due to the presence of the atom is then (Power 1966, see also Milonni 1994, Chap. 3)

$$\Delta\mathcal{E}_n = \sum_{\mathbf{k},\lambda} \frac{1}{2} \frac{\hbar\omega_k}{n(\omega_k)} - \sum_{\mathbf{k},\lambda} \frac{1}{2}\hbar\omega_k \simeq -\sum_{\mathbf{k},\lambda} [n(\omega_k) - 1] \frac{1}{2}\hbar\omega_k, \tag{6.54}$$

and the refractive index is given in this approximation by (Davydov 1965, Chap. 9)

$$n(\omega_k) \simeq 1 + \frac{4\pi}{3\hbar V} \sum_m \frac{|\boldsymbol{d}_{mn}|^2 \omega_{mn}}{\omega_{mn}^2 - \omega_k^2}, \tag{6.55}$$

where $\boldsymbol{d}_{mn} = e\boldsymbol{x}_{mn}$ is the electric dipole transition moment. After an integration over the solid angle $\hat{\boldsymbol{k}}$ and summation over the polarizations $\lambda = 1, 2$, Power obtains in the continuum limit for ω_k the formula

$$\Delta\mathcal{E}_n = -\frac{2}{3\pi c^3} \sum_m |\boldsymbol{d}_{mn}|^2 \omega_{mn} \int_0^\infty d\omega \, \frac{\omega^3}{\omega_{mn}^2 - \omega^2}, \tag{6.56}$$

which coincides with the previous result, Eq. (6.53).

The observable Lamb shift (called also Lamb shift proper) is obtained by subtracting from the total energy shift given by Eq. (6.53), the free-particle contribution, $\delta\mathcal{E}_{\text{fp}}$, represented by this same expression in the limit of continuous electron energies (when ω_{kn} can be ignored compared with ω in the denominator),

$$\delta\mathcal{E}_{\text{fp}} = \frac{2e^2}{3\pi c^3} \sum_m |\boldsymbol{x}_{nm}|^2 \omega_{mn} \int_0^\infty d\omega\, \omega = \frac{e^2\hbar}{\pi m c^3} \int_0^\infty d\omega\, \omega. \tag{6.57}$$

The last equality follows from the sum rule $\Sigma_m |\boldsymbol{x}_{nm}|^2 \omega_{mn} = 3\hbar/2m$. The Lamb shift proper of the energy level n is therefore given by

$$\delta\mathcal{E}_{\text{L}n} = \delta\mathcal{E}_n - \delta\mathcal{E}_{\text{fp}} = -\frac{2e^2}{3\pi c^3} \sum_k |\boldsymbol{x}_{nk}|^2 \omega_{kn}^3 \int_0^\infty d\omega \, \frac{\omega}{\omega_{kn}^2 - \omega^2}, \tag{6.58}$$

which again agrees with the nonrelativistic QED formula.[5] The logarithmic divergence of the integral calls for the introduction of the usual (nonrelativistic) regularizing cutoff $\omega_C = mc^2/\hbar$, which gives thus

$$\delta\mathcal{E}_{\text{L}n} = \frac{2e^2}{3\pi c^3} \sum_k |\boldsymbol{x}_{nk}|^2 \omega_{kn}^3 \ln \left| \frac{mc^2}{\hbar\omega_{kn}} \right|. \tag{6.59}$$

[5] In the denominator of Eq. (6.58) the term $\tau^2\omega^4$ due to Larmor radiation is missing (it has been neglected because the calculation is performed to lowest order in e^2). Its introduction is important for some applications, in particular to get a valid expression for the refractive index in Eqs. (6.55) and (6.58) (Sokolov and Tumanov 1956; de la Peña and Cetto 1977). This is a quite natural term in both QED and SED.

This is Bethe (1947) well known result. Note, however, that in the present approach (as in Power's) no mass renormalization was required.

The interpretation of the Lamb shift as a change of the atomic energy levels due to the interaction with the surrounding ZPF is fully in line with the general approach of the present theory. It constitutes one more manifestation of the influence of the particle on the near field, which is then fed back to the particle. An alternative way of looking at this reciprocal influence is by considering the general relation between the atomic polarizability α and the refractive index of the medium affected by it (for $n(\omega) \simeq 1$),

$$n(\omega) = 1 + 2\pi\alpha(\omega). \tag{6.60}$$

A comparison of this expression with Eq. (6.55) shows that

$$\alpha_n(\omega) = \frac{2}{3\hbar}\sum_m \frac{|\boldsymbol{d}_{mn}|^2\,\omega_{mn}}{\omega_{mn}^2 - \omega^2}, \tag{6.61}$$

which is the Kramers-Heisenberg formula (see Davydov 1965). This indicates that the Lamb shift can also be viewed as a Stark shift associated with the dipole moment $\boldsymbol{d}(\omega) = \alpha(\omega)\boldsymbol{E}$ induced by the electric component of the ZPF on the atom.

Let us recast Eq. (6.58) in a more familiar form, the one that is usual to find in textbooks and more adapted to direct calculation. In doing so it is convenient to introduce the energy $\mathcal{E}_n = \hbar\omega_n$ and the abbreviated notation

$$I_{nk} = \int_0^\infty d\mathcal{E}\, \frac{\mathcal{E}}{(\mathcal{E}_k - \mathcal{E}_n)^2 - \mathcal{E}^2}, \tag{6.62}$$

so that (6.58) rewrites as (with $\alpha = 3mc^2\tau/2\hbar$)

$$\begin{aligned}
\delta\mathcal{E}_{\mathrm{L}n} &= -\frac{2\alpha}{3\pi c^2}\sum_k I_{nk}\,|\omega_{kn}\boldsymbol{x}_{nk}|^2\,(\mathcal{E}_k - \mathcal{E}_n) \\
&= -\frac{2\alpha I_n}{3\pi m^2c^2}\sum_k |\boldsymbol{p}_{nk}|^2\,(\mathcal{E}_k - \mathcal{E}_n) \\
&= -\frac{2\alpha I_n}{3\pi m^2c^2} i\hbar \sum_k \boldsymbol{f}_{nk}\cdot\boldsymbol{p}_{kn},
\end{aligned} \tag{6.63}$$

since $\boldsymbol{p}_{nk} = i\omega_{nk}\boldsymbol{x}_{nk}$ and $\boldsymbol{f}_{nk} = i\omega_{nk}\boldsymbol{p}_{nk}$. We have assumed that I_{nk} depends so weakly on the index k that such dependence can be ignored; $\delta\mathcal{E}_{\mathrm{L}n}$ becomes, with $\hat{f} = -\nabla V$,

$$\begin{aligned}
\delta\mathcal{E}_{\mathrm{L}n} &= -\frac{2\alpha I_n}{3\pi m^2c^2} i\hbar \langle n|\, \boldsymbol{f}\cdot\boldsymbol{p}\, |n\rangle = \frac{\alpha I_n}{3\pi m^2c^2} i\hbar \langle n|\,[\nabla V, \boldsymbol{p}]\,|n\rangle \\
&= -\frac{\hbar^2\alpha I_n}{3\pi m^2c^2}\langle n|\,\nabla^2 V\,|n\rangle .
\end{aligned} \tag{6.64}$$

For the Coulomb potential, $\nabla^2 V = 4\pi Z e^2 \delta^3(\boldsymbol{r})$, so that only the wave function at the origin contributes to the Lamb shift in the present approximation.

The main interest of (6.64), apart from convenience of calculation, lies in the interpretation of the correction to the energy as due to fluctuations of the value of the potential V resulting from assumed fluctuations $\delta\boldsymbol{x}$ of the instantaneous position of the electron. Thus

$$V(\boldsymbol{x} + \delta(\boldsymbol{x})) = V(\boldsymbol{x}) + \delta x_i \frac{\partial V}{\partial x_i} + \frac{1}{2}\delta x_i \delta x_j \frac{\partial^2}{\partial x_i \partial x_j} V + \cdots \tag{6.65}$$

Assuming that the fluctuations are spherically symmetric on the average and statistically independent in orthogonal directions, so that in the mean $\overline{\delta x_i} = 0$, $\overline{\delta x_i \delta x_j} = \frac{1}{3}\overline{(\delta\boldsymbol{x})^2}\delta_{ij}$, the average (over the set of fluctuations) of the above equation becomes

$$\overline{V(\boldsymbol{x} + \delta(\boldsymbol{x}))} = V(\boldsymbol{x}) + \tfrac{1}{6}\overline{(\delta\boldsymbol{x})^2}\nabla^2 V + \cdots \tag{6.66}$$

Writing the energy shift (6.64) in terms of the expectation value of the deviation $\overline{V(\boldsymbol{x} + \delta(\boldsymbol{x}))} - V(\boldsymbol{x})$, and neglecting higher-order terms,

$$\delta\mathcal{E}_{\mathrm{L}n} = \left\langle \overline{V(\boldsymbol{x} + \delta(\boldsymbol{x}))} - V(\boldsymbol{x}) \right\rangle = \tfrac{1}{6}\overline{(\delta\boldsymbol{x})^2}\left\langle \nabla^2 V \right\rangle = -\frac{\hbar^2 \alpha I_n}{3\pi m^2 c^2}\left\langle \nabla^2 V \right\rangle \tag{6.67}$$

gives for the mean square displacement of $\boldsymbol{x}$

$$\overline{(\delta\boldsymbol{x})^2} = -\frac{2\hbar^2 \alpha I_n}{\pi m^2 c^2} = -\frac{2\alpha I_n}{\pi}\lambda_C^2, \quad \lambda_C = \frac{\hbar}{mc}, \tag{6.68}$$

where λ_C stands for the Compton wavelength. Thus

$$\frac{\sqrt{\overline{(\delta\boldsymbol{x})^2}}}{\lambda_C} = \sqrt{-\frac{4\alpha I_n}{\pi}} \sim 10^{-1}. \tag{6.69}$$

Qualitatively, this result is in line with the well-known proposal in Welton 1948 to identify the Lamb shift as a consequence of the fluctuations of the position of the electron due to the interaction with the vacuum field (Milonni 1994). We see that the fluctuations of $\boldsymbol{x}$ that give rise to the Lamb shift are smaller even in the mean than the Compton wavelength, which is by itself a small quantity in comparison with the Bohr radius $a_B = \hbar^2/me^2$, $\lambda_C = \alpha a_B$ ($\sim 10^{-9}$ cm for the electron). The Lamb shift is indeed a very small correction, so small that it requires the use of particularly refined spectroscopy to be detected.[6]

[6] There are several trembling motions of the electron, which should not be confused with each other. On one hand there is the zitterbewegung, a relativistic trembling of the order of λ_C, with a relatively well-defined frequency $\sim 2mc^2/\hbar$. This leads to the notion that an electron cannot be

It is interesting to observe that although the initial calculation of the Lamb shift on the basis of Eq. (6.45) exhibits it as an additional kinetic energy, the expression (6.64) describes it as the result of an extra potential energy. However, in both cases it is the fluctuations of $\boldsymbol{x}$ which give rise to the correction. The reason for this is that the fluctuations of the position variable generate both diffusion and extra potential energy. The two descriptions are therefore equivalent forms of accounting for the same process.

6.1.6 External Effects on the Radiative Corrections

By now it is clear that certain basic properties of the vacuum field—such as the intensity of its fluctuations or its spectral distribution—are directly reflected in the radiative corrections studied above. This means that a change in the properties of the vacuum should lead in principle to a corresponding modification of these corrections. The vacuum can be altered, for instance, by introducing material objects that modify the boundary conditions and hence affect the distribution of the normal modes of the field. Additionally, the background field can be modified by raising the temperature of the system or by introducing external radiation. These changes lead to observable effects on both the radiative lifetimes and the energy levels.

Such 'environmental' effects have been studied for more than 60 years, normally within the framework of quantum theory. However, some calculations have been made also within the framework of SED, in particular for the harmonic oscillator, leading to comparable results (see e.g. Cetto and de la Peña 1988a, b).[7] The formulas derived in the previous sections provide an opportunity to study the more general case, instead of restricting the calculations to the harmonic oscillator. The task is facilitated and becomes transparent by the use of the present theory, because the presence of the background radiation field is clear from the beginning.

6.1.6.1 External Effects on Atomic Lifetimes

In Sect. 6.1.3 we have already come across one observable effect of a change in the background field: according to Eq. (6.19) the rates of stimulated atomic transitions are directly proportional to the spectral distribution of the external (or additional) background field, be it a thermal field or otherwise. In the case of a thermal field at temperature T, in particular, with $g_a(\omega_{nk})$ given by Eq. (6.30), the (induced) transition rate from state n to state k becomes (with the help of Eqs. (6.4), (6.24) and (6.26))

---

confined to a region in space smaller than a Compton wavelength. As just discussed, the fluctuations of $\boldsymbol{x}$ associated with the Lamb shift are even smaller than λ_C.

[7] Related topics in which SED has been very successful are the Casimir effect and the van der Waals forces. The literature on this subject was iniciated with the old papers by Marshall (1965) and Boyer (1968, 1969), and followed by an extensive series of works by these authors and several others. A review of Boyer's work on the subject is Boyer (1980a); a more extensive list of references is given in *The Dice*. See also the literature at the end of the chapter.

$$\frac{dN_{nk}}{dt} = \rho_0(\omega_{nk}) g_a(\omega_{nk}) B_{nk}$$
$$= \frac{4e^2 \left|\omega_{nk}\right|^3 \left|x_{nk}\right|^2}{3\hbar c^3} \frac{1}{e^{\hbar|\omega_{nk}|/k_B T} - 1}. \tag{6.70}$$

This result shows that no single Hamiltonian eigenstate is stable at $T > 0$ (as is well known), because the thermal field induces both upward and downward transitions. For downward transitions ($\omega_{nk} > 0$) one can rewrite Eq. (6.70) for comparison purposes in terms of A_{nk} as given by (6.25),

$$\frac{dN_{nk}}{dt} = \frac{A_{nk}}{e^{\hbar\omega_{nk}/k_B T} - 1}. \tag{6.71}$$

This expression indicates that the effect of the thermal field on the decay rate is barely noticeable at room temperature ($k_B T \simeq 0.025$ eV), since for typical atomic frequencies, the denominator in Eq. (6.71) ranges between $\exp(-40)$ and $\exp(-400)$. In fact, the temperature would have to be raised by several orders of magnitude to obtain a noticeable effect on the lifetimes—but then other effects on atomic stability due to such high temperatures would be dominant. On the other hand, the simple addition of a high-intensity monochromatic field of frequency ω_{nk}—such as that produced by a laser—can have a visible effect on the emission rates, as is well known. Such effect in fact lies at the basis of the functioning of the laser.

Also when the geometry of the system or the spectral distribution of the background field are modified by the presence of nearby conducting objects, the transition rates are affected accordingly. Assume, for simplicity, that the modified field is still isotropic, with the density of modes of a given frequency ω_{nk} simply reduced by a (geometrical) factor $g(\omega_{nk}) < 1$. Then according to the results of Sect. 6.1.3 the corresponding spontaneous and induced transition rates are reduced by this factor, since both A and ρB are proportional to the density of modes. By enclosing the atoms in a high-quality cavity that excludes the modes of this frequency, one can therefore virtually inhibit the corresponding transition. For the more general, anisotropic case the calculations are somewhat more complicated, without however leading to a substantial difference from a physical point of view. These cavity effects have been the subject of a large number of experimental tests since the early works of Kleppner (1981), Goy et al. (1983), and others. In those cases where observable effects were predicted, the experiments have served to confirm the theoretical predictions.

Considering the essential role played by the background field in determining the basic structure and quantum behavior of the atom, according to SED, one may well wonder whether a geometric modification of this field (i.e., a change in the distribution of single modes of certain frequencies) would not have an impact on the basic behavior of the atom at the level of quantum mechanics. To respond to this question, let us look back at the energy-balance condition (6.5),

$$\tau \left\langle \dddot{\boldsymbol{x}} \cdot \boldsymbol{p} \right\rangle_n = \frac{e^2}{m} \left\langle \boldsymbol{p} \cdot \hat{\mathcal{D}} \right\rangle_n. \tag{6.72}$$

We observe that any such alteration of the field modes affects *both* the radiation reaction field and the ZPF on an equal footing. The reason is that on each side of this equation the field enters through its correlation function only (to lowest significative order in the approximation). Therefore, both sides of the equation are identically affected, and the equality continues to hold. This leads to an important conclusion, namely that the (quantum-mechanical) stationary solutions of the Schrödinger equation (which must comply with the balance equation) are immune to such environmental modifications of the background field.[8] This can be considered a sign of the robustness of the stationary quantum states.

6.1.6.2 External Effects on the Energy Levels

Let us now show how Eqs. (6.57) and (6.58) can be used to calculate the changes in the energy shift produced by the addition of an (external or thermal) background field. First we determine the shift $\delta\mathcal{E}_n(\rho)$ produced by the total field $\rho = \rho_0 + \rho_a$ (see Eq. (6.18)), following the same procedure that led to Eq. (6.53). Then, to the corresponding free-particle contribution $\delta\mathcal{E}_{\mathrm{fp}}(\rho)$ and the corresponding Lamb shift $\delta\mathcal{E}_{\mathrm{L}n}(\rho)$ we subtract the original shifts ($\delta\mathcal{E}_{\mathrm{fp}}(\rho_0)$ and $\delta\mathcal{E}_{\mathrm{L}n}(\rho_0)$) produced by the ZPF, thus obtaining the formulas for the variations of the (first-order) corrections. The calculation is straightforward using the cited equations, and the results are

$$\Delta\left(\delta\mathcal{E}_{\mathrm{fp}}\right) = \frac{4\pi e^2}{3\hbar}\sum_k |\boldsymbol{x}_{nk}|^2\,\omega_{kn}\int_0^\infty d\omega\,\frac{\rho_a}{\omega^2} = \frac{e^2\hbar}{\pi m c^3}\int_0^\infty d\omega\,\frac{\rho_a}{\rho_0}\omega, \tag{6.73}$$

$$\Delta\left(\delta\mathcal{E}_{\mathrm{L}n}\right) = -\frac{2e^2}{3\pi c^3}\sum_k |\boldsymbol{x}_{nk}|^2\,\omega_{kn}^3\int_0^\infty d\omega\,\frac{\rho_a}{\rho_0}\frac{\omega}{\omega_{kn}^2-\omega^2}, \tag{6.74}$$

for a homogeneous field. If, for instance, the additional field represents blackbody radiation at a temperature $T > 0$, i.e., if $\rho_a(\omega, T) = \rho_0(\omega) g_a(\omega, T)$ with $g_a(\omega, T)$ given by (6.30), then Eq. (6.73) gives

$$\Delta_T\left(\delta\mathcal{E}_{\mathrm{fp}}\right) = \frac{2\alpha}{\pi m c^2}(k_B T)^2\int_0^\infty dy\,\frac{y}{\exp y - 1}. \tag{6.75}$$

[8] A similar conclusion was arrived at in Cetto and de la Peña (1988a, b), where it was formulated in terms of a fluctuation-dissipation relation for SED.

With

$$\int_0^\infty dy \, \frac{y}{\exp y - 1} = \frac{\pi^2}{6} \tag{6.76}$$

equation (6.75) gives for the change of the free-particle energy the amount

$$\Delta_T \left(\delta\mathcal{E}_{\text{fp}}\right) = \frac{\pi\alpha}{3mc^2}(k_B T)^2. \tag{6.77}$$

Further, the formula for the change of the Lamb shift proper is given according to Eq. (6.74) by

$$\Delta \left(\delta\mathcal{E}_{\text{L}n}\right) = -\frac{4e^2}{3\pi c^3}\sum_k |\boldsymbol{x}_{nk}|^2 \, \omega_{kn}^3 \int_0^\infty d\omega \, \frac{\omega}{\omega_{kn}^2 - \omega^2}\left(\frac{1}{\exp(\hbar\omega/k_B T) - 1}\right). \tag{6.78}$$

These results coincide with those obtained within QED (Knight 1972; Zhou and Yu 2010), and the corresponding thermal shifts have been experimentally observed (see e.g. Hollberg and Hall 1984). From the point of view of SED (or QED) their interpretation is clear: they represent additional contributions to the kinetic energy impressed on the particle by the thermal field, according to the discussion at the beginning of Sect. 6.1.5.

6.2 The Spin of the Electron

A most fundamental problem in quantum theory relates to the origin and nature of the spin of the electron. Since this is a purely quantum phenomenon (albeit normally considered of relativistic origin), the present theory should be expected to provide an explanation for it, instead of merely taking it as one more intrinsic property of the particle such as its mass or its electric charge.

Despite its importance for QM, the question about the origin and nature of spin has received relatively little attention in SED. Indeed, during the initial period of SED the electron was considered mainly as a spinless particle; to our knowledge the only exception to this was the work of Braffort and Taroni (1967), showing the existence of some effects due to spin. The 1980s saw the publication of a series of phenomenological discussions by Moore and Ramírez (1982), Moore (1984), Cavalleri (1985) and Rueda (1993), focusing on the importance for quantum theory of the appearance of (helical) vibrations that might be identified with the zitterbewegung. By using an (otherwise classical) harmonic oscillator model for the electron and separating the ZPF into components of circular polarization, Jáuregui and de la Peña (1981), de la Peña and Jáuregui (1982), derived the mean squared electron spin angular momentum and its projections, within a numerical factor of order 1; see also Sachidanandam(1983). Similarly, Barranco and coworkers 1989 studied the spin and the magnetic moment of a (classical) particle subject to the ZPF, using a simple composite particle model with two constituents bound by a harmonic force. More recently, Muralidhar (2011) published a suggestive derivation of spin by assuming that the zero-point energy of the (free) electron, considered as a classical particle, is an energy of rotation within the region of space surrounding the particle. Specifically, by expressing the ZPF fluctuations as rotations on a complex plane, the spin angular momentum appears in this model connected with the imaginary part of the rotations.

The various SED calculations, though based on classical models, have led in one way or another to a result of order $\hbar^2$ for the mean square value of the spin and of order $\hbar$ for the spin projections. This strongly suggests the possibility of finding a reasonable explanation for the electron spin within the present approach. Our analysis of this possibility, presented in the following sections, exhibits the electron spin as one more emergent property arising from the interaction of the electron with the ZPF.

6.2.1 Unravelling the Spin

Let us again recall the equations for the averaged dynamical variables that ensue from the generalized Fokker-Planck equation, derived in Sect. 4.2.1. We are interested, in particular, in the balance equation for the angular momentum components

$$L_{ij} = x_i p_j - x_j p_i, \tag{6.79}$$

obtained from Eq. (4.28), namely

$$\frac{d}{dt}\langle L_{ij}\rangle_{\mathrm{Q}} = \langle M_{ij}\rangle_{\mathrm{Q}} + m\tau\langle x_i\dddot{x}_j - x_j\dddot{x}_i\rangle_{\mathrm{Q}} - e^2\left\langle x_i\hat{\mathcal{D}}_j - x_j\hat{\mathcal{D}}_i\right\rangle_{\mathrm{Q}}. \tag{6.80}$$

Here $M_{ij} = x_i f_j - x_j f_i$ is a component of the momentum of the external force (or torque), and the additional terms are due to radiation reaction and diffusion. For simplicity let us consider that only the ZPF is present and that the particle is in its ground state, $n = 0$. Then, since the state is (truly) stationary, $d\left\langle L_{ij}\right\rangle/dt = 0$, and Eq. (6.80) gives

$$\langle M_{ij}\rangle_0 + m\tau\langle x_i\dddot{x}_j - x_j\dddot{x}_i\rangle_0 - e^2\left\langle x_i\hat{\mathcal{D}}_j - x_j\hat{\mathcal{D}}_i\right\rangle_0 = 0. \tag{6.81}$$

To lowest order in τ one may take

$$m\langle x_i\dddot{x}_j - x_j\dddot{x}_i\rangle_0 = \frac{d}{dt}\langle M_{ij}\rangle_0 - \langle\dot{x}_i f_j - \dot{x}_j f_i\rangle_0, \tag{6.82}$$

which under stationarity reduces to

$$m\tau\langle x_i\dddot{x}_j - x_j\dddot{x}_i\rangle_0 = -\tau\langle\dot{x}_i f_j - \dot{x}_j f_i\rangle_0, \tag{6.83}$$

and Eq. (6.81) becomes thus

$$\left\langle M_{ij}\right\rangle_0 - \tau\left\langle \dot{x}_i f_j - \dot{x}_j f_i\right\rangle_0 = e^2\left\langle x_i\hat{\mathcal{D}}_j - x_j\hat{\mathcal{D}}_i\right\rangle_0. \tag{6.84}$$

For central forces, $M_{ij} = 0$ and one may write $f_i = g(r)x_i$ with $g(r)$ a radial function, whence

$$\tau\left\langle \dot{x}_i f_j - \dot{x}_j f_i\right\rangle_0 = \tau\left\langle g(r)\left(\dot{x}_i x_j - \dot{x}_j x_i\right)\right\rangle_0 = -\frac{\tau}{m}\left\langle g(r)L_{ij}\right\rangle_0, \tag{6.85}$$

which introduced in (6.84) leads to the balance equation for the angular momentum,

$$\tau\left\langle g(r)L_{ij}\right\rangle_0 = me^2\left\langle x_i\hat{\mathcal{D}}_j - x_j\hat{\mathcal{D}}_i\right\rangle_0. \tag{6.86}$$

Alternatively, in the Markovian limit equation (4.45) can be applied directly to $\mathcal{G} = L_{ij}$, thus obtaining

$$m\tau\left\langle x_i\dddot{x}_j - x_j\dddot{x}_i\right\rangle_0 = \left\langle D_{ij}^{px} - D_{ji}^{px}\right\rangle_0. \tag{6.87}$$

By combining this with Eqs. (6.83) and (6.85), we get a balance condition for the angular momentum components,

$$\tau\left\langle g(r)L_{ij}\right\rangle_0 = -m\langle D_{ji}^{px} - D_{ij}^{px}\rangle_0. \tag{6.88}$$

Alternatively, one may take directly the (stochastic) equation of motion for a central-force problem,

$$\dot{p}_i = f_i + m\tau\dddot{x}_i + eE_i(t), \tag{6.89}$$

where $p_i = m\dot{x}_i$, $E_i(t)$ is the electric component of the ZPF, and $f_i = g(r)x_i$. By multiplying this equation by x_j and anti-symmetrizing one gets

$$\dot{p}_i x_j - \dot{p}_j x_i = m\tau\left(\dddot{x}_i x_j - \dddot{x}_j x_i\right) - e\left(x_i E_j - x_j E_i\right). \tag{6.90}$$

Under stationarity, i.e. for the system in its ground state, one can perform the substitutions $\dot{p}_i \to f_i$ and $mx_i\dddot{x}_j \to -\dot{x}_i f_j$ (valid to zero order in τ), thus arriving at

$$\overline{\tau g(r)L_{ij}}^{(i)} = -me\overline{\left(x_i E_j - x_j E_i\right)}^{(i)} = -me\overline{(\boldsymbol{x}\times\boldsymbol{E})}_{ij}^{(i)}. \tag{6.91}$$

This expression, which is equivalent to (6.86), shows that there is an angular momentum $\boldsymbol{L}$ that results from the instantaneous torque exerted by the Lorentz force on the particle. Since only the fluctuating component of $\boldsymbol{x}$ can contribute to the average $\overline{(\boldsymbol{x}\times\boldsymbol{E})}^{(i)}$ (because $E_i(t)$ is purely random), it is clear that all the angular momentum thus generated is due to the random motion around the mean trajectory followed by the particle; thus, it is independent of the system of coordinates, and has an 'internal' (intrinsic) nature. This angular momentum will turn out to be a crucial phenomenon in our proposal for the origin of the electron spin.

That the Lorentz force due to the background field exerts indeed a torque on the particle, is explained by the following argument. It is experimentally observed that the interaction of the electron with the radiation field takes place via the circular polarized modes of the field (or modes of a certain helicity). This is known to be the case for the photonic field, which, from the present perspective, is the excited state of the radiation field, additional to the zero-point component. It is therefore natural to assume that the same is true for the interaction of the electron with the modes of the field in its ground state, i.e., the ZPF. To analyze the effect of such interaction, one should therefore consider the ZPF as composed of modes of both right- and left-handed circular polarization. Taken separately, these modes will indeed induce, through the Lorentz force, rotational (viz right- and left-handed) motions on the particle. The two effects will tend to conceal each other when the ensemble of modes and of particles is considered in its entirety; yet by focusing on one of the two subensembles of a given polarization of the field modes, the effective rotation induced on the corresponding particles should be disclosed. It is just this mean rotation present in each one of the two separate subensembles what will be identified below with the two degrees of freedom proper to each of the corresponding spin-1/2 states of the particle.

6.2.2 The Isotropic Harmonic Oscillator

As an example of application of the results of the previous section, let us consider the simplest case, namely a spherical isotropic harmonic oscillator of natural frequency ω_0, in its ground state. In this case $f_i = -m\omega_0^2 x_i$ so that $g(r) = -m\omega_0^2$, and Eq. (6.88) becomes

$$\left\langle L_{ij}\right\rangle_0 = \frac{1}{\tau\omega_0^2}\langle D_{ji}^{px} - D_{ij}^{px}\rangle_0. \tag{6.92}$$

The calculation of the *right-hand* side is carried out in appendix B. The result, in terms of the Cartesian components of $\hat{\boldsymbol{x}}$ and $\hat{\boldsymbol{p}}$, is

$$\left\langle L_{ij}\right\rangle_0 = \langle 0|\left(\hat{x}_i\hat{p}_j - \hat{x}_j\hat{p}_i\right)|0\rangle\,. \tag{6.93}$$

This convergence of results shows that indeed the fluctuations generate the angular momentum L_{ij}, as predicted by Eq. (6.92). Now, the numerical value $\left\langle L_{ij}\right\rangle_0 = 0$ for the ground state represents the net (average) angular momentum induced on the isotropic harmonic oscillator by the full ZPF. According to the discussion following Eq. (6.91), however, we should analyze separately the contributions arising from each of the two circular polarizations, characterized by the (circularly polarized) vectors

$$\epsilon_{k\pm} = \frac{1}{\sqrt{2}} \left(\epsilon_{ki} \pm i \epsilon_{kj} \right), \tag{6.94}$$

with ϵ_{ki}, ϵ_{kj} unit Cartesian vectors orthogonal to some axis $\boldsymbol{k}$, assumed to be fixed.

We will therefore now consider a situation in which the particle (the oscillator) is under the action of the background field of a given circular polarization with respect to an axis $\boldsymbol{k}$. Since according to the results in chapter 5 the response of the particle to the field is linear, when the particle is acted upon the (circularly) polarized field the appropriate variables to describe the system are the spherical variables (x^+, x^-, x_k), which are given, in accordance with (6.94), by,

$$x^{\pm} = \frac{1}{\sqrt{2}} \left(x_i \mp i x_j \right), \tag{6.95a}$$

$$x_i = \frac{1}{\sqrt{2}} \left(x^+ + x^- \right), \quad x_j = i \frac{1}{\sqrt{2}} \left(x^+ - x^- \right). \tag{6.95b}$$

The nonzero matrix elements of the oscillator are $x_{i01} = \left(x_{01}^+ + x_{01}^- \right) / \sqrt{2}$, and so on. Further, since $x_{10}^{\pm} = \left(x_{01}^{\mp} \right)^*$, Eq. (6.93) becomes

$$\left\langle L_{ij} \right\rangle_0 = m\omega_0 \left(x_{01}^+ x_{10}^- - x_{01}^- x_{10}^+ \right) = m\omega_0 \left(\left| x_{01}^+ \right|^2 - \left| x_{01}^- \right|^2 \right). \tag{6.96}$$

In the ground state, $\left\langle L_{ij} \right\rangle_0 = 0$; hence the two terms, $\left| x_{01}^+ \right|^2$ and $\left| x_{01}^- \right|^2$, contribute with equal magnitude and opposite sign to the k-th component of the total oscillator's angular momentum, as should be the case for a nonpolarized vacuum. These separate contributions are

$$\left\langle L_{ij} \right\rangle_0^+ = m\omega_0 \left| x_{01}^+ \right|^2, \quad \left\langle L_{ij} \right\rangle_0^- = -m\omega_0 \left| x_{01}^- \right|^2. \tag{6.97}$$

Using $x_{i01} = x_{j01} = \sqrt{\hbar/(2m\omega_0)}$ for the harmonic oscillator, one gets

$$m\omega_0 \left| x_{01}^{\pm} \right|^2 = \frac{\hbar}{2}, \tag{6.98}$$

whence the size of each separate contribution to the angular momentum in (6.96) is just $\hbar/2$. In order to distinguish this contribution from the (orbital) component of the angular momentum we write $\left\langle S_{ij} \right\rangle^{\pm}$ instead of $\left\langle L_{ij} \right\rangle_0^{\pm}$, so that

$$\left\langle S_{ij} \right\rangle^{\pm} = \pm \frac{\hbar}{2}. \tag{6.99}$$

Further, to study the square of the angular momentum we take into account that $\boldsymbol{L}^2$ corresponds to a 'classical' integral of motion for central forces; thus from equation (6.31) with $\xi = \boldsymbol{L}^2$, we get that for the (stationary) ground state, $\boldsymbol{L}^2$ satisfies the equation

$$m\tau\left\langle \dddot{x}_i \frac{\partial \boldsymbol{L}^2}{\partial p_i}\right\rangle_0 = e^2\left\langle \frac{\partial \boldsymbol{L}^2}{\partial p_i}\hat{\mathcal{D}}_i\right\rangle_0. \tag{6.100}$$

The approximation $\dot{p}_k = f_k$ allows us to write $m\dddot{x}_k = -\omega_0^2 p_k$ for the harmonic oscillator. Further, since $\boldsymbol{L}^2 = \boldsymbol{r}^2\boldsymbol{p}^2 - (\boldsymbol{r}\cdot\boldsymbol{p})^2$ is a homogeneous function of $\boldsymbol{p}$ of degree 2, we have

$$p_i \frac{\partial \boldsymbol{L}^2}{\partial p_i} = 2\boldsymbol{L}^2. \tag{6.101}$$

With these results Eq. (6.100) reduces to

$$\left\langle \boldsymbol{L}^2\right\rangle_0 = -\frac{3mc^3}{4\omega_0^2}\left\langle \frac{\partial \boldsymbol{L}^2}{\partial p_i}\hat{\mathcal{D}}_i\right\rangle_0. \tag{6.102}$$

The right-hand side of this equation is calculated in appendix C, and gives a result different from zero—even for s-states. Dividing again the full ensemble into two subensembles corresponding to different circular polarizations, one obtains

$$\left\langle \boldsymbol{L}^2\right\rangle_0 = \left\langle \boldsymbol{L}^2\right\rangle_0^+ + \left\langle \boldsymbol{L}^2\right\rangle_0^-, \tag{6.103}$$

where each separate contribution to the mean square angular angular momentum is given by

$$\left\langle \boldsymbol{L}^2\right\rangle_0^+ = \left\langle \boldsymbol{L}^2\right\rangle_0^- = \frac{3}{4}\hbar^2. \tag{6.104}$$

In terms of the notation introduced above (Eq. (6.99))—that distinguishes the (mean) orbital angular momentum from the (mean) angular momentum induced by the circular polarizations of the ZPF—Eq. (6.104) rewrites as

$$\left\langle \boldsymbol{S}^2\right\rangle^+ = \left\langle \boldsymbol{S}^2\right\rangle^- = \frac{3}{4}\hbar^2. \tag{6.105}$$

The fact that the results in Eqs. (6.99) and (6.105) do not depend on the oscillator's frequency ω_0, suggests that they hold in the general case, and for the free particle in particular (see Sect. 6.2.3). Therefore, we can conclude that when the transformation (6.95a, 6.95b) possesses physical meaning, so that the decompositions $\left\langle L_{ij}\right\rangle_0 = \left\langle L_{ij}\right\rangle_0^+ + \left\langle L_{ij}\right\rangle_0^-$ and $\left\langle \boldsymbol{L}^2\right\rangle_0^+ = \left\langle \boldsymbol{L}^2\right\rangle_0^+ + \left\langle \boldsymbol{L}^2\right\rangle_0^-$ make sense, equations (6.99) and (6.105) tell us that there exists an angular-momentum component that does *not* correspond to an orbital motion of the particle. It represents the angular momentum derived from the interaction of the particle with a given circular polarized mode of the ZPF. For an electron, which (as stated above) interacts with the radiation field via its circular polarized modes, the transformation (6.95) is indeed physically meaningful; thus the angular momentum induced by the ZPF can be identified with the electron's spin. The term 'intrinsic' usually attached to it, points to the permanence of this effect (the ZPF is always and everywhere present), although here appears as induced on (or acquired by) the particle.

We have thus disclosed the existence of the vector $\boldsymbol{S}$ (with k-component $\epsilon_{ijk} S_{ij}$) that stands for the intrinsic spin-angular momentum of a charged particle for which the circularly polarized modes of the ZPF are physically relevant. In this form we see that the theory does contain the elements to account for the spin of the electron. The extension of these results to the general case (arbitrary potential), as well as the construction of the quantum operator corresponding to the vector $\mathbf{S}$, are left for the following section.

One additional point merits consideration. According to these results a charged particle acquires spin 1/2; so what can be said about scalar bosons? A possible answer to this question is that the spin 1/2 is acquired by elementary particles, such as the electron, whereas composite particles may acquire it or not, depending on their structure and their specific interaction with the radiation field. Thus, bosons come out to be composite structures, with an even number of elements (if of fermion type).

6.2.3 General Derivation of the Electron Spin

Let us now present a general procedure to derive the spin of the electron, based on the Heisenberg formalism of QM developed in Chap. 5. According to the above discussion, our ansatz is that the electron spin is an angular momentum of the particle generated by its interaction with the ZPF, specifically with one of the circular polarizations of the field. In contrast with the previous section, here the electron is subject to an arbitrary external central force, and in any given stationary state. Denoting with the index n the set of quantum numbers that characterize the state of the particle, including the orbital angular momentum and its projection along the z axis, we have (for simplicity in the writing we use $x_i = x$, $x_j = y$, and $x_k = z$)

$$\begin{aligned} \left\langle \hat{L}_z \right\rangle_n &= \langle n | \hat{L}_z | n \rangle = \sum\nolimits_k \left(x_{nk} p_{ykn} - y_{nk} p_{xkn} \right) \\ &= im \sum\nolimits_k \omega_{kn} \left(x_{nk} y_{kn} - y_{nk} x_{kn} \right). \end{aligned} \tag{6.106}$$

To take into account that the electron responds to modes of a given circular polarization of the ZPF, we transform again to the variables defined in Eqs. (6.95). Following the same procedure that led to (6.96), Eq. (6.106) transforms into

$$\left\langle \hat{L}_z \right\rangle_n = m \sum\nolimits_k \omega_{kn} \left(\left| x_{nk}^{+} \right|^2 - \left| x_{nk}^{-} \right|^2 \right). \tag{6.107}$$

This expression can be rewritten as

$$\left\langle \hat{L}_z \right\rangle_n = \langle O_z \rangle_n^{+} + \langle O_z \rangle_n^{-}, \tag{6.108}$$

with $\langle O_z\rangle_n^{\pm}$ given by ($\sigma = \pm$)

$$\langle O_z\rangle_n^{\sigma} = \sigma m \sum\nolimits_k \omega_{kn} \left|x_{nk}^{\pm}\right|^2 . \tag{6.109}$$

On the other hand, from the commutator $\left[\hat{x}, \hat{p}_x\right] = i\hbar$ and $x_{kn} = x_{nk}^*$ it follows that

$$m \sum\nolimits_k \omega_{kn} \left|x_{nk}\right|^2 = \tfrac{1}{2}\hbar \tag{6.110}$$

and the same for $\left|y_{nk}\right|^2$. From (6.106) we obtain therefore

$$\hbar = m \sum\nolimits_k \omega_{kn} \left(\left|x_{nk}^{+}\right|^2 + \left|x_{nk}^{-}\right|^2\right) = \langle O_z\rangle_n^{+} - \langle O_z\rangle_n^{-} , \tag{6.111}$$

which combined with (6.108) gives

$$\langle O_z\rangle_n^{\sigma} = \tfrac{1}{2}\left\langle \hat{L}_z\right\rangle_n + \sigma\tfrac{1}{2}\hbar. \tag{6.112}$$

This result helps to avoid a possible confusion about the meaning of equation (6.108). It is simply

$$\left\langle \hat{L}_z\right\rangle_n = \tfrac{1}{2}\left\langle \hat{L}_z\right\rangle_n + \tfrac{1}{2}\hbar + \tfrac{1}{2}\left\langle \hat{L}_z\right\rangle_n - \tfrac{1}{2}\hbar. \tag{6.113}$$

Thus the spin projection is contained in the orbital angular momentum $\langle \hat{L}_z\rangle_n$ in a *dormant* form, so to say. It is the quantity $\langle O_z\rangle_n^{\sigma}$ what contains both a part of the orbital angular momentum and the spin associated with one or the other of the polarization states, $\sigma = \pm$.

To construct the (quantum) operator associated with the vector **S**, we proceed as follows. Clearly the mean value $\langle \hat{L}_z\rangle_n$ does not depend on σ, whereas the term $\sigma\hbar/2$ does not depend on the set n; this shows that the operator $\hat{L}_z$ and the operator to be associated with $\sigma\hbar/2$ (which we shall call $\hat{\Sigma}_z$) belong to different Hilbert spaces. Therefore, in order to express $\langle O_z\rangle_n^{\sigma}$ in (6.112) as the average of an operator, we must extend the Hilbert space to include the dichotomous variable σ in addition to the quantum index n. This we do by resorting to the product space $\mathcal{H} = \mathcal{H}_n \otimes \mathcal{H}_2$, with $\mathcal{H}_2$ a bidimensional vector space spanned by an orthonormal basis having as elements the vectors $\{|\sigma\rangle\} = \{|+\rangle, |-\rangle\}$. In terms of $|n\sigma\rangle = |n\rangle \otimes |\sigma\rangle$, Eq. (6.112) rewrites as

$$\langle O_z\rangle_n^{\sigma} = \tfrac{1}{2}\langle n\sigma|\, \hat{L}_z\, |n\sigma\rangle + \tfrac{1}{2}\hbar \langle n\sigma|\, \hat{\Sigma}_z\, |n\sigma\rangle , \tag{6.114}$$

with $\hat{\Sigma}_z$ an operator that has $|\sigma\rangle$ as eigenvector,

$$\langle n\sigma|\, \hat{\Sigma}_z\, |n\sigma\rangle = \langle\sigma|\, \hat{\Sigma}_z\, |\sigma\rangle = \sigma. \tag{6.115}$$

Expressing $\hat{\Sigma}_z$ in the general form in terms of the Pauli matrices gives

$$\hat{\boldsymbol{\Sigma}}_z = a_0\mathbb{I} + a_z\hat{\sigma}_z + a^+\hat{\sigma}_+ + a^-\hat{\sigma}_-, \tag{6.116}$$

where $a^\pm = \left(a_x \mp ia_y\right)/\sqrt{2}$, and $\hat{\sigma}_+ = \sqrt{2}\left|+\right\rangle\left\langle-\right|, \hat{\sigma}_- = \sqrt{2}\left|-\right\rangle\left\langle+\right|$ are ladder operators. Condition (6.115) imposed on $\hat{\boldsymbol{\Sigma}}_z$ gives $a_0 = 0$, $a_z = 1$. Further, since we are here considering the variables (x^+, x^-, z), the polarization vectors (6.94) fix $\hat{z}$ as the preferred axis, whence $a^\pm = 0$, $\hat{\boldsymbol{\Sigma}}_z = \hat{\sigma}_z$, and Eq. (6.114) becomes

$$\langle O_z\rangle_n^\sigma = \langle n\sigma|\left(\tfrac{1}{2}\hat{\boldsymbol{L}} + \hat{\boldsymbol{S}}\right)\cdot\hat{z}\left|n\sigma\right\rangle, \tag{6.117}$$

with $\hat{\boldsymbol{S}}$ the vector operator defined as $\hat{\boldsymbol{S}}_z = \hbar\frac{1}{2}\hat{\Sigma}_z$, i.e.,

$$\hat{\boldsymbol{S}} = \tfrac{1}{2}\hbar\hat{\boldsymbol{\sigma}}. \tag{6.118}$$

The identification of the operator $\hat{\boldsymbol{S}}$ with the spin of the electron is thus justified. The independence of $\langle\hat{L}_z\rangle_n$ from σ and of $\langle\hat{\Sigma}_z\rangle$ from n, indicates that under the present conditions, the fluctuations associated with the spin are not correlated with those that characterize the kinematics of the particle in the configuration space: $\boldsymbol{L}$ and $\boldsymbol{S}$ constitute independent dynamical variables. Of course the spaces of the spin and of the orbital angular momentum may become connected by the presence of magnetic fields, which here have been omitted.

It should be stressed that even if the orbital momentum $\hat{\boldsymbol{L}}$ and the spin $\hat{\boldsymbol{S}}$ are both contained in the same expression for the angular motions, Eq. (6.113), this does not mean that the spin is treated here as an orbital angular momentum. Indeed, as is well known, there are fundamental differences between $\hat{\boldsymbol{L}}$ and $\hat{\boldsymbol{S}}$. In particular, the mean value of $\hat{L}_z$, say, can be freely determined by adjusting external parameters, and may acquire a whole spectrum of values. However, only the sign of the projection $\hat{S}_z$ can be subject to external adjustment; its absolute value is determined by the fundamental commutator through (6.110), which in its turn is fixed by the ZPF. It is because of the *universal* value of the commutator that the spin of the electron is the same for all electrons under all circumstances, which reinforces its apparent 'intrinsic' nature.

The connection of the commutator $\left[\hat{x}, \hat{p}_x\right] = i\hbar$ with the spin of the electron deserves a couple of additional comments. As pointed out in chapter 4, Planck's constant $\hbar$ is a direct measure of the size of the fluctuations, both those of the ZPF and those impressed by it on the particle. Specifically, since the commutator implies that the fluctuations of x and p_x have a minimum value adjusted to the rule $\sigma_x^2\sigma_{p_x}^2\Big|_{\min} = \hbar^2/4$, one may write the *numerical* relation

$$\left|\langle\pm|\,\hat{S}_z\left|\pm\right\rangle\right| = \tfrac{1}{2}\hbar = \sigma_x\sigma_{p_x}\big|_{\min}, \tag{6.119}$$

which emphasizes the fact that the *value* of the electron spin is determined by the irreducible fluctuations of the phase-space variables x, p_x around the instantaneous position of the particle.

In QED the interaction of the electron with the vacuum leads to an effective radius of the order of Compton's wavelength λ_C. Similarly, in Chap. 9 we argue that the fluctuating motion of the electron assigns to it an effective structure, a result similar to that referred to in the discussion at the end of Sect. 6.1.5 above. From this point of view, the electron can still be considered as 'essentially' pointlike, but simultaneously possessing an 'effective' (measurable) size. The small, rapid trembling of the electron, taken as a (here nonrelativistic) zitterbewegung, suggests that it is tightly linked with the spin of the electron.[9] The appearance of an effective structure helps also to get an understanding of another important quality of the electron, namely its magnetic moment, a matter that is succinctly addressed in Sect. 6.2.5.

6.2.4 Angular Momentum of the Zero-Point Field

As a heuristic aid to our explanation of the spin of the electron, let us recall some of the most basic properties of the angular momentum of the radiation field. Since the detailed calculation is rather long, we present here a sketch of it (borrowing from Mandel and Wolf 1995, Sect. 10.6).

The total angular momentum operator of the electromagnetic field is usually defined as

$$\hat{\boldsymbol{J}} = \int_V \left(\boldsymbol{r} \times \hat{\boldsymbol{P}}\right) d^3r, \tag{6.120}$$

where $\hat{\boldsymbol{P}}$ stands for the linear momentum density operator (ε_0 is the electric permeability of the vacuum)

$$\hat{\boldsymbol{P}} = \tfrac{1}{2}\varepsilon_0 \left[\hat{\boldsymbol{E}}(\boldsymbol{r},t) \times \hat{\boldsymbol{B}}(\boldsymbol{r},t) - \hat{\boldsymbol{B}}(\boldsymbol{r},t) \times \hat{\boldsymbol{E}}(\boldsymbol{r},t)\right]. \tag{6.121}$$

The expression (6.120) for $\hat{\boldsymbol{J}}$ can be decomposed into a term $\hat{\boldsymbol{J}}_L$ that depends on $\boldsymbol{r}$ and can therefore be identified with the orbital angular momentum of the field, plus a second term, $\hat{\boldsymbol{J}}_S$, independent of $\boldsymbol{r}$ and thus interpreted as an intrinsic angular momentum. Specifically,

---

[9] The zitterbewegung is a phenomenon predicted by the Dirac equation for the electron. It consists of an oscillation (a trembling) around the relativistic motion of amplitude of order λ_C and frequency of order $2mc^2/\hbar$. The nonrelativistic 'zitterbewegung' discussed here differs from the relativistic jitter in that it involves nonrelativistic velocities and contains a wide spectrum of frequencies.

The often conjectured connection between spin and zitterbewegung was proposed for the first time by Schrödinger (1930) and investigated more deeply by Dirac (1958) (see also Maddox (1987)). Examples of related works, which include specific models of varying fortune, are Bhabha and Corben (1941), Huang (1952), Corben (1968), Barut and Zhangi (1984), Hestenes (1985), Hestenes (1990), Pavšič et al. (1993), Rodrigues et al. (1993), Rodrigues et al. (1998). Other mechanisms to generate zitterbewegung-like oscillations have been explored using relativistic models of the electron. For instance, by considering within SED that the structure of the particle is related to the difference between the centers of inertia and charge of the particle, it has been shown that the electron responds to the random field by performing a zitterbewegung (see e.g. Rueda 1993; also Cavalleri 1985, Cavalleri et al. 2010).

$$\hat{\boldsymbol{J}} = \hat{\boldsymbol{J}}_L + \hat{\boldsymbol{J}}_S, \tag{6.122}$$

where

$$\begin{aligned}\hat{\boldsymbol{J}}_L &= \frac{\varepsilon_0}{2}\int_V \sum_{i=1}^{3}\left\{\hat{E}_i\,(\boldsymbol{r}\times\nabla)\,\hat{A}_i + \left[(\boldsymbol{r}\times\nabla)\,\hat{A}_i\right]\hat{E}_i\right\} d^3r,\\ \hat{\boldsymbol{J}}_S &= \frac{\varepsilon_0}{2}\int_V \left(\hat{\boldsymbol{E}}\times\hat{\boldsymbol{A}} - \hat{\boldsymbol{A}}\times\hat{\boldsymbol{E}}\right) d^3r,\end{aligned} \tag{6.123}$$

and $\hat{\boldsymbol{A}}$ stands for the electromagnetic vector potential operator. Equation (6.123) shows that $\hat{\boldsymbol{J}}_S$ can give a result different from zero because the electromagnetic field is a vector field (i.e., because it has more than one spatial component); this endows the components $\hat{\boldsymbol{J}}_L$ and $\hat{\boldsymbol{J}}_S$ with very different properties and meaning. We use a discrete expansion in terms of plane waves, as is usual in QED (in the full three-dimensional reciprocal space $\boldsymbol{k}$, and with $\hat{a}^{\dagger}_{\boldsymbol{k}\sigma}$, $\hat{a}_{\boldsymbol{k}\sigma}$ creation and annihilation operators)

$$\hat{\boldsymbol{A}}\,(\boldsymbol{r},t) = \frac{1}{V^{1/2}}\sum_{\boldsymbol{k},\sigma}\left(\frac{\hbar}{2\omega\varepsilon_0}\right)^{1/2}\left[\boldsymbol{\epsilon}_{\boldsymbol{k}\sigma}\hat{a}_{\boldsymbol{k}\sigma}\,(0)\,e^{i\boldsymbol{k}\cdot\boldsymbol{r}-i\omega t} + \text{h.c.}\right], \tag{6.124}$$

with $\boldsymbol{\epsilon}_{\boldsymbol{k}\sigma}$ the orthogonal circular polarization vectors given by (6.94). Once the integration in (6.123) is carried out, $\hat{\boldsymbol{J}}_S$ reduces to

$$\hat{\boldsymbol{J}}_S = \sum_{\boldsymbol{k},\sigma=\pm1} \hbar\hat{\boldsymbol{k}}\sigma\left(\hat{n}_{\boldsymbol{k}\sigma} + \tfrac{1}{2}\right), \tag{6.125}$$

with $\hat{n}_{\boldsymbol{k}\sigma} = \hat{a}^{\dagger}_{\boldsymbol{k}\sigma}\hat{a}_{\boldsymbol{k}\sigma}$ the photon number operator in the basis of circular polarization, and $\hat{\boldsymbol{k}}$ a unit vector in the direction of the wave vector $\boldsymbol{k}$. Equation (6.125) is an expansion in states of definite helicity, which assigns to individual photons a spin angular momentum projection of value $\sigma\hbar = \pm\hbar$ along the direction $\hat{\boldsymbol{k}}$. Of major importance for what follows is that in the absence of photons, with only the ZPF present, each mode of the vacuum state still contains a component of angular momentum, with mean value given by

$$\left\langle\hat{\boldsymbol{J}}_S\right\rangle_{\text{vac}} = \sum_{\boldsymbol{k},\sigma=\pm1}\frac{\hbar}{2}\hat{\boldsymbol{k}}\sigma = \sum_{\boldsymbol{k}}\tfrac{1}{2}\hbar\hat{\boldsymbol{k}} - \sum_{\boldsymbol{k}}\tfrac{1}{2}\hbar\hat{\boldsymbol{k}}. \tag{6.126}$$

Therefore, a nonzero contribution to $\langle\hat{\boldsymbol{J}}_S\rangle_{\text{vac}} = \pm\frac{1}{2}\hbar\hat{\boldsymbol{k}}$ is associated with every mode $(\boldsymbol{k},\sigma)$ of the ZPF. For the unpolarized field the contributions of the right- and left-hand polarizations compensate each other (for each $\boldsymbol{k}$), and $\langle\hat{\boldsymbol{J}}_S\rangle_{\text{vac}}$ vanishes. This is the reason why the term $1/2$ in Eq. (6.125) is frequently omitted.

Consider now particles that couple with the right or left circularly polarized portions of the vacuum field. Equation (6.126) suggests that they acquire a component of angular momentum, just as they acquire energy or linear momentum, as a result of the coupling.

The above decomposition (6.122) into orbital and spinorial components of the radiation field possesses only a relative value. A detailed relativistic treatment of the field (recall that the radiation field *is* relativistic) shows that only the total angular momentum $\boldsymbol{J}$ satisfies a conservation law, but not its separate parts. The reason is that such decomposition is in general neither covariant nor gauge invariant, so it lacks a well-defined physical meaning. A detailed discussion of these matters can be seen in Rohrlich (1965) .

6.2.5 Gyromagnetic Factor for the Electron

It was discovered experimentally that the g-factor associated with the spin magnetic moment of the electron has an approximate value $g_S = 2$, whereas for the orbital magnetic moment the g-factor is $g_L = 1$. This characteristic value of g_S is incorporated into nonrelativistic quantum theory by hand, usually without further elaboration (it must be remarked that from the theory of Lande's factor it follows that for $L = 0$, $g = 2$; see Greiner 1998, Sect. 11.9). The issue is normally solved by resorting to the Dirac equation, which predicts the value $g_S = 2$ (plus corrections arising from QED). Since the present theory produces the electron spin, it becomes of interest to investigate the value predicted by it for the factor g_S.

Traditionally the gyromagnetic ratio of the electron has not been a subject for SED, due to the fact that the theory has paid little attention to the spin itself, as mentioned earlier. An exception to this is the (quite elaborate) calculation made in de la Peña and Jáuregui (1982) (see also *The Dice*, Sect. 8.3.4), using as a starting point the Fokker-Planck equation in the Markovian approximation. The problem studied was a spherical harmonic oscillator of natural frequency ω_0, subject to an external homogeneous magnetic field $\boldsymbol{B}$ in the z -direction. In terms of the Larmor frequency $\omega_L = |e|\, B/2mc$, the procedure led in the weak-field limit, when $\omega_L \ll \omega_0$, to a total average energy given (in the present notation) by

$$\mathcal{E}_\sigma = \frac{1}{2}\left[\frac{3}{2}\hbar\omega_0 + \omega_L\left(\langle L_z\rangle + 2S_z^\sigma\right)\right], \tag{6.127}$$

where the overall factor $1/2$ comes from the fact that the average is taken over half the ensemble, for a given value of σ. This expression contains already the correct result $g_S = 2$ in front of the term S_z^σ. To be precise, one should add that the referred calculations were made using the methods characteristically employed in SED during the eighties. Such methods led to the occurrence of some erroneous coefficients, although the correct Eq. (6.127) was derived.

With the tools developed so far one can now make a straightforward calculation of g_S. For this purpose consider the electron acted on, in addition to the external force $\boldsymbol{f}(\boldsymbol{x})$, by a static uniform magnetic field $\boldsymbol{B} = B\hat{\boldsymbol{z}}$. The contribution of the orbital angular momentum $\boldsymbol{L}$ to the Hamiltonian is given by

$$\hat{H} = -\hat{\boldsymbol{\mu}} \cdot \boldsymbol{B} = -\mu_z B, \tag{6.128}$$

where $\hat{\boldsymbol{\mu}} = -(g\mu_0 \hat{\boldsymbol{L}})/\hbar$ is the magnetic moment due to $\hat{\boldsymbol{L}}$, $\mu_0 = |e|\,\hbar/(2mc)$ is the Bohr magneton (with $-e = |e|$), and $g_L = 1$. Therefore the mean energy is

$$\mathcal{E} = -\frac{e}{2mc} B \langle \hat{L}_z \rangle = \frac{\mu_0}{\hbar} B \langle \hat{L}_z \rangle. \tag{6.129}$$

Consider a situation in which the spin projection along $\hat{z}$ has a well-defined value, say $\langle \hat{S}_z \rangle = +\hbar/2$. This means that one should take into account only the action of the subensemble of the ZPF that corresponds to $\sigma = +$. Resorting to Eq. (6.112) to write the contribution to $\langle L_z \rangle$ from the subensemble with $\sigma = +$ as $(\langle \hat{L}_z \rangle + \hbar)/2$, the component of $\mathcal{E}$ of interest is

$$\mathcal{E}^{+} = \frac{\mu_0}{\hbar} B \left(\frac{1}{2} \langle \hat{L}_z \rangle + \frac{\hbar}{2} \right) = \frac{\mu_0}{2\hbar} B \left(\langle \hat{L}_z \rangle + 2 \langle \hat{S}_z \rangle^{+} \right). \tag{6.130}$$

An analogous result holds for the subensemble with $\sigma = -$, for which $\langle \hat{S}_z \rangle = -\hbar/2$,

$$\mathcal{E}^{-} = \frac{\mu_0}{\hbar} B \left(\frac{1}{2} \langle \hat{L}_z \rangle - \frac{\hbar}{2} \right) = \frac{\mu_0}{2\hbar} B \left(\langle \hat{L}_z \rangle + 2 \langle \hat{S}_z \rangle^{-} \right). \tag{6.131}$$

The corresponding Hamiltonians describing each part of the magnetic interaction of the electron are therefore $\hat{H}^{+}_{LS}$, $\hat{H}^{-}_{LS}$, with

$$\hat{H}^{+}_{LS} = \frac{\mu_0}{2\hbar} B \left(\hat{L}_z + 2\hat{S}_z \right) = \hat{H}^{-}_{LS}; \tag{6.132}$$

thus the complete Hamiltonian (which includes both polarizations) reads

$$\hat{H}_{LS} = \hat{H}^{+}_{LS} + \hat{H}^{-}_{LS} = \frac{\mu_0}{\hbar} B \left(\hat{L}_z + 2\hat{S}_z \right). \tag{6.133}$$

This contains the correct g-factor of 2 for the spin of the electron. It is clear that such value derives from the two degrees of freedom associated with the polarization of the ZPF.

The result (6.133) gives a precise meaning to the operator appearing in Eq. (6.117). Indeed, from this latter equation one can write $\hat{\boldsymbol{O}} = (\hat{\boldsymbol{L}} + 2\hat{\boldsymbol{S}})/2$, whence

$$\hat{H}_{LS} = \frac{\mu_0}{\hbar} \boldsymbol{B} \cdot \left(\hat{\boldsymbol{L}} + 2\hat{\boldsymbol{S}} \right) = -\hat{\boldsymbol{\mu}} \cdot \boldsymbol{B}, \tag{6.134}$$

with

$$\hat{\boldsymbol{\mu}} = -\frac{2\mu_0}{\hbar} \hat{\boldsymbol{O}}. \tag{6.135}$$

This directly relates $\hat{O}$ with the total magnetic moment operator of the atomic electron.

Since with the present results we have at hand the usual theory of the electron spin, it is straightforward to incorporate it as usual to the Schrödinger equation and thus arrive at the Pauli equation.

6.3 Concluding Comments

We have found that the theory predicts radiative corrections to the results derived with the Schrödinger equation, which to lowest-order coincide with the corresponding ones of nonrelativistic QED. This is not a coincidence, since both theories are essentially equivalent in their physical content (to the order of approximation here studied), although very different in their conceptual perspective. In addition, just as the quantum-mechanical behavior at the Schrödinger (or Heisenberg) level ensues from the interaction with the ZPF, also the spin of the electron emerges as a result of such interaction. This is a most noteworthy outcome, since in quantum mechanics the spin is considered to be an innate property of the electron.

Appendix A

Contribution of Diffusion to the Energy Shift

In Sect. 6.1.5, Eq. (6.51) is obtained for the radiative shift of the energy level n. To calculate its value in the Markovian approximation (which amounts to taking $e^2\hat{\mathcal{D}}(t)Q$ to lowest order in e^2, as explained in Chap. 4) we start from the expression for $\hat{\mathcal{D}}$ in terms of the diffusion operators, Eq. (4.17),

$$e^2\hat{\mathcal{D}}_i = D_{ij}^{pp}\frac{\partial}{\partial p_j} + D_{ij}^{px}\frac{\partial}{\partial x_j}, \tag{A.1}$$

with

$$D_{ij}^{pp} = e^2\int_{-\infty}^{t} dt'\varphi(t-t')\frac{\partial p_j}{\partial p_i'}, \quad D_{ij}^{px} = e^2\int_{-\infty}^{t} dt'\varphi(t-t')\frac{\partial x_j}{\partial p_i'}. \tag{A.2}$$

We thus have (in one-dimensional notation)

$$e^2 x\hat{\mathcal{D}}Q = D^{pp}x\frac{\partial Q}{\partial p} + D^{px}x\frac{\partial Q}{\partial x}. \tag{A.3}$$

Performing the integration over phase space we get

$$e^2\left\langle x\hat{\mathcal{D}}\right\rangle = e^2\int x\hat{\mathcal{D}}Qdxdp = \int\left(D^{pp}x\frac{\partial Q}{\partial p} + D^{px}x\frac{\partial Q}{\partial x}\right)dxdp. \tag{A.4}$$

Upon an integration by parts this becomes

$$e^2\left\langle x\hat{\mathcal{D}}\right\rangle = -\left\langle D^{px} + x\left(\frac{\partial D^{pp}}{\partial p} + \frac{\partial D^{pp}}{\partial x}\right)\right\rangle = -\left\langle D^{px}\right\rangle, \tag{A.5}$$

where the last equality follows from Eq. (B.15) in Appendix 4B, namely

$$\frac{\partial D_{ij}^{pp}}{\partial p_j} + \frac{\partial D_{ij}^{px}}{\partial x_j} = 0. \tag{A.6}$$

For the calculation of $\langle D^{px}\rangle_n$ we resort to equation (B.14b) with $2\eta = \hbar$, which gives

$$\left\langle D^{px}\right\rangle_n = \frac{ie^2}{\hbar}\int_{-\infty}^{t} dt'\varphi(t-t')\left\langle\left[\hat{x}(t),\hat{x}(t')\right]\right\rangle_n, \tag{A.7}$$

so that

$$e^2\left\langle x\hat{\mathcal{D}}\right\rangle_n = -\frac{ie^2}{\hbar}\int_{-\infty}^{t} dt'\varphi(t-t')\left\langle\left[\hat{x}(t),\hat{x}(t')\right]\right\rangle_n. \tag{A.8}$$

The mean value of the commutator is

$$\left\langle\left[\hat{x}(t),\hat{x}(t')\right]\right\rangle_n = -2i\sum_k |x_{nk}|^2\sin\omega_{kn}(t-t').$$

Thus, with $\varphi(t-t')$ given by Eq. (4.10), i.e.,

$$\varphi(t-t') = \frac{4\pi}{3}\int_0^{\infty}\rho_0(\omega)\cos\omega(t-t')d\omega, \tag{A.9}$$

and ρ_0 given by (6.4), Eq. (A.8) gives

$$e^2\left\langle x\hat{\mathcal{D}}\right\rangle_n = -\frac{4e^2}{3\pi c^3}\sum_k |x_{nk}|^2\int_0^{\infty} d\omega\,\omega^3\int_{-\infty}^{t} dt'\cos\omega(t-t')\sin\omega_{kn}(t-t'). \tag{A.10}$$

Making the change of variable $t - t' = s$ and introducing

$$\int_0^\infty ds \cos \omega s \sin \omega_{kn} s = \tfrac{1}{2} \int_0^\infty ds \, [\sin(\omega_{kn} + \omega)s + \sin(\omega_{kn} - \omega)s] \qquad \text{(A.11)}$$
$$= \frac{\omega_{kn}}{\omega_{kn}^2 - \omega^2},$$

in Eq. (A.10) gives, finally,

$$\frac{e^2}{2} \left\langle x \hat{D} \right\rangle_n = -\frac{2e^2}{3\pi c^3} \sum_k |x_{nk}|^2 \, \omega_{kn} \int_0^\infty d\omega \, \frac{\omega^3}{\omega_{kn}^2 - \omega^2}. \qquad \text{(A.12)}$$

Appendix B

Angular-Momentum Components for the Harmonic Oscillator

In this appendix we calculate the right-hand side of Eq. (6.92) for the harmonic oscillator in its ground state,

$$\left\langle L_{ij} \right\rangle_0 = \frac{1}{\tau \omega_0^2} \langle D_{ji}^{px} - D_{ij}^{px} \rangle. \qquad \text{(B.1)}$$

With D_{ij}^{px} given by Eq. (B.14b) in appendix 4B (and $2\eta = \hbar$), we get

$$\langle D_{ji}^{px} - D_{ij}^{px} \rangle_{\mathrm{Q}} = \frac{-2ie^2}{3\pi c^3} \int_0^\infty d\omega \, \omega^3 \int_{-\infty}^t dt' \cos \omega(t - t') \left\langle \left[\hat{x}_j', \hat{x}_i \right] - \left[\hat{x}_i', \hat{x}_j \right] \right\rangle. \qquad \text{(B.2)}$$

For the ground state we have

$$\left\langle \left[\hat{x}_j(t'), \hat{x}_i(t) \right] - \left[\hat{x}_i(t'), \hat{x}_j(t) \right] \right\rangle_0 \qquad \text{(B.3)}$$
$$= -2 \sum_k \left(x_{i0k} x_{jk0} - x_{j0k} x_{ik0} \right) \cos \omega_{k0}(t - t'),$$

where the summation is performed over all possible excited states $k > 0$ connected to the ground state via the matrix elements x_{i0k}. Inserting this expression into (B.2) and resorting to Eq. (D.8) in appendix 4D, namely

$$\int_{-\infty}^t dt' \cos \omega(t - t') \cos \omega_{k0}(t - t') = \frac{\pi}{2} [\delta(\omega - \omega_{k0}) + \delta(\omega + \omega_{k0})], \qquad \text{(B.4)}$$

we obtain (with $\omega_{k0} > 0$)

$$
\begin{aligned}
\langle D_{ji}^{px} - D_{ij}^{px} \rangle_0 &= \frac{2e^2}{3mc^3} \sum_k im\omega_{k0}^3 \left(x_{i0k} x_{jk0} - x_{j0k} x_{ik0} \right) \\
&= \tau \sum_k \omega_{k0}^2 \left(x_{i0k} p_{jk0} - x_{j0k} p_{ik0} \right). \qquad \text{(B.5)}
\end{aligned}
$$

For the harmonic oscillator the only term that contributes to the sum is $k = 1$, with $\omega_{k0} = \omega_0$, whence Eq. (B.5) reads

$$
\langle D_{ji}^{px} - D_{ij}^{px} \rangle_0 = \tau\omega_0^2 \left(x_{i01} p_{j10} - x_{j01} p_{i10} \right) = \tau\omega_0^2 \left\langle L_{ij} \right\rangle_0. \qquad \text{(B.6)}
$$

Thus Eq. (B.1) reduces to an apparent tautology on account of the dynamics.

Appendix C

Calculation of $\langle S^2 \rangle$

To find $\left\langle \boldsymbol{L}^2 \right\rangle^{\pm}$ for the harmonic oscillator in its ground state we use Eq. (6.102), namely

$$
-2\omega_0^2\tau \left\langle \boldsymbol{L}^2 \right\rangle_0 = e^2 \left\langle \frac{\partial \boldsymbol{L}^2}{\partial p_i} \hat{\mathcal{D}}_i \right\rangle_0. \qquad \text{(C.1)}
$$

In order to calculate the right-hand side of this equation we proceed as in appendix A. Specifically, we resort to Eq. (A.1) to write

$$
e^2 \frac{\partial \boldsymbol{L}^2}{\partial p_i} \hat{\mathcal{D}}_i Q = \frac{\partial \boldsymbol{L}^2}{\partial p_i} D_{ij}^{pp} \frac{\partial Q}{\partial p_j} + \frac{\partial \boldsymbol{L}^2}{\partial p_i} D_{ij}^{px} \frac{\partial Q}{\partial x_j}. \qquad \text{(C.2)}
$$

Integration of this expression over phase space gives (after an integration by parts),

$$
\begin{aligned}
e^2 \left\langle \frac{\partial \boldsymbol{L}^2}{\partial p_i} \hat{\mathcal{D}}_i \right\rangle &= \int \left(\frac{\partial \boldsymbol{L}^2}{\partial p_i} D_{ij}^{pp} \frac{\partial Q}{\partial p_j} + \frac{\partial \boldsymbol{L}^2}{\partial p_i} D_{ij}^{px} \frac{\partial Q}{\partial x_j} \right) dxdp \qquad \text{(C.3)} \\
&= -\left\langle \frac{\partial}{\partial p_j} \left(\frac{\partial \boldsymbol{L}^2}{\partial p_i} D_{ij}^{pp} \right) + \frac{\partial}{\partial x_j} \left(\frac{\partial \boldsymbol{L}^2}{\partial p_i} D_{ij}^{px} \right) \right\rangle \\
&= -\left\langle D_{ij}^{pp} \frac{\partial^2 \boldsymbol{L}^2}{\partial p_j \partial p_i} + D_{ij}^{px} \frac{\partial^2 \boldsymbol{L}^2}{\partial x_j \partial p_i} \right\rangle,
\end{aligned}
$$

where in the third line we used Eq. (A.6). Taking into account that for the isotropic harmonic oscillator

$$
\frac{\partial p_j}{\partial p_i'} = \delta_{ij} \cos\omega_0(t - t'), \quad \frac{\partial x_j}{\partial p_i'} = \delta_{ij} \frac{1}{m\omega_0} \sin\omega_0(t - t'), \qquad \text{(C.4)}
$$

equation (A.2) shows that the diffusion coefficients are diagonal,

$$D_{ij}^{pp} = \delta_{ij} e^2 \int_{-\infty}^{t} dt' \varphi(t-t') \cos\omega_0(t-t') = \delta_{ij} D^{pp},$$
$$D_{ij}^{px} = \delta_{ij} \frac{e^2}{m\omega_0} \int_{-\infty}^{t} dt' \varphi(t-t') \sin\omega_0(t-t') = \delta_{ij} D^{px},$$

whence Eq. (C.3) reduces to

$$e^2 \left\langle \frac{\partial \boldsymbol{L}^2}{\partial p_i} \hat{\mathcal{D}}_i \right\rangle = -\left\langle D^{pp} \frac{\partial^2 \boldsymbol{L}^2}{\partial p_i \partial p_i} + D^{px} \frac{\partial^2 \boldsymbol{L}^2}{\partial x_i \partial p_i} \right\rangle, \tag{C.5}$$

where summation over i is understood.

Now, since $\boldsymbol{L}^2 = \boldsymbol{r}^2 \boldsymbol{p}^2 - (\boldsymbol{r} \cdot \boldsymbol{p})^2$, we have that

$$\frac{\partial^2 \boldsymbol{L}^2}{\partial p_i \partial p_i} = 4\boldsymbol{r}^2, \quad \frac{\partial^2 \boldsymbol{L}^2}{\partial x_i \partial p_i} = 0, \tag{C.6}$$

and (C.5) reads

$$e^2 \left\langle \frac{\partial \boldsymbol{L}^2}{\partial p_i} \hat{\mathcal{D}}_i \right\rangle = -4e^2 \left\langle \boldsymbol{r}^2 \right\rangle \int_{-\infty}^{t} dt' \varphi(t-t') \cos\omega_0(t-t'). \tag{C.7}$$

With $\varphi(t-t')$ given by (A.9), Eq. (C.7) becomes, for the ground state,

$$e^2 \left\langle \frac{\partial \boldsymbol{L}^2}{\partial p_i} \hat{\mathcal{D}}_i \right\rangle_0 = -\frac{8\hbar e^2}{3\pi c^3} \left\langle \boldsymbol{r}^2 \right\rangle_0 \int_0^{\infty} d\omega\, \omega^3 \int_{-\infty}^{t} dt' \cos\omega(t-t') \cos\omega_0(t-t'). \tag{C.8}$$

For the integral over t' we use Eq. (B.4), so that

$$e^2 \left\langle \frac{\partial \boldsymbol{L}^2}{\partial p_i} \hat{\mathcal{D}}_i \right\rangle_0 = -\frac{4\hbar e^2}{3c^3} \omega_0^3 \left\langle \boldsymbol{r}^2 \right\rangle_0 . \tag{C.9}$$

The factor $\langle \boldsymbol{r}^2 \rangle_0$ is calculated in accordance with the quantum methods, giving

$$\langle \boldsymbol{r}^2 \rangle_0 = 3\langle \hat{x}^2 \rangle_0 = 3x_{01}x_{10} = \frac{3\hbar}{2m\omega_0}.$$

Equation (C.1) becomes finally (Marshall 1965; de la Peña and Jáuregui 1982)

$$\left\langle \boldsymbol{L}^2 \right\rangle_0 = m\hbar\omega_0 \left\langle \boldsymbol{r}^2 \right\rangle_0 = \frac{3}{2}\hbar^2. \tag{C.10}$$

This value of $\langle \boldsymbol{L}^2 \rangle_0$ takes into account the action of the whole ZPF, with both states of circular polarization active. Considering the action of a single state of polarization $\sigma = \pm$, we get

$$\left\langle S^2 \right\rangle^{\sigma} \equiv \left\langle \boldsymbol{L}^2 \right\rangle_0^{\sigma} = \frac{3}{4}\hbar^2. \tag{C.11}$$

References

*Barranco, A.V., Brunini, S.A., Franca, H.M.: Spin and paramagnetism in classical stochastic electrodynamics. Phys. Rev. A **39**, 5492 (1989)

Barut, A.O., Zhangi, N.: Classical model of the Dirac electron. Phys. Rev. Lett. **52**, 2009 (1984)

Bethe, H.A.: The electromagnetic shift of energy levels. Phys. Rev. **72**, 339 (1947)

Bhabha, H.J., Corben, H.C.: General classical theory of spinning particles in a Maxwell field. Proc. Roy. Soc. A **178**, 273 (1941)

Bordag, M., Klimchitskaya, G.L., Mohideen, U., Mostepanenko, V.M.: Advances in the Casimir Effect. Oxford University Press, Oxford (2009)

*Boyer, T.H.: Quantum electromagnetic zero-point energy and retarded dispersion forces. Phys. Rev. **174**, 1631 (1968)

Boyer, T.H.: Derivation of the blackbody radiation spectrum without quantum assumptions. Phys. Rev. **182**, 1374 (1969)

Boyer,T.H.: A brief survey of stochastic electrodynamics. In: Barut, A.O. (ed.) Foundations of Radiation Theory and Quantum Electrodynamics, p. 49. Plenum Press, London (1980a)

Boyer, T.H.: Thermal effects of acceleration through random classical radiation. Phys. Rev. D **21**, 2137 (1980b)

*Braffort, P., Surdin, M., Taroni, T.: L'energie moyenne d'un oscillateur harmonique non relativiste en électrodynamique aléatoire. C. R. Acad. Sci. Paris **261**, 4339 (1965)

*Braffort, P., Taroni, A.: Mean energy of a free electron in a uniform magnetic field in aleatory electrodynamics. Compt. Rend. B **264**, 1437 (1967)

*Cavalleri, G.: Schrödinger's equation as a consequence of zitterbewegung. Lett. Nouvo Cim. **43**, 285 (1985)

*Cavalleri, G., Barbero, F., Bertazzi, G., Cesaroni, E., Tonni, E., Bosi, L., Spavieri, G., Gillies, G.T.: A quantitative assessment of stochastic electrodynamics with spin (SEDS). Front. Phys. China **5**(1), 107–122 (2010)

*Cetto, A.M., de la Peña, L.: Environmental effects on the Lamb shift according to stochastic electrodynamics. Phys. Rev. A **37**, 1952 (1988a)

*Cetto, A.M., de la Peña, L.: Environmental effects on radiative lifetime and mass renormalization according to SED. Phys. Rev. A **37**, 1960 (1988b)

*Cetto, A.M., de la Peña, L.: Environmental effects on spontaneous emission and Lamb shift according to stochastic electrodynamics. Phys. Scr. T. **21**, 27 (1988c)

*Cetto, A.M., de la Peña, L.: Radiative corrections for the matter-zeropoint field system: establishing contact with quantum electrodynamics. Phys. Scr. **T151**(014009), 1–7 (2012)

*Cetto, A.M., de la Peña, L., Valdés-Hernandez, A.: Quantization as an emergent phenomenon due to matter-zeropoint field interaction. J. Phys. CPCS. **361**, 012013 (2012).http://iopscience.iop.org/1742-6596/361/1/012013

*Cetto, A.M., de la Peña, L., Valdés-Hernandez, A.: Atomic radiative corrections without QED: role of the zero-point field. Rev. Mex. Fis. **59**, 433 (2013)

*Cetto, A.M., de la Peña, L., Valdés-Hernández, A.: Emergence of quantization: the spin of the electron. J. Phys.: JPCS. **504**, 012007 (2014)

Corben, H.C.: Classical and Quantum Theories of Spinning Particles. Holden-Day, San Francisco (1968)

Dalibard, J., Dupont-Roc, J., Cohen-Tannoudji, C.: Vacuum fluctuations and radiation reaction: identification of their respective contributions. J. Phys. (Paris) **43**, 1617 (1982)

Davies, B.: Note on the Lamb shift. Am. J. Phys. **50**, 331 (1982)

Davydov, A.S.: Quantum Mechanics. Addison-Wesley, Reading (1965) (English translation from the Russian edition with additions by ter Haar D., Moscow, 1963)

*de la Peña, L., Cetto, A.M.: The harmonic oscillator in a random electromagnetic field. Rev. Mex. Fís. **25**, 1 (1976)

*de la Peña, L., Cetto, A.M.: Derivation of quantum mechanics from stochastic electrodynamics. J. Math. Phys. **18**, 1612 (1977)

*de la Peña, L., Cetto, A.M.: The quantum harmonic oscillator revisited: a new look from stochastic electrodynamics. J. Math. Phys. **20**, 469 (1979)

de la Peña, L., Cetto, A.M.: The Quantum Dice. An Introduction to Stochastic Electrodynamics. Kluwer Academic, Dordrecht (1996) (Referred to in this book as *The Dice*)

*de la Peña, L., Cetto, A.M., Valdés-Hernández, A.: Atomic radiative corrections without QED: role of the zero-point field. arXiv:quant-ph:1301.6200v1 (2010)

*de la Peña, L., Cetto, A.M., Valdés-Hernández, A.: The emerging quantum. An invitation, advanced school on quantum foundations and open quantum systems. http://www.fisica.ufpb.br/asqf2012/index.php/invited-lectures 2012

*de la Peña, L., Jáuregui, A.: The spin of the electron according to stochastic electrodynamics. Found. Phys. **12**, 441 (1982)

Dirac, P.A.M.: The Principles of Quantum Mechanics. Clarendon, Oxford 1930 (1947, 1958)

Einstein, A.: Zur Quantentheorie der Strahlung. Mitteil. d. Phys. Ges. Zürich **18**, 47 (1916)

Einstein, A.: Zur Quantentheorie der Strahlung. Zeit. Phys. **18**, 121 (1917)

Fain, V.M.: Sov. Phys. JETP **23**, 882 (1966)

Fain, V.M., Khanin, Y.L.: Quantum Electronics: Basic Theory. Pergamon, Oxford (1969)

Feynman, R.P.: In: Stoops, R. (ed.), The Quantum Theory of Fields, Wiley, NY (1961)

Goy, P., Raimond, J.M., Gross, M., Haroche, S.: Observation of cavity-enhanced single-atom spontaneous emission. Phys. Rev. Lett. **50**, 1903 (1983)

Greiner, W.: Quantum Mechanics: Special Chapters. Springer, Berlin (1998)

Hestenes, D.: Quantum mechanics from self-interaction. Found. Phys. **15**, 63 (1985)

Hestenes, D.: The zitterbewegung interpretation of quantum mechanics. Found. Phys. **20**, 1213 (1990)

Hollberg, L., Hall, J.L.: Measurement of the shift of Rydberg energy levels induced by blackbody radiation. Phys. Rev. Lett. **53**, 230 (1984)

Huang, K.: On the zitterbewegung of the Dirac electron. Am. J. Phys. **20**, 479 (1952)

*Jáuregui, A., de la Peña, L.: The spin and the anomalous magnetic moment of the electron in stochastic electrodynamics. Phys. Lett. A **86**, 280 (1981)

Kalitsin, N.St.: On the interaction of an electron with the fluctuating electromagnetic field of the vacuum. Sov. Phys. JETP **25**, 407 (1953). (in Russian)

Kleppner, D.: Inhibited spontaneous emission. Phys. Rev. Lett. **47**, 233 (1981)

Knight, P.L.: Effects of external fields on the Lamb shift. J. Phys. A **5**, 417 (1972)

Louisell, W.H.: Quantum Statistical Properties of Radiation. Wiley, New York (1973)

Maddox, J.: Where zitterbewegung may lead. Nature **325**, 306 (1987)

Mandel, L., Wolf, E.: Optical coherence and quantum optics. Cambridge University Press, Cambridge (1995)

Marshall, T.W.: A classical treatment of blackbody radiation. Nuovo Cim. **38**, 206 (1965)

Milonni, P.W.: The Quantum Vacuum. An Introduction to Quantum Electrodynamics. Academic Press, Boston (1994)

*Moore, S.M.: Stochastic physics and the Abraham-Lorentz equation. Lett. Nuovo Cim. **20**, 676 (1977)

*Moore, S.M., Ramírez, J.A.: Electron spin in classical stochastic electrodynamics. Lett. Nuovo Cimento **33**, 87 (1982)

*Moore, S.M.: Spin relaxation and electromagnetic vacuum fluctuations. Lett. Nuovo Cim. **40**, 385 (1984)

Muralidhar, K.: Classical origin of quantum spin. Apeiron **18**, 146 (2011)

Pavšič, M., Recami, E., Rodrigues Jr, W.A., Maccarrone, G.D., Raciti, F., Salesi, G.: Spin and electron structure. Phys. Lett. B **318**, 481 (1993)

Power, E.A.: Zero-point energy and the Lamb shift. Am. J. Phys. **34**, 516 (1966)

Rodrigues Jr., W.A., Vaz, J., Recami, E., Salesi, G.: About zitterbewegung and electron structure. arXiv:quant-ph/9803037v1 (1998)

Rodrigues Jr, W.A., Vaz, J., Recami, E., Salesi, G.: About zitterbewegung and electron structure. Phys. Lett. B **318**, 623 (1993)

Rohrlich, F.: Classical Charged Particles. Addison-Wesley, Reading (1965)

*Rueda, A.: Stochastic electrodynamics with particle structure. Part I: zero-point Induced Brownian behavior. Found. Phys. Lett. **6**, 139 (1993)

Sachidanandam, S.: A derivation of intrinsic spin one-half from random electrodynamics. Phys. Lett. A **97**, 323 (1983)

*Santos, E.: The harmonic oscillator in stochastic electrodynamics. Nuovo Cim. B **19**, 57 (1974)

Schrödinger, E.: Über die kräftefreie Bewegung in der relativistischen Quantenmechanik. Sitzungsber. Preuss. Akad. Wiss. Phys.-Math. Kl. **24**, 418 (1930)

Sokolov, A.A., Tumanov, V.S.: The uncertainty relation and fluctuation theory. Soviet Phys. JETP **30**, 802. (Russian version); 3, 958, 1957 (English translation) (1956)

*Surdin, M.: L'état fondamental de l'oscillateur harmonique est-il un cycle limite? Ann. Inst. Henri Poincaré **13**, 363 (1970)

*Surdin, M.: The natural line-breadth in stochastic electrodynamics. Int. J. Theor. Phys. **9**, 185 (1974)

van der Waerden, B.L.: Exclusion principle and spin. In: Fierz, M., Weisskopf, V.F. (eds.) Theoretical Physics in the Twentieth Century, p. 202. Interscience, New York (1960)

Welton, T.A.: Some observable effects of the quantum-mechanical fluctuations of the electromagnetic field. Phys. Rev. **74**, 1157 (1948)

Zhou, W., Yu, H.: The Lamb shift in de Sitter spacetime, Phys. Rev. D **82**, 124067. arXiv:1012.4055v1 [hep-th], Dec 2010

Chapter 7
Disentangling Quantum Entanglement

The best possible knowledge of a whole does not necessarily include the best possible knowledge of all its parts, even though they may be entirely separate...

E. Schrödinger (1935)

The seminal paper by Einstein et al. (1935)—a milestone in the debates on the nature of quantum theory—led Schrödinger to see in the entangled states not *one* but *the* characteristic trait of quantum mechanics, that which makes it deviate completely from any classical notion (Schrödinger 1935). From that point on, entanglement has been recognized as a phenomenon of primary importance for our understanding of QM. Eventually the interest in it revived intensely, though pointing towards a new direction. The celebrated works of Bell as of 1964 (in particular 1987; see also Bell et al. 2001) seemed to put an end to the debate on the complete or incomplete nature of QM, and in the subsequent and extensive series of works focusing on entanglement, the emphasis shifted from the interpretative discussions to the new and promising perspectives offered by the applications of this phenomenon.[1] During recent times the investigations on entanglement have developed with increasing breakneck speed, mainly within the quantum information field. Nevertheless, and in spite of the progress made concerning the control, quantification, evolution, and distribution of entanglement, some fundamental aspects of its physical meaning remain as cryptic as they were 80 years ago. One central aspect, to which the present chapter is devoted, refers to the ascertainment of the mechanism that entangles two particles that do not interact by means of an external potential.

The mere mathematical structure of the (many-body) Schrödinger equation makes it possible to superpose two (or more) state vectors and create entangled state vectors for composite systems. From a physical point of view, it is natural for such superpositions to appear when the particles of the system are subject to mutual interactions: then the correlations (manifest in the form of entangled states) can be traced to this

[1] For recent literature on this subject see e.g. Horodecki et al. (2009). A different approach to entanglement is offered in Khrennikov (2010, 2011).

L. de la Peña et al., *The Emerging Quantum*,
DOI: 10.1007/978-3-319-07893-9_7,

interaction. However, for a system composed of two or more *noninteracting* particles things are different: in this case the standard quantum formalism *postulates* the symmetrization rule in the case of identical particles, thereby resorting to the mathematical possibility of superposing state vectors, but leaving aside the physical mechanism responsible for the ensuing correlations.[2] What is the *physical* agent that causes two (or more) noninteracting particles to get entangled? The answer (and even the very question) seems to be foreign to the quantum-mechanical domain. Here we have a dramatic instance of the failure of QM to identify, with its own tools, the physics behind one of its most far-reaching traits.

The theory developed in previous chapters is applied in this one to give due response to the question just raised. This is done specifically by extending the one-particle theory presented in Chap. 5 to a system of two particles that are embedded in a *common* background field. The dynamical variables of the particles are shown to become correlated whenever the particles resonate to a common frequency of the background field. When the description is reduced to one in terms of state vectors in the appropriate Hilbert space, the entangled states emerge naturally—the entanglement factor representing the correlation between the ZPF modes of common frequency—as the only ones that can reproduce such correlations. Further, for systems of identical particles the properties of invariance of the field correlations imply that entanglement is maximal and must be described by totally (anti)symmetric states.[3]

The results thus obtained are applied to the particular case of two electrons with spin, for which the total (orbital + spin) state vector turns out to be antisymmetric. States in which both particles are in the same orbital and spinorial state, are excluded because of the absence of a correlating field mode.

7.1 The Two-Particle System

7.1.1 The Field in the Vicinity of the Particles

The system of interest all along this chapter consists of a pair of particles with masses and electric charges m_1, m_2 and e_1, e_2, which are located at $\boldsymbol{x}_1$ and $\boldsymbol{x}_2$ and subject to external forces $\boldsymbol{f}_1(\boldsymbol{x}_1)$ and $\boldsymbol{f}_2(\boldsymbol{x}_2)$, respectively, with no interaction potential between them. Although the general arguments apply to systems of three or more

[2] The postulate of symmetrization of the wave function of the composite system is normally justified on the basis of the indistinguishability of the quantum corpuscles. It is within QFT that the spin-statistics theorem is derived as a relativistic result, detached from arguments about indistinguishability. Nevertheless, also QFT fails to provide elements to *unveil* the mechanism leading to this result —a fact that does not go without criticism (see e.g. Duck and Sudarshan 1997; Kaplan 2013; see also Tomonaga 1974).

[3] The material presented in this chapter is based to a large extent on Valdés-Hernández (2010), Valdés-Hernández et al. (2011), de la Peña et al. (2010, 2012).

particles, by focusing on a bipartite system only, multipartite entanglement (such as the 3-tangle, Coffman et al. 2000) is excluded from the present analysis. In a first instance, the particles are considered spinless; the consideration of spin, which is an important additional element, is left for Sect. 7.4.

The equations of motion for the two particles, again in the nonrelativistic approximation, are given by

$$m_1\ddot{\boldsymbol{x}}_1 = \boldsymbol{f}_1(\boldsymbol{x}_1) + m_1\tau_1\dddot{\boldsymbol{x}}_1 + \frac{e_1}{e_2}m_2\tau_2\dddot{\boldsymbol{x}}_2 + e_1\boldsymbol{E}(\boldsymbol{x}_1, t), \tag{7.1a}$$

$$m_2\ddot{\boldsymbol{x}}_2 = \boldsymbol{f}_2(\boldsymbol{x}_2) + m_2\tau_2\dddot{\boldsymbol{x}}_2 + \frac{e_2}{e_1}m_1\tau_1\dddot{\boldsymbol{x}}_1 + e_2\boldsymbol{E}(\boldsymbol{x}_2, t). \tag{7.1b}$$

Equation (7.1a) is just the Abraham-Lorentz equation for the particle located at $\boldsymbol{x}_1$, with the extra term $(e_1/e_2)m_2\tau_2\dddot{\boldsymbol{x}}_2 = 2e_1e_2\dddot{\boldsymbol{x}}_2/(3c^3)$ standing for the force that the radiation of the particle located at $\boldsymbol{x}_2$ exerts on charge e_1 (Landau and Lifshitz 1951), and similarly for Eq. (7.1b).

In line with the long-wavelength approximation made in the one-particle problem, we assume that noticeable changes in the field amplitudes of interest occur only for distances much greater than the deviations of each of the particles from their respective equilibrium positions $\boldsymbol{x}_i^0$ $(i = 1, 2)$, so that $\boldsymbol{E}(\boldsymbol{x}_i, t)$ can be replaced by $\boldsymbol{E}(\boldsymbol{x}_i^0, t)$. Therefore, rather than speaking of the background field at $\boldsymbol{x}_i$ we refer to the background field in the *vicinity* of the particle located at $\boldsymbol{x}_i$. To construct the expansions for these fields in a form analogous to the one used in Chap. 5, we proceed as follows.

Restricting the analysis to one-dimensional motions, we assume that the vectors $\boldsymbol{x}_1^0$ and $\boldsymbol{x}_2^0$ are colinear and write

$$\boldsymbol{x}_2^0 = \boldsymbol{x}_1^0 + \boldsymbol{R} = (x_1^0 + R)\hat{\boldsymbol{x}}_1^0, \tag{7.2}$$

where $\boldsymbol{R}$ is a constant vector that stands for the mean distance between the equilibrium positions of the particles, and $\hat{\boldsymbol{x}}_1^0$ is a unit vector. More generally, our description is valid for $\boldsymbol{R} = \boldsymbol{R}(t)$ whenever the characteristic times of the motions of each of the mechanical subsystems are much shorter than the times required for $\boldsymbol{R}(t)$ to change appreciably; here we are neglecting such (slow) changes. In analogy with Eq. (5.3), we write

$$\boldsymbol{E}(\boldsymbol{x}_1^0, t) = \sum_{\omega_k} \tilde{E}(\omega_k)\boldsymbol{a}_k(\boldsymbol{x}_1^0)e^{i\omega_k t} + \text{c.c.}, \tag{7.3a}$$

$$\boldsymbol{E}(\boldsymbol{x}_2^0, t) = \sum_{\omega_k} \tilde{E}(\omega_k)\boldsymbol{a}_k(\boldsymbol{x}_1^0 + \boldsymbol{R})e^{i\omega_k t} + \text{c.c.} \tag{7.3b}$$

Notice that by writing the *same* $\boldsymbol{a}_k$ for both $\boldsymbol{E}(\boldsymbol{x}_1^0, t)$ and $\boldsymbol{E}(\boldsymbol{x}_2^0, t)$, both particles are assumed to be subject to the same realization of the field. Therefore, systems composed of particles that are arbitrarily distant are excluded from the present description, because in such case the field in the vicinity of each particle is statistically independent from the other, and thus can be considered to correspond to different realizations. With this assumption we are tacitly describing a *single* (composite) system immersed in a *common* background field, rather than two independent systems.

Now we project the fields (7.3a, 7.3b) along the direction $\hat{\boldsymbol{R}}$ and fix the origin at $\boldsymbol{x}_1^0$, thus obtaining[4]

$$E_1(t) = \sum_{\omega_k} \tilde{E}(\omega_k)\boldsymbol{a}_k(0) \cdot \hat{\boldsymbol{R}} e^{i\omega_k t} + \text{c.c.} = \sum_k \tilde{E}(\omega_k) a_{1k} e^{i\omega_k t} + \text{c.c.}, \tag{7.4a}$$

$$E_2(t) = \sum_{\omega_k} \tilde{E}(\omega_k)\boldsymbol{a}_k(\boldsymbol{R}) \cdot \hat{\boldsymbol{R}} e^{i\omega_k t} + \text{c.c.} = \sum_k \tilde{E}(\omega_k) a_{2k} e^{i\omega_k t} + \text{c.c.}, \tag{7.4b}$$

where

$$a_{1k} = \boldsymbol{a}_k(0) \cdot \hat{\boldsymbol{R}}, \tag{7.5a}$$

$$a_{2k} = \boldsymbol{a}_k(\boldsymbol{R}) \cdot \hat{\boldsymbol{R}}. \tag{7.5b}$$

With $\hat{\boldsymbol{R}} = \hat{\boldsymbol{x}}$, the stochastic variable a_{1k} appearing here coincides with the one appearing in the expansion (5.5) for the one-particle case.

7.1.2 Looking for Stationary Solutions

7.1.2.1 The Coupled Equations of Motion

With the above approximations, Eqs. (7.1a, 7.1b) take the form

$$m_1\ddot{x}_1 = f_1(x_1) + m_1\tau_1 \dddot{x}_1 + e_1 E_1^{\text{eff}}, \tag{7.6a}$$

$$m_2\ddot{x}_2 = f_2(x_2) + m_2\tau_2 \dddot{x}_2 + e_2 E_2^{\text{eff}}, \tag{7.6b}$$

where each $f_i(x_i)$ can be expanded as a power series of its argument and ${E_i}^{\text{eff}}$ stands for the effective field

$$E_1^{\text{eff}} = E_1(t) + \frac{1}{e_2} m_2\tau_2 \dddot{x}_2, \quad E_2^{\text{eff}} = E_2(t) + \frac{1}{e_1} m_1\tau_1 \dddot{x}_1. \tag{7.7}$$

[4] In Blanco and Santos (1979) the problem of a system of several particles embedded in the ZPF is addressed, with a scope similar to the present one. However, and although that paper constitutes a significant and important effort to disclose the mechanism underlying entanglement, it differs essentially from the present one in that an identical background field is assumed for *all* particles, which leads to unconvincing results.

Equations (7.6a, 7.6b) thus show that the presence of the charge e_j modifies the net field acting on the charge e_i through the radiative term, proportional to $\dddot{x}_j$, which is superposed to the background field $E_i(t)$ in the vicinity of x_i. This extra term introduces a coupling —of radiative origin— between Eqs. (7.6a) and (7.6b), and consequently the particles cease to be independent. In particular, the coupling terms play an important role during the process leading to equilibrium, since, according to the discussion in Chap. 4, the particle located at x_i reaches a stationary state when the mean power radiated by it balances the mean power absorbed from the *effective* field E_i^{eff}, which is (partially) determined precisely by the coupling term $2e_ie_j\dddot{x}_j/3c^3$. Thus, the radiative effect of one of the particles on the other is initially as significant as the own radiation reaction. When the balance condition is eventually attained, these terms contribute only with radiative corrections that can be discarded in a first approximation, as was seen in Chaps. 4 and 5. However, the effect of the coupling between Eqs. (7.6a, 7.6b) endures, and has far-reaching consequences. To disclose and discuss such consequences is the main aim of this chapter.

7.1.2.2 Stationary Solutions

Suppose that once the stationary and radiationless regime has been reached, the particle located at x_1 is in state α whereas the particle located at x_2 is in state α' (in what follows, unprimed Greek indices denote states accessible to the particle at x_1 and primed Greek indices denote those accessible to the particle at x_2). Due to the coupling terms contained in the effective fields, each particle is in equilibrium with a net radiation field E_i^{eff} that bears information on the final state reached by *both* particles. Thus E_i^{eff}, and of course the stationary solutions of Eqs. (7.6a, 7.6b), must be labeled with a compound-state index $A = (\alpha, \alpha')$ that embodies both (final) states. Note that the index A can be understood in terms of a decomposition of the ensemble $\{i\}$ of realizations of the field in a similar way as in the one-particle case (see Sect. 5.2), where the index involved was α. In this sense, $A = (\alpha, \alpha')$ identifies the subensemble $\{i\}_A \subset \{i\}$ corresponding to those (bipartite) mechanical systems in which one particle has reached the state α, whereas the other has attained the state α'. This compels us to write $x_{iA}(t)$ instead of $x_i(t)$, thus generalizing the expression $x_\alpha(t)$ used in the one-particle case. Also the expansions for the (net) field and the external force must be generalized, so that [(cf. Eq. (5.31)]

$$x_{iA}(t) = \sum_B \tilde{x}_{iAB} b_{AB} e^{i\omega_{AB}t}, \tag{7.8a}$$

$$E_{iA}^{\text{eff}}(t) = \sum_B \tilde{E}_{iAB}^{\text{eff}} b_{AB} e^{i\omega_{AB}t}, \tag{7.8b}$$

$$f_{iA}(t) = \sum_B \tilde{f}_{iAB} b_{AB} e^{i\omega_{AB}t}. \tag{7.8c}$$

The index B plays the same role that β played in the one-particle expansions, i.e., it labels the elements of the set $\{\omega_{AB}\}$ of relevant frequencies for the given subensemble characterized by A. For the particle in state α, a set $\{\omega_{\alpha\beta}\}$ is defined; analogously, a set $\{\omega_{\alpha'\beta'}\}$ is defined for the second particle in state α'. Thus the index B is, like A, a compound index of the form $B = (\beta, \beta')$, with β and β' accesible states for the respective particles. As for $\{b_{AB}\}$, it stands for the set of random variables appropriate for describing the expansions of the dynamical variables in the bipartite case. Since the source of stochasticity in Eqs. (7.6a, 7.6b) are the two background fields $E_1(t)$ and $E_2(t)$ given by Eqs. (7.4a, 7.4b) clearly the variables $\{b\}$ depend on *both* families of variables, $\{a_1\}$ and $\{a_2\}$. The explicit relationship between them will be the subject of the next section.

Notice that, unlike the single-particle state, it may be that the stationary state A of the bipartite system is only *partially* characterized by its total (mechanical) energy, due to possible degeneracies. The index A is therefore in direct, though not necessarily univocal, correspondence with the total mechanical energy, denoted by $\mathcal{E}_A$. In the absence of an (external) interaction potential between the particles, $\mathcal{E}_A$ is given by

$$\mathcal{E}_A = \mathcal{E}_\alpha + \mathcal{E}_{\alpha'}, \tag{7.9}$$

and similarly for $B = (\beta, \beta')$, the total energy is $\mathcal{E}_B = \mathcal{E}_\beta + \mathcal{E}_{\beta'}$.

7.1.3 The Common Random Variable

Each of the coefficients of the form $\tilde{A}_{iAB}$ (with $A = x, E^{\text{eff}}, f, \ldots$) in Eqs. (7.8a–7.8c) is a quantity characterized by four state indices, $\tilde{A}_{iAB} = \tilde{A}_{i(\alpha,\alpha';\beta,\beta')}$, and its specific form requires investigation. However, we are interested in the time-asymptotic, radiationless limit only, in which Eqs. (7.6a, 7.6b) become decoupled and x_{1A} and x_{2A} acquire just the form of the single-particle expansions for the particles in states α and α', respectively. Therefore, introducing the superindex (0) in the coefficient $\tilde{x}_{iAB}$ appearing in Eq. (7.8b) to denote its value in the radiationless approximation, we have

$$\sum_B \tilde{x}^{(0)}_{1AB} b_{AB} e^{i\omega_{AB}t} = \sum_\beta \tilde{x}_{1\alpha\beta} a_{1\alpha\beta} e^{i\omega_{\alpha\beta}t}, \tag{7.10a}$$

$$\sum_B \tilde{x}^{(0)}_{2AB} b_{AB} e^{i\omega_{AB}t} = \sum_{\beta'} \tilde{x}_{2\alpha'\beta'} a_{2\alpha'\beta'} e^{i\omega_{\alpha'\beta'}t}. \tag{7.10b}$$

Since $\{a_{1\alpha\beta}\}$ and $\{a_{2\alpha'\beta'}\}$ are the sets of random variables that determine the background fields needed for the particles to reach separately the states α and α', each of these families satisfies the same statistical properties as the variables $\{a_{\alpha\beta}\}$ studied in the one-particle problem. Specifically, as a consequence of imposing the ergodic condition on each of the subsystems, both $\{a_{1\alpha\beta}\}$ and $\{a_{2\alpha'\beta'}\}$, as well as the corresponding frequencies $\{\omega_{\alpha\beta}\}$ and $\{\omega_{\alpha'\beta'}\}$, must satisfy the chain rule (Eqs. (5.59) and (5.54b), respectively),[5] whence in particular

$$a_{2\alpha'\alpha'}e^{i\omega_{\alpha'\alpha'}t} = a_{2\alpha'\alpha'} = 1, \quad a_{1\alpha\alpha}e^{i\omega_{\alpha\alpha}t} = a_{1\alpha\alpha} = 1. \tag{7.11}$$

We can now resort to these expressions to rewrite Eqs. (7.10a, 7.10b) as

$$\sum_B \tilde{x}^{(0)}_{1AB} b_{AB} e^{i\omega_{AB}t} = \sum_B \tilde{x}_{1\alpha\beta}\delta_{\alpha'\beta'}a_{1\alpha\beta}a_{2\alpha'\beta'}e^{i(\omega_{\alpha\beta}+\omega_{\alpha'\beta'})t}, \tag{7.12a}$$

$$\sum_B \tilde{x}^{(0)}_{2AB} b_{AB} e^{i\omega_{AB}t} = \sum_B \tilde{x}_{2\alpha'\beta'}\delta_{\alpha\beta}a_{1\alpha\beta}a_{2\alpha'\beta'}e^{i(\omega_{\alpha\beta}+\omega_{\alpha'\beta'})t}. \tag{7.12b}$$

Even though $x_{1A}(t)$ is formally equivalent to the expansion $x_{1\alpha}(t)$ (Eq. 7.10a), these expressions differ in an important sense, since Eq. (7.12a), by having α' fixed, constitutes the expansion for the position of one of the particles *in the presence of the other one.* Thus we have

$$\omega_{AB} = \omega_{\alpha\beta} + \omega_{\alpha'\beta'} = \hbar^{-1}\left(\mathcal{E}_A - \mathcal{E}_B\right), \tag{7.13}$$

where the last equality follows from Eq. (7.9) and the fact that in the radiationless approximation, Eq. (5.118) and its primed version hold. From Eqs. (7.12a, 7.12b) we also identify

$$\begin{aligned} \tilde{x}^{(0)}_{1AB} b_{AB} &= \tilde{x}_{1\alpha\beta}\delta_{\alpha'\beta'}a_{1\alpha\beta}a_{2\alpha'\beta'}, \\ \tilde{x}^{(0)}_{2AB} b_{AB} &= \tilde{x}_{2\alpha'\beta'}\delta_{\alpha\beta}a_{1\alpha\beta}a_{2\alpha'\beta'}. \end{aligned} \tag{7.14}$$

From here it follows that the nonstochastic coefficients $\tilde{x}^{(0)}_{iAB}$ are given by

$$\tilde{x}^{(0)}_{1AB} = \tilde{x}_{1\alpha\beta}\delta_{\alpha'\beta'}, \quad \tilde{x}^{(0)}_{2AB} = \tilde{x}_{2\alpha'\beta'}\delta_{\alpha\beta}, \tag{7.15}$$

whereas b_{AB} factorizes as

$$b_{AB} = a_{1\alpha\beta}a_{2\alpha'\beta'}. \tag{7.16}$$

This equation defines the joint random variable b_{AB} which is common to expansions (7.8a–7.8c) for both members of a bipartite system. The b_{AB} is determined by the random variables of the ZPF in the vicinity of the particles, and can therefore be used to expand the single *common* field in the form

$$E_{iA}(t) = E_A(x_i^0, t) = \sum_B \tilde{E}_{iAB} b_{AB} e^{i\omega_{AB}t}, \tag{7.17a}$$

[5] Notice that in Chap. 5, primed and unprimed indices were used to denote (stationary) states for the (single) particle, whereas here, primed and unprimed indices refer to *different* particles. Thus, care must be taken when 'translating' equations such as (5.59) and (5.54b) to the bipartite case.

with

$$\tilde{E}_{1AB} = \tilde{E}_{\alpha\beta}\delta_{\alpha'\beta'}, \quad \tilde{E}_{2AB} = \tilde{E}_{\alpha'\beta'}\delta_{\alpha\beta}. \tag{7.17b}$$

The chain rule (5.59) applied to the families of variables contained in Eq. (7.16) gives

$$\begin{aligned} b_{AB}b_{BG} &= (a_{1\alpha\beta}a_{2\alpha'\beta'})(a_{1\beta\gamma}a_{2\beta'\gamma'}) = \left(a_{1\alpha\beta}a_{1\beta\gamma}\right)(a_{2\alpha'\beta'}a_{2\beta'\gamma'}) \\ &= a_{1\alpha\gamma}a_{2\alpha'\gamma'} = b_{AG}. \end{aligned} \tag{7.18}$$

This relation can be easily generalized to any number of factors, therefore the chain rule holds also for the variables $\{b_{AB}\}$, and the following relations are satisfied,

$$b^*_{AB} = b_{BA}, \quad |b_{AB}|^2 = b_{AA} = b_{BB} = 1. \tag{7.19}$$

Moreover, according to Eq. (7.13), the chain rule (5.54b) applies to $\{\omega_{AB}\}$, whence

$$\omega_{AB} + \omega_{BG} = \omega_{AG}, \tag{7.20}$$

which can also be generalized immediately to an arbitrary number of terms.

7.1.4 Establishing Contact with the Tensor Product Hilbert Space

According to Eq. (7.15), the coefficient $\tilde{x}_{1AB}$ in the expansion for $x_{1A}(t)$ reduces in the radiationless regime to $\tilde{x}_{1\alpha\beta}\delta_{\alpha'\beta'}$, where $\tilde{x}_{1\alpha\beta}$ is the amplitude corresponding to the relevant frequency $\omega_{\alpha\beta}$ for the problem of a single particle located at x_1 in state α and subject to the external force $f_1(x_1)$. Since the same applies to $x_{2A}(t)$, we conclude that any pair of dynamical variables[6] $F(x_1, \dot{x}_1)$ and $G(x_2, \dot{x}_2)$ is represented in state A by expressions of the form

$$F_A(t) = \sum_B \tilde{F}_{AB}b_{AB}e^{i\omega_{AB}t}, \tag{7.21a}$$

$$G_A(t) = \sum_B \tilde{G}_{AB}b_{AB}e^{i\omega_{AB}t}, \tag{7.21b}$$

where

$$\tilde{F}_{AB} = \tilde{F}_{\alpha\beta}\delta_{\alpha'\beta'}, \quad \tilde{G}_{AB} = \tilde{G}_{\alpha'\beta'}\delta_{\alpha\beta}. \tag{7.22}$$

Here and in what follows we omit the superindex (0) (introduced in Sect. 7.1.3) in the coefficients, but keeping in mind that we are working in the radiationless regime.

Let us now find the appropriate Hilbert space for the composite system. In line with the results of Sect. 5.2.2, the coefficient $\tilde{F}_{\alpha\beta}$ in (7.22) stands for the element

[6] As discussed in Sect. 5.2.2, such variables are assumed to be functions of the form $h(x_i) + g(\dot{x}_i)$ with both h, g power series of their argument.

$\alpha\beta$ of a matrix $\hat{F}$ defined in a Hilbert space $\mathcal{H}_1$ spanned by the state vectors $\{|\alpha\rangle\}$ of the first particle, and similarly for $\tilde{G}_{\alpha'\beta'}$. Further, by including the time-dependent factor $e^{i\omega_{AB}t}$ to construct the amplitudes $\tilde{F}_{AB}(t)$, we obtain

$$\tilde{F}_{AB}(t) = \tilde{F}_{\alpha\beta}\delta_{\alpha'\beta'}e^{i(\omega_{\alpha\beta}+\omega_{\alpha'\beta'})t} = \tilde{F}_{\alpha\beta}(t)\delta_{\alpha'\beta'}e^{i\omega_{\alpha'\beta'}t}, \tag{7.23a}$$

so that $\tilde{F}_{\alpha\beta}(t) = \tilde{F}_{\alpha\beta}e^{i\omega_{\alpha\beta}t}$ stands now for the element $\alpha\beta$ of an evolving matrix $\hat{F}(t)$. Similarly,

$$\tilde{G}_{AB}(t) = \tilde{G}_{\alpha'\beta'}\delta_{\alpha\beta}e^{i(\omega_{\alpha\beta}+\omega_{\alpha'\beta'})t} = \tilde{G}_{\alpha'\beta'}(t)\delta_{\alpha\beta}e^{i\omega_{\alpha\beta}t}, \tag{7.23b}$$

where $\tilde{G}_{\alpha'\beta'}(t) = \tilde{G}_{\alpha'\beta'}e^{i\omega_{\alpha'\beta'}t}$ are the elements of a matrix $\hat{G}(t)$ defined in a Hilbert space $\mathcal{H}_2$. The one- and two-particle coefficients $\tilde{A}_{\alpha\beta}$, $\tilde{A}_{AB}$ are therefore replaced from now on by the corresponding matrix elements $A_{\alpha\beta}$, A_{AB}.

The above expressions for $\tilde{F}_{AB}(t) = F_{AB}(t)$ and $\tilde{G}_{AB}(t) = G_{AB}(t)$—or their time-independent version (7.22)—show that F_{AB} and G_{AB} are the elements of a matrix defined in the tensor product Hilbert space $\mathcal{H}_1 \otimes \mathcal{H}_2$. Notice that the description in the product space is here a consequence of expanding the dynamical variables in terms of random variables of the form $b_{AB} = a_{1\alpha\beta}a_{2\alpha'\beta'}$. The b_{AB}, in their turn, result from the presence of the coupling terms in Eqs. (7.6a, 7.6b). Hence the interaction of the particles via the common field lies at the root of the description in the Hilbert product space, even after the radiationless approximation has been taken.

By denoting with $\mathcal{F}$ and $\mathcal{G}$ the matrices with elements F_{AB} and G_{AB}, respectively, Eqs. (7.22) can be written in closed matrix notation as

$$\mathcal{F} = \hat{F} \otimes \mathbb{I}_2, \quad \mathcal{G} = \mathbb{I}_1 \otimes \hat{G}. \tag{7.24}$$

From this and Eqs. (7.21a, 7.21b) conclusions analogous to those reached in Sect. 5.2.2 are obtained. In particular, a variable $F(x_1, \dot{x}_1)$ has a matrix $\mathcal{F}$ associated with it, whereas a variable $G(x_2, \dot{x}_2)$ of the second particle can be put in correspondence with the matrix $\mathcal{G}$, and a variable of the form $F(x_1, \dot{x}_1)G(x_2, \dot{x}_2)$ is represented by[7]

$$\mathcal{F}\mathcal{G} = \hat{F} \otimes \hat{G} = \hat{F}\hat{G}, \tag{7.25}$$

where in the last equality the notation was simplified in the customary way. In a stationary state A the expansion for FG thus reads

$$(FG)_A = \sum_B \widehat{FG}_{AB} b_{AB} e^{i\omega_{AB}t}, \tag{7.26}$$

[7] More general functions $V(x_1, \dot{x}_1; x_2, \dot{x}_2)$ would be represented by linear combinations of elementary products FG.

with

$$\widehat{FG}_{AB} \equiv (\hat{F}\hat{G})_{AB} = \sum_{D} F_{AD}G_{DB}$$
$$= \sum_{\delta,\delta'} F_{\alpha\delta}\delta_{\alpha'\delta'}G_{\delta'\beta'}\delta_{\delta\beta} = F_{\alpha\beta}G_{\alpha'\beta'}. \tag{7.27}$$

Substitution of Eq. (7.27) in (7.26) gives, with the aid of (7.13) and (7.16),

$$(FG)_{A=(\alpha,\alpha')} = \sum_{\beta,\beta'} F_{\alpha\beta}G_{\alpha'\beta'}a_{1\alpha\beta}a_{2\alpha'\beta'}e^{i(\omega_{\alpha\beta}+\omega_{\alpha'\beta'})t} = F_A G_A = F_\alpha G_{\alpha'}. \tag{7.28}$$

Notice that according to Eq. (7.13), the term $F_{\alpha\beta}G_{\alpha'\beta'}$ in (7.28) for $F(t) = x_1(t)$, $G(t) = x_2(t)$ oscillates with a frequency ω_{AB} that is the sum of resonance frequencies of the corresponding particles. In this case ω_{AB} establishes the transition frequency of the *complete* system from state A to state B, with the total energy $\mathcal{E}_B$ determined by Eq. (7.13).

7.1.5 Implications of Ergodicity for the Common Random Field Variable

Decomposition of Eq. (7.26) into a time-independent contribution plus an oscillating term, here denoted as $\mathcal{O}(t)$, gives

$$(FG)_A = \sum_B \widehat{FG}_{AB}b_{AB}\Big|_{\omega_{AB}=0} + \mathcal{O}(t). \tag{7.29}$$

Hence, since $\overline{\mathcal{O}(t)}^t = 0$,

$$\overline{(FG)_A(t)}^t = \sum_B \widehat{FG}_{AB}b_{AB}\Big|_{\omega_{AB}=0}. \tag{7.30}$$

Now the ergodic demand applied to the bipartite system requires that

$$\overline{(FG)_A}^t = \overline{(FG)_A}^{(i)}, \tag{7.31}$$

so that the right-hand side of Eq. (7.30) must be independent of (i). As this must hold irrespective of A, it follows that

$$b_{AB}^{(i)} = b_{AB} \quad \text{whenever} \quad \omega_{AB} = 0. \tag{7.32}$$

Equation (7.32) is the generalization of our previous (one-particle) result $a_{\alpha\beta}^{(i)} = a_{\alpha\beta}$ for $\omega_{\alpha\beta} = 0$. In the one-dimensional single-particle case $\omega_{\alpha\beta}$ vanishes only for $\beta = \alpha$, and the ergodic condition reduces simply to $a_{\alpha\alpha} = 1$, Eq. (5.62). The same happens in the bipartite case when $A = B$. However, when there is energy degeneracy between two specific states A, B, the condition $\omega_{AB} = 0$ is satisfied also with $B \neq A$, i.e., [see Eq. (7.13)]

$$\omega_{\alpha\beta} = -\omega_{\alpha'\beta'} = \omega_{\beta'\alpha'} \neq 0. \tag{7.33}$$

This case, in which the two particles share a common relevant frequency $\omega_{\alpha\beta} = \omega_{\beta'\alpha'}$, will prove to be of particular relevance. Using (7.16) one can then rewrite Eq. (7.32) in terms of these frequencies as (recall that $a_{j\gamma\lambda} = a_j(\omega_{\gamma\lambda}),\ j = 1, 2$)

$$b_{AB}^{(i)} = a_{1\alpha\beta}^{(i)} a_{2\beta\alpha}^{(i)} = a_{1\beta'\alpha'}^{(i)} a_{2\alpha'\beta'}^{(i)} = b_{AB} \quad (\omega_{AB} = 0). \tag{7.34}$$

Since each $a_j(\omega_{\gamma\lambda})$ is of the form $a_j(\omega_{\gamma\lambda}) = \exp i(\varphi_{(j)\gamma\lambda})$, $a_{1\alpha\beta}^{(i)}$ and $a_{2\alpha\beta}^{(i)}$ differ only by a phase factor,

$$a_{1\alpha\beta}^{(i)} = a_{2\alpha\beta}^{(i)} e^{i\varsigma_{\alpha\beta}^{(i)}}. \tag{7.35}$$

Equation (7.34) thus becomes

$$b_{AB}^{(i)} = e^{i\varsigma_{\alpha\beta}^{(i)}} = e^{i\varsigma_{\beta'\alpha'}^{(i)}} = b_{AB}, \tag{7.36}$$

or rather

$$\begin{aligned} \lambda_{AB} &\equiv \left\langle a_{1\alpha\beta}^{(i)} a_{2\beta\alpha}^{(i)} \right\rangle = \langle b_{AB} \rangle = b_{AB} \\ &= e^{i\varsigma_{\alpha\beta}} = e^{-i\varsigma_{\alpha'\beta'}} \quad \text{for} \quad \omega_{\alpha\beta} = -\omega_{\alpha'\beta'}, \end{aligned} \tag{7.37}$$

where the notation λ_{AB} has been introduced to distinguish those b_{AB} that fulfill condition (7.32) from those that do not (hence are stochastic), and

$$\varsigma_{\alpha\beta} = -\varsigma_{\beta\alpha}, \tag{7.38}$$

as follows from taking the complex conjugate of (7.35) with $a_{(1,2)\alpha\beta}^{*(i)} = a_{(1,2)\beta\alpha}^{(i)}$ [cf. Eq. (5.61)]. Finally, from Eq. (7.19) we get

$$\lambda_{AB}^{*} = \lambda_{BA}, \quad |\lambda_{AB}|^2 = 1. \tag{7.39}$$

7.2 Correlations Due to Common Resonance Modes

7.2.1 Spectral Decomposition

An important feature of one-particle expansions such as (5.70) is that they represent spectral decompositions of the corresponding variable, so that each term of the sum oscillates with a different frequency. This ensures a one-to-one correspondence between the coefficients in the expansion and the corresponding relevant frequencies (meaning that the mode of frequency $\omega_{\alpha\beta}$ connects state α with state β only). However, bipartite expansions such as (7.26) do not necessarily constitute spectral decompositions of $(FG)_A$, since the (set of) relevant frequencies ω_{AB} may be degenerate, i.e., there may be two different states B (say, $B = D, K$ with $D \neq K$) such that $\omega_{AD} = \omega_{AK}$. According to Eq. (7.13), this of course corresponds to a degeneracy of the states D, K, so that they share the same energy $\mathcal{E}_D = \mathcal{E}_K$. In the following we investigate the implications of the existence of any frequency degeneracy for the spectral decomposition of the product $(FG)_A$.

Let us assume that in the expansion (7.26) or (7.28) there exist indeed two different pairs $\left(\beta, \beta'\right) = \left\{\left(\delta, \delta'\right), \left(\kappa, \kappa'\right)\right\}$ (i.e., two states $B = K, D$ with $K \neq D$) such that

$$\omega_{AD} = \omega_{\alpha\delta} + \omega_{\alpha'\delta'} = \omega_{\alpha\kappa} + \omega_{\alpha'\kappa'} = \omega_{AK}, \tag{7.40}$$

whence $\omega_{DK} = 0$ but

$$\omega_{\delta\kappa} = \omega_{\kappa'\delta'} \neq 0. \tag{7.41}$$

For simplicity we consider only one frequency to be (doubly) degenerate, which means that the only degeneracy present is the one explicitly written in (7.40). We then isolate the terms corresponding to $B = K, D$ in (7.26) and write

$$b_{AK} = b_{AD} b_{DK} = b_{AD} \lambda_{DK}, \tag{7.42}$$

where the second equality follows from Eq. (7.37) applied to b_{DK}, and $\lambda_{DK} = \lambda_{DK}^*$ is therefore a nonrandom coefficient of magnitude 1 [(see Eq. (7.39)]. The expansion (7.26) reads then

$$\begin{aligned}(FG)_A = {} & \left[\widehat{FG}_{AD} + \lambda_{DK} \widehat{FG}_{AK}\right] b_{AD} e^{i\omega_{AD}t} \\ & + \textstyle\sum_{B \neq D,K} \widehat{FG}_{AB} b_{AB} e^{i\omega_{AB}t},\end{aligned} \tag{7.43}$$

where the terms that oscillate with the same frequency have been separated from the sum and grouped in the first line. This represents the spectral decomposition of the product variable FG in state A. It is important to note that according to Eqs. (7.40) and (7.41), the degeneracy $\omega_{AK} = \omega_{AD}$ implies the existence of a nonzero relevant frequency common to both particles, $\omega_{\delta\kappa} = \omega_{\kappa'\delta'}$. By inverting the reasoning we conclude that a response of both particles to the same frequency of the background field implies a degeneracy ($\omega_{AD} = \omega_{AK}$) and gives rise to a nonfactorizable coefficient [the term within square brackets in (7.43)] in the spectral decomposition of $(FG)_A$.

Notice that owing to the ergodic condition, which implies that λ_{DK} in Eq. (7.43) is a nonrandom factor, every coefficient in the spectral decomposition of $(FG)_A$ is proportional to the joint random variable $b_{AB}^{(i)}$ (or $b_{AD}^{(i)}$). The linear response to the field, discussed at the end of Sect. 5.2 for the one-particle case, turns out to be a feature also of the composite system, but now with both particles responding to the field represented by $\{b_{AB}\}$.

Let us now take A to be one of the degenerate states, i.e., $A = D$. Then $\omega_{AK} = \omega_{DK} = \omega_{DD} = 0$, and Eq. (7.43) reads

$$(FG)_D = \widehat{FG}_{DD} + \lambda_{DK}\widehat{FG}_{DK} + \sum_{B \neq D,K} \widehat{FG}_{DB} b_{DB} e^{i\omega_{DB}t}. \tag{7.44}$$

Analogously, if $A = K$,

$$(FG)_K = \widehat{FG}_{KK} + \lambda_{KD}\widehat{FG}_{KD} + \sum_{B \neq D,K} \widehat{FG}_{KB} b_{KB} e^{i\omega_{KB}t}. \tag{7.45}$$

These equations give for the average value[8] of *FG* in states D and K, respectively (with $K \neq D$),

$$\langle (FG)_D \rangle = \overline{(FG)}^t_D = \widehat{FG}_{DD} + \lambda_{DK}\widehat{FG}_{DK}, \tag{7.46a}$$

$$\langle (FG)_K \rangle = \overline{(FG)}^t_K = \widehat{FG}_{KK} + \lambda_{KD}\widehat{FG}_{KD}, \tag{7.46b}$$

or in more explicit notation,

$$\langle (FG)_D \rangle = F_{\delta\delta} G_{\delta'\delta'} + \lambda_{DK} F_{\delta\kappa} G_{\delta'\kappa'}, \tag{7.47a}$$

$$\langle (FG)_K \rangle = F_{\kappa\kappa} G_{\kappa'\kappa'} + \lambda_{KD} F_{\kappa\delta} G_{\kappa'\delta'}. \tag{7.47b}$$

Equations (7.46a, 7.46b) show that $\langle (FG)_D \rangle$ and $\langle (FG)_K \rangle$ involve coefficients associated not only with their respective state $D = (\delta, \delta')$ or $K = (\kappa, \kappa')$, but also with the second state that shares the same energy, i.e., that satisfies

$$\mathcal{E}_D = \mathcal{E}_\delta + \mathcal{E}_{\delta'} = \mathcal{E}_\kappa + \mathcal{E}_{\kappa'} = \mathcal{E}_K, \tag{7.48}$$

[8] We recall that in the stationary regime, the principle of ergodicity holds and therefore time averages are equivalent to ensemble averages. Notice further that for the calculation of (7.46a, 7.46b), the system is considered to be either in state D or in state K. Hence, according to the discussion in Sect. 5.2, these averages are taken over *different* subensembles, namely those containing all realizations of the system in state D or K, respectively.

which is but another form of expressing Eq. (7.41).[9] This indicates that special attention must be paid to the expansion of dynamical variables in degenerate states, as will be shown in what follows.

7.2.2 State Expansion Versus Energy Expansion

As observed earlier, when there is no degeneracy the index A is univocally related to $\mathcal{E}_A$, so that (7.26) is both the expansion of FG when the system is in state A, and the expansion of FG when the system possesses energy $\mathcal{E}_A$. By contrast, in the presence of degeneracy both states D and K are linked through the condition $\mathcal{E}_D = \mathcal{E}_K$. So even if e.g. (7.44) represents FG for the system in state D, it does *not completely* represent this variable when the energy of the system is $\mathcal{E}_D = \mathcal{E}_K$. Because of this, it will be convenient to distinguish between *state expansions* (associated to a specific state), and *energy expansions* (associated to a specific energy).

Thus far we have focused on state expansions, characterized separately by the indices D, K. The energy expansions will be labeled with the energy index specifying the degeneracies; thus, according to the above we should write

$$\left.\begin{array}{r} (FG)_{\mathcal{E}_A} = (FG)_A \quad \text{for } A \text{ a nondegenerate state,} \\ (FG)_{\mathcal{E}_D=\mathcal{E}_K} \quad \text{for a (doubly) degenerate energy state, } D \neq K. \end{array}\right\} \tag{7.49}$$

The structure of $(FG)_{\mathcal{E}_D=\mathcal{E}_K}$ will be investigated below. At this point, it suffices to say that it represents FG in the *single* bipartite state characterized by the degeneracy $\mathcal{E}_D = \mathcal{E}_K$. Let us now find the state vector that corresponds to such state, and also to nondegenerate states.

7.2.3 State Vectors: Emergence of Entanglement

Equations (7.46a), (7.46b) for the average values of the product FG in states D, K with $\mathcal{E}_D = \mathcal{E}_K$ contain a feature that is distinctive of degenerate states, namely they consist of *linear combinations* of two products of matrix elements of F and G. This reflects the fact that the coefficients $\widehat{FG}_{AB}$ that oscillate with a degenerate frequency have ceased to be separable, as seen from the first line in Eq. (7.43). Such property

[9] The existence of common relevant frequencies and the condition of degeneracy are thus two faces of the same coin, yet it will become clear below that the former is more illuminating in disclosing the physical mechanism underlying entanglement. As stated after Eq. (5.135), focusing on the (relevant) frequencies rather than on the energy levels represents a shift from the Schrödinger to the Heisenberg approach. Here we find another example that shows that despite the formal equivalence of the two approaches, either one or the other is more suitable to understand certain aspects of the quantum phenomenon.

has important implications also for the structure of the state vectors associated with the degenerate states D, K, as is shown in what follows.

For this purpose, we apply the procedure of Sect. 5.4.2 to the bipartite case. In analogy with Eq. (5.119) for the one-particle case, we expand the matrix product $\widehat{FG}(t)$ in general as

$$\widehat{FG}(t) = \sum_{A,B} \widehat{FG}_{AB}(t)|e_A\rangle\langle e_B| \\ = \sum_{\alpha,\alpha',\beta,\beta'} F_{\alpha\beta}G_{\alpha'\beta'}e^{i(\omega_{\alpha\beta}+\omega_{\alpha'\beta'})t}\left(|e_\alpha\rangle\langle e_\beta|\right)_1\left(|e_{\alpha'}\rangle\langle e_{\beta'}|\right)_2, \tag{7.50}$$

where the canonical basis $\{|e_A\rangle\langle e_B| = \{(|e_\alpha\rangle\langle e_\beta|)_1 \otimes (|e_{\alpha'}\rangle\langle e_{\beta'}|)_2\}$ is constructed from the vectors $\{|e_A\rangle = |e_\alpha\rangle_1 |e_{\alpha'}\rangle_2\}$ spanning the product space $\mathcal{H}_1 \otimes \mathcal{H}_2$. The time evolution can be transferred from $\widehat{FG}(t)$ to the vectors of a new basis obtained from the original one by means of the unitary transformation

$$|e_A\rangle \to |A(t)\rangle = |\alpha(t)\rangle_1 \left|\alpha'(t)\right\rangle_2 = e^{-i(\mathcal{E}_A/\hbar)t}\,|e_A\rangle\,. \tag{7.51}$$

In analogy with Eq. (5.122), we get for the matrix elements of $\widehat{FG}(t)$

$$\widehat{FG}_{AB}e^{i\omega_{AB}t} = \langle A(0)|\,\widehat{FG}(t)\,|B(0)\rangle \\ = \langle A(t)|\,\widehat{FG}(0)\,|B(t)\rangle = \langle A|\,\widehat{FG}\,|B\rangle\,. \tag{7.52}$$

Now consider the case in which there are two degenerate states D, K, accessible from A, as in Sect. 7.2.1. Equation (7.43) shows that the spectral decomposition of $(FG)_A$ contains then coefficients of two kinds: those corresponding to all nondegenerate frequencies ω_{AB}, given by $\widehat{FG}_{AB}$ in the second line, and those associated with the degenerate frequency $\omega_{AK} = \omega_{AD}$, given by the first line. Expressed in terms of matrix elements, the former have the simple structure given in Eq. (7.52) (for $B \neq D,K$), whereas the latter take the form

$$\widehat{FG}_{AD}(t) + \lambda_{DK}\widehat{FG}_{AK}(t) = \langle A|\,\widehat{FG}\,(|D\rangle + \lambda_{DK}\,|K\rangle). \tag{7.53}$$

Therefore every coefficient in the spectral decomposition of $(FG)_A$ is contained in the A-th row (corresponding to the vector $\langle A|$) of the matrix $\widehat{FG}$. For nondegenerate states B (associated to $\widehat{FG}_{AB}$), the column vectors are simply $|B\rangle$, i.e.,

$$\text{nondegenerate } B \leftrightarrow |B\rangle = |\beta\rangle_1 \left|\beta'\right\rangle_2. \tag{7.54}$$

On the other hand, in the case of degeneracy, the column vector is given according to Eq. (7.53) by $|D\rangle + \lambda_{DK}\,|K\rangle$. This reveals the existence of a new type of vector that is not an element of the basis $\{|B\rangle\}$, but is given by the superposition

$$\text{degenerate } D, K \leftrightarrow |\varphi_{DK}\rangle \equiv |\delta\rangle_1 \left|\delta'\right\rangle_2 + \lambda_{DK} |\kappa\rangle_1 \left|\kappa'\right\rangle_2, \tag{7.55}$$

with $\omega_{\delta\kappa} = \omega_{\kappa'\delta'} \neq 0$. Therefore, whenever the two particles share a relevant frequency, a new class of vector arises naturally in the transition to the Hilbert-space description, which corresponds to a well-defined energy state of the bipartite system, is *nonfactorizable*, and gives rise to *entanglement*. To properly normalize this vector we consider that the original basis $\{|\beta\rangle_1 \left|\beta'\right\rangle_2\}$ is orthonormal and resort to Eq. (7.39), whence

$$|\psi_{DK}\rangle = \frac{1}{\sqrt{\langle\varphi_{DK}|\varphi_{DK}\rangle}} |\varphi_{DK}\rangle = \frac{1}{\sqrt{2}}\left(|\delta\rangle_1 \left|\delta'\right\rangle_2 + \lambda_{DK} |\kappa\rangle_1 \left|\kappa'\right\rangle_2 \right). \tag{7.56}$$

In the usual quantum formalism, the possibility of combining two (or more) state vectors to construct a third one—which is key for the existence of entanglement—is understood as a (mathematical) result of the linearity of the Schrödinger equation. Here we find that this combination has a deeper physical origin in the relations between the random variables of the common background field, expressed through the chain rule. In fact, it is the product Eq. (7.42) containing the nonrandom phase factor λ_{DK} what allows to rewrite the original expansion (7.26) in the form of (7.43), leading to the combination of terms given by (7.53), with information about the two degenerate states ($|\delta\rangle_1 \left|\delta'\right\rangle_2$ and $|\kappa\rangle_1 \left|\kappa'\right\rangle_2$).

7.2.4 Entanglement as a Vestige of the ZPF

We recall from Eq. (7.37) that λ_{DK} is defined by the correlation of the field variables $a_1(\omega_{\delta\kappa})$, $a_2(\omega_{\delta'\kappa'})$. The entanglement factor λ_{DK} in (7.56) is thus exhibited as a further vestige of the ZPF in quantum mechanics, reminding us of the active role of this field as a member of the whole system. Because of this correlation between field variables, the state of the system becomes correctly described by the single state $|\psi_{DK}\rangle$ given by Eq. (7.56) instead of the separate-particle states $|D\rangle$, $|K\rangle$, which contain only partial information about the system, as discussed above. This observation contrasts with the widespread interpretation of states of the form (7.56) as meaning that the system can be simultaneously in two different (degenerate) states $|D\rangle$, $|K\rangle$.[10] Instead, the structure of $|\psi_{DK}\rangle$ means that the original one-particle states $|\delta\rangle_1 \left|\delta'\right\rangle_2$, $|\kappa\rangle_1 \left|\kappa'\right\rangle_2$, have become entangled, giving rise to a *single* new state.

[10] In fact, as follows from the remark in note 8, the averages over states $|D\rangle$ and $|K\rangle$ imply an averaging over different (incompatible) subesembles, hence they cannot be taken simultaneosly over one and the same system.

7.2.5 Emergence of Correlations

The fact that two particles embedded in a common background field may share relevant frequencies, suggests possible correlations between those dynamical variables F and G (corresponding to different particles) to which the common frequencies contribute in the expansions (7.21a, 7.21b). In order to analyse such possible correlations with the usual matrix tools developed here and in Chap. 5, we calculate the covariance defined as

$$\Gamma_{(FG)_{\mathcal{E}}} = \langle\psi|\,\widehat{FG}\,|\psi\rangle - \langle\psi|\,\hat{F}\,|\psi\rangle\,\langle\psi|\,\hat{G}\,|\psi\rangle\,, \tag{7.57}$$

where $|\psi\rangle$ stands for the state vector of the bipartite system when its energy is $\mathcal{E}$.

According to the first line in (7.49) and Eq. (7.54), for a nondegenerate state A (i.e. $\mathcal{E} = \mathcal{E}_A$) one gets

$$\Gamma_{(FG)_{\mathcal{E}_A}} = \langle A|\,\widehat{FG}\,|A\rangle - \langle A|\,\hat{F}\,|A\rangle\,\langle A|\,\hat{G}\,|A\rangle\,. \tag{7.58}$$

Direct calculation gives

$$\Gamma_{(FG)_{\mathcal{E}_A}} = 0. \tag{7.59}$$

By contrast, when there is a degeneracy (i.e. $\mathcal{E}_D = \mathcal{E}_K$), the second line in (7.49) and (7.55) imply that the covariance is given by

$$\Gamma_{(FG)_{\mathcal{E}_D=\mathcal{E}_K}} = \langle\psi_{DK}|\,\widehat{FG}\,|\psi_{DK}\rangle - \langle\psi_{DK}|\,\hat{F}\,|\psi_{DK}\rangle\,\langle\psi_{DK}|\,\hat{G}\,|\psi_{DK}\rangle\,, \tag{7.60}$$

where $|\psi_{DK}\rangle$ is the state vector (7.56).The first term gives

$$\langle\psi_{DK}|\,\widehat{FG}\,|\psi_{DK}\rangle = \tfrac{1}{2}(F_{\delta\delta}G_{\delta'\delta'} + F_{\kappa\kappa}G_{\kappa'\kappa'}) + \operatorname{Re}\lambda_{DK}\,F_{\delta\kappa}G_{\delta'\kappa'}, \tag{7.61}$$

because of the hermiticity of the matrices $\hat{F}$, $\hat{G}$ and Eq. (7.39). The second term is obtained by successively taking $\hat{F}, \hat{G} = \mathbb{I}$ in (7.61),

$$\langle\psi_{DK}|\,\hat{F}\,|\psi_{DK}\rangle\,\langle\psi_{DK}|\,\hat{G}\,|\psi_{DK}\rangle = \tfrac{1}{4}(F_{\delta\delta} + F_{\kappa\kappa})(G_{\delta'\delta'} + G_{\kappa'\kappa'}), \tag{7.62}$$

whence

$$\Gamma_{(FG)_{\mathcal{E}_D=\mathcal{E}_K}} = \tfrac{1}{4}(F_{\delta\delta} - F_{\kappa\kappa})(G_{\delta'\delta'} - G_{\kappa'\kappa'}) + \operatorname{Re}\lambda_{DK}\,F_{\delta\kappa}G_{\delta'\kappa'}. \tag{7.63}$$

This result shows that the existence of common relevant frequencies is crucial for the emergence of nonclassical correlations between the particles, or more properly, between (some of) their dynamical variables.[11] Which variables become correlated

[11] According to the results in Chap. 5 (see also Appendix B), the relevant frequencies are constructed via linear combinations of the resonance frequencies, hence the existence of common relevant

depends of course on the frequencies $\omega_{\delta\kappa}$ and $\omega_{\delta'\kappa'}$ that satisfy $\omega_{\delta\kappa} = \omega_{\kappa'\delta'} \neq 0$, since they determine the matrix elements appearing in Eq. (7.63). To see this more clearly, let us first note that the covariance is composed of two contributions of entirely different nature (we write $\Gamma_{(FG)\mathcal{E}_D=\mathcal{E}_K} = \Gamma_{(FG)_{DK}}$),

$$\Gamma_{(FG)_{DK}} = \Gamma^0_{(FG)_{DK}} + \Gamma^\lambda_{(FG)_{DK}}, \tag{7.64}$$

with

$$\begin{aligned} \Gamma^0_{(FG)_{DK}} &= \tfrac{1}{4}(F_{\delta\delta} - F_{\kappa\kappa})(G_{\delta'\delta'} - G_{\kappa'\kappa'}), \\ \Gamma^\lambda_{(FG)_{DK}} &= \mathrm{Re}\,\lambda_{DK} F_{\delta\kappa} G_{\delta'\kappa'}. \end{aligned} \tag{7.65}$$

The first one is different from zero only when both $\hat{F}$ and $\hat{G}$ have diagonal terms, with different values for states δ and κ, and δ' and κ', respectively. In particular, when these matrices commute with $\hat{H}_1$ and $\hat{H}_2$, which means that they represent constants of motion for particles 1 and 2, respectively, they become diagonal matrices in the energy representation and *only* contribute to $\Gamma^0_{(FG)_{DK}}$; such matrices do not serve to exhibit entanglement. For illustration purposes take the simple case $\hat{F} = \hat{H}_1$ and $\hat{G} = \hat{H}_2$: then

$$\Gamma^0_{(FG)_{DK}} = \tfrac{1}{4}(\mathcal{E}_\delta - \mathcal{E}_\kappa)(\mathcal{E}_{\delta'} - \mathcal{E}_{\kappa'}) = -\left(\tfrac{1}{2}\hbar\omega_{\delta\kappa}\right)^2. \tag{7.66}$$

This result, as should be expected, simply expresses the negative (classical) correlation between the individual energy differences for particle 1 ($\Delta\mathcal{E}_{\delta\kappa} = \hbar\omega_{\delta\kappa}$) and for particle 2 ($\Delta\mathcal{E}_{\delta'\kappa'} = \hbar\omega_{\delta'\kappa'}$) when the energies are subject to the restriction $\mathcal{E}_\delta + \mathcal{E}_{\delta'} = \mathcal{E}_\kappa + \mathcal{E}_{\kappa'}$.

For F, G to exhibit entanglement, the second term in Eq. (7.65) must be different from zero. This happens only when both $\hat{F}$ and $\hat{G}$ have nondiagonal terms (and such that $\lambda_{DK} F_{\delta\kappa} G_{\delta'\kappa'}$ is not purely imaginary). In particular, for purely nondiagonal matrices $\hat{F}$ and $\hat{G}$ this is the only contribution to $\Gamma_{(FG)_{DK}}$. The term $\Gamma^\lambda_{(FG)_{DK}}$ clearly depends on the correlation λ_{DK} between the random field amplitudes when there is a common relevant frequency [$\omega_{\delta\kappa} = -\omega_{\delta'\kappa'}$; see Eq. (7.37)]. In other words, the correlation of the common modes of the background field in the vicinity of the particles can manifest itself on the mechanical system in the form of an entanglement-related correlation between the particles. It is impossible to understand the genesis of such correlation when the presence of the common field is ignored.

(Footnote 11 continued)
frequencies goes back to the existence of common resonance frequencies. We can therefore rephrase the above statement by saying that the existence of (nontrivial) common *resonance* frequencies lies at the root of the emergence of correlations between the particles.

7.2.5.1 Statistical Meaning of the Covariance

Above we resorted to the matrix formalism and the state vectors derived in Sect. 7.2.3 to study the emergence of correlations between the particles when there is a common relevant frequency. Now we shall establish contact with the statistical language more proper of LSED, and recover the previous covariances with the tools of the latter theory. For this purpose we focus on the covariance [cf. Eq. (7.57)]

$$\Gamma_{(FG)_{\mathcal{E}}} = \langle (FG)_{\mathcal{E}} \rangle - \langle F_{\mathcal{E}} \rangle \langle G_{\mathcal{E}} \rangle , \tag{7.67}$$

when the composite system has a total (mechanical) energy $\mathcal{E}$.

As seen in Sect. 7.2.2, for a nondegenerate state A, $(FG)_{\mathcal{E}_A}$ is simply $(FG)_A$, so one gets

$$\Gamma_{(FG)_{\mathcal{E}_A}} = \langle (FG)_A \rangle - \langle F_A \rangle \langle G_A \rangle . \tag{7.68}$$

A direct calculation using Eq. (7.26) gives in this case

$$\langle (FG)_A \rangle = F_{\alpha\alpha} G_{\alpha'\alpha'} = \langle F_A \rangle \langle G_A \rangle , \tag{7.69}$$

whence F and G are uncorrelated, in line with Eq. (7.59). Thus, for a nondegenerate state A, the covariance of F and G can be calculated indistinctly as a statistical average [right-hand side of Eq. (7.68)], or as a quantum average [right-hand side of Eq. (7.58)].

We now focus on the degenerate case, with $\mathcal{E}_D = \mathcal{E}_K$ $(D \neq K)$. According to the discussion in Sect. 7.2.2, the covariance (7.67) must be calculated with the energy expansion $(FG)_{\mathcal{E}_D = \mathcal{E}_K}$. Since this latter represents FG in the state characterized by the energy $\mathcal{E}_D = \mathcal{E}_K$, its mean value

$$\left\langle (FG)_{\mathcal{E}_D = \mathcal{E}_K} \right\rangle \tag{7.70}$$

will be calculated resorting to Eqs. (7.47a, 7.47b) under the assumption that both states D and K contribute to it on an equal footing. Thus,

$$\left\langle (FG)_{\mathcal{E}_D = \mathcal{E}_K} \right\rangle = \tfrac{1}{2} \left[\langle (FG)_D \rangle + \langle (FG)_K \rangle \right] . \tag{7.71}$$

Explicitly, we obtain

$$\left\langle (FG)_{\mathcal{E}_D = \mathcal{E}_K} \right\rangle = \tfrac{1}{2} \left[F_{\delta\delta} G_{\delta'\delta'} + F_{\kappa\kappa} G_{\kappa'\kappa'} + 2\mathrm{Re}\, \lambda_{DK} F_{\delta\kappa} G_{\delta'\kappa'} \right] . \tag{7.72}$$

To calculate the corresponding covariance [see Eq. (7.67)],

$$\Gamma_{(FG)_{\mathcal{E}_D = \mathcal{E}_K}} = \left\langle (FG)_{\mathcal{E}_D = \mathcal{E}_K} \right\rangle - \left\langle F_{\mathcal{E}_D = \mathcal{E}_K} \right\rangle \left\langle G_{\mathcal{E}_D = \mathcal{E}_K} \right\rangle , \tag{7.73}$$

we write $\langle F_{\mathcal{E}_D = \mathcal{E}_K} \rangle$ (and $\langle G_{\mathcal{E}_D = \mathcal{E}_K} \rangle$) by taking $G = 1$ (and $F = 1$) in (7.72),

$$\left\langle F_{\mathcal{E}_D=\mathcal{E}_K}\right\rangle = \tfrac{1}{2}(F_{\delta\delta} + F_{\kappa\kappa}), \quad \left\langle G_{\mathcal{E}_D=\mathcal{E}_K}\right\rangle = \tfrac{1}{2}(G_{\delta'\delta'} + G_{\kappa'\kappa'}). \tag{7.74}$$

Equation (7.73) gives thus, using (7.72) and (7.74),

$$\Gamma_{(FG)_{\mathcal{E}_D=\mathcal{E}_K}} = \tfrac{1}{4}(F_{\delta\delta} - F_{\kappa\kappa})(G_{\delta'\delta'} - G_{\kappa'\kappa'}) + \mathrm{Re}\,\lambda_{DK} F_{\delta\kappa} G_{\delta'\kappa'}, \tag{7.75}$$

which is exactly Eq. (7.63). This confirms the equivalence of the matrix method and the statistical method, under the (natural) assumption that the partial (statistical) averages over the degenerate states D, K contribute an equal amount.

7.3 Systems of Identical Particles

7.3.1 Natural Entanglement

We have seen in Sect. 7.2.3 that the existence of a common relevant frequency—due to a degeneracy in the energy of the composite system—is a necessary and sufficient condition for the state of a bipartite system to get entangled. Now, in the case of two identical, noninteracting particles subject to the same external force, there is an intrinsic degeneracy that makes entanglement unavoidable. This degeneracy arises from the fact that the particles satisfy the same equation of motion, namely [cf. Eqs. (7.6a, 7.6b)]

$$m\ddot{x}_i = f_i(x_i) + m\tau\dddot{x}_i + eE_i^{\mathrm{eff}}. \tag{7.76}$$

It is clear that in this case the sets of states that are accessible to each particle are identical,

$$\{\beta\} = \{\beta'\}, \tag{7.77}$$

and hence all the relevant frequencies are common to both.[12]

From Eq. (7.77) it follows that for each and every pair $B_I = (\beta, \beta')$ of states of the two particles (with $\beta \neq \beta'$) there is another one, $B_{II} = (\beta', \beta)$, having the same total energy $\mathcal{E}_B = \mathcal{E}_\beta + \mathcal{E}_{\beta'}$, such that the degeneracy condition (7.41) [with $(\delta,\delta') = (\beta,\beta')$, $(\kappa,\kappa') = (\beta', \beta)$] holds, i.e,

$$\omega_{\beta\beta'} = -\omega_{\beta'\beta} \neq 0. \tag{7.78}$$

Although the first equality follows immediately from the antisymmetry of the relevant frequencies, it is important to stress its physical meaning by observing that the left-

[12] In the above lines we treat as identical particles those having the same intrinsic properties relevant for the present description, namely m and e. No explicit reference to their spin or other properties is made, since the description developed so far [rooted in Eq. (7.76)] leaves out any spinorial effect, thus treating the particles as spinless systems. However, the equality of their spins is tacitly assumed by considering that the Hilbert spaces in which the matrices $\hat{F}$ and $\hat{G}$ are defined coincide, hence possess the same dimension. In Sect. 7.4, we extend the analysis to particles with spin 1/2.

hand side stands for the relevant frequency of one of the particles, whereas the right-hand side stands for the relevant frequency of the other one.

According to the discussion in Sect. 7.2.3, the double degeneracy of all the states (β, β') as well as of all the frequencies $\omega_{\beta\beta'}$ compels us to describe every state of the bipartite system in terms of an entangled vector of the form given by (7.56) [with $(\delta,\delta') = (\beta,\beta')$, $(\kappa,\kappa') = (\beta',\beta)$],

$$|\psi_B\rangle = \frac{1}{\sqrt{2}}\Big(|\beta\rangle_1 \left|\beta'\right\rangle_2 + \lambda_B \left|\beta'\right\rangle_1 |\beta\rangle_2 \Big), \tag{7.79}$$

where

$$\lambda_B \equiv \langle a_{1\beta\beta'} a_{2\beta'\beta}\rangle = e^{i\varsigma_{\beta\beta'}} \quad (\beta \neq \beta'), \tag{7.80}$$

in accordance with Eq. (7.37). The covariance of F and G, Eq. (7.63), reduces in this case to (we use the shorthand notation $\Gamma_{\beta\beta'} = \Gamma_{(FG)\varepsilon_{B_I}=\varepsilon_{B_{II}}}$)

$$\Gamma_{\beta\beta'} = \tfrac{1}{4}\left(F_{\beta\beta} - F_{\beta'\beta'}\right)\left(G_{\beta'\beta'} - G_{\beta\beta}\right) + \mathrm{Re}\,\lambda_B F_{\beta\beta'} G_{\beta'\beta}. \tag{7.81}$$

7.3.2 The Origin of Totally (Anti)symmetric States

In the case of two identical particles, the covariance of F and G is invariant under an exchange of single-particle states, as can be readily verified by applying the operator $I_s\colon \beta \leftrightarrow \beta'$ to $\Gamma_{\beta\beta'}$ in Eq. (7.81),

$$I_s \Gamma_{\beta\beta'} = \Gamma_{\beta\beta'}. \tag{7.82}$$

Further, using Eq. (7.38) the state vector (7.79) becomes

$$\left|\psi'_B\right\rangle = I_s |\psi_B\rangle = \frac{1}{\sqrt{2}}\Big(\left|\beta'\right\rangle_1 |\beta\rangle_2 + \lambda_B^* |\beta\rangle_1 \left|\beta'\right\rangle_2 \Big) = \lambda_B^* |\psi_B\rangle, \tag{7.83}$$

which is the same as the original $|\psi_B\rangle$ multiplied by an overall phase factor.

Now, we recall that each of the particles is in correspondence with the stochastic field $E_i = E(x_i^0, t)$ in its vicinity. By making this field explicit, we may carry out a physical transformation having no counterpart in the usual quantum description, which consists of an exchange of the field in the vicinity of the particles, $E_1 \leftrightarrow E_2$. This operation is equivalent to the exchange of the field variables $a_1 \leftrightarrow a_2$ and therefore will be denoted as $I_f\colon 1 \leftrightarrow 2$. Since the particles are identical, such permutation is equivalent to an exchange of the states of the particles, and hence, according to what has been said above, it must leave the covariance (7.81) invariant

$$I_f \Gamma_{\beta\beta'} = \Gamma_{\beta\beta'}. \tag{7.84}$$

In order to calculate the left-hand side of this equation, we first recall that $a_{i\beta\beta'} = a^*_{i\beta'\beta}$, whence

$$I_f \lambda_B = \lambda_B^*. \tag{7.85}$$

As for the matrix elements in $\Gamma_{\beta\beta'}$, they have no trace of the field variables, so they remain invariant under I_f.[13] With this and Eq. (7.85), the condition (7.84) becomes

$$\left(\lambda_B^* - \lambda_B\right) \operatorname{Im} F_{\beta\beta'} G_{\beta'\beta} = 0. \tag{7.86}$$

Since this equation must hold irrespective of β, β', $\hat{F}$, and $\hat{G}$, its solution is λ_B real,[14] i.e.,

$$\lambda_B = \pm 1. \tag{7.87}$$

This means that for two identical (noninteracting) particles subject to the same external potential, the states (7.79) are (anti)symmetric, hence maximally entangled,

$$|\psi_B\rangle = \frac{1}{\sqrt{2}}\left(|\beta\rangle_1 |\beta'\rangle_2 \pm |\beta'\rangle_1 |\beta\rangle_2 \right), \tag{7.88}$$

for $\beta' \neq \beta$. From Eq. (7.80) we see that in this case the field variables $a_{1\beta\beta'}$ and $a_{2\beta'\beta}$ are maximally (anti)correlated and λ_B is symmetric in the field variables, as should be expected in view of the symmetry of the problem. Further, Eqs. (7.65) and (7.81) give for the entanglement-related covariance

$$\Gamma^{\lambda}_{\beta\beta'} = \lambda_B \operatorname{Re} F_{\beta\beta'} G_{\beta'\beta} = \pm \operatorname{Re} F_{\beta\beta'} G_{\beta'\beta}. \tag{7.89}$$

7.3.3 Comments on Particle Exchange

The effect of interchanging particle states is identical to the effect of interchanging the fields in the vicinity of the particles; therefore, the product of both operations, $I_s I_f = I_f I_s$, leaves the description invariant. Further, since the field variables are in direct correspondence with the positions of the particles, the effect of I_f can be seen as an interchange of positions (I_x), whence the operation $I_s I_f = I_s I_x$ is equivalent to

---

[13] As follows from the analysis in Sect. 7.1.4, $\hat{F}$ ($\hat{G}$) stands for the matrix defined in $\mathcal{H}_1$ ($\mathcal{H}_2$) associated with a variable $F(G)$ proper of the particle immersed in the background (not effective!) field $E_1(E_2)$, with random amplitudes $a_1(a_2)$. Now, in the case of identical particles subject to the same external potential, $\mathcal{H}_1$ and $\mathcal{H}_2$ coincide, and therefore the matrices $\hat{F}$ and $\hat{G}$ are invariant under I_f.

[14] If $\hat{F} = \hat{G}$, then $F_{\beta\beta'} G_{\beta'\beta} = \left|F_{\beta\beta'}\right|^2$, thus $\operatorname{Im} F_{\beta\beta'} G_{\beta'\beta} = 0$ and nothing can be said about $\lambda_B^* - \lambda_B$. However, there exist $\hat{F}$ and $\hat{G}$ that have the same relevant frequencies and give nevertheless a complex value for $F_{\beta\beta'} G_{\beta'\beta}$. An obvious example is $\hat{F} = \hat{p}$ and $\hat{G} = \hat{x}$, for which $\operatorname{Im} F_{\beta\beta'} G_{\beta'\beta} = m\omega_{\beta\beta'} \left|x_{\beta\beta'}\right|^2$.

an interchange of particles, denoted as I_p.[15] However, I_p does not coincide with the corresponding particle-exchange transformation $I_p^{\rm qm}$ made within the usual quantum formalism, because the field (or position)-exchange operation I_f is foreign to the usual scheme. It is precisely the alleged impossibility to distinguish trajectories what is usually considered the basis for the *indistinguishability* of the particles, a property that is identified as responsible for the essential differences between quantum systems of identical particles and their classical analogue [see e.g. Cohen-Tannoudji et al. (1977)]. Consequently, $I_p^{\rm qm}$ is simply I_s; that is, in the Hilbert-space description the exchange of particles amounts to an exchange of their actual states only (this is the so-called *exchange degeneracy*). Such discrepancy in the definition of the particle-exchange operation causes the symmetry (or invariance) arguments in these approaches to differ conceptually.

In the quantum formalism the operation $I_f{:}1 \leftrightarrow 2$ has the mathematical effect of interchanging the Hilbert spaces of the one-particle vectors in products such as $|\beta\rangle_1|\beta'\rangle_2$. Thus, it still holds that I_f and $I_s = I_p^{\rm qm}$ produce identical effects, but here I_f is merely a mathematical operation that switches the labels attached to the particles, without these labels providing any additional physical information, or characterizing any intrinsic property of the particles. In contrast, the formalism developed here puts these labels into correspondence with the fields in the vicinity of the particles, and in this sense they are endowed with a physical meaning.

7.4 Spin-Symmetry Relations

The derivation of the symmetry or antisymmetry of the stationary states for systems of two identical particles made in Sect. 7.3.2 represents a step towards the spin-statistics theorem of quantum field theory, which is adopted by QM. Let us now take a further step in connection with this problem, by focusing on a system composed of two noninteracting electrons with their spin taken into account.[16] The analysis that follows rests on the main result in Sect. 7.3.2, namely that whenever condition (7.78) holds, the state of the bipartite system is described by the entangled vector (7.88).

The system to be studied consists, therefore, of a pair of noninteracting electrons subject to the same external potential, such that the 'orbital' (i.e., nonspinorial) part of the total Hamiltonian decomposes as

$$\hat{H}_{\rm o} = \hat{H}_{1{\rm o}} + \hat{H}_{2{\rm o}}, \tag{7.90}$$

[15] The interchange of particles thus defined is valid only for identical particles subject to the same external potential. This is so because to ensure that upon the exchange I_s the resulting state continues to be stationary, both β and β' must be accessible states for both particles, for any pair (β, β').

[16] In Chap. 6 the electron spin was obtained as another quantum property that emerges from the particle-ZPF interaction. In the transition to the Heisenberg description, this new variable appears represented by a two-dimensional matrix operator $\hat{\mathbf{S}}$, which is expressed in terms of the Pauli matrices as $\hat{\mathbf{S}} = \hbar\hat{\boldsymbol{\sigma}}/2$ and satisfies the usual angular-momentum commutation rules.

with $\hat{H}_{io}$ having eigenvalues $\mathcal{E}_n$. In addition, and in order to include the spinorial degrees of freedom, the system is immersed in an external magnetic field $\boldsymbol{B} = B\hat{z}$, so that the following extra term is added to the Hamiltonian,

$$\hat{H}_{\text{spin}} = \mu_0 B \hat{\sigma}_{1z} + \mu_0 B \hat{\sigma}_{2z}, \tag{7.91}$$

where μ_0 stands for Bohr's magneton ($\mu_0 = -e\hbar/2mc$) and $\hat{\sigma}_{iz}$ is the z-Pauli matrix of particle i. The total Hamiltonian of the composite system is then

$$\hat{H} = \hat{H}_1 + \hat{H}_2 = \hat{H}_{1o} + \hat{H}_{2o} + \mu_0 B \hat{\sigma}_{1z} + \mu_0 B \hat{\sigma}_{2z}, \tag{7.92}$$

and the energy levels of each particle are

$$\mathcal{E}_\beta = \mathcal{E}_n + \mu_0 B s_z, \tag{7.93}$$

with $s_z = \pm$ the eigenvalue of $\hat{\sigma}_{iz}$, i.e., in the usual notation,

$$\hat{\sigma}_{iz} \left|\pm\right\rangle_i = \pm \left|\pm\right\rangle_i, \qquad \text{or} \qquad \hat{\sigma}_{iz} \left|s_{zi}\right\rangle = s_{zi} \left|s_{zi}\right\rangle. \tag{7.94}$$

Equation (7.93) thus shows that the (single-particle) state index β is now a compound index $\beta = (n, s_z)$ that includes both the orbital and the spinorial states. With this equation, condition (7.78) reads

$$\omega_{\beta\beta'} = \hbar^{-1}(\mathcal{E}_\beta - \mathcal{E}_{\beta'}) = \hbar^{-1}(\mathcal{E}_n - \mathcal{E}_{n'}) + \mu_0 B \hbar^{-1}(s_z - s_z') \neq 0. \tag{7.95}$$

7.4.1 Two Electrons in the Singlet State

Let us first apply the above results to a system of two electrons assuming for simplicity that both particles are in the same orbital state $|n\rangle$,

$$|\varphi\rangle = |n\rangle_1 \, |n\rangle_2 \,, \tag{7.96}$$

so that $n = n'$ in the above expressions. The common relevant frequency is then determined solely by the difference between the magnetic energy terms; according to Eq. (7.95),

$$\omega_{\beta\beta'} = \mu_0 B \hbar^{-1}(s_z - s_z'). \tag{7.97}$$

For this frequency to be different from zero we must have $s_z \neq s_z'$, hence $|\beta\rangle = |n\rangle \left|\pm\right\rangle$ and $\left|\beta'\right\rangle = |n\rangle \left|\mp\right\rangle$. From Eq. (7.88) we thus obtain the entangled state vector

$$|\psi\rangle = \frac{1}{\sqrt{2}} |n\rangle_1 \, |n\rangle_2 \left(|+\rangle_1 \, |-\rangle_2 + \lambda_S \, |-\rangle_1 \, |+\rangle_2 \right), \tag{7.98}$$

with $\lambda_S = \pm 1$. To determine the sign of λ_S we use Eq. (7.81) with $\lambda_B = \lambda_S$, $\hat{F} = \hat{\boldsymbol{\sigma}}_1 \cdot \boldsymbol{a}$, and $\hat{G} = \hat{\boldsymbol{\sigma}}_2 \cdot \boldsymbol{b}$, where $\boldsymbol{a}$, $\boldsymbol{b}$ are two unitary vectors in arbitrary directions. The calculation is straightforward and gives

$$\Gamma_{\beta\beta'} = \Gamma_S = -a_z b_z + \lambda_S (a_x b_x + b_y a_y). \tag{7.99}$$

We now assume that the two electrons are in the singlet state (total spin $S = 0$). In this case, for $\boldsymbol{b} = \boldsymbol{a}$ the spin components must be antiparallel, which means that

$$\Gamma_{S=0}(\boldsymbol{a} = \boldsymbol{b}) = -a_z^2 + \lambda_{S=0}(a_x^2 + a_y^2) = -a_z^2(1 + \lambda_{S=0}) + \lambda_{S=0} = -1 \tag{7.100}$$

irrespective of a_z. This fixes the value of the correlation factor,

$$\lambda_{S=0} = -1, \tag{7.101}$$

whence

$$\Gamma_{S=0} = -\boldsymbol{a} \cdot \boldsymbol{b} \tag{7.102}$$

in agreement with the quantum formula. This calculation has served to univocally determine the sign of λ_S for $S = 0$, which introduced into (7.98) leads to the antisymmetric spin state vector:

$$|\chi\rangle_{S=0} = \frac{1}{\sqrt{2}}\Big(|+\rangle_1 |-\rangle_2 - |-\rangle_1 |+\rangle_2 \Big), \tag{7.103}$$

and to the complete antisymmetric state vector

$$|\psi\rangle_{S=0} = \frac{1}{\sqrt{2}} |n\rangle_1 |n\rangle_2 \Big(|+\rangle_1 |-\rangle_2 - |-\rangle_1 |+\rangle_2 \Big). \tag{7.104}$$

It is clear from the way it has been constructed, that (7.103) encodes the (maximum) available statistical information of the spin state of the entire bipartite system. This means that it encodes the probabilities of the outcomes of experiments that may be performed on the two particles, whether jointly or individually, whether close to each other or a distance apart—as long as the correlation between the common relevant field modes is not broken. When the latter occurs, the value of the correlation parameter λ_S becomes zero, the system ceases to be correctly described by the entangled state vector, and the bipartite system must be considered as two separate systems, each described by its own state function.

7.4.2 The Helium Atom

An important example of a system having two electrons is of course the Helium atom. Let us consider, as in the previous example, that the direction of z is fixed by an external magnetic field $\boldsymbol{B}$ of arbitrary size that aligns the spins and introduces a spin-dependent energy term. For the present calculation, the motion of the nucleus and the direct interaction between the particles can be neglected, so that the Hamiltonian takes the form (7.92).

Take first the case in which both electrons are in the same orbital state, say n, and the spins are antiparallel, i.e., $S = 0$. As seen above, the complete state is described by the antisymmetric state vector (7.104). Now the question arises: is it possible for the two electrons to have parallel (instead of antiparallel) spins (i.e., $S = 1$) when they are in the same orbital state? We know of course that the answer is in the negative; let us look at it from the approach developed in the preceding section.

As remarked under Eq. (7.97), for correlating modes to exist when the particles are in the same orbital state, their spin projections along the direction $\hat{z}$ must be antiparallel (i.e., $s_z \neq s_z'$). This led us to write the general form of the bipartite state vector as (7.98) with $\lambda_S = \pm 1$. This means that only two states can describe the composite system. Since the case with $\lambda_S = -1$ has been shown to correspond to $S = 0$, the only possibility left is that the vector corresponding to $S = 1$ has $\lambda_{S=1} = 1$. Equation (7.99) would thus give

$$\Gamma_{S=1} = -a_z b_z + (a_x b_x + b_y a_y), \tag{7.105}$$

which for $\boldsymbol{a} = \boldsymbol{b}$ reduces to

$$\Gamma_{S=1}(\boldsymbol{a} = \boldsymbol{b}) = -a_z^2 + a_x^2 + a_y^2 = 1 - 2a_z^2. \tag{7.106}$$

However, the fact that the condition $\omega_{\beta\beta'} \neq 0$ allows only for *one* state vector corresponding to $S = 1$, namely (7.98) with $\lambda_{S=1} = 1$, creates a problem, since a single state does not suffice to represent the three spin-projection configurations consistent with $S = 1$, that conform the triplet state. In other words, when the total spin is $S = 1$ three vectors are required to describe the system, which is not allowed by the condition $\omega_{\beta\beta'} \neq 0$. In fact, the restriction $s_z \neq s_z'$ that ensues from this latter already indicates that the bipartite state cannot be of the form

$$|\psi\rangle = |n\pm\rangle_1 \, |n\pm\rangle_2 \,. \tag{7.107}$$

Thus the orbital states must be different, and they become entangled due to the fact that there is a common resonance frequency $\omega_{\beta\beta'} \neq 0$. In other words, no two electrons can be in exactly the same (spin and orbital) state. This is, in essence, an expression of the Pauli principle, which appears here as a consequence of the existence of the ZPF connecting the particles and correlating their motions via the common relevant frequencies.

In connection with this result, it is in place to emphasize the key role played by the constraint imposed on the variables characterizing the state of the bipartite system for entanglement to appear—as already remarked by Schrödinger in his celebrated 1935 papers. Consider for simplicity a bipartite state defined by the total spin and its projection along an arbitary axis z (i.e., $S = 0$, or $S = 1$ with $m = 0$); all these cases correspond to an entangled state. Assume now that the constraint on the total spin is lifted (e.g., we focus only on the orbital, identical states); then the electrons become disentangled and must be treated as two separate systems. This observation can be extended to other variables characterizing a quantum system. For example, in the case studied in Sect. 7.3.1, the state of the bipartite system was defined by the total energy only, $\mathcal{E}_B = \mathcal{E}_\beta + \mathcal{E}_{\beta'}$, with $\beta \neq \beta'$. Having the total energy fixed led to $\omega_{\beta\beta'} = -\omega_{\beta\beta'} \neq 0$, and to the consequent entanglement of the particles through the respective field modes. If the system is perturbed by a force that does not conserve the (total) energy, the particles get disentangled.

Applied to the Helium atom, we may draw a further consequence from the above discussion. Whether the atom is in the ground state or in any one of the excited states, for the (composite) state vector to represent a stationary state it must be defined by the *total* energy of the bipartite system. This means that all stationary states of the atom are entangled, either through the spin vectors, or the orbital vectors, or both. Now, our analysis of the singlet spin case in Sect. 7.4.1 led us to conclude that the complete state vector must be antisymmetric; see (7.104). This property should hold regardless of whether the electrons are in the same orbital state $|n\rangle$ or in different states $|n\rangle$, $|n'\rangle$, since the action of any operator (satisfying the superselection rule of univalence)[17] acting on the state vector does not alter its symmetry or antisymmetry property. This means that *any* vector describing the complete, stationary state of the system of two electrons must be antisymmetric.

Again, this is a well-known property in QM; however, instead of being introduced here by hand, it appears as a further consequence of the correlations established (via the ZPF) between the electrons (with spin) that form a bipartite system.

7.5 Final Comments

We have seen that when the components of a bipartite system resonate to a common frequency of the background field, it is no longer possible to describe them in a separate and independent way; particles and field constitute a single physical system, which explains why, when the description focuses on the mechanical subsystem only, the state is described by an entangled state vector. By ignoring the presence of the background field, however, entanglement appears as an elusive feature, and (apparent) nonlocal and noncausal effects enter into the picture.

The importance of the ZPF for the understanding of entanglement goes well beyond the quantum-mechanical case studied here. As an important example, in a series of works related to stochastic optics (see e.g. Santos 2002, 2008; Marshall and Santos 1988, 2002; Casado et al. 1998 and references therein), the 'nonclassical' entanglement of photons is shown to arise as a consequence of the extra correlations involving the ZPF. Taken together, those results and the present ones point towards a fundamental mechanism responsible for entanglement, one that rests on the extra correlations—

[17] The superselection rule of univalence forbids transtions from symmetric into antisymmetric vector states, and conversely.

unknown to classical physics and normally concealed in quantum theory—due to the presence of the ZPF.

By now the mediating action of the environment in the process of entanglement is well recognized. The environment appears as a physical entity capable not only of degrading entanglement—as happens in decoherence processes—but also of producing it. The ability of the environment to generate correlations among the constituents of a composite system has been largely studied within the quantum formalism (see, e.g., Braun 2002; Kim et al. 2002; Benatti et al. 2003; Ficek and Tanaś 2006; Hor-Meyll et al. 2009, and references therein).

The picture built along this chapter suggests a remarkable parallelism between (quantum) entanglement and superconductivity. In this latter case a passing electron modifies the phonon field due to electromagnetic interaction, so a second electron passing by feels the modified phonon field and reacts accordingly, thus establishing a correlation between the two electrons to form a Cooper pair (see e.g. Greiner 1998). The background phonon field due to the vibrations in the crystal lattice plays a role similar to that of the background electromagnetic field in the present case, as the medium that transfers information of the state of one electron (by inducing a change in its mechanical energy) to another one (by inducing a corresponding change).

It should be mentioned also that a result analogous to quantum entanglement can be obtained for a system of Brownian particles (Allahverdyan et al. 2005). For this to happen, the particles must however have interacted in the past via an external potential. This essential difference between the *Brownian entanglement* and the quantum entanglement presented here rests on the specific nature of the stochastic environment acting in each case. The thermal bath represents a white noise and thus is completely uncorrelated and incoherent, so that it is capable of maintaining the pre-existent correlations for some limited time, but is unable to originate them.

Finally, we have seen that the ZPF, precisely through the mechanism of entanglement, becomes an important piece in producing states of definite permutation symmetry for systems composed of electrons. This provides a physical clue for a better understanding of the relation between spin and statistics in quantum mechanics.

References

Allahverdyan, A.E., Khrennikov, A., Nieuwenhuizen, ThM: Brownian entanglement. Phys. Rev. A **72**, 032102 (2005)

Bell, J.S.: On the Einstein Podolsky Rosen paradox. Physics **1**, 195 (1964)

Bell, J.S.: Speakable and Unspeakable in Quantum Mechanics. Cambridge University Press, Cambridge (1987)

Bell, M., Gottfried, K., Veltman, M.: John S. Bell on the Foundations of Quantum Mechanics. World Scientific, Singapore (2001)

Benatti, F., Floreanini, R., Piani, M.: Environment induced entanglement in Markovian dissipative dynamics. Phys. Rev. Lett. **91**, 070402 (2003)

Blanco, R., Santos, E.: Coupled harmonic oscillators in stochastic electrodynamics. Lett. Nuovo Cim. **25**, 360 (1979)

Braun, D.: Creation of entanglement by interaction with a common heat bath. Phys. Rev. Lett. **89**, 277901 (2002)

*Casado, A., Marshall, T.W., Santos, E.: Type II parametric downconversion in the Wigner-function formalism: entanglement and Bell's inequalities. J. Opt. Soc. Amer. B **15**, 1572 (1998)

Coffman, V., Kundu, J., Wootters, W.K.: Distributed entanglement. Phys. Rev. A **61**, 052306 (2000)

Cohen-Tannoudji, C., Diu, B., Laloë, F.: Quantum Mechanics. Wiley, New York (1977)

*de la Peña, L., Valdés-Hernández, A., Cetto, A.M.: Entanglement of particles as a result of their coupling through the common background zero-point radiation field. Physica E **42**, 308 (2010)

*de la Peña, L., Cetto, A.M., Valdés-Hernández, A.: The emerging quantum. An invitation. In: Proceedings of the Advanced School on Quantum Foundations and open Quantum Systems, João Pessoa, Brazil, 16–28 Jul 2012

Duck, I., Sudarshan, E.C.G.: Pauli and the Spin-Statistics Theorem. World Scientific, Singapore (1997)

Einstein, A., Podolsky, B., Rosen, N.: Can quantum-mechanical description of physical reality be considered complete? Phys. Rev. **47**, 777 (1935)

Ficek, Z., Tanaś, R.: Dark periods and revivals of entanglement in a two-qubit system. Phys. Rev. A **74**, 024304 (2006)

Greiner, W.: Quantum Mechanics. Special Chapters. Springer, Berlin (1998)

Hor-Meyll, M., Auyuanet, A., Borges, C.V.S., Aragão, A., Huguenin, J.A.O., Khoury, A.Z., Davidovich, L.: Environment-induced entanglement with a single photon. Phys. Rev. A **80**, 042327 (2009)

Horodecki, R., Horodecki, P., Horodecki, M., Horodecki, K.: Quantum entanglement. Rev. Mod. Phys. **81**, 865 (2009)

Kaplan, I.G.: The Pauli exclusion principle. Can it be proved? Found. Phys. **43**, 1233 (2013)

Kim, M.S., Lee, J., Ahn, D., Knight, P.L.: Entanglement induced by a single-mode heat environment. Phys. Rev. A **65**, 040101 (2002)

Khrennikov, A.: Description of composite quantum systems by means of classical random fields. Found. Phys. **40**, 1051 (2010)

Khrennikov, A.: Prequantum classical statistical field theory: Schrödinger dynamics of entangled systems as a classical stochastic process. Found. Phys. **41**, 317 (2011)

Landau, L., Lifshitz, E.: The Classical Theory of Fields. Addison-Wesley, Cambridge (1951)

*Marshall, T.W., Santos, E.: Stochastic optics: a reaffirmation of the wave nature of light. Found. Phys. **18**, 185 (1988)

*Marshall, T.W., Santos, E.: Semiclassical optics as an alternative to nonlocality. Recent Res. Devel. Opt. **2**, 683 (2002)

Santos, E.: What is entanglement? arXiv:quant-phys/0204020 (2002)

*Santos, E.: Photons are fluctuations of a random (zeropoint) radiation field filling the whole space. In: Roychoudhuri, C., Kracklauer, A.F., Creath, K. (eds.) The nature of light. What is a photon? CRC Press, Florida (2008)

Schrödinger, E.: Die gegenwärtige Situation in der Quantenmechanik I, II, III, Naturwiss 23, 807, 823, 844. English translation: Quantum Theory and Measurement (Trans: Wheeler, J.A., Zurek, W.H. (eds.)) (Princeton University Press, Princeton, 1983) (1935)

Tomonaga, S.I.: The Story of the Spin. The University of Chicago Press, Chicago (1974)

Valdés-Hernández, A.: Investigación del origen del enredamiento cuántico desde la perspectiva de la Electrodinámica Estocástica Lineal. Ph. D. Thesis, Universidad Nacional Autónoma de México, México (2010)

*Valdés-Hernández, A., de la Peña y, L., Cetto, A.M.: Bipartite entanglement induced by a common background (zero-point) radiation field. Found. Phys. **41**, 843 (2011)

Chapter 8
Causality, Nonlocality, and Entanglement in Quantum Mechanics

> *Why is the pilot-wave picture ignored in the text books? Should it not be taught, not as the only way, but as an antidote to the prevailing complacency? To show that vagueness, subjectivity, and indeterminism, are not forced on us by experimental facts, but by deliberate theoretical choice?*
>
> Bell (1987, page 160)

The material presented in previous chapters has dealt with some fundamental elements of QM such as (non)locality, (non)causality, (in)determinism and the (non)existence of trajectories. These elements have been at the core of some of the most longstanding controversies, and also of the most thorough attempts to develop alternative interpretations of the quantum formalism.

Among such reformulations, perhaps the best known one is the de Broglie-Bohm theory. This represented the first serious attempt to recover for QM two notions that are dear to (classical) physics, namely those of causality and of trajectory (or realism, in a by now extended usage)—although at the high price of nonlocality. It seems therefore appropriate to initiate the discussion on these subjects with a brief, critical review of the de Broglie-Bohm theory. This will set the framework for a fresh analysis of quantum (non)locality from the perspective of the SED theory as developed in the preceding chapters.

8.1 Causality at Stake

Quantum indeterminism, as represented by the Heisenberg inequalities or the inherently unpredictable specific outcome of a measurement, is for some a cause of discomfort with the theory. When the status of QM as a fundamental theory of physics is at stake, such trait is indeed hardly acceptable from a realist standpoint. The simplest solution to the problem of quantum indeterminism consists in assuming that there exist some variables that specify, along with the quantum-mechanical state

L. de la Peña et al., *The Emerging Quantum*,
DOI: 10.1007/978-3-319-07893-9_8,

vector, a *dispersion-free state*, i.e. a state for which the specific outcomes of a measurement are fully determined. These hypothetical variables that restore determinism are the so-called 'hidden variables' or 'hidden parameters'. Of course, one must assume that their existence is compatible with the quantum laws. And here sprouts the problem...

8.1.1 Von Neumann's Theorem

In 1932 the mathematician J. von Neumann published what constituted the first formal textbook on quantum mechanics (English version: von Neumann 1932). The book contains a theorem that was quickly accepted, and which half a century ago still seemed to be in full force. This theorem—which stands even today as the archetype of the 'impossibility proofs', more popularly known as 'no-go' theorems— asserts that one cannot add hidden variables to render QM deterministic.[1] Since QM is supported by a vast amount of empirical evidence, a deterministic description of it happens to be refuted by such evidence.

In 1952 a second round started with the publication by D. Bohm of two now classical papers (Bohm 1952a, b; see also Bohm 1953) providing a specific counterexample to von Neumann's theorem. It was just a causal and deterministic description completely consistent with the quantum-mechanical formalism, built in terms of *nonlocal* hidden variables. One should have expected such a result to immediately draw the attention from physicists (and mathematicians and philosophers of science) eager to find the error in von Neumann's demonstration. However, this was not the case; the large majority simply ignored Bohm's results and took von Neumann's conclusion for granted.

In the proof of his theorem, von Neumann had introduced some postulates that he considered of general validity. Starting with the expression

$$\left\langle \hat{C} \right\rangle = \mathrm{tr}\,(\rho C) \tag{8.1}$$

that defines the mean value of an operator $\hat{C}$ (associated with the observable C) in the quantum state described by the density matrix ρ, it is straightforward to arrive at

$$\left\langle \hat{A} + \hat{B} \right\rangle = \left\langle \hat{A} \right\rangle + \left\langle \hat{B} \right\rangle, \tag{8.2}$$

irrespective of the operators $\hat{A}$ and $\hat{B}$. Von Neumann then assumed that this additive relation holds *also* when the mean values are calculated in dispersion-free states. And since such states must yield one (well-defined) result from among the possible ones predicted by the quantum state, the expectation values of $\hat{C}$ in a dispersion-free state

[1] It is much less known that almost simultaneously and independently, a similar result was published by Solomon (1933).

are just the eigenvalues of $\hat{C}$ (Bell 1987, articles 1 and 4). Yet for noncommuting $\hat{A}$ and $\hat{B}$ the eigenvalue of the sum $\hat{A}+\hat{B}$ is not necessarily equal to the sum of the corresponding eigenvalues,[2] whence the relation (8.2) ceases to be true for dispersion-free states.

Two years after the original publication of von Neumann's theorem, Hermann (1935) observed that this additivity assumption was too restrictive within the quantum domain.[3] Hermann's criticism was largely ignored, as were the few other criticisms raised much later, notably by Feyerabend (1956), and Mugur-Schächter (1964).[4] The widely accepted proof that von Neumann's theorem demanded revision came only with the work of Bell (1966), who again observed that the theorem was mathematically correct but not of general applicability, as the additivity postulate was too restrictive.

The generally accepted conclusion at present is that the theorem is not general enough to eliminate all kinds of hidden-variable theories (see e.g. Bub 2010); in particular, it does not exclude nonlocal hidden variables, as Bohm's work evinced. Alternative versions of von Neumann's theorem have been devised in the meantime that escape from the previous criticisms, by Gleason (1957) for Hilbert spaces of dimensionality greater than 2, Bell (1966), Kochen and Specker (1967), Belinfante (1973), and Peres (1996), among others. In their turn, Gudder (1970) and Santos (1975) have proved that QM does accept contextual hidden variables.

---

[2] Take for example the spin projections along three different directions: $\hat{A}=\hat{S}_x$, $\hat{B}=\hat{S}_y$ and $\hat{C}=(\hat{A}+\hat{B})/\sqrt{2}=(\hat{S}_x+\hat{S}_y)/\sqrt{2}$. If the system possesses spin 1/2, the eigenvalues of each of these operators are the same and equal to ± 1; clearly the eigenvalues of $\hat{C}$ are not the linear combination $(\pm 1 \pm 1)/\sqrt{2}$.

[3] Since Hermann's argument is little known, and is just the same discovered by Bell 30 years later, it seems of interest to transcribe it here: "Suppose we have an ensemble of physical systems, with $\mathfrak{R}$ and $\mathfrak{S}$ physical quantities that can be measured on this ensemble; the expectation value of $\mathfrak{R}$ (Expt($\mathfrak{R}$)) is the average value of all measurement outcomes that will be obtained when measuring $\mathfrak{R}$ on all systems of the ensemble, and is also the value that is expected to be obtained when measuring $\mathfrak{R}$ on an arbitrary element of this ensemble. Von Neumann requires that for this expectation value-function Expt($\mathfrak{R}$), defined using an ensemble of physical systems and producing a number for every physical quantity, Expt($\mathfrak{R}+\mathfrak{S}$) = Expt($\mathfrak{R}$) + Expt($\mathfrak{S}$). In words: The expectation value of a sum of physical quantities is equal to the sum of the expectation values of both quantities. With this assumption the proof of von Neumann either succeeds or fails."

"For classical physics this requirement is trivial and also for those quantum mechanical observables that [commute]... Not trivial however is the relation for quantum mechanical quantities for which indeterminacy relations hold. In fact the sum of two such quantities is not even defined: Because a sharp measurement of one of them excludes sharp measurement of the other one and thus because both quantities cannot have sharp values at the same time, the commonly used definition of the sum of two quantities breaks down."

[4] Feyerabend noticed that the postulates used in von Neumann's derivation did not exclude dispersive hidden variables. Now if the hidden variables added to QM had an irreducible dispersion, the quantum variables themselves should continue to be dispersive and things remained essentially the same, except that the theorem needed some reformulation. Mugur-Schächter, on her part, argued that the demonstration was not as general as assumed, since it presupposes that the distribution of the hidden variables (once more, distributed variables) has properties similar to those of the quantum distribution.

8.1.2 Bohm's Counterexample

For historical fairness, the theory proposed by Bohm—*causal quantum mechanics*, as he called it—should be named after de Broglie and Bohm, since L. de Broglie proposed his *pilot-wave theory* already during the construction of quantum mechanics as an alternative to both Heisenberg's and Schrödinger's routes to the new theory (de Broglie 1926a, b, 1927a, b, c; see also de Broglie 1963).[5] The pillars on which de Broglie's and Bohm's theories rest are quite similar, although from the outset the former was more ambitious. The theory proposed by de Broglie was an attempt to arrive at QM through the study of the trajectories followed by electrons or photons 'guided' by the quantum field (hence the name of pilot-wave theory). De Broglie's initial work preceded the theories of Heisenberg and Schrödinger, and can therefore be considered to be the first attempt to find a formulation for QM. Unfortunately the author abandoned it around 1930, due to several unresolved difficulties, and with this the search for quantum trajectories was left aside for a long time. A detailed discussion of de Broglie's theory, covering both the technical and the historical aspects, can be seen in the excellent book by Bacciagaluppi and Valentini (2009); see also Cushing (1992) and Bohm and Hiley (1995). In 1956, de Broglie himself published a general discussion of his theory

As for Bohm's theory, it is derived from QM—just the converse of what de Broglie's theory was designed for—so one can say that it is QM supplemented with a guidance formula—a pilot formula that can be derived from within QM—which is the entry point for the hidden variables. Thus Bohm's theory represents an alternative reading of QM rather than a different theory. Bohm himself insisted on this point: his theory is totally consistent with QM, it is QM seen from a causal and deterministic stance. The theory was born precisely out of the drive to demonstrate that, in defiance of von Neumann's theorem, QM accepts a hidden-variable description that makes it causal and deterministic. Yet even though Bohm's formulation did provide at its time a real counterexample to von Neumann's theorem, it was largely left aside, as mentioned earlier. An eloquent testimony of the reaction generated by Bohm's theory is given by F. Bopp in his summary of the discussions at a 1957 Conference in Bristol (Jammer 1974, p. 256):

> '...we say that Bohm's theory cannot be refuted, adding... that we don't believe in it.'.

Hostility towards Bohm's theory continued and still continues in many circles, although much attenuated. An illustrative example is due to Pauli (1952), who in the book in homage to de Broglie dismissed the de Broglie-Bohm approach as 'artificial metaphysics' because this theory breaks the symmetric treatment of canonically conjugate variables. The theory developed in Chap. 4 shows that this violation is merely

[5] Without pretending to undermine de Broglie's credit for his seminal contribution, in most of this book we shall refer to *Bohm's theory*, for short, as is customary in present-day literature. An alternative form of QM, similar to Bohm's, had been proposed many years earlier by Madelung (1927).

apparent, since the description in configuration space is the result of a free choice rather than of necessity.

Since Bohm's formulation is derived just from QM, its novelty might be questioned; however, this question would be out of place. Bohm's approach has contributed in various important ways to our understanding of quantum theory. Firstly, it opened a door to the idea of the feasibility of hidden-variable theories. This stimulated work on the subject, which led to important results and to a weakening of the extended conviction that the quantum world is elusory. A most important contribution is the insistence on a *causal* interpretation of QM, openly confronting the orthodox interpretation in terms of quantum fluctuations and selections without an underlying cause. A further merit is that it does not resort to the observer, as orthodox QM does. If you want to perform a measurement, you should incorporate the instrument into the description just as you would do in a classical context (an example of this is given below in Sect. 8.2.3). Moreover, the variables used in Bohm's formulation allow for the introduction into QM of the notion of trajectory. In fact, the electron remains all the time a corpuscle following a trajectory and never becoming a wave. This additional touch of realism is another important virtue that is in want in the usual renderings of QM. In discussing Bohm's formulation more at depth in what follows, we will find a new opportunity to cogently establish the meaning of several of the most characteristic features of QM from a realist point of view, and to show the many coincidences and some divergences with the theory discussed in this book.

As Bohm's formulation gained acceptance with time, the emphasis shifted towards a full-fledged alternative to the orthodox interpretation, and the theory expanded its aims. From a counterexample to von Neumann's theorem it evolved into a causal and realistic interpretation of QM that acquired some popularity.[6] It has even been presented as a quantum theory by itself, not based on the Schrödinger equation but on its own principles [see e.g. Dürr and Teufel (2009)] and leading to the Schrödinger equation as one of its major results.[7] A very important extension of the theory is its generalization to the many-body problem. Here also de Broglie was the originator, having presented his many-body theory already at the 1927 Solvay Conference (de Broglie 1928). Also since its 1952 revival by Bohm, the theory was presented as a many-body corpus. There are works studying further possibilities, such as the introduction of spin, the extension to the relativistic domain, (see e.g. Nikolić 2007; Hernández-Zapata and Hernández-Zapata 2010), or the generalization to a statistical

[6] The well-known book by Bell containing the collection of his articles on the foundations of QM (Bell 1987) was very influential in the revival of Bohm's theory. Bell appreciated the objective, deterministic and causal aspects of the pilot-wave theory. It was the search for an answer to the question: Is it that any hidden-variables theory is by necessity nonlocal? what prompted Bell's work leading to his now famous inequalities.

[7] This is achieved by simply inverting the reasoning in the derivations. Two crucial postulates are needed: one is of course the guidance equation; the second demand serves to introduce the quantum potential V_Q into Eq. (8.9) on the basis of an appropriately contrived argument. The simplest procedure is to consider the quantum potential as an empirical—and thus phenomenological—expression, and to proceed from there on. There exist all sorts of interpretations and 'derivations' of the quantum potential, as commented in footnote 4.14.

situation (Bohm and Hiley 1996), usually treated by means of the density matrix, and into other modern topics (Oriols and Mompart 2012). Extensive monographs on the subject are Bohm and Hiley (1995), Holland (1993), Dürr and Teufel (2009), or the introductory course Towler (2009); see also Passon (2005) and Thiounn (1965). For another 'branch' of the theory see Floyd (2000). A related critical work on Bohm's theory is Dürr et al. (1992).

8.2 Essentials of the de Broglie-Bohm Theory

8.2.1 The Guiding Field

A straightforward derivation of Bohm's theory starts with the introduction into the Schrödinger equation

$$i\hbar\frac{\partial\psi}{\partial t} = -\frac{\hbar^2}{2m}\nabla^2\psi + V\psi \tag{8.3}$$

of the wave function $\psi(\boldsymbol{x}, t)$ written in polar form

$$\psi(\boldsymbol{x}, t) = \sqrt{\rho(\boldsymbol{x}, t)}e^{iS(\boldsymbol{x},t)}, \tag{8.4}$$

with $\rho(\boldsymbol{x}, t)$ and $S(\boldsymbol{x}, t)$ real functions. By separating Eq. (8.3) into its real and imaginary parts one obtains the couple of equations

$$\frac{\partial\rho}{\partial t} + \nabla\cdot\rho\boldsymbol{v} = 0, \tag{8.5a}$$

$$\hbar\frac{\partial S}{\partial t} + \frac{\hbar^2}{2m}(\nabla S)^2 + V - \frac{\hbar^2}{2m}\frac{\nabla^2\sqrt{\rho}}{\sqrt{\rho}} = 0, \tag{8.5b}$$

where $\rho\boldsymbol{v} = \boldsymbol{j}$ is the probability current or particle flux, with [see Eqs. (4.57) and (4.127)]

$$\boldsymbol{v}(\boldsymbol{x}, t) = \frac{\hbar}{m}\nabla S(\boldsymbol{x}, t). \tag{8.6}$$

Clearly the content of Eqs. (8.5a) and (8.5b) is the same as that of Schrödinger's equation. However, in the causal interpretation a formal analogy with classical mechanics is established, by taking $\boldsymbol{v}(\boldsymbol{x}, t)$ as the velocity field of a *single* particle located at $\boldsymbol{x}$,

$$\frac{d\boldsymbol{x}}{dt} = \frac{\hbar}{m}\nabla S(\boldsymbol{x}, t)|_{\boldsymbol{x}=\boldsymbol{x}(t)}, \tag{8.7}$$

and thus interpreting $\hbar S$ as the action function of the quantum problem. This action differs from the classical one (S_c) in that the latter solves a true (classical) Hamilton-Jacobi equation

$$\frac{\partial S_c}{\partial t} + \frac{1}{2m}(\nabla S_c)^2 + V = 0, \tag{8.8}$$

whereas the (dimensionless) action S is governed by Eq. (8.5b), which can be rewritten as

$$\hbar\frac{\partial S}{\partial t} + \frac{\hbar^2}{2m}(\nabla S)^2 + V_{\text{eff}} = 0, \quad V_{\text{eff}} = V + V_Q, \tag{8.9}$$

with V_Q the *quantum potential* or *Bohm's potential* [already found, for example, in Eqs. (2.80) and (4.86)],

$$V_Q = -\frac{\hbar^2}{2m}\frac{\nabla^2\sqrt{\rho}}{\sqrt{\rho}}. \tag{8.10}$$

In terms of the 'effective potential' $V_{\text{eff}} = V + V_Q$, Eq. (8.9) takes the *form* of a Hamilton-Jacobi equation for the principal function S (Goldstein et al. 2002). Equation (8.9) is thus interpreted as the quantum Hamilton-Jacobi equation, that is, the quantum version of Eq. (8.8). According to this interpretation, the difference between a classical problem and the corresponding quantum one is due the presence of the function V_Q (cf. the discussion in Sect. 4.4.1).[8]

Notice, however, that in contrast to the external (classical) potential V, V_Q is not a preestablished function of $\boldsymbol{x}$; it depends on the evolution of $\rho = \psi\psi^*$, the dynamics of which is in its turn determined by the quantum potential itself. Hence V_Q applies a kind of feedback on the particle, dependent on the distribution of particles, a feature that endows the theory with highly nonclassical properties. It should come as no surprise that the implications of this potential for the dynamics are far reaching, as we know is just the case. In particular, the dependence of V_Q on ρ endows Eq. (8.9)

[8] The kinetic origin of the quantum potential is discussed in de la Peña et al. (2011). To the varied proposals to derive the quantum potential cited in footnote 4.14, one should add those of Dürr et al. (1992), and Ván and Fülöp (2003), as well as.the thermodynamic approach of Grössing (2008, 2009). A somewhat bolder one is that of Floyd (2002), who proposes a trajectory description based on a peculiar quantum potential containing derivatives of third order. An interesting point of this theory is that it contains extra parameters that allow for a distribution of the velocity $\boldsymbol{v}$, resulting in a more realistic description. Salesi (1996) and Recami and Salesi (1998) propose that the quantum potential can be derived by considering the energy associated with the internal zitterbewegung (considered as the antecedent of the spin). A similar proposal is made by Esposito (1999), who associates the quantum potential with the (internal) kinetic energy due to a generalized spin; see also Yang (2006). For these authors, the notions of spin, zitterbewegung and quantum potential are intimately related. Garbaczewski (1992) offers a nice derivation of the quantum potential as due to the fluctuations of the momentum. In Carroll (2007, 2010), additional arguments are introduced about the origin of the quantum potential, related to Fisher information.

with a statistical content, and is responsible for the essential difference between (8.5b) and a true Hamilton-Jacobi equation, which by definition describes the motion of a congruency of (single) particles acted on by local potentials (a congruency refers to a single-valued trajectory field).

The fact that in Bohm's theory $\boldsymbol{v}$ stands for the velocity of a single particle, allows for the introduction of the notion of trajectory into the description. Such trajectory is described by the velocity field, determined by ψ according to Eq. (8.6) [see also Eq. (4.125)],

$$\boldsymbol{v}(\boldsymbol{x},t) = \frac{i\hbar}{2m}\left(\frac{\nabla\psi^*}{\psi^*} - \frac{\nabla\psi}{\psi}\right). \tag{8.11}$$

The wave function ψ is taken here as a physically real field—just as real as, say, the electromagnetic field—that pervades the entire available space and guides the particle according to (8.11), which is therefore known as the *guidance* (or *pilot*) equation.[9] The two basic physical elements of the theory are thus the wave (guiding or pilot) field, determined by the Schrödinger equation, and the particle, with its motion determined by the solution of Eq. (8.7), or rather by the solution of

$$m\left(\frac{\partial}{\partial t} + \boldsymbol{v}\cdot\nabla\right)\boldsymbol{v} = m\frac{d\boldsymbol{v}}{dt} = -\nabla(V + V_Q), \tag{8.12}$$

obtained by applying the operator ∇ to Eq. (8.9). Notice that the operator

$$\mathcal{D}_c = \frac{\partial}{\partial t} + \boldsymbol{v}\cdot\nabla, \tag{8.13}$$

frequently called co-moving derivative (introduced in Sect. 2.3.1), coincides with the the total time derivative when the latter is taken along the path determined by the velocity field $\boldsymbol{v}$ (whence in the Bohmian approach $\mathcal{D}_c\boldsymbol{v}$ is the actual acceleration of the particle).

The solution of Eq. (8.7) depends on the initial condition $\boldsymbol{x}_0 = \boldsymbol{x}(t=0)$. When different values are assigned to $\boldsymbol{x}_0$, an ensemble of motions (for a given ψ) is determined; a statistical meaning is therefore ascribed to the field ψ by *postulating* that the probability for a particle to be in the vicinity $d\boldsymbol{x}$ around $\boldsymbol{x}$ is given by

$$|\psi(\boldsymbol{x},t)|^2\, d\boldsymbol{x}. \tag{8.14}$$

This allows to establish contact with the probability interpretation of the density $\rho(\boldsymbol{x},t)$. Moreover, it can be shown that given the initial distribution $\rho(\boldsymbol{x},0)$, Eq.

[9] The field ψ differs in essence from those known to classical physics. In contrast to the gravitational or the electromagnetic field, for example, it does not have a generating source. Moreover, it affects the particle (by guiding it) but is not affected by it. This lack of reciprocity in the field-particle influence led de Broglie (1956) (and afterwards Bohm himself) to regard the pilot-wave theory as just a step towards a necessarily more developed theory. [See item 15 in Bell (1987)].

(8.11) holds at any future time. The fact that the (ensemble of) initial positions $\boldsymbol{x}_0$ is the only information required by the theory that is not contained in $\psi(\boldsymbol{x}, t)$, has led to refer to such positions as the set of hidden variables of the theory. They are here, but remain hidden to usual QM.

The pilot equation was introduced early in de Broglie's work (1927, 1928) as a law that unifies the Maupertuis principle of least action for the path of a particle and Fermat's principle of least time for a ray in optics; pt thus represents a law that unifies mechanics and optics. That both principles led to the same Eq. (8.11) allowed de Broglie to bring together the mechanical and undulatory aspects of the behavior of a quantum particle, following his fundamental proposal of the dual behavior of both, particles and photons. Equation (8.11) was therefore the starting point of de Broglie's theory (the pilot-wave theory). For Bohm, on the other hand, the point of departure was Eq. (8.12), which represents the guidance principle applied to the acceleration, not to the velocity, and which can be identified with Newton's equation of motion for a particle subject to the potential V_{eff}. Of course, the presence of the quantum potential in this latter modifies drastically the strictly classical (Newtonian) interpretation of Eq. (8.12).

What is sometimes called *Bohmian mechanics* is the theory based on Eqs. (8.11) and (8.3); it is a kind of mixture of the two theories (de Broglie's and Bohm's) that leaves aside the initial motivations of both authors in favor of the formal aspects of the theory. Notwithstanding its shortcomings (some of which will be discussed below) Bohmian mechanics should be recognized for its success in restoring realism, objectivity, determinism and causality for QM.

8.2.2 *Quantum Trajectories*

In the de Broglie-Bohm theory of motion a particle possesses an exact position $\boldsymbol{x}$ and an exact velocity $\boldsymbol{v}$ at any given time. This endows the theory with one of its main virtues, namely that of allowing for a phase-space trajectory (Holland 1993, Vasudevan et al. 2008). Since the notion of trajectory is foreign to the usual quantum description, it seems appropriate to comment on it here.

Textbooks on QM usually appeal to physical intuition to introduce some concepts, yet very soon the physical intuition dissapears in favour of a dynamical description carried out in an appropriate Hilbert space. In this new (highly mathematical) context, the notion of trajectory in physical space is dispensed of, avoided, or even negated.[10]

[10] This assertion requires some qualification. It is not too difficult to find (both in orthodox textbooks and in research papers, and of course also in popular works), arguments that bear implicitly or explicitly on the notion of trajectory. For example, in discussions on van der Waals or molecular forces a drawing is sometimes made of atoms with well-localized orbiting (point) electrons, and the Hamiltonian is written accordingly. True, at some moment an average is taken, but nevertheless the discussion refers, or at least seems to refer, to orbiting point particles. Another example is an atom or a particle in a Stern-Gerlach experiment, which in every analysis is considered to follow a definite trajectory.

In general, the denial of quantum trajectories is rooted on a certain reading of the uncertainty relations, so this kind of assertions is interpretation-dependent; valid only within the Copenhagen interpretation broadly understood, and particularly embraced by Bohr, as revealed in the following:

> Bohr was also at the meeting. After I had tried many times to explain what I was doing and didn't succeed, I talked about trajectories, then I would swing back –I was being forced back all the time. I said that in quantum mechanics one could describe the amplitude of each particle in such and such a way. Bohr got up and said 'Already in 1925, 1926, we knew that the classical idea of a trajectory or a path is not legitimate in quantum mechanics: one could not talk about the trajectory of an electron in the atom, because it was something not observable.' In other words, he was telling me about the uncertainty principle. It became clear to me that there was no communication between what I was trying to say and what they were thinking. Bohr thought that I didn't know the uncertainty principle, and was actually not doing QM right either. He didn't understand at all what I was saying. I got a terrible feeling of resignation. R. P. Feynman, taken from Towler (2009), lecture 6.

In contrast to the Copenhagen interpretation (where the quantum description is taken to refer to a single particle), for the ensemble interpretation the Heisenberg inequality $\Delta x \Delta p \geq \hbar/2$ says nothing about the impossibility for a single particle to have a definite position and momentum. It represents, instead, a statistical expression relating the simultaneous dispersions of position and momentum in the ensemble. Further, acknowledging the statistical essence of QM, the issue is not whether both x and p are simultaneously distributed, but *why* their distributions are conditioned by the Heisenberg inequality. We have found in Chaps. 3–5 an answer to this question, involving the action of the zero-point field. Indeed, in the present approach the original equation of motion refers to a stochastic process with a well-defined trajectory for every realization of the field. However, when an ensemble of systems is considered, the possibility to *identify* the single trajectories is lost.[11] Our corollary is thus that QM is unable to explicitly allude to the notion of trajectory, without this however implying a negation of the existence of trajectories.

Historically speaking, the denial of quantum trajectories was not always the case. Quite the opposite, as stated in Sect. 8.1.2, in his very first attempts to construct modern quantum mechanics—some ten years after Bohr's model of the H-atom—de Broglie was trying to describe the quantum trajectories when he proposed the guidance formula $\boldsymbol{v} = \hbar \nabla S/m$. In the course of time, the notion of quantum trajectory has found invigoration with the emergence of novel ideas from the consistent-histories interpretation of quantum mechanics (see e.g. Griffiths 1993, Omnès 1994), the quantum-trajectory method (Lopreore and Wyatt 1999, 2000), quantum optics (see e.g. Brun 2002), and even from a study of the Dirac equation (Gull et al. 1993). The Workshops on Quantum Trajectories held in 2008 and 2010 attest to the current interest on the subject. Indeed, the possibility of tracing quantum trajectories has attracted the attention of a vast number of investigators due to the valuable information that can be effectively gained from their study. The related literature shows a growing trend,

[11] It is this statistical treatment what engenders 'indistinguishability', and this occurrs regardless of whether the system is classical or quantum. This, for instance, explains the use of the notion of indistinguishability to solve the Gibbs paradox in classical statistical physics (see e.g. Mandl 1988).

and eventually such endeavour should bear important fruits. Hence, even though the notion of trajectory in QM continues to meet opposition among strongly Copenhagen-minded physicists, it seems that we are still in a period of search and definition about the physical principles of quantum theory, the trajectory issue being only a small part of the picture.

From within SED, refined studies have been carried out to determine the trajectories as predicted by the theory. Such studies demand a detailed simulation of the ZPF, instead of the hazard being introduced by hand through the use of distributed initial velocities (Bohm's theory) or of a white noise (stochastic mechanics). For this purpose a background noise with spectrum proportional to ω^3 has to be generated and the full equation of motion (4.2) must be used. A most important example of this kind of calculations is the work of Cole and Zou (2003, 2004a, b, c, d, 2009) on the H-atom, which (nearly) reproduces the stationary quantum predictions as the result of a statistical analysis of a long run of the ground-state orbit. It should be stressed that these results are obtained by averaging the orbital motions, that is, by tracking individual trajectories of electrons for a long time interval, in correspondence with the discussions in Chaps. 4 and 5. Another most interesting example is given in Huang and Batelaan (2012); here the authors study the details and statistics of the trajectories followed by a classical oscillator immersed in the ZPF, and exhibit the mechanism by which the classical distribution of positions gets transformed into the corresponding quantum distribution.

That quantum trajectories may differ greatly from classical trajectories should be expected in advance. Popular illustrations such as the stylised atoms used in postage stamps, logos and comic strips are of course very rude representations, far from reality. The atomic orbits do not resemble planetary orbits; they are something much more complex due to the complex dynamics they obey. The best-known graphs of quantum trajectories are perhaps those obtained within Bohm's theory.[12] Even if they cannot provide a detailed description, the studies of electron trajectories from Bohm's perspective satisfactorily reproduce some central features of the experimental results and offer an intuitive picture of the underlying dynamics. For some physicists the notion of quantum trajectory is natural and is an interesting consequence of the quantum potential (see e.g. Dewdney et al. 1993, Dürr et al. 1993); for others, their weirdness makes them unacceptable, or even surrealistic (Englert et al. 1992, Aharonov and Vaidman 1996).

The trajectories predicted by Bohm's theory are in general very complex, since the quantum potential is an intricate function of ψ and ψ^* that can vary very fast with the coordinates, and thus along the path followed by the particle. Moreover, the trajectories are normally quite sensitive to the initial conditions and to the full distribution of particles. A most popular example of a Bohmian trajectory is presented in Fig. 8.1, which shows the paths followed by electrons diffracted by two Gaussian slits (taken from Philippidis et al. 1979). Another example is given in Fig. 8.2 (taken

[12] Also Nelson's theory and more generally the stochastic description of QM have been successfully used to investigate quantum trajectories, as shown by the examples in Chap. 2.

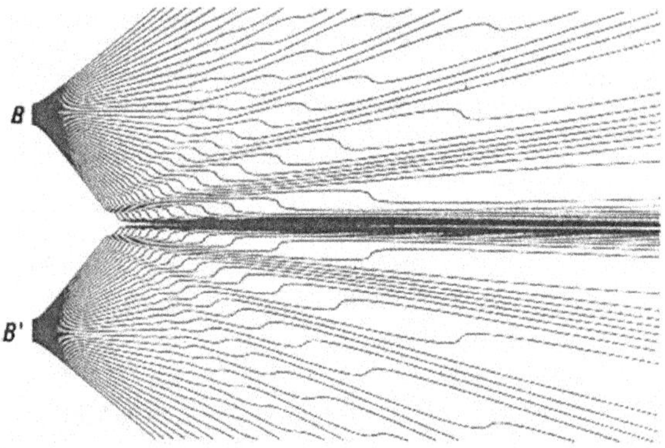

Fig. 8.1 Trajectories scattered by two Gaussian slits, according to Bohm's theory. The initial beams of particles have a uniform distribution at each slit. Reprinted from Philippidis et al. (1979) with permission from Springer

from Dewdney and Hiley 1982), which shows the trajectories of electrons from a Gaussian packet incident on a semitransparent barrier with a transmission ratio of one-half. The pattern shows clearly how the outer particles are reflected *before* reaching the barrier, whilst the ones closer to the barrier are mostly transmitted. In the next section we will have opportunity to discuss this (nonlocal) feature in more detail.[13] Notice that *individual* particles do not conserve their *mechanical* energy, but change speed before reaching the barrier, as shown also in Chap. 2. The total energy is of course conserved, but since QM does not consider the energy of interaction with the ZPF, it loses accountability in instances as the present one.

As mentioned earlier, the Bohmian trajectories are described by individual particles having an instantaneous velocity $\boldsymbol{v}(\boldsymbol{x})$ as given by Eq. (8.6). Such interpretation differs from the one developed in this book, according to which (8.6) refers to a local *mean* velocity, obtained as a result of a (partial) averaging over the momentum space, and hence containing statistical information about all the individual instantaneous velocities at $\boldsymbol{x}$. From the perspective adopted here, the stochastic field is responsible for fluctuations around the mean motion, as shown for example in Figs. 2.1 and 2.2, so that the actual velocity of a single particle varies at random from case to case and should be expressed as[14]

$$\boldsymbol{v}_{\text{actual}}(\boldsymbol{x},t) = \frac{\hbar}{m}\nabla S(\boldsymbol{x},t) + \Delta\boldsymbol{v}, \tag{8.15}$$

with $\Delta\boldsymbol{v}$ a stochastic deviation that averages to zero. Consequently the SED approach admits (random) trajectories that can mutually cross, an effect that the de Broglie-Bohm theory, by its deterministic nature, cannot take into account. Indeed, a well-

[13] Further examples can be seen in Holland (1993), Lopreore and Wyatt (1999, 2000)—who have generated what they call the 'quantum trajectory method'—; Suñé and Oriols (2000), Matzkin and Nurock (2008), Sanz et al. (2002), Philippidis et al. (1982), and Kumar Chattaraj (2010).

[14] In a variant of Bohm's theory the idea of a fluid *à la* Madelung is entertained. By considering this fluid to be subject to fluctuations, a random element is then added; see Bohm and Vigier (1954). See also Wang (2006) for related work.

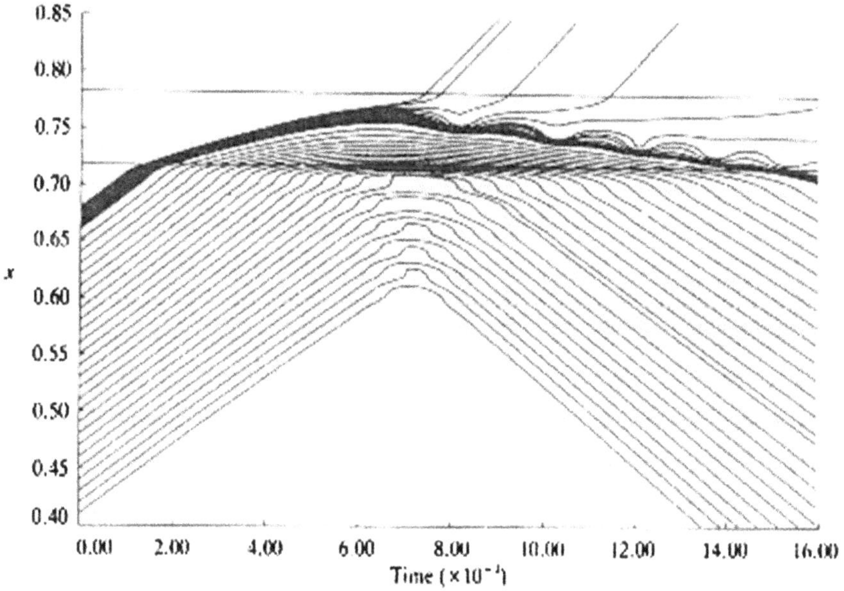

Fig. 8.2 A collimated beam of particles scattered by a barrier according to Bohm's theory. Notice that many particles are scattered before reaching the barrier. Reprinted from Dewdney and Hiley (1982) with permission from Springer

known prediction of Bohm's theory is that the (deterministic) trajectories never cross. This is foreseeable, for particles that follow the (by form) classical rule, Eq. (8.12): the crossing of classical trajectories at a point $\boldsymbol{x}$ would mean that at that point the velocity (the local tangent) would not be a single-valued function. Illustrative examples of crossing trajectories that go in line with the stochastic perspective can be seen in the cited figures of Chap. 2.

The aforementioned difference between the meanings ascribed to $\boldsymbol{v}$ is reflected also in the way Eq. (8.9) is read. Whereas in Bohm's theory the statistical content of this equation is encoded in the quantum potential V_Q (via its dependence on the probability density ρ), within our approach the analogy with the classical case is more distant, due to the intrinsic statistical sense of the kinetic terms involved. In addition, the fact that in Bohm's theory the velocity $\boldsymbol{v}$ is identified with $d\boldsymbol{x}/dt$ runs into a problem for stationary states with real (spatial) wave function, since in such case $\nabla S = 0$ and there is no flow of particles, as argued by Heisenberg (1955). This sounds unconvincing, especially considering that in general (even for stationary states) a nontrivial distribution of momenta is to be expected, so that a static image seems incongruent with the theory. Of course, there still exists the diffusive velocity $\boldsymbol{u}$ (see Sect. 4.5.1), although it is not recognized as a velocity in the usual quantum account, nor in the causal interpretation. By contrast, if $\boldsymbol{v}$ refers to a mean local velocity, the result $\boldsymbol{v} = 0$ only indicates that there are equally many particles travelling in opposite directions. This restores consistency with the distribution of the momenta.[15]

8.2.3 The Measurement Task in the Pilot Theory

The causal and realistic posture that characterizes the pilot theory can be extended to cope with the measurement problem in a natural, objective and rational way: if a measurement is to be performed, that means the system has been prepared by inserting the measuring instrument, which has thus become an integral part of it. The next step is therefore to solve the Schrödinger equation with the apparatus and its gear included; the solution should give the statistical answer, and so the measurement problem should dissolve.

Let us illustrate how this is achieved, with the help of a simple example proposed by Bell (1987, article 17) and derived from the original theory presented by de Broglie (1928) in the Fifth Solvay Conference 1927 (other, more elaborate and realistic examples can be seen in the suggested literature). Suppose that a spin component is to be measured, and that the 'measurement' is represented by an interaction Hamiltonian similar to the one proposed by von Neumann (1932) in his famous textbook (one-dimensional notation is used for simplicity),

$$-i\hbar g\hat{O}\frac{\partial}{\partial x}. \tag{8.16}$$

Here g is a coupling constant and $\hat{O}$ an appropriate spin operator (the observable to be measured). Assume also that the particle is sufficiently massive so that the complete Hamiltonian is very nearly represented by its interaction part. The Schrödinger equation for the state $|\psi\rangle$ reads then

$$i\hbar\frac{\partial}{\partial t}|\psi\rangle = -i\hbar g\hat{O}\frac{\partial}{\partial x}|\psi\rangle . \tag{8.17}$$

Introduction of the spin eigenvectors $|\alpha_n\rangle$ such that $\hat{O}|\alpha_n\rangle = O_n|\alpha_n\rangle$, allows to express the state function in the form

[15] Here it is in place to recall the argument against the Copenhagen interpretation raised by Einstein (1953), considering the stationary states of an infinite one-dimensional square well potential. The spatial part of the wave function can be written in the form $\varphi = N\sin kx$. From Eq. (8.11) it follows that $v = 0$, hence there is no flow velocity. However, by writing the wave function in the form $\psi = (N/2i)(e^{ikx} - e^{-ikx})$, it can be interpreted as referring to two similar subensembles of particles, traveling to the right and to the left, with velocities $\pm\hbar k/m$. Thus, it is the net (mean) velocity that is null. Einstein used this example to argue that the statistical reading is the single one that can be made in the limit of high energies. Since passing to this limit does not change the nature of the problem, Einstein concluded that one should consider the wave function as describing an ensemble, not an individual particle.

$$|\psi(x,t)\rangle = \sum_n \Phi_n(x,t)\,|\alpha_n\rangle\,. \tag{8.18}$$

Substituting into Eq. (8.17) leads to

$$\frac{\partial \Phi_n}{\partial t} + gO_n \frac{\partial \Phi_n}{\partial x} = 0, \tag{8.19}$$

with solution

$$\Phi_n(x,t) = \Phi_n(x - v_n t) \tag{8.20}$$

with $v_n = gO_n$. Therefore the state is given by

$$|\psi(x,t)\rangle = \sum_n \Phi_n(x - v_n t)\,|\alpha_n\rangle\,. \tag{8.21}$$

The result speaks for itself: the wavepackets $\Phi_n(x - v_n t)$ travel each with its own velocity gO_n, and thus move apart with time. Eventually the overlap between packets will be negligible so that they can be considered isolated from one another. A position measurement corresponds then to a specific O_n and thus to a particular spin state $|\alpha_n\rangle$, which contributes with the probability $|\Phi_n|^2$. For example, when x is taken to represent a radial direction, different positions correspond to different angles and the example becomes an elementary model of a Stern-Gerlach experiment.

When the overlap between the different Φ_n becomes negligible, there is no appreciable interference between them. It thus appears as if the state has suffered an effective, practically irreversible collapse, although subject at all times to a unitary evolution. It is the approximation of treating the final packets as truly independent which breaks the unitary dynamics, thus evoking a kind of 'collapse' (introduced by hand) that does not correspond to any physical process, yet is useful as a practical procedure to deal with the final state.[16] One merit of the theory becomes thus clear: there is no observer that 'induces' the wave collapse; no need for partial tracings; no two laws of evolution but only the one that governs the dynamics all along the process. The notion of collapse dissolves and no measurement theory is needed, since now the measurement becomes a normal experiment, softly merged into the conventional theory of quantum evolution. The conclusion is in full agreement with the stance promoted by van Kampen (1988) and others: "The measuring act is fully described by the Schrödinger equation for object and apparatus together..." In plain words, quantum mechanics is more fitting without the addendum of the weighty theory of measurement.

The rather elementary model just discussed illustrates well various fundamental aspects of the meaning of 'measurement of an observable'. The first one, remarkable

[16] In the usual quantum theory of measurement, the process of extracting from (8.21) the observed result Φ_n (the reduction or collapse of the after-measurement state) is referred to as the problem of objectification. See e.g. Mittelstaedt (2009).

enough, is that every measurement (here of a spin component) ends up being a position measurement. Thus, the breaking of the initial wave packet into several ones as a product of the interaction, which separate in the course of time (until becoming almost noninterfering) allows to identify the presence or absence of any component in the initial wave packet. A second, fundamental aspect is that the description is entirely objective and avoids dividing the world into undefined observed and observer, object and subject—which is one of the most discomforting aspects of the usual quantum theory of measurement (see e.g. Bell 1987; Wick 1995).

The amplitude of the nth component in Eq. (8.21) depends on the instrument through the factor gO_n. Thus one is tempted to conclude that the result of the measurement is determined by both the system and the measuring apparatus (which here is part of the enlarged system). In this direct sense, it is true. The probability with which the packet Φ_n contributes to $|\psi(x,t)\rangle$ is $|\Phi_n(x - gO_nt)|^2$, which for $t \neq 0$ is instrument-dependent. However, it is only in the relative weights of the states that there is a dependence on the measuring device: if initially $|\Phi_n(x)|^2$ is zero for a given component, this component will never be registered in the output. Only those components that contribute at $t = 0$ will have a chance to show up at later times. In this sense, the end result depends exclusively on the system itself. In other words, only predefined values are eventually observed, assuming the measuring instrument does not directly affect the system itself.[17]

One of the reasons for the success of Bohm's approach with the measurement problem is the fact that it contains and uses the notion of quantum trajectory, as in the above example, where the packets representing particles tend to separate. We have here a nice and important instance of both the possibility of introducing this notion into conventional QM , as already discussed, and the usefulness of such endeavor. Even if, according to our perspective, a more detailed description exists that contains fluctuations that are absent in Bohm's' theory, for many purposes such local mean description suffices to give an approximate idea of how the quantum system behaves.

8.3 The Quantum Potential

As stated in Sect. 8.2.1, a key element in Eq. (8.5b) is the term $V_Q(\boldsymbol{x}, t)$. It stands as a sui generis potential, essentially different from any classical one in many ways, on some of which we comment in the present section. Our purpose is both to show

[17] The possibility that the result of a measurement depends on both the system under observation and the measuring apparatus is also present at the classical level. A common example of the class of nondisturbing classical procedures is a photocell detector that checks the presence or absence of somebody before closing the door of an elevator. An example of the second class could be a 'tail or head' detector for tossed coins which operates by inserting a card to stop and receive the coins. Of course this second mechanism can be replaced with more elaborate optical procedures that do not disturb the observed coins. This is a matter of the measurer's skills and of the existing technical possibilities.

that essential features of the quantum system are linked to V_Q, and to provide new insights into its physical origin.[18]

Rather than starting from Eq. (8.4), we resort to the more general Eq. (2.81),

$$ia\frac{\partial\psi}{\partial t} = -\frac{a^2}{2m}\nabla^2\psi + V\psi + (1-\lambda)\frac{a^2}{2m}\frac{\nabla^2\sqrt{\rho}}{\sqrt{\rho}}\psi, \text{ with } \psi = \sqrt{\rho}e^{iS}, \tag{8.22}$$

which is valid provided the flux momentum of the system has the structure $m\boldsymbol{v}(\boldsymbol{x},t) = a\nabla S(\boldsymbol{x},t)$, with a constant. The procedure used at the beginning of Sect. 8.2.1 leads then to a couple of equations involving the variables S and ρ: one is the continuity Eq. (8.5a), the other is a generalized form of Eq. (8.5b),

$$a\frac{\partial S}{\partial t} + \frac{a^2}{2m}(\nabla S)^2 + V - \lambda\frac{a^2}{2m}\frac{\nabla^2\sqrt{\rho}}{\sqrt{\rho}} = 0. \tag{8.23}$$

The case of interest here corresponds evidently to $a = \hbar$; however, the specific value of a is irrelevant for the present purposes. As discussed in Sect. 2.4, what is important is that Eq. (8.22) is linear in ψ if and only if $\lambda = 1$, that is, if Eq. (8.23) involves an additional 'potential' given precisely by $-\left(a^2/2m\right)(\nabla^2\sqrt{\rho}/\sqrt{\rho})$ (which for $a = \hbar$ is the quantum potential). This establishes the following concomitance: for a given purpose we can resort to the (Schrödinger) equation for ψ, or to the equivalent (Hamilton-Jacobi-type) equation for (ρ, S)—the continuity equation is merely a constraint, not a dynamical equation—, and the effects that arise from the linearity of the former will be attributable to the term V_Q in the latter. In the next two sections we apply this parallelism—exploiting the linearity of the Schrödinger equation—to draw conclusions about the role played by V_Q in the dynamics of the system. We shall thereafter inquire further into the physical meaning of V_Q.

8.3.1 Linearity and Nonlocality

The most recognized feature of Bohm's theory is its essential nonlocality. However, almost every analysis on nonlocality that one finds in the physical literature is carried out within the context of composite systems, the discussions on nonlocal effects in single-particle systems being rather scarce. A possible explanation for this rests on the

[18] A first peculiarity of V_Q is that it is independent of the field's strength, or rather of the intensity ($\sim\rho$) of the wave. This follows from the fact that $V_Q(\rho) = V_Q(A\rho)$ for any constant A, and indicates that the effects due to the particles do not depend on the number of particles present; but on their distribution. That there are forces within the classical realm, particularly in the hydrodynamical analogy, with similar peculiarities does not suffice to surmount the problem, since in the hydrodynamical case there is a medium that supports and transmits the presure and the stresses. By contrast, in the quantum single-particle problem we are not dealing with a collective system; the 'collection' may be a conceptual ensemble, devoid of physical existence.

fact that since the EPR paper, and mostly from Bell's theorem on, the debates regarding nonlocality have centered on bipartite (or multi-partite) quantum systems, to the extent that the notion of nonlocality has become widely understood as synonymous with violation of Bell's inequalities,[19] or as a manifestation of some sort of action at a distance between the constituents of the system. Another reason may be that for single-particle systems the notion of nonlocality is somewhat more blurred and subtle. In particular, there is apparently no room for actions at a distance simply because a single particle has no partner to interact with! However, a look at Fig. 8.2—which shows that as an effect of the barrier some particles are reflected before reaching it—indicates that it does make sense to talk about nonlocal effects in single-particle systems. As the alien element in the otherwise classical (single-particle) Eq. (8.5b), V_Q is to be blamed for the characteristic quantum properties of the system: hence it must account also for its nonlocal properties.

The fact that V_Q depends explicitly on the spatial distribution of particles could be considered indeed as a sufficiently strong argument to ascribe to it the nonlocal properties characteristic of quantum systems. However, a more detailed argument goes as follows. Expressing the Schrödinger equation in the form $\mathcal{L}\psi = 0$, with $\mathcal{L}$ a linear operator, the determination of the (causal) Green function for $\mathcal{L}$, call it $K(\boldsymbol{x}, t|\boldsymbol{x}', t')$, suffices to express the solution in the form

$$\psi(\boldsymbol{x}, t) = \int K(\boldsymbol{x}, t|\boldsymbol{x}', t')\psi(\boldsymbol{x}', t')\, d^3x', \quad t \geq t', \tag{8.24}$$

with $K = 0$ for $t < t'$. Equation (8.24) shows that ψ at any point $\boldsymbol{x}$ and time t carries information regarding its previous value at *all* points of the available space. A consequence of this is that every quantity that is determined by ψ bears in general information about the whole setup. Therefore, the trajectory defined by the guidance Eq. (8.11) reflects the presence of boundary conditions and of (possibly distant) external potentials, for example. This accounts for some of the nonclassical effects manifested in numerous Bohm trajectories; in particular, it explains why the particles begin to gain or lose energy or deviate before reaching the barrier, as in Fig. 8.2, as if 'perceiving' its effect in advance.[20] This kind of behavior is the one we identify as the quantum single-particle nonlocality, where the term 'nonlocal' means that the dynamics of the particle in some region is affected by what happens in regions that may be far away from it, without an intermediate recognized (external) agent.

In the de Broglie-Bohm interpretation the underlying cause for such nonlocal behaviour is found in the quantum potential, which transmits, to each point, information about the wave field in the entire space. Thus, a particle at a point where V is constant (e.g., located at $x < a$ in the barrier example) is not a free particle, but

[19] Bell inequalities is a collective name referring to a number of inequalities (such as the CHSH-type inequalities) that involve correlations between variables of the constituents of a composite system and are violated by QM, reputedly due to the nonlocal properties of the quantum description.

[20] The transmitted particles are among those that gain enough energy to travel not *through* the barrier, but *over* it (see e.g. Lopreore and Wyatt 1999).

is acted upon by a *quantum force* $-\nabla V_Q(\boldsymbol{x}, t)$. This force, being dependent on the point $\boldsymbol{x}$ where it is exerted, acts of course locally; what is nonlocal is the information it carries. There is no room for superluminal action at a distance. Moreover, since in Bohm's interpretation the field ψ is considered to be a physical entity, the nonlocality appears as entirely admissible: the anticipated reflection of the particles is as natural (and local) as it would be if there existed a *real* fluid, as was conceived by Madelung. In this scenario some incident particles are reflected before reaching the barrier simply because the reflected wave acts upon them (Holland 1993).

This explanation does not hold in the SED framework, where the correct description of the entire (field plus particle) system has a local structure and quantum nonlocality appears as a feature of the reduced quantum-mechanical description, rather than an ontological property. At the end of Sect. 8.3.3 we comment on this point in more detail.

8.3.2 Linearity and Fluctuations

Equation (8.24) is a direct consequence of the linearity of the Schrödinger equation. Another most important property of the solutions of linear equations, is that they satisfy the superposition principle. Let us apply it here to a free-particle system, by superposing plane waves of different momenta to construct a Gaussian wave packet, and use it to show that V_Q is also linked to the presence of quantum fluctuations, which appear as irreducible.

Consider the following one-dimensional packet of free particles of mass m,

$$\psi(x, t) = \frac{1}{\sqrt{2\pi}} \int \phi(k) e^{-i(\hbar k^2 t/2m)+ikx} dk, \tag{8.25}$$

where $\hbar k = p$. If the initial wave function is

$$\psi(x, 0) = \left(\frac{1}{2\pi\sigma_0^2}\right)^{1/4} \exp(-x^2/4\sigma_0^2), \tag{8.26}$$

the distribution $\rho(x, t)$ is a Gaussian centered at the origin,

$$\rho(x, t) = \frac{1}{\sqrt{2\pi\sigma^2}} \exp\left(-x^2/2\sigma^2\right) \tag{8.27}$$

with

$$\sigma^2(t) = \sigma_x^2(t) = \sigma_0^2 + \left(\frac{\hbar t}{2m\sigma_0}\right)^2. \tag{8.28}$$

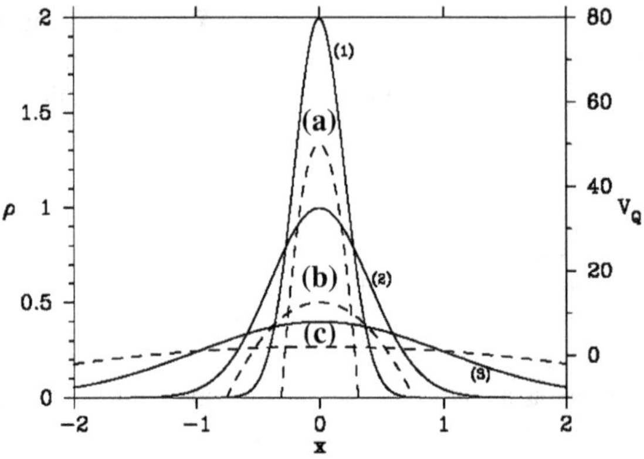

Fig. 8.3 Distribution of particles ρ (*solid lines*, marked with numbers) and quantum potential V_Q (*dotted lines*, marked with letters) of Eqs. (8.27) and (8.29) for three different times $t_1 < t_2 < t_3$. Curves (1) and (a) correspond to t_1, (2) and (b) to t_2, and (3) and (c) to t_3. Arbitrary units

A direct calculation gives for the quantum potential

$$V_Q = \frac{\hbar^2}{8m\sigma^4}\left(2\sigma^2 - x^2\right). \tag{8.29}$$

Figure 8.3 shows ρ (solid line) and V_Q (dotted line) for three different times, as the spread of the packet is seen to increase. A first conspicuous property of V_Q is that it does not approach a constant value at the boundaries ($x \to \pm\infty$), where the field intensity, or rather the distribution of particles ρ, tends to zero.[21] On the contrary, the effect of the potential (8.29), as measured by the quantum force $-\partial V_Q/\partial x$, increases in those regions where the particles have a smaller probability to be. Indeed, it follows from Eq. (8.29) that the exerted quantum force is repulsive and linear in x; those particles in the region $x > 0$ are 'pushed' to the right by such force, and those in $x < 0$ are 'pushed' to the left, so the packet spreads continuously. In the frame of the Hamilton-Jacobi-type equation, the dispersion is explained in mechanistic terms as an effect of the quantum potential.

Further, from Eq. (8.29) it follows that

$$2m\left\langle x^2\right\rangle\langle V_Q\rangle = \frac{\hbar^2}{4}, \tag{8.30}$$

since $\sigma^2 = \left\langle x^2\right\rangle$. This result (to be analyzed later for the general case) suggests a strong relationship between the mean quantum potential and the irreducible momentum fluctuations that lie at the root of the Heisenberg inequalities. It also exhibits quantitatively what was said above in qualitative terms about the dispersive effect of the quantum potential. Thus, in Bohm's approach, the inherent dispersive nature of quantum systems becomes causal, the quantum potential being the physical element that causes it by exerting a (here) repulsive force on the particles.

---

[21] This feature is not exclusive of the present example, but rather the general rule: V_Q does not decay as $x \to \pm\infty$, i.e., at far distances from the particles.

If at variance with Bohm's interpretation, ψ is taken as a mathematical entity that bears statistical information, but is unable to produce direct physical effects on the particles, the spread of the wave packet cannot be understood in mechanistic terms. It becomes explained instead in statistical terms: the ensemble is composed of subensembles of particles traveling with different velocities [each contributing with a probability $|\phi(k)|^2$, as follows from (8.25)]; the faster ones move forward whereas the slower ones are left behind, so the packet spreads. In this scenario each particle of the ensemble is always a true free particle, since no physical force acts upon it. The quantum potential is the bearer of the statistical information regarding the momentum dispersion, so that not only the spread of the packet becomes natural, but so does Eq. (8.30), which implies a relation between the quantum potential and the momentum fluctuations.

The spread of the packet induced by V_Q given by Eq. (8.29) can be compensated by an external harmonic potential, so as to ensure that the net force $-\partial V_{\text{eff}}/\partial x$ vanishes. Writing the external potential as $V = m\omega^2 x^2/2$, such condition is guaranteed provided the oscillator frequency is exactly $\omega = \hbar/\left(2m\sigma^2\right)$. The effective potential becomes thus

$$V_{\text{eff}} = V + V_Q = \tfrac{1}{2}\hbar\omega. \tag{8.31}$$

Under this condition there is no net force on the particle, the packet (8.5b) does not spread and a stationary state is reached. This is the simplest example of a coherent state, and serves also to explain the stability of the distribution of the ground state of the harmonic oscillator (which is a Gaussian): it is an effectively free particle, in the sense the no *net* force acts upon it. Moreover, Eq. (8.31) fixes a natural reference energy level—an observation that can be used to infer the existence of the ZPF underlying the Schrödinger (or Hamilton-Jacobi-type) description.

8.3.3 The Quantum Potential as a Kinetic Term

In Bohm's theory the quantum potential is accepted as a natural entity that needs no further explanation. Just as is the case with fundamental laws of nature, it is taken as an expression of the structure and workings of the physical world, which physics has the duty to discover and describe. Yet clarification of the deeper meaning of V_Q *does* have importance because in it resides a fundamental ingredient of the quantum description. All quantum problems involve the quantum potential, even if normally it remains concealed behind the veil of the Schrödinger equation.

The discussion in Chap. 4 helps us elucidate the nature of the quantum potential and understand, from a more fundamental perspective, why it is so intimately related with the dispersive and nonlocal effects studied in the previous section. In particular, by disclosing the origin of V_Q it will become clear that when dealing with a quantum problem one is (knowingly or unknowingly) taking into account much more information about the behavior of the system in momentum space than is explicitly acknowledged.

Let us for this purpose recall an alternative derivation of Eqs. (8.5a) and (8.5b) that serves to emphasize the significance of both velocities $\boldsymbol{v}$ and $\boldsymbol{u}$. We start from Eq. (4.128), namely

$$\hat{\boldsymbol{p}}\psi = -i\hbar\nabla\psi = m(\boldsymbol{v} - i\boldsymbol{u})\psi, \tag{8.32}$$

with $\boldsymbol{u}$ the diffusive velocity given by

$$\boldsymbol{u} = \frac{\hbar}{2m}\nabla \ln \rho. \tag{8.33}$$

From Eq. (8.32) it follows that the kinetic energy operator $(1/2m)\,\hat{\boldsymbol{p}}^2$ applied to ψ results in

$$\frac{\hat{\boldsymbol{p}}^2}{2m}\psi = \frac{1}{2}\left[\left(m\boldsymbol{v}^2 - m\boldsymbol{u}^2 - \hbar\nabla\cdot\boldsymbol{u}\right) - i\left(\hbar\nabla\cdot\boldsymbol{v} + 2m\boldsymbol{v}\cdot\boldsymbol{u}\right)\right]\psi. \tag{8.34}$$

On the other hand, Eq. (8.4) gives

$$i\hbar\frac{\partial\psi}{\partial t} = \left(-\hbar\frac{\partial S}{\partial t} + i\frac{\hbar}{2}\frac{\partial \ln \rho}{\partial t}\right)\psi. \tag{8.35}$$

Combining the last two expressions with Schrödinger's equation

$$\frac{\hat{\boldsymbol{p}}^2}{2m}\psi + V(\boldsymbol{x})\psi = i\hbar\frac{\partial\psi}{\partial t} \tag{8.36}$$

gives an expression whose imaginary and real parts are, respectively, the continuity Eq. (8.5a) and the dynamical law

$$\hbar\frac{\partial S}{\partial t} + \frac{1}{2}\left(m\boldsymbol{v}^2 - m\boldsymbol{u}^2 - \hbar\nabla\cdot\boldsymbol{u}\right) + V = 0. \tag{8.37}$$

A comparison with Eq. (8.5b) (with $\left(\hbar^2/2m\right)(\nabla S)^2 = (m/2)\,\boldsymbol{v}^2$) allows to write the quantum potential in a form that reveals its kinetic nature,

$$V_Q = -\tfrac{1}{2}\left(m\boldsymbol{u}^2 + \hbar\nabla\cdot\boldsymbol{u}\right). \tag{8.38}$$

In Bohm's theory the term $(m/2)\,\boldsymbol{v}^2$ alone is identified with the total kinetic energy, whereas the remaining terms in Eq. (8.5b) are taken as a 'potential' energy. This latter form of separating the kinetic and potential terms is a consequence of reading (8.5b) in a classical fashion, as if it were a true Hamilton-Jacobi equation. Equations

(8.34) and (8.38), by contrast, identify V_Q with a contribution to the kinetic energy due to the diffusive velocity $\boldsymbol{u}$. These equations give for the mean kinetic energy

$$\frac{1}{2m}\left\langle \hat{\boldsymbol{p}}^2 \right\rangle = \int \left(\frac{1}{2}m\boldsymbol{v}^2 + V_Q\right)\rho d^3x = \frac{1}{2}m\left\langle \boldsymbol{v}^2 \right\rangle + \left\langle V_Q \right\rangle, \tag{8.39}$$

so with the help of Eq. (4.120), namely

$$\left\langle \hat{\boldsymbol{p}}^2 \right\rangle = m^2\left\langle \boldsymbol{v}^2 + \boldsymbol{u}^2 \right\rangle, \tag{8.40}$$

one gets

$$\left\langle V_Q \right\rangle = \tfrac{1}{2}m\left\langle \boldsymbol{u}^2 \right\rangle = \tfrac{1}{2}m\sigma_{\boldsymbol{u}}^2. \tag{8.41}$$

This is an interesting result: the mean quantum potential coincides with the mean kinetic energy of diffusion. Since $\boldsymbol{u}(\boldsymbol{x}) \neq 0$ whenever $\rho(\boldsymbol{x})$ is not constant (which happens in all cases of interest, when there is finite spatial dispersion), Eq. (8.41) implies that $\left\langle V_Q \right\rangle$ is strictly positive. From Eq. (8.40) we find $\sigma_{\hat{\boldsymbol{p}}}^2 = m^2\sigma_{\boldsymbol{v}}^2 + m^2\sigma_{\boldsymbol{u}}^2$ [which is Eq. (4.122)], whence

$$\sigma_{\hat{\boldsymbol{p}}}^2 = \sigma_{\boldsymbol{p}}^2 = \sigma_{m\boldsymbol{v}}^2 + 2m\left\langle V_Q \right\rangle > \sigma_{m\boldsymbol{v}}^2, \tag{8.42}$$

where $\sigma_{m\boldsymbol{v}}^2$ stands for the variance of the flux momentum $m\boldsymbol{v}$. Classically, this latter coincides with the total momentum dispersion, $\sigma_{\boldsymbol{p}}^2 = \sigma_{m\boldsymbol{v}}^2$. Therefore, (8.42) states that the quantum momentum dispersion (normally) exceeds the classical one. The result $\sigma_{\boldsymbol{p}}^2 > \sigma_{m\boldsymbol{v}}^2$ is immediate from (4.122), yet the inequality (8.42) is expressed in terms of the (mean) quantum potential, thus confirming that the 'quantumness' of the system is indeed encoded in V_Q. From (8.42) it follows that the minimum value of $\sigma_{\boldsymbol{p}}^2$ is

$$(\sigma_{\hat{\boldsymbol{p}}}^2)_{\min} = 2m\left\langle V_Q \right\rangle_{\min}, \tag{8.43}$$

a result that exhibits the existence of irreducible momentum fluctuations of value $2m\left\langle V_Q \right\rangle_{\min}$. Equation (8.30) corresponds to the particular case $\boldsymbol{v} = 0$ (a stationary bounded s state), when $\sigma_{m\boldsymbol{v}}^2$ vanishes.

From the previous results and the Schwartz inequality it follows that

$$\sigma_{\boldsymbol{x}}^2\sigma_{\boldsymbol{p}}^2 \geq m^2\sigma_{\boldsymbol{x}}^2\sigma_{\boldsymbol{u}}^2 \geq \tfrac{1}{4}\hbar^2. \tag{8.44}$$

The diffusive velocity is therefore the one that determines the Heisenberg inequality, by expressing the presence of diffusion in the quantum system.

With Eq. (8.40) rewritten in terms of local mean values (with $\langle \boldsymbol{p} \rangle_{\boldsymbol{x}} = m\boldsymbol{v}$) one obtains, after rearrangements,

$$\int V_Q \rho d^3x = \frac{1}{2m} \int \left\langle \left(\boldsymbol{p} - \langle \boldsymbol{p} \rangle_x\right)^2 \right\rangle_x \rho d^3x. \tag{8.45}$$

From here it follows that up to an arbitrary term that averages to zero, $2mV_Q$ plays the role of $\langle (\boldsymbol{p} - \langle \boldsymbol{p} \rangle_x)^2 \rangle_x = \langle \boldsymbol{p}^2 \rangle_x - \langle \boldsymbol{p} \rangle_x^2$, the local mean deviation of the momentum from its local mean value. This assigns a more fundamental meaning to V_Q: it bears information, at each point $\boldsymbol{x}$, of the local fluctuations impressed upon the momentum of the particle.

With this, Eq. (8.43) becomes natural from the point of view of SED. Indeed, unavoidable fluctuations exist due to the ZPF , the information about which is contained in V_Q. The quantum potential is the element that (re)incorporates the momentum fluctuations impressed by the ZPF into the dynamics governed by the Hamilton-Jacobi-type equation.

In addition, the fact that V_Q plays the role of a partially averaged quantity that results from restricting the description to the configuration subspace of the particle, explains the origin of its nonlocal effects, since at each point $\boldsymbol{x}$, V_Q bears statistical information about the entire momentum space. The single-particle quantum nonlocality, rather than an ontological property, appears thus as a semblance, an artifact of the reduced statistical description, which would dissolve by going back to the full, original phase-space description.

8.4 Nonlocality in Bipartite Systems

As mentioned at the beginning of Sect. 8.3.1, nonlocality is normally discussed in relation with composite systems. Nonlocality in such context has gained so much attention in the last decades, that huge numbers of papers and entire volumes have been devoted to its study with different purposes and in many directions. The issue has evidently not been exhausted, and it therefore seems pertinent to contribute to its clarification from the perspective of the present theory. In this section we focus on a two-particle system as the simplest example that can be used to study the complexities arising in composite systems, and resort to the tools developed so far to gain further insight into (bipartite) entanglement and nonlocality.

The two-particle system is described by the Schrödinger equation

$$i\hbar \frac{\partial \psi}{\partial t} = \left(-\frac{\hbar^2}{2m_1} \nabla_1^2 - \frac{\hbar^2}{2m_2} \nabla_2^2 + V \right) \psi, \tag{8.46}$$

with a general (time-independent) external potential of the form $V = V(\boldsymbol{x}_1, \boldsymbol{x}_2)$. Substitution of

$$\psi(\boldsymbol{x}_1, \boldsymbol{x}_2, t) = \sqrt{\rho(\boldsymbol{x}_1, \boldsymbol{x}_2, t)} e^{iS(\boldsymbol{x}_1, \boldsymbol{x}_2, t)} \tag{8.47}$$

in (8.46) leads again to a couple of equations (corresponding to its real and its imaginary part). The continuity equation is

$$\frac{\partial \rho}{\partial t} + \frac{\hbar}{m_1}\nabla_1 \cdot (\rho \nabla_1 S) + \frac{\hbar}{m_2}\nabla_2 \cdot (\rho \nabla_2 S) = 0, \tag{8.48a}$$

whereas the dynamical law reads

$$\hbar\frac{\partial S}{\partial t} + \frac{\hbar^2}{2m_1}(\nabla_1 S)^2 + \frac{\hbar^2}{2m_2}(\nabla_2 S)^2 - \frac{\hbar^2}{2m_1}\frac{\nabla_1^2\sqrt{\rho}}{\sqrt{\rho}} - \frac{\hbar^2}{2m_2}\frac{\nabla_2^2\sqrt{\rho}}{\sqrt{\rho}} + V = 0. \tag{8.48b}$$

Comparison of Eq. (8.48a) with its one-particle version (8.5a), allows to identify the flow velocity associated to particle i ($i = 1, 2$) with

$$\boldsymbol{v}_i = \frac{\hbar}{m_i}\nabla_i S, \tag{8.49}$$

so that the continuity equation for the bipartite case reads

$$\frac{\partial \rho}{\partial t} + \nabla_1 \cdot (\rho \boldsymbol{v}_1) + \nabla_2 \cdot (\rho \boldsymbol{v}_2) = 0, \quad \text{or} \quad \frac{\partial \rho}{\partial t} + \nabla \cdot (\rho \boldsymbol{v}) = 0, \tag{8.50}$$

where the last equation is written in the six-dimensional configuration space. On comparing Eq. (8.48b) with Eq. (8.5b) one obtains for the quantum potential V_{Qi} associated with particle i

$$V_{Qi} = -\frac{\hbar^2}{2m_i}\frac{\nabla_i^2\sqrt{\rho}}{\sqrt{\rho}} = -\frac{\hbar^2}{4m_i}\left[\frac{\nabla_i^2\rho}{\rho} - \frac{1}{2}\left(\frac{\nabla_i\rho}{\rho}\right)^2\right]. \tag{8.51}$$

Equation (8.48b) takes thus the form

$$\hbar\frac{\partial S}{\partial t} + \frac{1}{2}m_1\boldsymbol{v}_1^2 + \frac{1}{2}m_2\boldsymbol{v}_2^2 + V_{Q1} + V_{Q2} + V = 0. \tag{8.52}$$

Application of the operator ∇_i to Eq. (8.52) gives, with the aid of (8.49) (unless explicitly stated, from now on we assume $i, j = 1, 2$ with $i \neq j$ when both indices appear in the same expression),

$$m_i\mathcal{D}_c^{(i)}\boldsymbol{v}_i = -\nabla_i V - \nabla_i\left(V_{Qi} + V_{Qj}\right) - \tfrac{1}{2}m_j\nabla_i\boldsymbol{v}_j^2, \tag{8.53}$$

where $\mathcal{D}_c^{(i)}$ stands for the co-moving derivative of particle i [see Eq. (8.13)],

$$\mathcal{D}_c^{(i)} = \frac{\partial}{\partial t} + (\boldsymbol{v}_i \cdot \nabla_i). \tag{8.54}$$

In the Bohmian approach, where $\boldsymbol{v}_i = d\boldsymbol{x}_i/dt$, $\mathcal{D}_c^{(i)}\boldsymbol{v}_i$ equals the total derivative $d\boldsymbol{v}_i/dt$ and is taken as the actual acceleration of the particle describing the trajectory $\boldsymbol{x}_i(t)$. From the present perspective instead, according to which $\boldsymbol{v}_i$ stands for a (partially) averaged velocity, $\mathcal{D}_c^{(i)}\boldsymbol{v}_i$ constitutes a statistical acceleration characterizing the local flow. Hence $m_i\mathcal{D}_c^{(i)}\boldsymbol{v}_i$ stands for a (local mean) representative force, already much smoother than the actual (stochastic) one. Let us denote this force by $\boldsymbol{F}_i$; then Eq. (8.53) reads

$$\begin{aligned} \boldsymbol{F}_i &= -\boldsymbol{\nabla}_i V - \boldsymbol{\nabla}_i V_{Qi} - \boldsymbol{\nabla}_i V_{Qj} - \tfrac{1}{2} m_j \boldsymbol{\nabla}_i \boldsymbol{v}_j^2 \\ &= \boldsymbol{f}_i^{\mathrm{c}} + \boldsymbol{f}_{ii}^{\mathrm{q}} + \boldsymbol{f}_{ij}^{\mathrm{q}} + \boldsymbol{f}_{ij}^{\mathrm{f}}. \end{aligned} \tag{8.55}$$

The second equality identifies the different forces that contribute to $\boldsymbol{F}_i$. The first one, $\boldsymbol{f}_i^{\mathrm{c}} = -\boldsymbol{\nabla}_i V$, is the classical force due to the external potential V, whereas the three remaining terms

$$\begin{aligned} \boldsymbol{f}_{ii}^{\mathrm{q}} &= -\boldsymbol{\nabla}_i V_{Qi}, \\ \boldsymbol{f}_{ij}^{\mathrm{q}} &= -\boldsymbol{\nabla}_i V_{Qj}, \\ \boldsymbol{f}_{ij}^{\mathrm{f}} &= -\frac{1}{2} m_j \boldsymbol{\nabla}_i \boldsymbol{v}_j^2, \end{aligned} \tag{8.56}$$

are of kinetic origin. This is obvious for $\boldsymbol{f}_{ij}^{\mathrm{f}}$—which originates in the flux kinetic energy proportional to $\boldsymbol{v}_j^2$—and becomes clear for the 'quantum' forces $\boldsymbol{f}_{ii}^{\mathrm{q}}$ and $\boldsymbol{f}_{ij}^{\mathrm{q}}$ once the quantum potential (8.51) is written in the form

$$V_{Qi} = -\tfrac{1}{2}\left(m_i \boldsymbol{u}_i^2 + \hbar \boldsymbol{\nabla}_i \cdot \boldsymbol{u}_i\right), \tag{8.57}$$

with $\boldsymbol{u}_i$ the diffusive velocity associated wtih particle i,

$$\boldsymbol{u}_i = \frac{\hbar}{2m_i} \boldsymbol{\nabla}_i \ln \rho. \tag{8.58}$$

Equations (8.57) and (8.58) generalize the single-particle expressions (8.38) and (8.33), and show that both $\boldsymbol{f}_{ii}^{\mathrm{q}}$ and $\boldsymbol{f}_{ij}^{\mathrm{q}}$ are due to a diffusive velocity.

The fact that $\boldsymbol{f}_{ij}^{\mathrm{q}}$ and $\boldsymbol{f}_{ij}^{\mathrm{f}}$ represent forces that are exerted at $\boldsymbol{x}_i$ but originate in quantities (velocities) associated with an (arbitrarily distant) point $\boldsymbol{x}_j$, suggests the emergence of further nonlocal effects that add to the nonlocal features characteristic of the single-particle case (which ensued from the term $\boldsymbol{f}_{ii}^{\mathrm{q}}$). In the following we shall investigate some aspects of this new kind of nonlocality.

8.4.1 Nonlocality and Entanglement

For simplicity, in what follows we shall assume that each particle is restricted to a one-dimensional motion. The above equations thus give for the two contributions to F_i due to the presence of particle j,

$$f_{ij}^{\mathrm{q}} = -\frac{\partial}{\partial x_i} V_{Qj} = \left(m_j u_j + \frac{\hbar}{2} \frac{\partial}{\partial x_j} \right) \frac{\partial u_j}{\partial x_i}, \tag{8.59}$$

$$f_{ij}^{\mathrm{f}} = -m_j v_j \frac{\partial v_j}{\partial x_i}. \tag{8.60}$$

Clearly, a necessary condition for the force f_{ij}^{q} to exist is that $\partial_i u_j \neq 0$, and similarly for f_{ij}^{f}, that $\partial_i v_j \neq 0$. Let us determine the properties of those states $\psi(x_1, x_2, t)$ for which these conditions hold.

We start by writing the density $\rho(x_1, x_2, t)$ and the phase $S(x_1, x_2, t)$ of the wave function (8.47) in the general form

$$\rho(x_1, x_2, t) = r_1(x_1, t) r_2(x_2, t) r(x_1, x_2, t), \tag{8.61}$$

$$S(x_1, x_2, t) = s_1(x_1, t) + s_2(x_2, t) + s(x_1, x_2, t). \tag{8.62}$$

With this, and using the one-dimensional version of Eqs. (8.49) and (8.58), we arrive at

$$\frac{\partial u_j}{\partial x_i} = \frac{\hbar}{2m_j} \frac{\partial^2}{\partial x_j \partial x_i} \ln r(x_1, x_2, t), \tag{8.63a}$$

$$\frac{\partial v_j}{\partial x_i} = \frac{\hbar}{m_j} \frac{\partial^2}{\partial x_j \partial x_i} s(x_1, x_2, t). \tag{8.63b}$$

Notice that $m_j \partial_i u_j = m_i \partial_j u_i$ and $m_j \partial_i v_j = m_i \partial_j v_i$, two properties that will be (and had been) freely used without explicit mention. According to Eq. (8.63a), $\partial_i u_j = 0$ if and only if r has the form $r(x_1, x_2, t) = R_1(x_1, t) R_2(x_2, t)$, i.e., if and only if ρ factorizes as

$$\rho(x_1, x_2, t) = \rho_1(x_1, t) \rho_2(x_2, t), \tag{8.64}$$

with ρ_i the marginal distribution functions,

$$\rho_i(x_i, t) = \int \rho(x_i, x_j, t) dx_j. \tag{8.65}$$

Analogously, Eq. (8.63b) implies that $\partial_i v_j = 0$ if and only if s decomposes as $s(x_1, x_2, t) = \Sigma_1(x_1, t) + \Sigma_2(x_2, t)$, which in its turn means that S has the additive structure

$$S(x_1, x_2, t) = S_1(x_1, t) + S_2(x_2, t). \tag{8.66}$$

It is evident that with ρ and S given by (8.64) and (8.66), respectively, $\psi = \sqrt{\rho}\exp(iS)$ factorizes as

$$\psi(x_1, x_2, t) = \psi_1(x_1, t)\psi_2(x_2, t), \tag{8.67}$$

with $\psi_i(x_i, t) = \sqrt{\rho_i}\exp(iS_i)$. We are thus led to conclude that

$$\psi = \psi_i\psi_j \Longleftrightarrow \frac{\partial}{\partial x_i}u_j = 0 \text{ and } \frac{\partial}{\partial x_i}v_j = 0, \tag{8.68}$$

and consequently, that for a factorizable (separable) state the forces (8.59) and (8.60) exerted at x_i due to the presence of particle j vanish.

On the other hand, with $\partial_i u_j = 0$, V_{Qi} reduces to

$$V_{Qi} = -\frac{\hbar^2}{2m_i}\frac{\nabla_i^2\sqrt{\rho_i}}{\sqrt{\rho_i}} = V_{Qi}(x_i, t). \tag{8.69}$$

If in addition $\partial_i v_j = 0$, consistency with Eq. (8.52) demands the external potential to be of the form $V(x_1, x_2) = V_1(x_1) + V_2(x_2)$. This shows that a separable state is a consistent solution only for a system of noninteracting particles, and allows us to write F_i as

$$F_i(x_i, t) = -\frac{\partial}{\partial x_i}V_i(x_i) - \frac{\partial}{\partial x_i}V_{Qi}(x_i, t). \tag{8.70}$$

Therefore, whenever ψ factorizes as in (8.67), no force arises in the composite system additional to those found in the single-particle case. In other words, in a factorizable state the dynamics corresponds to that of a couple of independent particles, each following its own laws. The nonlocalities are, then, those that correspond to single-particle systems.

However, if at least one of the conditions for the velocities in (8.68) fails, the wave function can no longer be factorized and hence it describes a nonseparable or entangled state. According to the statement following Eq. (8.60), only in this case the forces f_{ij}^{q} and f_{ij}^{f} may be different from zero. In other words, f_{ij}^{q} and f_{ij}^{f} are conditioned by the existence of a nonfactorizable ρ or a nonfactorizable $\exp(iS)$, respectively. This serves to identify the separate physical effects of the magnitude and the phase of the wave function on the dynamics of the bipartite system. The entanglement may be encoded either in the amplitude of ψ (if $\partial_i u_j \neq 0$), or in its phase (if $\partial_i v_j \neq 0$), or in both; occasionally we will refer to amplitude entanglement or phase-entanglement, respectively. The fact that either u_i, or v_i, or both, may depend on x_i *and* x_j, precludes the possibility of determining such velocities by focusing on the subsystem i only. Instead, these dynamical variables pertain to the bipartite system as a whole; they are associated with one of the subsystems but cannot be defined separately from the second one. We will come back to this point in Sect. 8.4.3.

Equations (8.61) and (8.62) allow us to write (for simplicity in what follows we omit the time dependence)

$$v_i = \frac{\hbar}{m_i}\left[\frac{\partial s_i(x_i)}{\partial x_i} + \frac{\partial s(x_1, x_2)}{\partial x_i}\right], \tag{8.71a}$$

$$u_i = \frac{\hbar}{2m_i}\left[\frac{\partial \ln r_i(x_i)}{\partial x_i} + \frac{\partial \ln r(x_1, x_2)}{\partial x_i}\right]. \tag{8.71b}$$

These expressions display the extra contributions to v_i and u_i that are exclusively due to entanglement, and which introduce the two-point dependence. Whenever there is amplitude-entanglement ($\partial_i \ln r \neq 0$) the quantum potential (8.57) takes the form $V_{Q_i} = V_{Q_i}(x_1, x_2)$, the (total) quantum potential $V_Q = V_{Q_1} + V_{Q_2}$ acquires an additional contribution that formally plays the role of an interaction potential, and a generally nonzero term $f_{ij}^{\mathrm{q}} = -\partial_i V_{Qj}$ depending on both x_1 *and* x_2 adds to F_i, which becomes a function of the form $F_i(x_1, x_2, t)$.[22] Analogouity, when there is phase entanglement ($\partial_i s \neq 0$), the flux kinetic energy associated with particle j affects the particle located at x_i, and a force f_{ij}^{f} results which also depends on the position of both particles, leading again to a total force $F_i(x_1, x_2, t)$. While giving rise to the forces f_{ij}^{f} and f_{ij}^{q}, the entanglement may also modify the force $f_{ii}^{\mathrm{q}} = -\partial_i V_{Qi}$, which for a nonfactorizable ρ will depend in general on the position variables of both particles.

The ensuing two-point dependence of F_i brings out naturally the notion of nonlocal effects, particularly when there is no (external) interaction between the particles. This is the case, for example, if the particles interacted (got entangled) in the past, or if the system is composed of identical noninteracting parties in a state described by the superposition

$$\psi(x_1, x_2, t) = \frac{1}{\sqrt{2}}\left[\phi_n(x_1, t)\phi_m(x_2, t) \pm \phi_m(x_1, t)\phi_n(x_2, t)\right], \tag{8.72}$$

despite the fact that $V = V_1 + V_2$. In the absence of an external interaction potential, the issue of nonlocal effects due to entanglement has become so befogged that even 'spooky' actions at a distance—of unexplained physical origin, of course—have been invoked. Nevertheless, considering for example that the Bohm particle that follows the trajectory $x_i(t)$ with a velocity $v_i = dx_i/dt$ is not the real, physical particle, but rather a representative particle that at each point reproduces the statistical dynamics of the appropriate subensemble, the spooky-action problem disappears. The nonlocalities due to the two-point dependence of v_i are not the result of a direct

[22] The effective interaction potential introduced via $V_Q(\boldsymbol{x}_1, \boldsymbol{x}_2)$, which remains 'hidden' in the depths of the Schrödinger equation, formally transforms the original noninteracting system into an interacting one [see Eq. (8.52)]. By contrast, the possible nonfactorizability of $\exp(iS)$ does not manifest itself as a formal interaction potential in Eq. (8.52). The nonlocal effect of this kind of entanglement is manifested when a description in terms of forces is made, as we have seen.

physical action between the real particles, but only an imprint (in configuration space) of the presence of correlated fluxes. In the identical-particle case, as follows from the results of Chap. 7, it is clear that such correlations are rooted in the coupling of both particles through common modes of the field. Of course, a similar mechanism takes place, although less effectively, for nonidentical particles provided they have common relevant frequencies.

It is clear from the above results that for a noninteracting system ($V = V_1 + V_2$), the force F_i depends on both x_i and x_j only if at least one of the conditions (8.68) fails. In such case the correlations that ensue from the two-point dependence of F_i are ascribed to the entanglement of ψ. In particular, correlations between the diffusive and flux velocities denote entanglement. To see this, observe that given a function $h(x_1, x_2)$, an integration by parts leads to

$$\left\langle \frac{\partial h}{\partial x_i} \right\rangle = \int \frac{\partial h}{\partial x_i} \rho dx_1 dx_2 = -\int h \frac{\partial \rho}{\partial x_i} dx_1 dx_2 = -\frac{2m_i}{\hbar} \langle h u_i \rangle , \tag{8.73}$$

under the assumption that $h\rho$ vanishes at infinity. This result is particularly useful when h is one of the velocities v_j or u_j, since in such case, according to Eq. (8.68) a nonzero value of any of the covariances[23]

$$\langle u_i u_j \rangle, \quad \langle u_i v_j \rangle \tag{8.74}$$

implies entanglement of the state ψ. Moreover, depending on which of the covariances is nonzero, one can determine whether there is entanglement encoded in either the modulus of the wave function ($\langle u_i u_j \rangle \neq 0$), or its phase ($\langle u_i v_j \rangle \neq 0$), or both.

Quantum mechanics does not recognize in u a quantity with a particular physical meaning (let alone a velocity with a diffusive connotation!), and something similar goes frequently for the flux velocity v. Yet the above results, particularly (8.68), indicate that these velocities play a significant role not only for the understanding of several properties of single-particle quantum systems (as follows, for example, from Chap. 4), but also in the bipartite case in connection with entanglement. In the following sections the velocities v and u are used for an analysis of certain aspects of entanglement and nonlocality. Even though some of the conclusions are well known, the method used to reach them is not, and this may help to get a fresh look at them.

8.4.2 Momentum Correlations

The entry point for nonlocality in the present description has been the two-point dependence of u_i and v_i. Now, these velocities are connected with the momentum

---

[23] We use here the term covariance to refer to a two-point momentum $\langle \hat{F}_1 \hat{G}_2 \rangle$, even if the product $\langle \hat{F}_1 \rangle \langle \hat{G}_2 \rangle$ differs from zero. In the literature the term 'correlation' is frequently used for $\langle \hat{F}_1 \hat{G}_2 \rangle$, so we use it here when convenient.

operator $\hat{p}_i$ by means of Eq. (8.32), which in the one-dimensional case reads

$$\hat{p}_i \psi = -i\hbar \frac{\partial}{\partial x_i} \psi = m_i (v_i - i u_i) \psi = \pi_i \psi, \tag{8.75}$$

where π_i stands for the complex variable [cf. Eq. (4.130)]

$$\pi_i = m_i (v_i - i u_i). \tag{8.76}$$

This suggests that the correlations (8.74) may be contained in the expectation value of $\hat{p}_i \hat{p}_j$. Applying the operator $\hat{p}_j$ to Eq. (8.75) one obtains

$$\hat{p}_j \hat{p}_i \psi = \left(\pi_j \pi_i - i\hbar \partial_j \pi_i \right) \psi. \tag{8.77}$$

The properties $m_j \partial_i u_j = m_i \partial_j u_i$ and $m_j \partial_i v_j = m_i \partial_j v_i$ imply that $\partial_i \pi_j = \partial_j \pi_i$, whence Eq. (8.77) is symmetrical in i, j, as expected. Multiplying this equation from the left by ψ^* and integrating leads to

$$\left\langle \hat{p}_i \hat{p}_j \right\rangle = m_i m_j \left\langle v_i v_j + u_i u_j \right\rangle. \tag{8.78}$$

So even though the imaginary part of π_i, u_i, does not contribute to the mean value of $\hat{p}_i$, it plays a central role in higher-order moments. This has already been pointed out when calculating $\left\langle \hat{p}_i^2 \right\rangle$ [see also Eq. (8.40), or Sect. 4.5.2],

$$\left\langle \hat{p}_i^2 \right\rangle = m_i^2 \left\langle v_i^2 + u_i^2 \right\rangle. \tag{8.79}$$

As discussed in Sect. 8.3.3, a nonzero value of $\left\langle u_i^2 \right\rangle$ indicates the presence of irreducible momentum fluctuations; on the other hand, according to the discussion following Eq. (8.73), a nonzero value of $\left\langle u_i u_j \right\rangle$ reflects entanglement and hence a nonzero correlation $\left\langle \hat{p}_i \hat{p}_j \right\rangle$. Thus, the same physical entity, namely the velocity u, bears information about two of the most characteristic quantum features, which now appear as intimately related.[24]

Coming back to Eq. (8.78), the fact that for a factorizable ψ the mean value $\langle u_1 u_2 \rangle$ vanishes implies that any discrepancy between $\left\langle \hat{p}_i \hat{p}_j \right\rangle$ and $m_i m_j \left\langle v_i v_j \right\rangle$ is due to entanglement, specifically due to the nonfactorizability of ρ. Moreover, (8.78) leads to

$$\frac{1}{m_i m_j} \left| \left\langle \hat{p}_i \hat{p}_j \right\rangle \right| \leq \left| \left\langle v_i v_j \right\rangle \right| + \left| \left\langle u_i u_j \right\rangle \right|, \tag{8.80}$$

which shows that a nonfactorizable ρ increases the upper limit of $\left| \left\langle \hat{p}_i \hat{p}_j \right\rangle \right|$. This result is in line with one of the main conclusions related with the violation of Bell's

[24] From this perspective, the conclusions reached regarding the dispersive and nonlocal features of the quantum potential (a quantity that depends on u only) become evident.

inequalitites, namely that entanglement introduces extra contributions to the correlations, with no classical analogue. Here the source of such additional terms are fluctuation-related velocities.

Before ending this section, it seems in place to add that the correspondence

$$\hat{p}_i \rightarrow \pi_i = m_i \left(v_i - i u_i\right) \tag{8.81}$$

discussed in Sect. 4.5.2, besides establishing the equivalence between the average of the quantum operator $\hat{p}_i$ and the average of the complex scalar function π_i, works also for the mean value $\left\langle \hat{p}_i \hat{p}_j \right\rangle$. Indeed, direct calculation shows that the covariance of π_i and π_j, defined as

$$C(\pi_i, \pi_j) = \tfrac{1}{2}\left\langle \pi_i^* \pi_j + \pi_i \pi_j^* \right\rangle, \tag{8.82}$$

is just Eq. (8.78), that is,

$$C(\pi_i, \pi_j) = \left\langle \hat{p}_i \hat{p}_j \right\rangle = m_i m_j \left\langle v_i v_j + u_i u_j \right\rangle. \tag{8.83}$$

This shows that the local mean value of $p_i p_j$ coincides, up to a term that averages to zero, with $m_i m_j \left(v_i v_j + u_i u_j\right)$.

8.4.3 The Whole and the Parts

The procedure that led to Eq. (8.77), introduced in Sect. 4.5.2, can be used to write the mean value of the operator $\hat{p}_i^n$ as an average of a scalar function, in the form

$$\left\langle \hat{p}_i^n \right\rangle = \left\langle p_i^n \right\rangle = \int P_i(\pi_i, \partial_i^{k_n} \pi_i) \rho(x_1, x_2) dx_1 dx_2, \tag{8.84}$$

where $P_i(\pi_i, \partial_i^{k_n} \pi_i)$, a real function of π_i and its first n derivatives ($k_n = 0, 1, \ldots n$), plays the role of the local—at point (x_1, x_2)—mean value of $\hat{p}_i^n$. It follows from Eq. (8.84) that for any operator $g_i(\hat{p}_i)$ that can be expanded as a power series of its argument, the quantum average $\left\langle g_i(\hat{p}_i) \right\rangle$ can also be obtained by averaging a c-number $G_i(\pi_i, \partial_i^k \pi_i)$, the local mean value of the variable $g_i(p_i)$, as

$$\left\langle g_i(\hat{p}_i) \right\rangle = \left\langle g_i(p_i) \right\rangle = \int G_i(\pi_i, \partial_i^{k_n} \pi_i) \rho(x_1, x_2) dx_1 dx_2. \tag{8.85}$$

It is clear that if $\pi_i = \pi_i(x_1, x_2)$, G_i will in general be also a two-point function; consequently any dynamical variable defined through it will in general display nonlocal features.

According to Eq. (8.68), if the state is factorizable $\partial_j \pi_i = 0$, i.e. $\pi_i = \pi_i(x_i)$ and Eq. (8.85) reduces to

$$\langle g_i(\hat{p}_i)\rangle = \langle g_i(p_i)\rangle = \int G_i(\pi_i, \partial_i^{k_n}\pi_i)\rho_i dx_i, \tag{8.86}$$

that is, an average weighted with the marginal probability distribution $\rho_i(x_i)$. For (8.85) to reduce to (8.86) irrespective of $g_i(\hat{p}_i)$, necessarily $\partial_j \pi_i = 0$, which means that the state is nonentangled. Therefore, the mean value of an arbitrary $g_i(\hat{p}_i)$ is just the average of the local mean value of $g_i(p_i)$ at point x_i weighted with the marginal distribution $\rho_i(x_i)$, if and only if the state is nonentangled. Otherwise stated, all the information required to determine $\langle g_i(\hat{p}_i)\rangle$ can be obtained from observations on system i *only*, if and only if the state is separable.

If, on the contrary, the state ψ is an entangled one, then the dependence of G_i on x_j precludes the possibility of expressing $\langle g_i(\hat{p}_i)\rangle$ in the form (8.86). In this case, determining the mean value of an arbitrary variable $g_i(p_i)$ of *one* of the particles requires information about the *whole* system; in particular, the marginal distribution ρ_i is not enough and we must resort to the joint distribution function $\rho(x_1, x_2)$.

The above observations provide an alternative way of looking at nonlocality in composite systems without the notion of action at a distance: *nonlocality, as a property encoded in the entanglement of the state, reflects the impossibility of considering each of the constituents of the system separately one from another, i.e., as nonseparability*. It forces us to consider the system as a whole, rather than as composed of two separate parts, well within the spirit of Bohm's interpretation. By considering the presence of the ZPF, as was done in Chap. 7, this point of view is the single natural one: there is a unique system, composed of field and particles, with certain field modes playing a correlating function between particles.

8.4.4 Nonlocality and Noncommutativity

At variance with what occurred with $\langle g_i(\hat{p}_i)\rangle$, the mean value of a function of the position operator only, $f_i(\hat{x}_i)$, is *blind* to the nonfactorizability of $\psi(x_1, x_2, t)$. This follows from the fact that any $f_i(\hat{x}_i)$ has a local mean value that is just the function $f_i(x_i)$, and hence the average

$$\langle f_i(\hat{x}_i)\rangle = \langle f_i(x_i)\rangle = \int f_i(x_i)\rho(x_1, x_2)dx_1dx_2 = \int f_i(x_i)\rho_i(x_i)dx_i \tag{8.87}$$

does not exhibit nonlocal features, irrespective of the state. Therefore, a variable A_i proper to a *single* particle exhibits nonlocality only when such variable is momentum-dependent.[25]

[25] When the operator $\hat{A}$ does not correspond to a *single* particle, this statement ceases to be true. For example, for $\hat{A} = A_1(\hat{x}_1)A_2(\hat{x}_2)$, the entanglement is revealed in the covariance $\langle A_1 A_2\rangle$ even

This particular asymmetry between momentum and position variables is due to the fact that the description is carried out in the configuration-space representation. If the momentum representation is used instead, with

$$\psi_p(p_1, p_2, t) = \sqrt{\rho_p(p_1, p_2, t)}e^{iS_p(p_1,p_2,t)}, \tag{8.88}$$

the p-local mean value associated with $g_i(\hat{p}_i)$ becomes the real variable $g_i(p_i)$,[26] whereas the p-local mean value for $f_i(\hat{x}_i)$ is defined, in analogy with (8.75), via the equation

$$\hat{x}_i\psi_p = i\hbar\frac{\partial}{\partial p_i}\psi_p = (\xi_i - i\zeta_i)\psi_p = \chi_i\psi_p, \tag{8.89}$$

where $\xi_i = \hbar(\partial S_p/\partial p_i)$ and $\zeta_i = (\hbar/2)(\partial \ln \rho_p/\partial p_i)$. An analysis entirely analogous to the one carried out above leads to conclude that the complex function χ_i, which now plays the role of the previous π_i, depends on both momenta p_1 and p_2 if and only if ψ_p does *not* factorize as $\psi_p = \psi_1(p_1, t)\psi_2(p_2, t)$. Under these circumstances, conclusions entirely similar to the previous ones apply, *mutatis mutandi*: nonlocal effects in momentum space are manifested in connection with the p-local mean value associated with $f(\hat{x}_i)$.

The above observations lead us to assert that once a representation in terms of the eigenvalues of the operator $\hat{x}_i$ is chosen to describe the evolution of the system, the nonlocal features become manifest through those variables $\hat{A}_i$ that are functions of the corresponding noncommuting operator $\hat{p}_i$, and vice versa. Thus, considering that (for a spinless system) any dynamical variable is a function of the fundamental variables x and p, it follows that the x-local mean value of A_i will exhibit nonlocal features when $[\hat{x}_i, \hat{A}_i] \neq 0$, whereas its p-local mean value will exhibit nonlocal features when $[\hat{p}_i, \hat{A}_i] \neq 0$.[27]

The requirement of noncommutativity of operators for the disclosure of nonlocality can alternatively be shown as follows. Let $\{|\alpha\beta\rangle \equiv |\alpha\rangle_i \otimes |\beta\rangle_j\}$ be an orthonormal

(Footnote 25 continued)
though none of the variables is momentum-dependent. In fact, the point here is to show that the present approach allows to reach conclusions about entanglement by focusing on single-particle variables, rather than on correlations between variables of the two subsystems, as is customarily done [see discussion following Eq. (8.96)].

[26] The p-local mean value of a dynamic variable g is defined, in analogy with Eq. (4.50), as its partial average over the configuration space, using the distribution Q,

$$\langle g\rangle(p_1, p_2) = \langle g\rangle_p = \frac{1}{\rho_p}\int gQ(x_1, x_2, p_1, p_2)dx_1dx_2.$$

[27] The fact that the kind of variables that may exhibit nonlocality is representation-dependent does not mean that the very existence of nonlocality is representation-dependent. Indeed, for any entangled state $|\psi\rangle$ there will always be some variable exhibiting nonlocal features; which one depends on the representation used to project $|\psi\rangle$.

basis of the product Hilbert space $\mathcal{H}_1 \otimes \mathcal{H}_2$. One may write the expectation value of $\hat{A}_i$ in the pure state $|\psi\rangle$ as

$$\langle\psi| \hat{A}_i |\psi\rangle = \int \langle\alpha\beta| \psi\rangle \langle\psi| \alpha'\beta'\rangle \langle\alpha'\beta'| \hat{A}_i |\alpha\beta\rangle \, d\alpha d\beta d\alpha' d\beta'$$
$$= \int \langle\alpha\beta| \psi\rangle \langle\psi| \alpha'\beta\rangle \langle\alpha'| \hat{A}_i |\alpha\rangle \, d\alpha d\beta d\alpha'. \tag{8.90}$$

If the basis $\{|\alpha\rangle_i\}$ is selected such that its elements are the eigenvectors of an operator $\hat{a}_i$ satisfying $[\hat{a}_i, \hat{A}_i] = 0$, then $\langle\alpha'| \hat{A}_i |\alpha\rangle \sim \delta(\alpha - \alpha')$, and (8.90) reduces to

$$\langle\psi| \hat{A}_i |\psi\rangle = \int \rho(\alpha, \beta) \langle\alpha| \hat{A}_i |\alpha\rangle \, d\alpha d\beta, \tag{8.91}$$

where $\rho(\alpha, \beta) = |\psi(\alpha, \beta)|^2$ and $\psi(\alpha, \beta) = \langle\alpha\beta| \psi\rangle$ is the wave function in the (α, β)-representation. Integration over β of the joint probability $\rho(\alpha, \beta)$ gives the marginal probability $\rho_i(\alpha)$, so that Eq. (8.91) simplifies into

$$\left\langle \hat{A}_i \right\rangle = \int \langle\alpha| \hat{A}_i |\alpha\rangle \, \rho_i(\alpha) d\alpha. \tag{8.92}$$

Equation (8.92) is usually read as saying that when projective measurements are performed corresponding to a physical variable a such that $\hat{a}_i |\alpha\rangle = \alpha |\alpha\rangle$ with $[\hat{a}_i, \hat{A}_i] = 0$, $\langle \hat{A}_i \rangle$ can be obtained from the sole (local) inspection of system i, and any nonlocality due to the possible entanglement of $\psi(\alpha, \beta)$ remains hidden.[28] By contrast, if a different basis $\{|\gamma\rangle\}$ is chosen for the representation, such that $[\hat{\gamma}_i, \hat{A}_i] \neq 0$, the above reduction cannot be made. Now, it is always possible to find a basis $\{|\alpha\rangle_1\}$ in which $\hat{A}_1$, say, is diagonal; but then the basis in which $\hat{B}_2$ is diagonal will not be $\{|\alpha\rangle_2\}$ unless $[\hat{A}_1, \hat{B}_1] = 0$; therefore, entanglement shows up only when dynamical variables $\hat{A}_1$, $\hat{B}_2$ are considered such that $[\hat{A}_i, \hat{B}_i] \neq 0$.[29,30]

[28] Equation (8.87) is just Eq. (8.92) with $\hat{A}_i = f(\hat{x}_i)$, yet Eq. (8.86) differs from the structure of (8.92). To see this consider in particular Eq. (8.84) with $n = 1$; then $P_i(\pi_i, \partial_i^k \pi_i) = m_i v_i(x_i) \neq \langle x_i| \hat{p}_i |x_i\rangle$, so indeed (8.86) is not Eq. (8.92) for $\hat{A} = \hat{g}_i$ and $\alpha = x_i$.

[29] Notice that the use of a fixed representation for both elements of the composite system, i.e. $\{|\alpha\rangle_1\}$, $\{|\beta\rangle_2\}$, is a matter of necessity when discussing entanglement. The same applies when considering measurements on a system. In fact, given an (α, β)-representation, the distribution function $\rho(\alpha, \beta)$ is defined as the joint probability density that determines the probability of obtaining the values α and β when performing the projective measurements corresponding to the proyectors $\Pi_{A_1}^{\alpha} = |\alpha\rangle\langle\alpha| \in \mathcal{H}_1$ and $\Pi_{B_2}^{\beta} = |\beta\rangle\langle\beta| \in \mathcal{H}_2$, respectively. Thus the representation used is linked with the variables that are measured in a certain experiment.

[30] This conclusion is in line with the results obtained in Sect. 7.2.5. Specifically, the discussion following Eq. (7.63) tells us that for entanglement to become manifest through a correlation, both dynamical variables involved (i.e., F, G, the equivalent of A_1, B_2 in the present case) must have nondiagonal elements in a given representation (the energy representation, in that case).

The results obtained above disclose the tight relation between noncommutativity and nonlocality that is well recognized in the literature (see e.g. Tsirelson 1980, Landau 1987, Revzen et al. 1997), though generally limited to the context of dichotomic operators. In particular, Landau shows that given two operators $\hat{a}$ and $\hat{A}$ in $\mathcal{H}_1$, and two operators $\hat{b}$ and $\hat{B}$ in $\mathcal{H}_2$, the (CHSH) operator defined as

$$\begin{aligned} \hat{C} &= \hat{a}\hat{b} + \hat{a}\hat{B} + \hat{A}\hat{b} - \hat{A}\hat{B} \\ &= \hat{a}(\hat{b} + \hat{B}) + \hat{A}(\hat{b} - \hat{B}), \end{aligned} \tag{8.93}$$

satisfies the following inequality,[31]

$$\begin{aligned} \langle\hat{C}\rangle^2 \leq & \left\langle\left(\hat{a}^2 + \hat{A}^2\right)(\hat{b}^2 + \hat{B}^2)\right\rangle + \left\langle\{\hat{b}, \hat{B}\}\left(\hat{a}^2 - \hat{A}^2\right)\right\rangle \\ & + \left\langle\{\hat{a}, \hat{A}\}(\hat{b}^2 - \hat{B}^2)\right\rangle + \left\langle\left[\hat{a}, \hat{A}\right][\hat{B}, \hat{b}]\right\rangle. \end{aligned} \tag{8.94}$$

This result ensues exclusively from the fact that $\sigma_C^2 = \langle\hat{C}^2\rangle - \langle\hat{C}\rangle^2 \geq 0$. Landau, in line with the usual treatments, considered operators $\hat{a}$, $\hat{A}$, $\hat{b}$ and $\hat{B}$ such that

$$\hat{a}^2 = \hat{A}^2 = \mathbb{I}_1, \quad \hat{b}^2 = \hat{B}^2 = \mathbb{I}_2 \tag{8.95}$$

(the typical example being the Pauli matrices). In such case Eq. (8.94) reduces to

$$\langle\hat{C}\rangle^2 \leq 4 + \left\langle\left[\hat{a}, \hat{A}\right][\hat{B}, \hat{b}]\right\rangle, \tag{8.96}$$

an expression that exhibits the significant role of (the covariance of) the commutators in determining the maximum possible value of $\langle\hat{C}\rangle^2$. For commuting $\hat{a}$ and $\hat{A}$, or $\hat{b}$ and $\hat{B}$, the inequality gives $\left|\langle\hat{C}\rangle\right| \leq 2$, which is the limit established in Bell's theorem (see e.g. Bell 1966, 1987). Thus, according to Eq. (8.96), the noncommutativity of the operators involved is a necessary condition for the violation of Bell's inequalities (in the form of the CHSH inequality, Clauser et al. 1969), as well as the nonnull correlation between both commutators.

This observation that noncommutativity and covariance unequivocally signal the nonlocal feature of a state goes nicely with our previous exposition. Our conclusions above were drawn by focusing not on covariances between operators of the form $\hat{F}_i\hat{G}_j$, but on the single-particle operators $\hat{A}_i$, thus showing that some aspects of

[31] Note that, as already remarked in connection with von Neumann's theorem [see Eq. (8.2)], the equality

$$\langle\hat{C}\rangle = \langle\hat{a}\hat{b}\rangle + \langle\hat{a}\hat{B}\rangle + \langle\hat{A}\hat{b}\rangle - \langle\hat{A}\hat{B}\rangle$$

does not hold in general if the operators in the terms of the sum do not commute. This important restriction needs to be borne in mind when attempting to apply (8.94) [or (8.96)] to draw conclusions about correlations. See e.g. Accardi (1984).

nonlocality can be unveiled without resorting to nonlocal (i.e. i and j-dependent) operators. In addition, our exposition applies to continuous-variable systems, contrary to the more usual approach that focuses on dichotomic variables, such as spin projections, when discussing these matters. The natural question arises as to why noncommutativity is required for entanglement (or nonlocality) to become evident. To give an answer, let us put $[\hat{a}, \hat{A}] = i\hat{F}_1$, and $[\hat{B}, \hat{b}] = -i\hat{G}_2$, with $\hat{F}_1$ and $\hat{G}_2$ two Hermitian operators. Equation (8.96) thus reads

$$\langle \hat{C} \rangle^2 \leq 4 + \langle \hat{F}_1 \hat{G}_2 \rangle, \tag{8.97}$$

and the inequality now states that $\left| \langle \hat{C} \rangle \right|$ will exceed the value 2 only if the variables $\hat{F}_1$ and $\hat{G}_2$ are correlated. If one of the commutators in (8.96) vanishes, then $\langle \hat{F}_1 \hat{G}_2 \rangle$ is trivially zero, no information at all is obtained regarding any possible correlation between the systems, and no conclusion can be drawn about entanglement. In other words, (non)commutativity *per se* says nothing about nonlocality; it is required merely as a useful way to specify which are the operators that may give evidence of entanglement, via the correlation between $\hat{F}_1$ and $\hat{G}_2$. As stated above, in Chap. 7 a similar definition of the appropriate operators was made, by specifying that only those variables F_1 and G_2 that share relevant frequencies become correlated in such a way as to disclose entanglement [see the discussion following Eq. (7.63)].

8.5 Final Remarks

Nonlocality is a weird trait of quantum mechanics that is considered to have been revealed (and has been made popular) by the Bell inequalities. Bohm, on his side, used it to develop an entire line of philosophy of nature based on a holistic picture of the world. As a result, dominant voices today affirm that 'Nature *is* nonlocal'. But, is it? Or is it merely the (quantum) *description* that is nonlocal? There are of course physicists (and even a few philosophers of science, see e.g. Brown and Harré 1988) who cannot accept nonlocality as a trait of fundamental science, arguing that it is an artifact of the formalism and our reading of it. The derivations presented in previous chapters and lines above, add their own share.

Let us briefly elaborate on this point. There exist today several derivations of the Bell inequalities; one that has become standard with time started with the famous paper by Clauser et al. (1969) cited above and based on the expression (8.93). A careful consideration of the known derivations allows one to verify that the Bell inequalities are merely statistical relations: no physics is involved in their derivation, just as no physics is involved in Eq. (8.96). It suffices to take the mean value of (8.93) for a series of trials[32] and determine its bounds, to arrive at the CHSH inequality.

[32] With an eye put on note (31), making sure that the average is taken over the same distribution in each term.

In short, the Bell (or CHSH) inequalities by themselves say nothing about Nature. The physics enters when the theorem is applied to a given physical system. Any experimental violation means that at least some postulate used for the derivation, is not satisfied by that system. It is usual to blame the physical demand of locality for the violation of the inequalities.

Let us consider the case of a pair of noninteracting particles, discussed in Chap. 7. The results obtained there indicate that the nonlocality of the description—which is made evident in the case of an entangled state—arises from the neglect of the ZPF.[33] Since quantum mechanics lacks of a fundamental explanation for the origin of entanglement—it is the result of a basic postulate—, the physics behind the associated nonlocality remains hidden. Even if our results on this problem are still limited, our analysis suggests there is a real chance that a more refined description—in phase space, for example— would allow to recover locality.

References

Accardi, L.: The probabilistic roots of the quantum mechanical paradoxes. In: Diner, S., Fargue, D., Lochak, G., Selleri, F. (eds.) The Wave-Particle Dualism. Springer, Dordrecht (1984)

Aharonov, Y., Vaidman, L.: Bohmian Mechanics and Quantum Theory: An Appraisal, p. 141. Kluwer, Dordrecht (1996)

Bacciagaluppi, G.,Valentini, A.: Quantum Theory at the Crossroads. Reconsidering the 1927 Solvay Conference. Cambridge University Press, Cambridge (2009)

Bell, J.S.: On the problem of hidden variables in quantum mechanics. Rev. Mod. Phys. **38**, 447 (1966). Reprinted in Bell 1987

Bell, J.S.: Speakable and Unspeakable in Quantum Mechanics. Cambridge University Press, Cambridge (1987)

Belinfante, F.J.: A Survey of Hidden-Variables Theories. Pergamon, Oxford (1973)

Bohm, D.: A suggested interpretation of the quantum theory in terms of "hidden" variables. I. Phys. Rev. **85**, 166 (1952a)

Bohm, D.: A suggested interpretation of the quantum theory in terms of "hidden" variables. II. Phys. Rev. **85**, 180 (1952b)

Bohm, D.: Comments on an article of Takabayasi concerning the formulation of quantum mechanics with classical pictures. Progr. Theor. Phys. **9**, 273 (1953)

Bohm, D., Hiley, D.J.: The Undivided Universe. An Ontological Interpretation of Quantum Theory. Routledge, London (1995)

Bohm, D., Hiley, B.J.: Statistical mechanics and the ontological interpretation. Found. Phys. **26**, 823 (1996)

Bohm, D., Vigier, J.-P.: Model of the causal interpretation of quantum theory in terms of a fluid with irregular fluctuations. Phys. Rev. **96**, 208 (1954)

Brown, H., Harré, R. (eds.): Philosophical Foundations of Quantum Field Theories. Clarendon Press, Oxford (1988)

Brun, T.A.: A simple model of quantum trajectories. Am. J. Phys. **70**, 719 (2002) quant-ph/0108132v1

Bub, J.: Von Neumann's 'no hidden variables' proof: a re-appraisal. Found. Phys. **40**, 1333 (2010)

[33] The extra correlations that lead to the violation of a Bell inequality exist also in the case of photons, due to the correlations between the excitations of the field and the corresponding modes of the ZPF; see e.g. Casado et al. (1998).

Carroll, R.: Some remarks on Ricci flow and the quantum potential. (2007). arXiv:math-ph/0703065v2

Carroll, R.: On the Emergence Theme of Physics. World Scientific, Singapore (2010)

Casado, A., Marshall, T.W., Santos, E.: Type II parametric downconversion in the Wigner-function formalism: entanglement and Bell's inequalities. J. Opt. Soc. Am. B **15**, 1572 (1998)

Clauser, J.F., Horne, M.A., Shimony, A., Holt, R.A.: Proposed experiment to test local hidden-variable theories. Phys. Rev. Lett. **23**, 880 (1969)

*Cole, D.C., Zou, Y.: Quantum mechanical ground state of hydrogen obtained from classical electrodynamics. Phys. Lett. A **317**, 14 (2003)

*Cole, D.C., Zou, Y.: Analysis of orbital decay time for the classical hydrogen atom interacting with circularly polarized electromagnetic radiation. Phys. Rev. E **69**, 016601 (2004a)

*Cole, D.C., Zou, Y.: Simulation study of aspects of the classical hydrogen atom interacting with electromagnetic radiation: circular orbits. J. Sci. Comput. **20**, 43 (2004b)

*Cole, D.C., Zou, Y.: Simulation study of aspects of the classical hydrogen atom interacting with electromagnetic radiation: elliptical orbits. J. Sci. Comput. **20**, 379 (2004c)

*Cole, D.C., Zou, Y.: Perturbation analysis and simulation study of the effects of phase on the classical hydrogen atom interacting with circularly polarized electromagnetic radiation. J. Sci. Comput. **21**, 145 (2004d)

*Cole, D.C., Zou, Y.: Subharmonic resonance behavior for the classical hydrogen atomic system. J. Sci. Comput. **39**, 1 (2009)

Cushing, J.T.: Causal quantum theory: why a nonstarter? In: Selleri, F. (ed.) The Wave-Particle Duality. Plenum, New York (1992)

de Broglie, L.: Sur la possibilité de relier les phénomènes d'interférences et de diffraction à la théorie des quanta de lumière. C. R. Acad. Sci. Paris **183**, 447 (1926a)

de Broglie, L.: Les principes de la nouvelle Mécanique ondulatoire, série VI. J. Phys. **7**(11), 321–333 (1926b)

de Broglie, L.: La structure atomique de la matière et du rayonnement et la Mécanique ondulatoire. C. R. Acad. Sci. Paris **184**, 273 (1927a)

de Broglie, L.: Sur le rôle des ondes continues en Mécanique ondulatoire. C. R. Acad. Sci. Paris **185**, 380 (1927b)

de Broglie, L.: La Mécanique ondulatoire et la structure atomique de la matière et du rayonnement. J. Phys. Rad. **8**, 225 (1927c)

de Broglie, L.: Électrons et Photons: Rapports et discussions du Cinquième Conseil de Physique. Gauthier-Villars, Paris (1928)

de Broglie, L.: Une tentative d'interprétation causale et non linéare de la Mécanique Ondulatoire: La Théorie de la Double Solution. Gauthier-Villars, Paris (1956)

de Broglie, L.: Recherches sur la Théorie des Quanta. Réédition du texte de 1924. Masson, Paris (1963)

*de la Peña, L., Cetto, A.M., Valdés-Hernández, A., França, H.M.: Genesis of quantum nonlocalities. Phys. Lett. A **375**, 1720 (2011). arXiv:quant-phys:1110.4641v1

Dewdney, C., Hardy, L., Squires, E.J.: How late measurements of quantum trajectories can fool a detector. Phys. Lett. A **184**, 6 (1993)

Dewdney, C., Hiley, B.J.: A Quantum potential description of one-dimensional time-dependent scattering from square barriers and square wells. Found. Phys. **12**, 27 (1982)

Dürr, D., Fusseder, W., Goldstein, S., Zanghi, N.: Comment on "Sur-realistic Bohm trajectories". Z. Naturforsch. **48a**, 1261 (1993)

Dürr, D., Goldstein, S., Zanghi, N.: Quantum mechanics, randomness, and deterministic reality. Phys. Lett. A **172**, 6 (1992)

Dürr, D., Teufel, S.: Bohmian Mechanics. The Physics and Mathematics of Quantum Theory. Springer, Dordrecht (2009)

Einstein, A.: Elementare Überlegungen zur Interpretation der Grundlagen der Quanten-Mechanik. In: Scientific Papers, Presented to Max Born. Oliver & Boyd, Edinburgh (1953)

Englert, B.G., Scully, M.O., Süssmann, G., Walther, H.: Surrealistic Bohm Trajectories. Z. Naturforsch. **47a**, 1175 (1992)
Esposito, S.: Photon wave mechanics: a de Broglie-Bohm approach. Found. Phys. Lett. **12**, 533 (1999)
Feyerabend, P.K.: Eine Bemerkung zum Neumannschen Beweis. Zeit. Phys. **145**, 421 (1956)
Floyd, E.R.: Classical limit of the trajectory representation of quantum mechanics, loss of information and residual indeterminacy, Int. J. Mod. Phys. A **15**, 1363 (2000). arXiv:quant-ph/9907092v3
Floyd, E.R.: The philosophy of the trajectory representation of quantum mechanics. In: Amoroso, R.L., et al. (eds.) Gravitation and Cosmology: From the Hubble Radius to the Planck Scale, Fundamental Theories of Physics, vol. 126, Part IV, p. 401. Kluwer Academic, Netherlands (2002)
Garbaczewski, P.: Derivation of the quantum potential from realistic Brownian particle motions. Phys. Lett. A **162**, 129 (1992)
Gleason, A.M.: Measures on the closed subspaces of a Hilbert space. J. Math. Mech. **6**, 885 (1957)
Goldstein, H., Poole, Ch., Safko, J.: Classical Mechanics, 3rd edn. Pearson, New Jersey (2002)
Griffiths, R.B.: Consistent interpretation of quantum mechanics using quantum trajectories. Phys. Rev. Lett. **70**, 2201 (1993)
Grössing, G.: A classical explanation of quantization. Phys. Lett. A **372**, 4556 (2008)
Grössing, G.: On the thermodynamic origin of the quantum potential. Physica A **388**, 811 (2009)
Gudder, S.: On hidden-variable theories. J. Math. Phys. **11**, 431 (1970)
Gull, S., Lasenby, A., Doran, Ch.: Electron paths, tunnelling and diffraction in the spacetime algebra. Found. Phys. **23**, 1329 (1993)
Heisenberg, W.: The development of the interpretation of the quantum theory. In: Pauli, W. (ed.) Niels Bohr and the Development of Physics. Pergamon Press, London (1955)
Hermann, G.: Die naturphilosophischen Grundlagen der Quantenmechanik, section 7. Abhandlungen der Fries'schen Schule 6 (1935) (English translation: Seevinck, M.: www.phys.uu.nl/igg/seevinck/trans.pdf)
Hernández-Zapata, S., Hernández-Zapata, E.: Classical and non-relativistic limits of a Lorentz-Invariant Bohmian model for a system of spinless particles. Found. Phys. **40**, 532 (2010)
Holland, P.R.: The Quantum Theory of Motion. An Account of the de Broglie-Bohm Causal Interpretation of Quantum Mechanics. Cambridge University Press, Cambridge (1993)
Huang, W.Ch.-W., Batelaan, H.: Dynamics underlying the gaussian distribution of the classical harmonic oscillator in zero-point radiation. (2012). arXiv:quant-ph:1206.5323v1
Jammer, M.: The Philosophy of Quantum Mechanics. Wiley, NY (1974)
Kochen, S., Specker, E.P.: The problem of hidden variables in quantum mechanics. J. Math. Mech. **17**, 59 (1967)
Kumar Chattaraj, P.: Quantum Trajectories. CRC Press, NY (2010)
Landau, L.J.: On the violation of Bell's inequality in quantum theory. Phys. Lett. A **120**, 54 (1987)
Lopreore, C.L., Wyatt, R.E.: Quantum wave packet dynamics with trajectories. Phys. Rev. Lett. **82**, 5190 (1999)
Lopreore, C.L., Wyatt, R.E.: Lopreore and Wyatt reply. Phys. Rev. Lett. **85**, 895 (2000)
Madelung, E.: Quantentheorie in hydrodynamischer form. Zeit. Phys. **43**, 354 (1927)
Mandl, F.: Statistical Physics. Wiley, NY (1988)
Matzkin, A., Nurock, V.: Classical and Bohmian trajectories in semiclassical systems: mismatch in dynamics, mismatch in reality? Stud. Hist. Philos. Sci. B **39**, 17 (2008)
Mittelstaedt, P.: Objectification. In: Greenberger, D., Hentschel, K., Weinert, F. (eds.) Compendium of Quantum Physics. Springer, Berlin (2009)
Mugur-Schächter, M.: Étude du Caractère Complet de la Théorie Quantique. Gauthier-Villars, Paris (1964)
Nikolić, H.: Bohmian mechanics in relativistic quantum mechanics, quantum field theory and string theory. J. Phys. Conf. Ser. **67**, 012035 (2007)
Omnès, R.: The Interpretation of Quantum Mechanics. Princeton University Press, Princeton (1994)

Oriols, X., Mompart, J.: Applied Bohmian Mechanics. From Nanoscale Systems to Cosmology. Pan Stanford Publishing Pte. Ltd., Singapore (2012)
Passon, O.: Why Isn't Every Physicist a Bohmian? (2005). arXiv:quant-ph/0412119v2
Pauli, W., George, A. (ed.) Louis de Broglie Physicien et Penseur, vol. 33. Albin Michel, Paris (1952)
Peres, A.: Generalized Kochen-Specker theorem. Found. Phys. **41**, 807 (1996)
Philippidis, C., Dewdney, C., Hiley, B.J.: Quantum interference and the quantum potential. Nuovo Cimento **52B**, 15 (1979)
Philippidis, C., Dewdney, C., Kaye, R.D.: The Aharonov-Bohm effect and the quantum potential. Nuovo Cimento **71B**, 75 (1982)
Recami, E., Salesi, G.: Kinematics and hydrodynamics of spinning particles. Phys. Rev. A **57**, 98 (1998)
Revzen, M., Lokajíček, M., Mann, A.: Bell's inequality and operators' noncommutativity. Quantum Semiclass. Opt. **9**, 501 (1997)
Salesi, G.: Spin and Madelung fluid. Mod. Phys. Lett. A **11**, 1815 (1996). arXiv:0906.4147v1
Santos, E.: On the possibility of local hidden-variable theories. Phys. Lett. A **53**, 432 (1975)
Sanz, A.S., Borondo, F., Miret-Artés, S.: Particle diffraction studied using quantum trajectories. J. Phys.: Condens. Matter **14**, 6109 (2002)
Solomon, J.: Sur l'indéterminisme de la Mécanique quantique. J. Phys. **4**, 34 (1933)
Suñé, J., Oriols, X.: Comment on "quantum wave packet dynamics with trajectories". Phys. Rev. Lett. **85**, 894 (2000)
Thiounn, M.: Doctorate thesis, Cahiers de Physique **174**, 53 (1965); for further works on the theory of the double solution see C. R. Acad. Sci. Paris **262B**, 657 (1966), Portugaliae, Physica **4**, 85 (1966)
Towler, M.D.: De Broglie-Bohm Pilot Wave Theory and the Foundations of Quantum Mechanics. A Graduate Lecture Course. Cambridge University, Cambridge (2009). www.tcm.phy.cam.ac.uk/mdt26/pilot_waves.html
Tsirelson, B.S.: Quantum generalizations of Bell's inequality. Lett. Math. Phys. **4**, 93 (1980)
Ván, P., Fülöp, T.: Stability of stationary solutions of the Schrödinger-Langevin equation (2003). arXiv:quant-ph/0304190v4
van Kampen, N.G.: Ten theorems about quantum mechanical measurements. Physica A **153**, 97 (1988)
Vasudevan, R., Parthasarathy, K.V., Ramanathan, R.: Quantum Mechanics. A Stochastic Approach. Alpha Science, Oxford (2008)
von Neumann, J.:Mathematische Grundlagen der Quantenmechanik. Springer, Berlin (1932). English Trans: Mathematical Foundations of Quantum Mechanics. Princeton University Press, Princeton (1932/1955)
Wang, X.-S.: Derivation of the Schrödinger equation based on a fluidic continuum model of vacuum and a sink model of particles. (2006). arXiv:0610224
Wick, D.: The Infamous Boundary. Seven Decades of Heresy in Quantum Physics. Copernicus-Springer, New York (1995)
Yang, C.D.: Modeling quantum harmonic oscillator in complex domain. Chaos Solitons Fractals **30**, 342 (2006)

Chapter 9
The Zero-Point Field Waves (and) Matter

Students should not be taught to doubt that electrons, protons and the like are particles... The waves cannot be observed in any way than by observing particles.

Mott (1964)

The electron is either here, or there, or somewhere else, but wherever it is, it is a point charge.

Feynman et al. (1965)

So far in our exploration of the fundamentals of QM we have paid null attention to the very concept that gave rise to wave mechanics, the de Broglie wavelength and the associated undulatory behavior of matter—certainly one of the most mysterious properties of the quantum world. The notion of the quantum corpuscle as something that possesses intrinsic wave properties, which preclude the possibility of describing it as a localized entity, is widely extended. What we intend to show in the present chapter, by contrast, is that the fundamental wave properties associated with quantum particles can be understood without renouncing the notion of localized corpuscles.

Of course, formal manipulations of the results obtained in previous chapters allow to ascertain the wave content of quantum mechanics and eventually arrive at de Broglie's wavelength. But such procedure would appear to reduce it to a mere mathematical artifact, without providing a clue about its physical content, and more deeply, without throwing light about the nature of the de Broglie wave. It seems therefore obligatory to pay closer attention to this most significant entity.

In line with the spirit of the theory exposed in the present volume, the ZPF should be expected to play an important role in the elucidation of de Broglie's wave, and more generally in the explanation of the undulatory properties of matter. It would be even surprising if the ZPF did not in some way or another impress its wave properties on the particles embedded in it. However, such possibility has only been occasionally explored within SED; therefore, this chapter contains the results of some of the initial

L. de la Peña et al., *The Emerging Quantum*, DOI: 10.1007/978-3-319-07893-9_9,

investigations into the territory of the quantum waves, which obviously deserves further exploration.[1]

9.1 Genesis of de Broglie's Wave

We recall that in de Broglie's theory of matter waves,[2] an oscillation of Compton's frequency

$$\omega_C = \frac{m_0 c^2}{\hbar} \tag{9.1}$$

is associated with a corpuscle at rest (m_0 stands for the rest mass). If the particle moves with respect to the laboratory with velocity v along some axis, the frequency ω in this latter reference frame is Doppler-shifted according to the formula

$$\omega = \gamma\omega_C(1+\beta), \tag{9.2}$$

with

$$\gamma = \left(1-\beta^2\right)^{-1/2}, \quad \beta = v/c. \tag{9.3}$$

The shift $\gamma\beta\omega_C$ in Eq. (9.2) can be rewritten as ($m = \gamma m_0$)

$$\gamma\beta\omega_C \equiv \omega_B = 2\pi c\frac{mv}{h} = \frac{2\pi c}{\lambda_B}, \tag{9.4}$$

and therefore

$$\lambda_B = \frac{h}{mv} = \frac{h}{p}. \tag{9.5}$$

Equation (9.5) is the well-known expression for the de Broglie wavelength, which originates in the Doppler shift of the frequency ω_C. In de Broglie's theory, a physical

[1] Previous versions of the material presented in the first part of this chapter can be found in de la Peña and Cetto (1992, 1994), Cetto and de la Peña (1955a, b), and *The Dice*.

[2] Detailed, first-hand expositions of de Broglie's theory can be found in de Broglie (1926, 1956, 1963). Modern presentations by one of its advocates made in Selleri (1990). A most elaborate development of a variant of de Broglie's theory for the relativistic electron is the *geometrical mechanics* developed by Synge (1954). An informed historical discussion of de Broglie's work up to the 1927 Solvay conference is given in Bacciagaluppi and Valentini (2009). MacKinnon (1976) presents a detailed analysis and improvement of de Broglie's derivation in his thesis. Another detailed discussion of de Broglie's phase waves is presented in Espinosa (1982).

wave with the wavelength λ_B becomes a central entity, directly related with the moving particle; yet the nature of such wave remains unspecified.[3]

Within the quantum formalism it is customary to introduce the expression (9.5) as a means to assign wave properties to the quantum corpuscle. Practical applications of the de Broglie wavelength are contained in almost any textbook, largely in the form of restrictions on λ_B associated with atomic stationarity conditions, from which (quantized) spectra are extracted. De Broglie's wavelength appears also in connection with particle diffraction patterns, notably the electron equivalent of Young's double-slit experiment, and in the optics of electron microscopy. However, discussions on the nature and origin of the de Broglie *wave* (not just the wavelength λ_B) are found only rarely. In the following sections we dig into such matters, with the intention to throw some light on the concept of de Broglie's wave.

9.1.1 The de Broglie 'Clock'

The first point that deserves attention in any attempt to understand the de Broglie wave relates to the physical origin of the oscillations of frequency ω_C associated with the particle in its rest frame, which constitute a sort of 'clock' in de Broglie's theory. In this regard we recall that according to QED (see e.g. Milonni 1994, Chap. 11), the interaction of an electron with the electromagnetic vacuum dresses the particle and endows it with an effective size, estimated between $(\lambda_C r_c)^{1/2} = \lambda_C(\alpha/2\pi)^{1/2} \simeq \lambda_C/30$ and λ_C, where $r_c = e^2/(mc^2) = (\alpha/2\pi)\lambda_C$ is the classical electron radius and λ_C is the Compton wavelength

$$\lambda_C = \frac{2\pi c}{\omega_C} = \frac{h}{m_0 c}. \tag{9.6}$$

In terms of λ_C, Eq. (9.5) takes the form

$$\lambda_B = \frac{\lambda_C}{\gamma\beta} = \lambda_C\sqrt{\frac{c^2}{v^2} - 1}, \tag{9.7}$$

which means that for nonrelativistic motions λ_B is usually much larger than Compton's wavelength.

From the point of view of SED, it is also natural to consider the charged particle immersed in the vacuum field as endowed with an effective size of the order of the Compton wavelength λ_C.[4] As a result, the particle decouples from the components

---

[3] In Surdin (1979) it is proposed to consider that de Broglie's wave is of electromagnetic nature, in some undefined way associated with the electromagnetic ZPF.

[4] A crude way to reach the same conclusion is the following. From the Heisenberg inequality one obtains $\sigma_x^2 \geq (\hbar^2/4\sigma_p^2)$, whence the minimum dispersion in the position variable determines an

of the radiation field with wavelengths smaller than λ_C (and frequencies larger than ω_C), so that the Compton frequency appears as a cutoff frequency. Any specific model for the charge with structure (real or effective) would be arbitrary at this stage, but also unnecessary, since our present purpose is limited to the introduction of the appropriate cutoff, which we accept to be of the order of ω_C.

The characteristic equation of motion for a free particle with structure (real or effective) acted on by the ZPF and radiation reaction has complex roots, giving rise to oscillations of a very high frequency.[5] This frequency is determined basically by the size of the particle rather than by the details of its structure, so the phenomenon is quite general; for an (effective) radius of order λ_C the frequency is of the order of ω_C. In a classical context, these high-frequency oscillations are transient, related to initial motions, momentary disturbances and the like. However, when the particle is in permanent interaction with the random background field, as is the present case, things change essentially. The electromagnetic environment not only puts the particle into resonance and makes it radiate, but it is also constantly knocking the particle, so that the high-frequency oscillations become continuously renewed and acquire a permanent (though fluctuating) character. It is appealing to identify these fine oscillations of frequency ω_C with the zitterbewegung, of which we have here an informal rendering.

In short, even if the particle is initially conceived of as pointlike—which sounds somewhat extreme for a physical, rather than mathematical element— it behaves as an object with some structure that performs, in addition to any other motion, a sustained oscillation with a frequency of about ω_C. In this way the vacuum field provides the physical sustenance for the de Broglie clock.

Because of its oscillating behavior, the particle at rest is continuously radiating at the frequency ω_C, a process that in a stationary state must be compensated by absorption from the vacuum field. This means that the particle interacts intensely with the modes of frequency ω_C, as measured in its proper frame, and that these modes sustain the jitter. The specific mechanism of this interaction is irrelevant for the kinematics that follow; what is important is that the particle interacts selectively with a narrow band of modes of the field of frequencies around ω_C.

Let us assume for simplicity that the particle motion is restricted to one dimension, along some axis $\hat{\boldsymbol{x}}'$.This means that in its proper frame (denoted with S') the components of the ZPF of interest are the two plane waves of frequency ω_C travelling in opposite directions. The resulting (standing) wave is thus the superposition

$$\varphi'(x', t') = e^{-i(\omega_C t' - \boldsymbol{k}'_+ \cdot \boldsymbol{x}' + \theta_+)} + e^{-i(\omega_C t' - \boldsymbol{k}'_- \cdot \boldsymbol{x}' + \theta_-)} + \text{c.c.}, \tag{9.8}$$

(Footnote 4 continued)
effective radius $a \sim (\sigma_x)_{\min}$. Such minimum value is achieved for the largest σ_p^2, which in the nonrelativistic regime can be limited by $m_0^2 c^2$. This results in $a \sim (\hbar/m_0 c)$.

[5] A detailed discussion can be seen in de la Peña et al. (1982), and *The Dice*, Sects. 3.4 and 7.3.3. In this latter it is shown that the selfcorrelation of the position coordinate of a harmonic oscillator contains a permanent oscillatory contribution of a frequency determined by the cutoff (Eq. 7.101), and with a value that is not too far from the Compton frequency.

where

$$\boldsymbol{k}'_{\pm} = \pm k_C \hat{\boldsymbol{x}}', \quad k_C = \omega_C / c, \tag{9.9}$$

and $\theta_{\pm}$ are statistically independent random phases, in accordance with the results of Chap. 4. In the laboratory frame (denoted by S) where the particle is seen to move with velocity $\boldsymbol{v} = v\hat{\boldsymbol{x}}'$, the frequency ω_C and the wave vectors $\boldsymbol{k}'_{\pm}$ transform in such a way that the phases appearing in (9.8), being a relativistic scalar, remain the same. This means that if $\omega_{\pm}$ and $\boldsymbol{k}_{\pm}$ stand, respectively, for the frequency and the wave vector as seen in S of the plane waves traveling in the positive and negative direction along the axis $\hat{\boldsymbol{x}}'$, then

$$\omega_C t' - \boldsymbol{k}'_{\pm} \cdot \boldsymbol{x}' = \omega_{\pm} t - \boldsymbol{k}_{\pm} \cdot \boldsymbol{x}. \tag{9.10}$$

The expressions for the frequencies $\omega_{\pm}$ and the wave vectors $\boldsymbol{k}_{\pm}$ read (see, e.g., Jackson 1975, Sect. 11.3)

$$\begin{aligned} \omega_{\pm} &= \gamma \omega_C (1 \pm \beta), \\ \boldsymbol{k}_{\pm} &= \pm \gamma k_C (1 \pm \beta) \hat{\boldsymbol{x}} = \pm k_{\pm} \hat{\boldsymbol{x}}, \end{aligned} \tag{9.11}$$

and the standing wave $\varphi'(x', t')$ in S' has therefore the following form in S,

$$\varphi_v(x, t) = e^{-i(\omega_+ t - k_+ x + \theta_+)} + e^{-i(\omega_- t + k_- x + \theta_-)} + \text{c.c.} \tag{9.12}$$

In terms of the frequencies

$$\omega_A = c k_A = \tfrac{1}{2}(\omega_+ + \omega_-) = \gamma \omega_C, \tag{9.13a}$$

$$\omega_B = c k_B = \tfrac{1}{2}(\omega_+ - \omega_-) = \gamma \beta \omega_C = \beta \omega_A, \tag{9.13b}$$

Equation (9.12) becomes

$$\varphi_v(x, t) = 4 \cos(\omega_A t - k_B x + \theta_1) \cos(\omega_B t - k_A x + \theta_2), \tag{9.14}$$

with $\theta_{1,2} \equiv \frac{1}{2}(\theta_+ \pm \theta_-)$. This result, to which we shall return below, represents the standing wave of the ZPF that activates the de Broglie clock, as seen from the laboratory frame.

9.1.2 Energy, Frequency and Matter Waves

In order to relate $\varphi_v(x, t)$ with the de Broglie wave, let us resort to the relativistic expression for the energy

$$\mathcal{E}^2 = m_0^2 c^4 + c^2 p^2. \tag{9.15a}$$

From Eqs. (9.1) and (9.4) we obtain

$$\hbar\omega_C = m_0 c^2, \tag{9.15b}$$

$$\hbar\omega_B = cp, \tag{9.15c}$$

which together with (9.13a) and (9.13b) allows us to recast Eq. (9.15a) as

$$\mathcal{E}^2 = \hbar^2 \left(\omega_C^2 + \omega_B^2\right) = \hbar^2\omega_C^2 \left(1 + \gamma^2\beta^2\right) = \hbar^2\omega_C^2\gamma^2 = \hbar^2\omega_A^2. \tag{9.16}$$

It follows that

$$\mathcal{E} = \hbar\omega_A = \hbar\gamma\omega_C, \tag{9.17}$$

and the relation for the energy (9.15a) becomes equivalent to

$$\omega_A^2 = \omega_B^2 + \omega_C^2. \tag{9.18}$$

Formula (9.17) exhibits the energy as a manifestation of a vibration of very high frequency, so that energy and frequency become two aspects of the same reality, as is strongly expressed by Eq. (9.18). This suggests that all forms of energy are essentially the same thing, namely vibrations (energy *is* motion!). Under the consideration that ω_B refers to an electromagnetic wave, the successive discoveries by Planck (captured in the quantum relation $\mathcal{E} \sim \omega$), by Einstein (Eq. (9.15a)) and by de Broglie become integrated into the general law (9.18), which is simultaneously relativistic and quantum. In addition, this equation shows that de Broglie's frequency can be understood as a measure of the deviation of the actual frequency of vibration of the particle in the laboratory (ω_A) from its reference value (the Compton frequency ω_C), i.e., $\omega_B = \left(\omega_A^2 - \omega_C^2\right)^{1/2}$.

Taken together, Eqs. (9.15c) and (9.17) associate the wave number $k_B = p/\hbar$ and the frequency $\omega_A = \mathcal{E}/\hbar$ with a particle having momentum p and energy $\mathcal{E}$. Such quantities are thus the natural ones to characterize a 'matter' wave associated with the moving corpuscle. The dispersion relation for such wave is therefore given by the relation $\mathcal{E} = \mathcal{E}(p)$, whence from Eqs. (9.15a) and (9.17) it follows that the group velocity v_g of the matter wave is

$$v_g = \frac{\partial \mathcal{E}}{\partial p} = v. \tag{9.19a}$$

On the other hand, the phase velocity is just

$$v_p = \frac{\mathcal{E}}{p} = \frac{mc^2}{p} = \frac{c^2}{v}. \tag{9.19b}$$

Notice that the *sole* specification of the wave number and the frequency of the matter wave (k_B and ω_A, respectively) could suggest to identify it with a simple wave of the form $\cos(\omega_A t - k_B x + \alpha)$ (with α constant). However, such wave does not comply with the above expression for v_g. This stresses the importance of the correct dispersion relation, and clearly indicates that the matter wave must be more complex than a simple oscillation.

9.1.3 The de Broglie Wave

We see that the intimate connection between energy and frequency not only brings in the notion of a matter wave associated with the moving corpuscle, but also determines its group and phase velocities. Two immediate questions arise, about its identification and about its physical reality. Is the matter wave simply a mathematical artifact, conveniently put in correspondence with the physical corpuscle, or is it a truly physical wave? In this section we briefly tackle this issue.

From the above discussion we know that the matter wave is not simply $\cos(\omega_A t - k_B x + \alpha)$, but this wave modulated so that there is a wave traveling with velocity v; hence it must be a wave of the form

$$\cos(\omega_A t - k_B x + \alpha) \times f(x - vt). \tag{9.20}$$

In Eq. (9.14) we have precisely this kind of wave. Indeed, with $\omega_B/k_A = v$, $\varphi_v(x,t)$ is found to have just the structure of (9.20),

$$\varphi_v(x,t) = 4\cos\left(\omega_A t - k_B x + \theta_1\right)\cos\left[k_A\left(x - vt\right) - \theta_2\right]. \tag{9.21}$$

Taking a snapshot of (9.21) at $t = 0$ gives

$$\varphi_v(x,0) = 4\cos\left(k_A x - \theta_2\right)\cos\left(k_B x - \theta_1\right). \tag{9.22}$$

Since $k_B = \beta k_A < k_A$, $\varphi_v(x,0)$ represents a rapid spatial oscillation with an amplitude that is modulated by a wave of wavelength $\lambda_B = 2\pi/k_B$; that is, the wavelength of the (spatial) modulation (envelope) *is precisely de Broglie's* λ_B. Let us now assume that instead of a snapshot we take a video with the position fixed at $x = 0$; this gives

$$\varphi_v(0,t) = 4\cos\left(\omega_A t + \theta_1\right)\cos\left(\omega_B t + \theta_2\right). \tag{9.23}$$

The fact that $\omega_B = \beta\omega_A < \omega_A$, implies that the amplitude of the higher-frequency wave (the carrier) is modulated by an oscillation of frequency ω_B. In other words, the frequency of the (temporal) envelope *coincides with the de Broglie frequency*.

We are now in a position to identify the whole structure $\varphi_v(x,t)$ with the 'matter wave', or de Broglie wave. Recognizing the origin of $\varphi_v(x,t)$ in the ZPF, we conclude that the de Broglie wave represents a physically real wave, as 'seen' from the laboratory. Since the spatial modulation of $\varphi_v(x,t)$ travels with velocity v, to an observer in S it appears to keep company to the particle, as if surrounding and 'guiding' it along its motion—thus calling to mind the idea behind the guidance formula in de Broglie's theory. Both entities, particle and wave, appear thus as an indissoluble couple, yet each of them has a well-defined and complementary nature; in particular, the particle remains always a corpuscle, a nonextended object (though with some structure), in contrast with the always extended $\varphi_v(x,t)$. Notice that, even though from this perspective the particle is an intrinsically localizable object, its specific position within the matter wave's wavelength is not determined.

Consideration of the ZPF seems thus to be a natural means to incorporate not only the de Broglie wavelength, but also the de Broglie *wave*, into the narrative of quantum mechanics.[6] An additional relation between λ_B, the vacuum field, and the dynamics of the particle, can be obtained rewriting Eq. (9.4) in the form

$$\omega_B \lambda_B = 2\pi c. \tag{9.24}$$

This relation characterizes an electromagnetic wave in vacuum, with de Broglie's wavelength and with a linear momentum equal to $p_B = \hbar\omega_B/c$, which, according to Eq. (9.15c), $\hbar\omega_B = cp$, coincides with the momentum p of the particle,

$$p = p_B. \tag{9.25}$$

Consequently, while the particle travels 'sitting' on the de Broglie wave, it bears the same momentum as the ZPF modes of frequency ω_B; such modes thus acquire special relevance for the moving particle. In this sense it is natural to associate the ZPF modes of wavelength λ_B also to the moving corpuscle—bearing in mind, however, that de Broglie's wavelength λ_B does actually originate in the background field. De Broglie's formula should then be recast in the form

$$\lambda_B = \frac{h}{p_B}, \tag{9.26}$$

representing a genuine wave formula written in terms of parameters pertaining to a wave only, *without reference at all to the particle*. From this perspective, it is via the condition (9.25) that the wave property is *transferred* to the particle, so that $\lambda_B = h/p$. That the modes of the ZPF having frequency ω_B (and wavelength λ_B) turn out to be of particular importance for the dynamics of the particle will be further discussed in Sect. 9.3, in relation with matter diffraction.

---

[6] Or rather, into the ontology of quantum mechanics. We see in the wave *function* of quantum mechanics an abstract object that lives in a mathemathical configuration space. By contrast, the de Broglie wave associated with the ZPF modulations should be understood as a real wave in three-dimensional space. They are therefore two objects of an entirely different nature.

9.2 An Exercise on Quantization à la de Broglie

In this section we resort to the de Broglie wave constructed above to show by means of an example how it can be applied to analyze some properties of stationary, bounded, one-dimensional quantum motions. With this aim let us consider a benchmark case and examine the stationary description of a particle trapped in an infinite square potential well of width a. In this case there is no net flux and the particles will be performing periodical back and forth motions inside the box. In order to construct the de Broglie description for this situation, one must take into account not only the $\varphi_v(x,t)$, representing the wave associated with a particle that travels in the $+x$ direction with velocity v, but also the reflected wave $\varphi_{-v}(x,t)$ that travels in the $-x$ direction with the same speed. We therefore take the superposition

$$\varphi(x,t) = \varphi_v(x,t) + \varphi_{-v}(x,t). \tag{9.27}$$

As follows from Eq. (9.11), with the substitution $v \to -v$ the frequency $\omega_\pm$ becomes $\omega_\mp$, and similarly for $k_\pm = \omega_\pm/c$. We shall assume that the phases $\theta_\pm$ in Eq. (9.12) are the same in both components (they both refer to the same wave). Taking all this into account, Eq. (9.27) reads

$$\varphi(x,t) = e^{-i\theta}[e^{-i(\omega_+ t - k_+ x)} + e^{-i(\omega_+ t + k_+ x)} + e^{-i(\omega_- t - k_- x)} + e^{-i(\omega_- t + k_- x)}] + \text{c.c}, \tag{9.28}$$

which reduces to

$$\varphi(x,t) = 4\left[\cos(\omega_+ t + \theta)\cos k_+ x + \cos(\omega_- t + \theta)\cos k_- x\right]. \tag{9.29}$$

This standing wave inside the well is consistent with the condition of zero flux velocity. Unlike the de Broglie wave, the superposition $\varphi(x,t)$ does not travel with the particle, but reflects the periodicity of the motion. Further, since $\varphi(x,t)$ corresponds to a stationary situation, it means that it is periodic in x with period a,

$$\varphi(x,t) = \varphi(x+a,t). \tag{9.30}$$

This stationarity condition applied to Eq. (9.29) leads to

$$k_\pm = \frac{2\pi}{a} n_\pm, \quad n_\pm = 0, 1, \ldots \tag{9.31}$$

Notice that for $v \neq 0$ we have $k_+ > k_-$ (see Eq. (9.11)), whence $n_+ > n_-$. From here and Eqs. (9.13a), (9.13b) it follows that

$$\tfrac{1}{2}(k_+ - k_-) = k_B = \tfrac{\pi}{a}(n_+ - n_-) \equiv \tfrac{\pi}{a} n, \quad n = 1, \ldots, \tag{9.32a}$$

$$\tfrac{1}{2}(k_+ + k_-) = k_A = \tfrac{\pi}{a}(n_+ + n_-) \equiv \tfrac{\pi}{a} N, \quad N = 1, \ldots \tag{9.32b}$$

Equation (9.32b), together with (9.15c), gives

$$p = \hbar k_B \rightarrow p_n = \frac{\hbar\pi}{a} n, \tag{9.33}$$

whence

$$n\lambda_B = 2a. \tag{9.34}$$

One can recognize here the well-known statement that the well can accommodate an integer number of half-de Broglie's wavelengths under stationarity, in agreement with usual phenomenology. Notice that the result arises as a consequence of imposing the stationarity condition on the *wave* $\varphi(x,t)$ that reflects the periodicity of the *corpuscle's* motion. Equations (9.33) and (9.34) mean that the dynamics and the de Broglie wave have become conformed to the geometry of the system.

Notice that Eq. (9.33) follows also from (9.25), under conditions of stationarity of the standing waves of the ZPF inside the well. In other words, the quantization implied by Eq. (9.33) can be seen as a result of the presence of the vacuum field and the identification $p = p_B$, a relation that plays thus the role of a quantization rule.

Let us now turn to Eq. (9.32b), which together with $\lambda_A = 2\pi/k_A$ gives

$$N\lambda_A = 2a. \tag{9.35}$$

According to this expression, also an integer number of half-wavelengths λ_A must be accommodated inside the well to attain stationarity. However, since $k_B/k_A = \beta = n/N$, in the nonrelativistic regime $n \ll N$. Comparison between Eqs. (9.34) and (9.35) thus indicates that the wave with λ_A inside the well has many more nodes that the wave with λ_B. In terms of the de Broglie wave, this is explained by recalling that at any given time, $\varphi_v(x, t_0)$ represents a rapid oscillation of wavelength λ_A modulated by an oscillation of wavelength $\lambda_B \gg \lambda_A$ (cf. Eq. (9.22)). Physically, this reflects the fact that the particle inside the box is not simply performing a uniform motion with (mean) velocity v (like a classical particle would do), but that such motion is superposed to a vibration at the high frequency $\omega_A \sim \omega_C$. As mentioned earlier, this oscillation, the zitterbewegung, constitutes an echo—the laboratory frame—of de Broglie's clock.

The above results can be somewhat completed to get a more detailed picture of what is happening inside the well. The formula for the energy $\mathcal{E}_n$ associated with the smooth motion of the particles follows directly from Eq. (9.33),

$$\mathcal{E}_n = \frac{p_n^2}{2m} = \frac{\pi^2\hbar^2}{2ma^2} n^2, \quad n = 1, ... \tag{9.36}$$

Since the particles are being perfectly reflected at the walls of the well, it has sense to define the period of the (mean) motion in state n as

$$\mathcal{T}_n \equiv \frac{2a}{v_n} = 2\pi n \frac{\hbar}{2\mathcal{E}_n}, \tag{9.37}$$

which suggests to introduce a *mechanical* frequency ω_n^{mec} such that

$$\omega_n^{mec}\mathcal{T}_n = 2\pi n. \tag{9.38}$$

With this definition, Eq. (9.37) gives

$$\mathcal{E}_n = \frac{\hbar}{2}\omega_n^{mec}, \quad \omega_n^{mec} = \frac{\pi^2\hbar}{ma^2}n^2, \quad v_n = \frac{\pi\hbar}{ma}n = v_1 n. \tag{9.39}$$

In state n the particle surveys in the mean n times the distance that corresponds to the fundamental state $n = 1$. This result relates the index n in p_n to the number of complete cycles performed by the component n during the time that the slowest component (for $n = 1$) completes one cycle.

The first relation in (9.39) looks akin to the substance of SED, and states that the energy of the particle in the state n coincides with the energy of the modes of the ZPF of frequency ω_n^{mec}. Further, direct calculation gives

$$\lambda_{Bn}\omega_n^{mec} = 2\pi v_n, \tag{9.40}$$

with $\lambda_{Bn} = h/p_n$. This relation defines a geometric wave inside the well, moving with the particle; it has the de Broglie wavelength and the mechanical frequency ω_n^{mec}. This latter can be related with the de Broglie frequency by a comparison of Eqs. (9.24) and (9.40), resulting in

$$\omega_n^{mec} = \beta_n \omega_{B_n}. \tag{9.41}$$

For nonrelativistic motions, we verify that the de Broglie frequency is very high compared with the frequency of the dominant motion. In contrast, the de Broglie wavelength is considerably larger than the Compton wavelength (see Eq. (9.7)).

Unlike the resonance frequencies studied in Chap. 5, which are the frequencies of transition between states, the ω_n^{mec} $(n = 1, 2, ...)$ are related directly to the *permanence* in the respective state n. They remain hidden to QM, not being part of its ontology, its epistemology, or its semantics. Equation (9.39) together with Bohr's rule shows that there exists a relation between the transition and the permanence frequencies,

$$\omega_{nm} = \hbar^{-1}(\mathcal{E}_n - \mathcal{E}_m) = (\omega_n^{mec} - \omega_m^{mec})/2. \tag{9.42}$$

To the extent to which the definition $\mathcal{E}_n = \hbar\omega_n^{mec}/2$ has a meaning, such relationship may be significant. The factor 2 is specific of the present example, of course.

9.3 Undulatory Properties of Matter

Because of their wave-like properties, quantum corpuscles are often not 'seen' as particles—or waves—but as another sort of entity, such as 'wavicles' (Eddington 1928), 'microparticles' (Blokhinsev 1953/1964), 'quantons' (Bunge 1967, 1973), 'smearons' (Maxwell 1981), 'wavelets' (Barut 1993), and what not.[7] A few examples, taken from among scores of them, of the kind of contrasting points of view to which the consideration of the wave properties of matter is prone to lead, may be seen in Diner et al. (1983), Agazzi (1988), and Combourieu and Rauch (1992). Leaving aside particular details, what the existence of so many and diverse approaches evinces is that the undulatory properties of matter are among the most perplexing and least understood aspects of the quantum world.

To pay attention to the wave-like properties of matter, let us appeal to one of the simplest and at the same time most revealing quantum experiments, that of electrons passing through two parallel slits made on a screen. The amazing result is well known, and popularized by the Tonomura et al 1989 experiment, which has been seen by many thanks to the web (www.hitachi.com/rd/portal/research/em/movie.html; see also Bach et al. 2013). It is important to draw attention to this experiment (and earlier ones, such as those described in Jönsson (1961), and Matteucci and Pozzi (1978), since all of them reveal that a *single* electron does not give rise to the diffraction pattern: it merely produces a spot (seemingly at random) on the screen. The diffraction pattern, a wave-like phenomenon, results from the addition of tens of thousands of events, and hence depicts the statistical distribution of electrons on the screen. The Schrödinger equation, which refers to the wave properties of particles, describes just this statistical behavior. It cannot provide in general a detailed description of the wanderings of an individual electron. It is devised to describe the multitude, not what each and every electron is doing. And it certainly does not provide a physical *explanation* for the diffraction pattern.

Let us now look from the present SED perspective at the problem of particle diffraction by the pair of parallel slits. One should start by considering that the the ZPF is not immune to the presence of the slits. The Casimir effect, as well as the cavity effects on atomic lifetimes and energy levels, are well-known instances that remind us that ZPF must satisfy the same boundary conditions as any other electromagnetic field in the presence of matter [see e.g. Boyer (1980) and references therein; Cetto and de la Peña (1988a, b)]. And indeed, Fig. 9.1 shows an image of the ZPF diffracted by two parallel slits, opened on an infinite, totally reflecting plate; the wavelentgth of the field modes has been chosen to be of the order of the distance between slits. This is the kind of field that the electrons 'feel' when traveling in the neighborhood of the screen. The partially reorganized electric forces act on the particles, and one

[7] The term *wavelet* refers to localized nonspreading solutions of massless wave equations that move like massive quantum particles. Wavelets are seen as a bridge between classical point particles and the waves of QM; the mass of the particle is determined by the internal frequency of the wavelet, much as the 'internal clock' in the Broglie's theory.

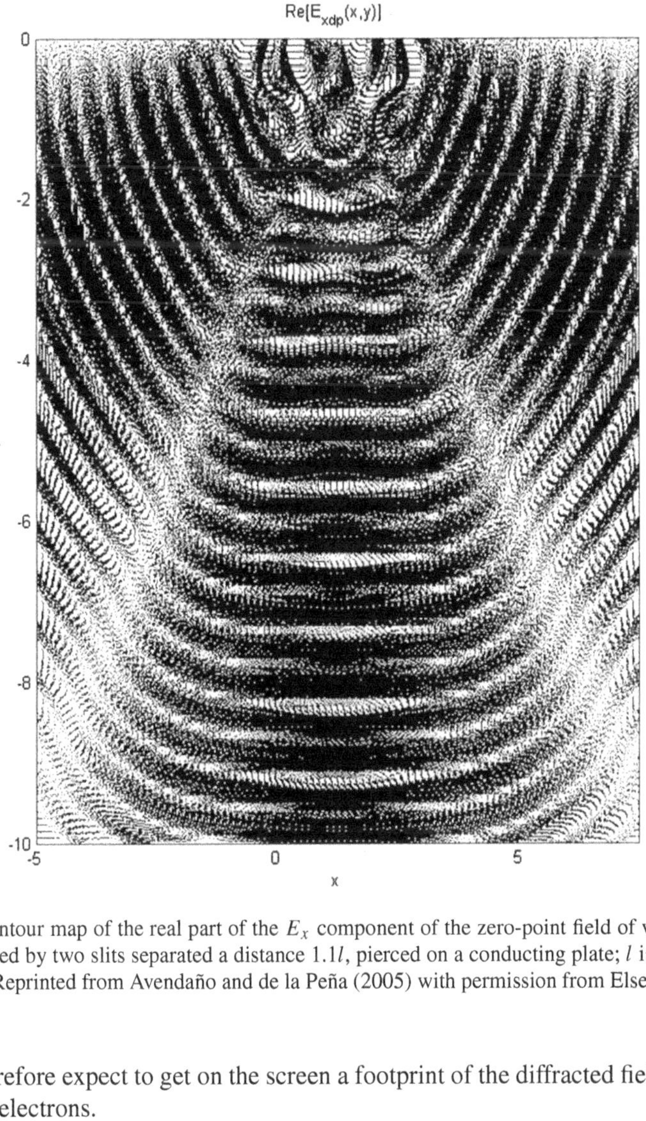

Fig. 9.1 Contour map of the real part of the E_x component of the zero-point field of wavelength $0.9l$, diffracted by two slits separated a distance $1.1l$, pierced on a conducting plate; l is the width of the slits. Reprinted from Avendaño and de la Peña (2005) with permission from Elsevier

should therefore expect to get on the screen a footprint of the diffracted field, traced out by the electrons.

The detailed dynamics of the particles travelling in a diffracted field like the one in figure reffig1 needs still to be worked out. However, according to the discussion in the previous section, one may reasonably assume that the electrons with a momentum p will be particularly affected by the diffracted modes that satisfy the condition (9.25), $p_B = p$ (i.e., the modes of wavelength $\lambda_B = h/p$), and guided by them towards the screen along the preferred directions determined by Bragg's law. This would give shape to an interference pattern superimposed on the noisy background. Thus, the particle needs to 'know' nothing about the existence of the slits: it is the background field what carries the required information and operates accordingly *on* the particles.

The picture that emerges reminds us of the image suggested by J. Clauser some time ago: "If a bunch of surfers pass through a breakwater with two entrances, you'll see the two-slit pattern later on the beach in surfer flesh!" (quoted in Wick (1995), p. 116). And indeed, for over 80 years we have been observing interference patterns in electron flesh. The electrons maintain their corpuscular identity all along the experiment, and there is no need of particle self-interference. Since the diffracted field exists even in the absence of the electrons, it might be possible to put this explanation to experimental test. The observation of the diffracted field with independence from the presence or absence of the electrons would demonstrate that it is the field, not matter, what is diffracted. An initial exposition of these matters is given in Avendaño and de la Peña (2010). An example of the kind of results that such an explanation can afford is shown in Fig. 9.2. This figure shows some preliminary results obtained again by numerical calculation (Avendaño and de la Peña 2005, and work in preparation), for the trajectories followed by electrons in the double-slit experiment. The fluctuating component of the diffracted field has been suppressed to highlight the guiding effect of the field. The momentum p of the particles has been selected according to the law $p = p_B$ and the kinetic energy of the electrons has been assumed to remain constant, i.e., the particles are deflected without changing their speed. Even if in the real situation the trajectories may be not as smooth, the figure offers a clear image of the behavior one may expect for the particles under the action of the diffracted ZPF. In particular, it is distinctly seen that on its way to the distant screen, each particle crosses a single time a single slit, and behaves as a localized corpuscle all along its journey.[8]

Particle diffraction patterns have been obtained experimentally also with neutrons and other neutral particles, as is well known from crystallography and has been confirmed by the famous experiments by Rauch and colleagues [Rauch et al. (1974); a detailed review is Greenberger (1983)]. This means that the SED explanation should not be restricted to charged particles. As suggested in Chap. 4, a possible answer to this observation is that all known particles, including the neutral ones, have electromagnetic interactions. An interaction with the ZPF through the coupling of the electric dipole moment, the magnetic moment, or any other multipole, is able in principle to lead to results that are similar to the ones obtained from electric charge coupling, although details such as the relaxation times may vary; such differences, however, are irrelevant for the performed experiments, which proceed very slowly in comparison.

The double-slit experiment affords an opportunity to give a more precise meaning to the assertion that the Schrödinger equation predicts *only* the wavelike behavior of quantum corpuscles. That such statement requires qualification can be illustrated with the aid of Fig. 9.3. This figure shows the numerical solution of the Schrödinger equation for the problem of two ideal slits, at different distances from the plane

[8] The results obtained with this numerical experiment are similar to those obtained by Couder and Fort (2006) in their macroscopic Young-type experiment, showing clearly that the bouncing droplet goes through either one of the two slits but the associated wave passes through both slits, and the interference of the resulting waves is responsible for the trajectory of the walker.

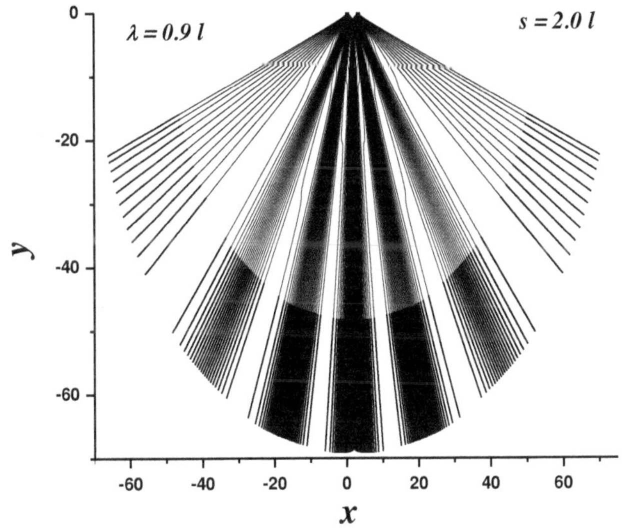

Fig. 9.2 Trajectories followed by electrons in a realistic simulation of a two-slit experiment. The particles are uniformly distributed in the beam behind the slits. The diffracted modes of the field have momentum p_B and the momentum of the particles is p, with $p = p_B$. Figure courtesy of J. Avendaño, adapted from Avendaño and de la Peña (2010)

that contains the slits. The solution shows that sufficiently close to the slits the Schrödinger equation predicts a corpuscular behavior of the particles, whereas the wavelike behavior corresponds to the Fraunhofer (far) region. For intermediate distances both aspects are simultaneously manifested. Here we have an example of coexistence of corpuscular and wavelike manifestations, which reminds us of Einstein's reading of Eq. (3.71) as an expression of these two complementary aspects in the case of light.

9.4 Cosmological Origin of Planck's Constant

Let us make a detour from the line of inquiry followed so far in this chapter, and direct our attention to another interesting question directly related with the ZPF, namely: what fixes the scale $\hbar$ of the ZPF fluctuations? Being the ZPF of cosmological origin, it sounds natural to assume that $\hbar$ should be linked in any way to other universal constants.

To find an answer to this question let us consider a world made of just harmonic oscillators, representing both matter and the modes of the zero-point radiation field. Such a crude model should be appropriate for the purpose of performing an order-of-magnitude estimate of certain quantities of interest for our present intent. Since the thermal (photonic) background radiation is of no interest here, we assume that all

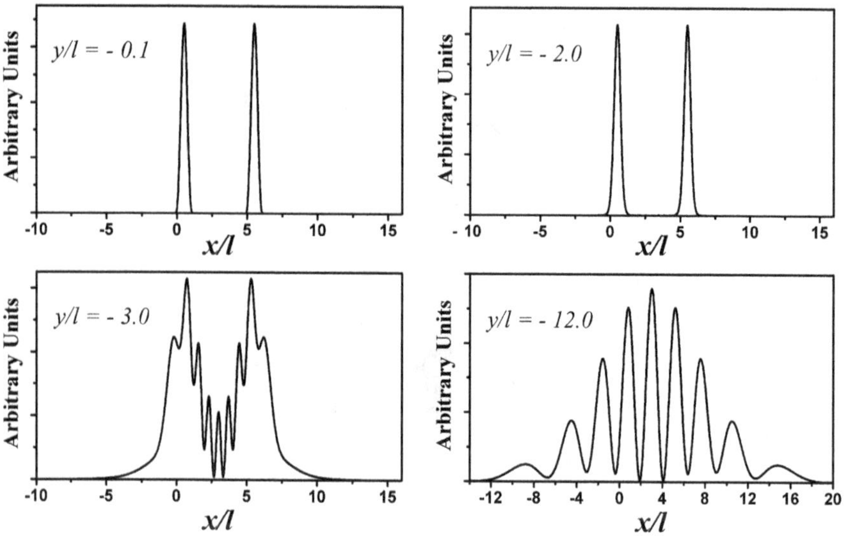

Fig. 9.3 Numerical solution of the Schrödinger equation for two slits, shown as function of the distance from the plane of the slits. At very short distances the solution resembles the one that corresponds to particles, whereas at very far distances a Fraunhofer diffraction pattern is observed. The distance from the slits to the screen is given by $-y$. Figure courtesy of J. Avendaño

elementary oscillators of a given frequency, irrespective of their nature, are in their ground state. According to the image developed in this book, the charged (matter) oscillators (electric dipoles) are radiating and contributing to the background field, but they are also absorbing energy from the field. The random field so regenerated should coincide with the vacuum field at each frequency under the assumption of a universe in equilibrium. We herewith espouse a sort of cosmological principle associated with SED, or, if preferred, a kind of *electromagnetic Mach principle*: the field produced at a given point by all dipoles in the Universe should equal the random field acting at that point on the particles themselves. This requirement establishes a relationship between cosmological and atomic constants; in other words, it establishes the scale of quantum fluctuations. Planck's constant becomes thus determined by cosmological parameters (de la Peña and Cetto 1984, 1997).[9]

A parallel reasoning, but dealing entirely with the gravitational field, has been discussed by Calogero (1997) in an interesting essay restricted to order-of-magnitude considerations. In that work, the identification of the unavoidable gravitational fluctuations with the quantum fluctuations of atomic systems is shown to lead to a relationship between atomic and cosmological constants. One should recall also a similar attempt made in Puthoff (1991) within SED, also on the basis of a self-regenerating model. The basic idea in Puthoff's paper is much in line with the one studied here,

[9] Recently we have become aware of a similar proposal by Mavrychev (1967), in which the author reaches a comparable result.

although in our rough estimate we refrain from resorting to any specific cosmological model. Still we leave aside any problem related with the infinite gravitational effects of the zero-point field, just as is done in QED (and in cosmology) with all vacuum fields, simply because nobody knows yet how to solve this so-much studied and poorly understood problem (see e.g. Weinberg 1989). By equating the radiation field predicted by the model at a given point with the corresponding component of the zero-point field, Puthoff obtains a prediction for the baryonic mass density of the Universe, which establishes a relation between atomic constants and the Hubble constant. This relation happens to correspond essentially to the one discussed in Weinberg (1972) book (Sect. 16.4), which is usually taken as a numerical coincidence, of unknown origin and meaning.

Consider, then, the radiating dipole α ($\alpha = 1, 2, \ldots, N$) of frequency ω located at the position $\boldsymbol{r}_\alpha$; we take the origin of coordinates at our place. The Fourier amplitude of the electric field produced by this oscillator at the origin is

$$\boldsymbol{E}_\alpha(\omega) = -k^2 \boldsymbol{n}_\alpha \times (\boldsymbol{n}_\alpha \times \boldsymbol{p}_\alpha) \frac{e^{ikr_\alpha}}{r_\alpha}, \quad k = \frac{\omega}{c}, \tag{9.43}$$

where $\boldsymbol{p}_\alpha = (e/2)(\boldsymbol{q}_{0\alpha} + i\dot{\boldsymbol{q}}_{0\alpha}/\omega)$ is the (complex) amplitude of the dipole moment $\boldsymbol{p}_\alpha e^{-i\omega t}$, and $\boldsymbol{n}_\alpha = \boldsymbol{r}_\alpha / r_\alpha$ is the unit vector in the direction of $\boldsymbol{r}_\alpha$. Since the mean energy of the oscillator is $\hbar\omega/2$, one has $\left\langle \boldsymbol{q}_{0\alpha}^2 + \dot{\boldsymbol{q}}_{0\alpha}^2/\omega^2 \right\rangle = \hbar/m\omega$, where the average is taken over the set of oscillators of frequency ω, so that we write

$$\boldsymbol{p}_\alpha = \frac{e}{2}\sqrt{\frac{\hbar}{m\omega}} \boldsymbol{B}_\alpha \tag{9.44}$$

and consider the components $B_{i\alpha}$ of $\boldsymbol{B}_\alpha$ to be statistically independent complex random variables with zero mean and second moments given by

$$\left\langle B_{i\alpha} B_{j\beta}^* \right\rangle = \delta_{\alpha\beta}\delta_{ij}, \quad \left\langle B_{i\alpha} B_{j\beta} \right\rangle = 0. \tag{9.45}$$

With these assumptions the mean square of Eq. (9.43) is (omitting the index α)

$$\langle |\boldsymbol{E}(\omega)|^2 \rangle = \frac{\hbar e^2}{4mc^4} \frac{\omega^3}{r^2} \langle |\boldsymbol{n} \times (\boldsymbol{n} \times \boldsymbol{B})|^2 \rangle = \frac{\hbar e^2}{2mc^4} \frac{\omega^3}{r^2}, \tag{9.46}$$

since $\langle |\boldsymbol{n} \times (\boldsymbol{n} \times \boldsymbol{B})|^2 \rangle = \left\langle \boldsymbol{B} \cdot \boldsymbol{B}^* - (\boldsymbol{n} \cdot \boldsymbol{B})(\boldsymbol{n} \cdot \boldsymbol{B}^*) \right\rangle = 2$. We have taken into account that the field amplitudes produced by statistically independent oscillators are uncorrelated.

To evaluate the average energy content $\langle \mathcal{E}(\omega) \rangle$ of the radiation field of frequency ω at the origin, we integrate Eq. (9.46) over a spherical volume of radius R, assuming an isotropic and homogeneous distribution of oscillators, of which there are $n(\omega)$ of frequency ω and a total number $N = \sum_\omega n(\omega)$:

$$\langle \mathcal{E}(\omega) \rangle = \frac{n(\omega)}{4\pi} \int_V \langle |\boldsymbol{E}(\omega)|^2 \rangle dV = \frac{\hbar e^2 \omega^3 R}{2mc^4} n(\omega). \tag{9.47}$$

The cosmological postulate asserts that this energy should correspond to the ZPF energy of a mode of frequency ω, i.e., $\hbar\omega/2$; one thus obtains

$$n(\omega) = \frac{mc^4}{e^2 \omega^2 R}. \tag{9.48}$$

To estimate the total number of oscillators we integrate over all frequencies, using the rule $V^{-1} \sum_\omega \rightarrow (2\pi^2 c^3)^{-1} \int d\omega\, \omega^2$, which gives

$$N = \sum_\omega n(\omega) = \frac{mc^4}{e^2 R} \sum_\omega \frac{1}{\omega^2} \rightarrow \frac{mcV}{2\pi^2 e^2 R} \int_0^\Omega d\omega = \frac{mc\Omega V}{2\pi^2 e^2 R}. \tag{9.49}$$

Since the integral is divergent we have introduced a cutoff frequency Ω for the material oscillators. Indeed, the material oscillators are transparent at arbitrarily high frequencies; one can consider a cutoff around the pair-creation (Zitterwebegung) frequency $\Omega = 2mc^2/\hbar$ as physically meaningful (see also the discussion in Sect. 9.1.1), so that (9.49) becomes

$$\frac{N}{V} = \frac{m^2 c^2}{\pi^2 \alpha \hbar^2 R}, \tag{9.50}$$

where $\alpha = e^2/\hbar c$ stands for the fine-structure constant. Here V must be taken as the volume of the visible part of the Universe, as this is the part that contributes to the radiation field, and thus N/V is to be identified with the cosmological density of charged particles, which multiplied by m_N (the nucleon mass or any typical baryon mass) gives for the baryonic density of the Universe the estimate

$$\rho \simeq \frac{m^2 m_N c^2}{\pi^2 \alpha \hbar^2 R}. \tag{9.51}$$

Before proceeding further let us add a couple of remarks with regard to this expression. Firstly, we have not taken into account any absorption process, the reason being that we are dealing with the ZPF, which is not absorbed by matter in equilibrium with it. Of course it is scattered by matter, but for a uniform and homogeneous universe the final distribution remains the same. Therefore Eq. (9.51) needs *no* correction from Thomson scattering.

The second comment refers to the simplicity of the model. The intention here is to make a qualitative test of the SED cosmological principle, and for such purpose the present rough estimate should suffice. For example, a somewhat more realistic model would take into account the expansion of the Universe, which produces a redshift, so that instead of the original frequency ω radiated when the Universe had a scale factor $R(t)$, the red-shifted frequency

$$\omega_0 = \frac{R(t)}{R_0}\omega \tag{9.52}$$

should be used, where the subindex 0 refers to the present moment and place of observation. Thus, if $\nu(\omega)$ represents the spatial density of oscillators of local frequency ω at a distance r from us, instead of Eq. (9.47) one should write

$$\langle \mathcal{E}(\omega)\rangle = \frac{e^2\hbar}{2mc^4}\omega_0^3 R_0^3 \int_0^{R_0} \frac{\nu(\omega_0 R_0/R(t))}{R^3(t)r^2} r^2 dr, \tag{9.53}$$

where, using Weinberg's (1972) notation, one must put $dr = (\sqrt{1-kr^2}/R(t))dt$. To go further one would have to specify the cosmological model; however, any reasonable choice for $R(t)$ would only change the numerical factors, without altering the essential contents of Eq. (9.51). Thus, up to such numerical factors we take the former result (9.51) as a reasonable relation among the relevant constants of nature.

Let us try to draw some conclusion from Eq. (9.51). For this purpose we first introduce an auxiliary (representative) mass defined as

$$\bar{m} = \left(\frac{m^2 m_N}{\pi^2 \alpha}\right)^{1/3} \backsimeq 30m. \tag{9.54}$$

where m is the mass of the electron. Equation (9.51) can then be rewritten in the form (we put $R = R_0$, and add the subindex 0 to mark the present values of the cosmological parameters)

$$\frac{\rho_0 R_0^3}{\bar{m}} = \frac{\bar{m}^2 c^2 R_0^2}{\hbar^2} = \left(\frac{R_0}{\lambda_{\bar{m}}}\right)^2, \tag{9.55}$$

where $\lambda_{\bar{m}}$ is the Compton wavelength (divided by 2π) associated with the mass $\bar{m}, \lambda_{\bar{m}} = \hbar/\bar{m}c$. We recognize in each side of Eq. (9.55) one of the 'large numbers' of cosmology, which are (H_0 is the present value of Hubble constant, $H_0 = c/R_0$, and G stands for the gravitational coupling constant)

$$N_1 = \frac{\hbar c}{G m_N^2} \sim \frac{1}{6}10^{39}, \tag{9.56}$$

$$N_2 = \frac{mc^2}{\hbar H_0} = \frac{mcR_0}{\hbar} = \frac{R_0}{\lambda_m} \sim \frac{1}{3}10^{39}, \tag{9.57}$$

$$N_3 = \frac{\rho_0 c^3}{m_N H_0^3} = \frac{\rho_0 R_0^3}{m_N} \sim 10^{79}. \tag{9.58}$$

Except for the differences in the masses, Eq. (9.55) reads

$$N_3 = N_2^2, \tag{9.59}$$

which is one of the well-known numerical coincidences among these large numbers. The surprising content of this expression is that it relates cosmological parameters with Planck's constant, which is a highly nontrivial result (remember that Weinberg (1972) qualifies Eq. (9.59) as mysterious). The second independent relation among these numbers, which can be taken to be $N_1 N_2 \simeq N_3$, does not involve Planck's constant and can be obtained from cosmological models, such as the Friedmann model.

We conclude that the SED Cosmological Principle, namely that the energy of the vacuum fluctuations corresponds to the energy radiated by all dipoles of the Universe in a self-regenerating process, seems to hold and serves to explain the relation $N_3 = N_2^2$ up to a constant factor of at most a few orders of magnitude.

Let us now recast Eq. (9.51) in a different form. In terms of the dimensionless gravitational coupling constant $\alpha_G = Gmm_N/\hbar c$ it reads

$$\alpha_G R_0 \simeq \frac{3\pi}{8}\alpha\lambda_m, \tag{9.60}$$

which we write simply as

$$\alpha\lambda_m \simeq \alpha_G R_0, \tag{9.61}$$

where the value of the common length $l = \alpha\lambda_m = e^2/mc^2 = r_0$ equals the classical electron radius. It seems interesting to observe that Eq. (9.61) can be extended to include nuclear forces by taking the coupling constant of order 1, $\alpha_N \simeq 1$, and a characteristic lenght $R_N \simeq \hbar/m_\pi c$, (m_π is the pion mass), which gives a numerical value $\simeq \alpha\lambda_m/2$, so that

$$\alpha_G R_0 \simeq \alpha\lambda_m \simeq \alpha_N R_N \simeq r_0. \tag{9.62}$$

Equation (9.61) explains why Calogero's gravitational arguments and the present electromagnetic ones lead to equivalent results. This is another form of saying that it should be feasible to represent the effects of the zero-point field as a fluctuating metric field, a possibility that was already studied by Einstein himself (1924). It is interesting to observe that Eq. (9.59) (or Eq. (9.61)) cannot be obtained solely from the usual quantum formalism; it is within the conceptual frame of SED where the cosmological principle leading to Eq. (9.59) finds its natural place.[10]

To end this detour, we note that Eq. (9.55) can be written in the form

$$\hbar = \left(\frac{\bar{m}^3 c^2}{\rho_0 R_0}\right)^{1/2}. \tag{9.63}$$

---

[10] For some enriching comments of differing nature on the ZPF see Ibison (2003), and Dasgupta and Roy (2007).

This suggests speculating that the fluctuations of matter density at cosmological scales may produce local fluctuations on the value of $\hbar$. Whether this has any sense at all is a question that we leave to cosmologists.

References

Agazzi, E.: The nature of quantum paradoxes. In: Tarozzi, G., van der Merwe, A. (eds.) Kluwer, Netherlands (1988)

*Avendaño, J., de la Peña, L.: Reordering of the scarlets of stochastic electromagnetic field by diffraction due to an ideal slit. Phys. Rev. E **72**, 066605 (2005)

*Avendaño, J., de la Peña, L.: Matter diffraction through a double slit obtained by numerical simulation using a diffracted random electromagnetic field. Phys. E **42**, 313 (2010)

Bacciagaluppi, G., Valentini, A.: Quantum Theory at the Crossroads. Reconsidering the 1927 Solvay Conference. Cambridge University Press, Cambridge (2009)

Bach, R., Pope, D., Liou, S.-H., Batelaan, H.: Controlled double-slit electron diffraction. New J. Phys. **15**, 033016 (2013)

Barut, A.O: Diffraction and interference of single de Broglie wavelets. Deterministic wave mechanics, document IC/93/106 (ICTP, Trieste). Reprinted in fondation de Broglie (1993)

Blokhintsev. D.I.: Grundlagen der Quantenmechanik (Deut. Verlag d. Wissenschaften, Berlin). From the Russian original of 1949. English translation: Principles of Quantum Mechanics Reidel, Dordrecht (1953/1964)

Boyer, T.H.: A brief survey of stochastic electrodynamics. In: Barut, A.O. (ed.) Foundations of Radiation Theory and Quantum Electrodynamics Plenum, New York (1980)

Bunge, M.: Foundations of Physics. Springer, New York (1967)

Bunge, M.: Philosophy of Physics. Reidel, Dordrecht (1973)

Calogero, F.: Cosmic origin of quantization. Phys. Lett. A **228**, 335 (1997)

*Cetto, A.M., de la Peña, L.: Environmental effects on the Lamb shift according to stochastic electrodynamics. Phys. Rev. A **37**, 1952 (1988a)

*Cetto, A.M., de la Peña, L.: Environmental effects on radiative lifetimes and mass renormalization, according to stochastic electrodynamics. Phys. Rev. A **37**, 1960 (1988b)

*Cetto, A.M., de la Peña, L. Zeropoint waves and quantum particles. In: Ferrero, M., van der Merwe, A. (eds.) Fundamental Problems in Quantum Physics, p. 47. Kluwer, Amsterdam (1995a)

*Cetto, A.M., de la Peña, L.: Wave mechanics: the interplay between stochastics and quanta. In: Garbaczewski, P., Wolf, M., Weron, A. (eds.) Chaos: The Interplay Between Stochastic, Classic and Quanta, p. 51. Springer, Berlin (1995b)

Combourieu, M.-C., Rauch, H.: The wave-particle dualism in 1992: a summary. Found. Phys. **22**, 1403 (1992)

Couder, Y., Fort, E.: Single-particle diffraction and interference at a macroscopic scale. Phys. Rev. Lett. **97**, 154101 (2006)

Dasgupta, R., Roy, S.: Spectrum of zero-point field and statistical stability. Commun. Appl. Anal. **11**, 23 (2007)

de Broglie, L.: Les principes de la nouvelle Mécanique ondulatoire. J. de Physique, série VI, tome VII no. **11**(321), 337 (1926)

de Broglie, L.: Une Tentative d'interprétation Causale et Non Linéare de la Mécanique Ondulatoire: La Théorie de la Double Solution. Gauthier-Villars, Paris (1956)

de Broglie, L.: Recherches sur la Théorie des Quanta. Réédition du texte de 1924, Masson, Paris (1963)

*de la Peña, L., Cetto, A.M.: A cosmological implication of stochastic electrodynamics, preprint IFUNAM, Mexico. (Unpublished) (1984).

*de la Peña, L., Cetto, A.M.: Role of the zeropoint field in wave mechanics, In: Fondation de Broglie (1992)
*de la Peña, L., Cetto, A.M.: The wave properties of matter and the zeropoint radiation field. Found. Phys. **24**, 753 (1994)
**de la Peña, L., Cetto, A.M.: The Quantum Dice. An Introduction to Stochastic Electrodynamics, Kluwer, Dordrecht (1996) (Referred to in the book as *The Dice*)
*de la Peña, L., Cetto, A.M.: Estimate of Planck's constant from a cosmological Mach principle. Found. Phys. Lett. **10**, 591 (1997)
de la Peña, L., Jiménez, J.L., Montemayor, R.: The classical motion of an extended charged particle revisited. Il Nuovo Cimento B **69**, 71 (1982)
Diner, S., Fargue, D., Lochak, G., Selleri, F. (eds.): The Wave Particle Dualism. Reidel, Netherlands (1983)
Eddington, A.S.: The Nature of the Physical World. Cambridge University Press, Cambridge (1928)
Einstein, A.: Über den Äther. Schweiz. Naturforsch. Gesells. Verh. **105**, 85 (1924)
Espinosa, J.M.: Physical properties of de Broglie's phase waves. Am. J. Phys. **50**, 357 (1982)
Feynman, R.P., Leighton, R.B., Sands, M.: The Feynman Lectures on Physics, vol. III. Addison-Wesley, Reading (1965)
Greenberger, D.M.: The neutron interferometer as a device for illustrating the strange behavior of quantum systems. Rev. Mod. Phys. **55**, 875 (1983)
*Ibison, M.: Electrodynamics in the zero-point field: on the equilibrium spectral energy distribution and the origin of inertial mass. Found. Phys. Lett. **16**, 83 (2003)
Jackson, J.D.: Classical Electrodynamics, 2nd edn. Wiley, New York (1975)
Jönsson, C.: Zeit. Phys. **161**, 454. (Partial) English translation: Electron diffraction at multiple slits. Am. J. Phys. **42**, 4 1974 (1961)
MacKinnon, E.: De Broglie's thesis: a critical retrospective. Am. J. Phys. **44**, 1047 (1976)
Matteucci, G., Pozzi, G.: Two further experiments on electron interference. Am. J. Phys. **46**, 619 (1978)
Mavrychev, Yu.S.: Derivation of the relation between cosmological parameters and microphysical constants. Izvestiya VUZOV. Fizika **10**, 148 (1967)
Maxwell, N.: Instead of particles and fields: a micro realistic quantum "Smearon" theory. Found. Phys. **12**, 607 (1981)
Milonni, P.W.: The Quantum Vacuum. An Introduction to Quantum Electrodynamics. Academic Press, San Diego (1994)
Mott, N.F.: On teaching quantum phenomena. Contemp. Phys. **5**, 401 (1964)
*Puthoff, H.E.: Source of vacuum electromagnetic zero-point energy. Phys. Rev. A **40**, 4857 (1989), Reply to Comments on "Source of vacuum electromagnetic zero-point energy", Puthoff, H.E.: Phys. Rev. A **44**, 3385 (1991)
Rauch, H., Treimer, W., Bonse, U.: Test of a single crystal neutron interferometer. Phys. Lett. A **47**, 369 (1974). For further references see Greenberger 1983
Selleri, F.: Quantum Paradoxes and Physical Reality. Kluwer, Dordrecht (1990)
Surdin, M.: L'onde de de Broglie et l'électrodynamique stochastique. Ann. Fond. L. de Broglie **4**, 139 (1979)
Synge, J.L.: Geometrical Mechanics and de Broglie Waves. Cambridge University Press, Cambridge (1954)
Weinberg, S.: Gravitation and Cosmology. Wiley, New York (1972)
Weinberg, S.: The cosmological constant problem. Rev. Mod. Phys. **61**, 1 (1989)
Wick D.: The Infamous Boundary. Seven Decades of Controversy in Quantum Physics. Birkhäuser, Boston (1995)

Chapter 10
Quantum Mechanics: Some Answers

Nothing in Nature is random.... A thing appears random only through the incompleteness of our knowledge.

Spinoza (2005)

10.1 The Genetic Gist of the Zero-Point Field

Our investigations started with a recurrent theme of SED, namely the consideration of the fluctuating zero-point radiation field as a basis to arrive at the Planck distribution for the thermal equilibrium radiation field. The quantum of energy introduced by Planck to obtain his law was not a necessary hypothesis[1]; instead, the essential ingredient turned out to be the zero-point component of the field—just the one that Planck discovered in his 1912 paper. This second discovery by Planck acquires, from the present perspective, an importance at least as high as his first one.

The cornerstone for this remarkable result is shown to be contained already in Wien's law, which opens the door to the ZPF—traditionally put aside as an article of faith—by allowing the zero-point energy $\mathcal{E}_0$ of the field oscillators to be different from zero. The conventional (classical) selection $\mathcal{E}_0 = 0$ is clearly unfit for the complexity of a physical world full of electric charges, following all kinds of unending motions and constantly emitting radiation. One should expect in advance the physical reality of the random zero-point field to lead to a statistical behavior of submicroscopic matter that departs from the classical one. And indeed, the presence of this ZPF in an otherwise classical system turns out to radically transform its behavior, leading to the

[1] Something similar happens with the photoelectric effect. This just was the example that Einstein 1905 used to argue in favor of the quantun of radiation. But it is a matter of fact that the cause of this effect can be attributed to the quantization of matter (see e.g. Schiff 1955, Chap. X; Lamb and Scully 1969; Mandel 1976).

L. de la Peña et al., *The Emerging Quantum*,
DOI: 10.1007/978-3-319-07893-9_10,

quantization not only of the energy of the field oscillators in interaction with matter, but also to that of matter in interaction with the field, under appropriate conditions.

The specific properties of this background field have very much to do with the peculiarities of quantum systems; they are, so to say, the ultimate source of 'quantum weirdness'. First, the wave nature of the ZPF is responsible for a fundamental difference between the effects of this field on matter and those of an incoherent source of white noise, characteristic of classical, dissipative, Brownian-motion systems—as already anticipated in Chap. 2. A number of results obtained in Chap. 4 through 9 show that a field with highly (spatially and temporally) coherent independent modes, gives rise to an entirely different behavior of the system. Further to leading to nondissipative states and sustaining them—a property remarkable by itself, being at the root of the atomic stability as envisaged by Nernst—it impresses on the mechanical system the 'wavelike' properties encoded in the Schrödinger equation. Moreover, as shown in Chap. 6, the two degrees of polarization of the ZPF modes induce helicoidal motions of the electron, giving rise to an angular momentum that is independent of the orbital motion. The electron spin is thus identified as a further emergent quantum property, with the correct gyromagnetic factor of 2 for its magnetic moment.

The ergodic properties of the stationary mechanical system—one of the many features that are concealed in usual quantum theory—turn out to be of primary significance, as revealed in Chap. 5. Linked to them is the linear resonant response of the mechanical system to selected modes of the radiation field, encoded in the Heisenberg formulation of quantum mechanics. These are the modes involved in atomic transitions; they are *selected* by the mechanical system, as expressed in the well-known selection rules for atomic transitions. Such rules are here endowed with a physical mechanism (absent in QM) that explains the notorious quantum jumps. According to the results in Chap. 5, for a given state of the atomic electron there is a defined set of resonance frequencies to which it may respond. Which will be the one selected in each instance is a matter of chance, but there is no 'guessing' on the part of the electron about the frequency once the resonance is established. Of course, 'chance' should be here understood to mean that the result depends on a multitude of (unknown and uncontrollable) factors, such as the instantaneous state of the atom and the specific realization of the active mode of the random vacuum field, of which we have neither control nor knowledge.

Further, under certain conditions the ZPF modes have the capacity to induce nonclassical correlations between the parties of a multipartite system, impressing on the latter an effective nonlocal behaviour, encoded in the ensuing entanglement. This observation helps to demystify the mechanism of entanglement by eliminating the spooky nonlocality that seems to characterize it, say, in the case of (classically) noninteracting particles. It is the absence of the ZPF from the usual quantum description what makes of entanglement, and all its peculiar properties, an abstruse feature, difficult to reconcile with the rest of physics. Particularly interesting is the fact that the entanglement of the spin variables in a system of identical particles introduces the demand of antisymmetric wave functions.

The usual quantum-mechanical formalism is known to provide a very accurate description of the time-reversible regime. But strictly speaking, as shown here, it provides only an approximate description for at least two reasons: firstly, the initial transient phase, when field and particle have just started to interact, is not considered; secondly, the radiative terms that were essential in taking the system to the stationary regime, have been dropped from the equations in the derivations leading to quantum mechanics, by taking the radiationless approximation. The first remark should not be a cause for concern (yet), as the initial relaxation time is estimated to be of the order of or shorter than 10^{-20} s. The second limitation is normally lifted by using QED to calculate, perturbatively, the corrections produced by the neglected radiative terms. In SED, by contrast, advantage is taken of the fact that these radiative terms are embedded in the original formulation, as shown in Chap. 6.

Moreover, as follows from Chap. 4, the fact that the initial continuity equation for the phase-space probability density is reduced to an equation in configuration space for the local averaged dynamical variables, indicates that the ensuing (quantum) description is incomplete, this incompleteness being due not solely to its statistical nature, but also to the partial information it bears about the dynamics of the system. Quantum mechanics provides the maximum possible information through the statistical description of (an ensemble of) systems in a certain (radiationless) state. The lost information about the system cannot be reconstructed by going in reverse order, from the reduced x- or p-(or any other equivalent-) representation back to the full phase-space description. An immediate consequence of this is that characteristic quantum features—such as the existence of *irreducible* fluctuations (as encoded in the Heisenberg inequalities), the inexistence of joint probabilities for noncommuting operators, the negative probabilities,[2] the apparent nonlocalities, the nonclassical correlations due to entanglement, and so on—become difficult (if not impossible) to understand from within QM. The physical mechanisms that account for the quantumness are *irreversibly* concealed when the ZPF, the main culprit, has been left out of the picture.

From the present analysis we conclude that even if the description of Nature afforded by present-day quantum mechanics is nonlocal, indeterministic, in some instances noncausal,[3] and free of trajectories and physical images, it strictly speaking does not negate either physical or philosophical realism. It is when one assigns the characteristics of the (partial, asymptotic and approximate) quantum description

[2] We have in mind true probabilities. Negative 'probabilities' are consubstantial to the technique of the so-called weak measurements, i.e., non-projective measurements (Aharonov et al. 1988).

[3] There are other cases in theoretical physics where approximations transform an otherwise causal theory into one that violates causality. Perhaps one of the best known examples is the Abraham-Lorentz equation of motion. This equation is derived from a perfectly causal combination of Maxwell's theory and classical mechanics. The end result, the Abraham-Lorentz equation, can however give rise to noncausal phenomena as preacceleration, the anticipated response to a future force [see e.g. Eq. (4.42)]. Again in this case, the root of such noncausal behavior is to be found in the approximations leading from the original causal full description to the final simplified (and noncausal) one. Approximate physical theories are not bound to satisfy the same rigorous requirements that fundamental theories are supposed to fulfil; this is particularly true in what refers to consistency with first principles.

to Nature itself, by assuming one or another of the (free) interpretative hypotheses of present-day QM, that realism seems to stall. Reality is richer than that: both QM and realism are alive, each one in its own province. Arguments as those that follow from the present theory show that the frequently repeated dicta about quantum noncausality, essential indeterminism, (irreducible) quantum fluctuations, (unexplained) nonlocality, and the like, normally attributed to Nature, are indeed an attribute of the description. The most essential properties of the classical world maintain their force at the quantum level, although a very specific not-less-powerful-statistical quantum lawfulness emerges from the complexity of the situation.

10.1.1 Origin of Quantization

The fact that from an essentially stochastic theory in which dynamical variables can acquire any value from among a continuum we arrived at a description in which such variables may attain sure, i.e., nonstochastic, and discrete values, seems to be a contradiction, or at least an obscure property. However, there is an explanation for it, and a multifaceted one indeed. The more fundamental reason for the appearance of sharp values for some dynamical variables (in particular the energy) is the highly peaked resonant response of the mechanical system to certain field modes, when energy balance and ergodicity are in place; this leads to a (radiationless) description in terms of eigenvalues of Hilbert-space operators, corresponding to stable stationary motions. Moreover, the quantum description leaves aside small fluctuations around the corresponding eigenvalues (as explained in Chap. 5), which thus appear as nonfluctuating quantities. In other words, Nature is noisier than what the theoretical description (in the quantum-mechanical, radiationless approximation) asserts.

By fixing the reduced set of stationary solutions that are robust with respect to the fluctuations of the field and correspond to a local extremum (minimum) of the mean energy, which makes them particularly stable as discussed in Chap. 4, the resulting ergodicity can be considered the source of quantization in the present theory. At the same time, it becomes intuitively clear that the demand of detailed energy balance can be satisfied only by a selected (frequently discrete) set of motions. By its physical content, the present explanation of quantization stands in sharp contrast to the usual one related to the mathematical properties of the wave function—although they are both *formally* equivalent, as seen in Chaps. 4 and 5. This issue acquires relevance if the wave function is taken as a mathematical object, because even as such it must satisfy conditions of continuity and single-valuedness. In fact these conditions on ψ arise from physical demands (such as that of energy balance in the mean) and, very importantly, from those imposed on the spatial distribution, the flux currents, and so on. This stresses the importance of the fact that the theory leads naturally to the Born rule, which assigns to the wave function its probabilistic meaning and related properties.

10.1.2 Recovering Realistic Images

As discussed in Chap. 8, the notion of trajectory is foreign to QM. This correct conclusion, however, is frequently translated to mean that in the real quantum systems of Nature trajectories do not exist. This would force us to renounce the possibility of constructing a realist image of what a quantum corpuscle is actually doing in space and time, a ban that goes contrary to the precepts and results of the theory here developed.

Since the origins of quantum mechanics, particularly through the foundational work of Heisenberg (see e.g. Jammer 1966), the statement about the inexistence of quantum trajectories has permeated almost every textbook on the subject. The argument is founded on the Heisenberg inequalities, interpreted as saying that non-commuting observables (e.g., position and momentum) imply not simultaneously existing values. However, according to our exposition in previous chapters, the Heisenberg inequalities express an acquired property of the statistical description—ultimately originating in the fluctuations impressed on the particle by the ZPF—once the system reaches the time-asymptotic quantum regime. In other words, they set a limit to the applicability of present-day theory, not an (ontic or epistemic) final limit to our capability of apprehension of Nature, nor to the capacity of a single corpuscle to follow a given (yet unknown) trajectory. Even if the individual trajectories become unrecoverable from the quantum-mechanical description, they exist in nature; they are in fact present in the very initial description in terms of Eq. (4.2), say, but disappear from the narrative in the quantum description as a result of its (reduced) statistical nature. The absence of trajectories in the quantum description is an obstacle of significance for the account of individual events, such as the deflection of single particles by a potential barrier or by the atoms of a crystal, and more importantly, for the construction of a realistic image of the microscopic world.

Among the attempts to introduce hidden variables into quantum theory to recover the hidden trajectories, the best known one is Bohm's causal theory. According to the view here advocated such attempts can have at most a partial success, since the individual behavior of a particle becomes irretrievable once its stochastic motion has been smoothed out by the local averaging process. The only sensible way to follow the real trajectories would be to go back to the original equation of motion (4.2), but even then there is the intrinsic problem of any stochastic description, namely the specific realization of the field is unknown and with it also the specific particle trajectory. The best and only thing one can do in any real situation is, therefore, to resort to the statistical treatment of the problem.

In summary, this means that to the extent that the present theory is a sensible one, the mere addition of hidden variables to the usual (reduced) quantum mechanical description to recover determinism or realism is a very limited recourse. Even if one completes the (quantized) theory by adding the (quantized) background field, as is done in QED, the trajectory of a specific particle remains unknown. In other words, within present-day knowledge an indeterministic *description* of the quantum system results unavoidable. It is interesting to compare this conclusion with the old eagerness, expressed so many times by Einstein as his most tenacious devotee, for a final description free of statistical elements. Unfortunately (for some), that is a goal that seems difficult to attain.

10.2 Some Answers

In the following we list a number of conclusions that can be drawn from the developments, discussions and results presented in the body of this book. The list is intended to address some of the main (physical and conceptual) difficulties associated with the usual interpretations of quantum mechanics mentioned in Chap. 1, and to provide a succinct picture of the way out of such difficulties as offered by present SED. Those readers who have accompanied us throughout this long journey, will appreciate that the assertions made emanate from the theory here advanced and do not ensue from personal preconceptions or prejudices. The list contains what we consider are the most relevant points, but of course it is by no means exhaustive.

- The quantum phenomenon is not intrinsic, neither to matter nor to the field, but emerges from a complex process of matter-field interaction.
- The stochastic zero-point radiation field is the physical entity ultimately responsible for such emergency.
- This applies in particular to the description of the thermal equilibrium radiation field. No assumption of discreteness is required to arrive at the Planck distribution.
- The equilibrium eventually attained by matter and field is such that a (detailed) balance exists between mean absorbed and radiated power at each frequency of the background field. Radiative effects can then be neglected.
- In such time-reversible regime, the mechanical system satisfies ergodic properties. Taken together, detailed energy balance and ergodicity define the *quantum regime*.
- Once in this regime, the evolution of the mechanical system is governed by the quantum laws, in the formulations of Schrödinger and Heisenberg. Both formulations are formally equivalent, yet they disclose different and complementary aspects of the quantum machinery.
- The Schrödinger equation provides a reduced statistical description of an ensemble of particles immersed in a stochastic environment. It is unable to afford in general a detailed account of a single element of the ensemble.
- The Schrödinger equation contains the core of the stochastic source in the form of Planck's constant, which is a measure of the fluctuations impressed by the ZPF.
- The spectral energy density of the ZPF, proportional to ω^3—and hence consistent with special relativity and with the principle of inertia—is essential in ensuring the stability of the ground state. This singles out the ZPF as the physical field responsible for atomic stability.
- The relatively high coherence of the modes of the ZPF that sustain the stationary solutions contrasts markedly with an uncorrelated noise, such as that leading to a single asymptotic (Brownian) solution with purely stochastic motion.

- The transitions between 'stationary' states (hence the states themselves) are determined by the extremely sharp resonances of the mechanical system to the background field.
- The operators and vectors in a Hilbert space are a powerful tool for the statistical description in terms of possible states of the system—though they conceal the physical mechanism sustaining such states, and give no idea of what is happening in real three-dimensional space.
- In a given quantum state, the dynamical variables are controlled by certain modes of the ZPF. The noncommuting operators and the canonical quantum commutator $[\hat{x}, \hat{p}]$ are an imprint of this field.
- The Heisenberg inequalities refer to the minimum statistical variances of the variables, induced by the action of the ZPF. The quantum fluctuations are causal fluctuations due to the (random) field.
- The particles remain particles all the time and follow definite and causal (stochastic) trajectories, but the description of these is beyond the possibilities of the limited statistical treatment offered by quantum mechanics.
- The reduction of the complete statistical description from phase space to the configuration subspace is the source of several of the most characteristic and deceitful properties of the quantum systems.
- In particular, such reduction precludes the possibility to start from QM and arrive, by purely logical steps, at a genuine (Kolmogorovian) phase-space distribution. Such distribution exists, but is beyond the realm of QM.
- The local averaging process conceals the detailed dynamics in phase space and leads to apparent nonlocalities. Therefore nonlocality exists, in the restricted sense that it a feature of the quantum *description*, but it does not refer to an ontological property of Nature. The reduction process leads to the artful transformation of the original local description into a nonlocal one.
- In particular, the concealed momentum fluctuations reappear in configuration space in the form of a strange contribution, conventionaly taken as an additional (quantum) potential, despite its kinetic origin. This 'potential' is an entry point of nonlocality and irreducible fluctuations in the quantum description.
- The resonant response to selected field modes explains the emergence of nonclassical correlations between noninteracting particles. Partners that share common relevant frequencies become entangled, and the high spatial coherence of the field modes involved gives stability to the ensuing states.
- For identical particles subject to the same potential, the entanglement induced by the ZPF common modes is maximal and leads to the (anti)symmetry of the wave function. The antisymmetry of the total electron wave function is rooted in the entanglement of the spin states.
- There are no spooky actions at a distance. It is the ZPF what connects the parties and transforms the bipartite system into a single entity .
- The electron spin is an acquired property that ultimately emerges from the (rotational) fluctuations impressed on the particle by the ZPF. Its magnitude is hidden in the fundamental quantum commutator (the signature of the ZPF).

- Associated with the spin (unavoidable, thus in a sense intrinsic) angular momentum is a gyromagnetic g-factor of value 2, derived from the two degrees of freedom of the circular polarization of the ZPF.
- A moving quantum corpuscle has a physical (de Broglie) wave associated to it. The de Broglie wave possesses an electromagnetic origin, linked with the ZPF. Yet both entities keep their individual corpuscle and wavelike natures, respectively.
- The diffraction pattern formed by a beam of electrons sent through an array of slits reflects the action on them by the diffracted ZPF. The trajectories of individual particles convey and reproduce statistically the wave nature of the diffracted background field.
- The theory defined by the starting Eq. (4.2) derives from a fundamental description that has all the properties that a fundamental realist and objective physical theory should possess: it is causal, deterministic and local. It accepts depictive images and offers a space-time description, without the need to bring the observer into the picture.
- The loss, total or partial, of any of the above mentioned traits, is a result of the approximations, reductions and simplifications made in arriving at the ultimate quantum description. Of particular relevance is the disappearance of the zero-point field from the picture.

10.3 The Photon

The photon is the quantum of electromagnetic interactions. This commonplace statement refers to the photon in interaction with matter—not to the photon itself. If we try to find an answer to the simple question "What is a photon itself?" things become blurred. This is fittingly reflected in the well-known confession by Einstein (1951) to a life-long friend:

> All these fifty years of conscious brooding have brought me not closer to the answer to the question, 'What are light quanta?' Of course today every Tom, Dick and Harry thinks he knows the answer, but he is deluding himself.

This popular sentence, written more than sixty years ago, is still very much alive. As a token of this, we recall the discussion organized in 2003 by The Optical Society (OSA) "to bring together different views regarding a question asked over the course of centuries: 'What is the nature of light?'" The editors of the meeting materials (C. Roychoudhuri and R. Roy) write in response: "Despite significant progress in our understanding, it remains an open question", and with a similar bent A. Zajonc, one of the contributors to the issue, declares "We are today in the same state of 'learned ignorance' with respect to light as was Einstein."[4]

[4] The contributions to the cited supplement to Optics & Photonic News are collected in Roychoudhuri and Roy (2003). The book Roychoudhuri et al. (2008) and the proceedings of subsequent SPIE meetings under the same title "The Nature of Light: What are Photons?"contains an ample collection of papers on the nature of the photon, showing the broad diversity of views and deep contradictions still existing on this matter.

The theory developed in Chap. 3 suggests a view of the quantized radiation field that is closer to Planck's than to Einstein's, in the sense that the quantization of the energy of each mode of the radiation field ensues as a consequence of matter-field interaction, the field alone remaining continuous. This contrasts with the intrinsic discrete structure of the radiation field, which is the reading that evolved into the notion of photon (Kuhn 1978)—yet quantum field theory has generally adopted and extended the heuristic point of view proposed by Einstein (1905). We recall that Eq. (3.84)

$$\langle f(\mathcal{E})\rangle = \underbrace{\int_0^\infty W_g(\mathcal{E}) f(\mathcal{E}) d\mathcal{E}}_{\text{continuous}} = \underbrace{\frac{1}{Z_g}\sum_{n=0}^{\infty} f(\mathcal{E}_n) e^{-\beta\mathcal{E}_n}}_{\text{discrete}} \tag{10.1}$$

for the mean equilibrium value of a function $f(\mathcal{E})$ was derived without any quantum assumption, but can be read both in terms of a continuous energy distribution or a discrete set of energy eigenstates. It is also true that such energy eigenvalues are never captured *exactly* in the lab, due to the unavoidable fluctuations (line breadth, and so on), which tend to give a more continuous structure to the energy distribution.

It seems interesting to make a short review of the most usual 'definitions' of the photon. This is a complex task, given their diversity, so we restrict ourselves to some representative examples. Probably the most extended one is precisely the early concept proposed by Einstein himself in (1909) of the photon as a "singular point just like the occurrence of electrostatic fields according to the electron theory", i.e., a singularity surrounded by the electromagnetic field of light of Maxwell's theory. The present-day QED definition of the photon as the unit of excitation associated with a quantized mode of the radiation field (a state of Fock space) does not throw much light on the structure of such quantum, normally understood—or wanted to be understood—as a localized 'particle', but described as an infinitely extended entity. Again, this goes somewhat beyond wave-particle duality, since the definition includes momentum, energy and polarization, but eludes position and time. As Zajonc (2003) puts it "Location in space and in time is no longer a means for theoretically distinguishing photons as elementary particles". Indeed it is well known, since the work of Newton and Wigner (1949), that there is no Hermitian operator that befittingly corresponds to position for photons.[5,6] As Bohm (1951) puts it in an introductory

[5] Newton and Wigner showed that it is not possible to define any position operator for a massless free particle with a nonzero spin, in sharp contrast to the case of massive particles, which can be localized. This is clearly in contradiction to the almost familiar notion of 'position of a photon', as one basic ingredient of the intended theoretical description.

[6] Einstein's photon of (1909) is defined by $E = cp = \hbar\omega$. It was in 1916 that he added a well-defined direction to the photon, transforming it into a 'needle of radiation'. On the other hand, Wigner's
(Footnote 6 continued)
photon is its helicity, which is a Lorentz-invariant concept coming from a subgroup of the Lorentz group for massless particles.

account: "There is, strictly speaking, no function that represents the probability of finding a light quantum at a given point. If we choose a region large compared with a wavelength, we obtain approximately $P(x) \sim \left(E^2(x) + H^2(x)\right)/8\pi h\nu(x)$, but if this region is defined too well, $\nu(x)$ has no meaning."

The single photon is frequently defined not by its intrinsic properties, but by its capabilities, such as its ability to trigger a single photodetection event. There is then no word about the spatial distribution, because the photon becomes defined by the entire excited optical system, which may be a closed or an open one. For the closed system a single standing-wave mode happens to be enough, whereas in the open case a traveling wavepacket is required. In either case the same energy of one quantum of light, $\hbar\omega$, is considered distributed over the entire apparatus, without a hint as to the localization of the photon. Despite this and other similar obscurities, it is usually asserted that both wave and particle properties are present in the quantum description, considering that the corpuscle aspect is fully described by the language of particle creation and annihilation. The definition reduces then to a description of the mathematics: we are invited to conceive the photon as a discrete excitation of a mode or a set of modes $\{k\}$ of the electromagnetic field in some cavity (see e.g. Muthukrishan et al. 2008). Going to the extreme along this direction of defining the photon by its capabilities, we find the jocund pragmatic definition by Glauber "A photon is what a photodetector detects." And in the most instrumentalist vein Muthukrishan et al. (2008) add "A photon is where the photodetector detects it.", whatever the size of the detector. Thus the photon extends in the literature from at least the size of the interferometer to that of an atom, or smaller. It would perhaps be more suitable to say that a photon ends its life where and when the photodetector detects it. On the other hand, one can read here and there of the photon as if it were already well established and experimentally confirmed that it is something like a minuscule localized object with well-defined frequency and wavelength (however, see e.g. Raymer and Srinivasan 2012). We find nevertheless that the photon of QED continues to be something radically different from a localized dimensionless corpuscle, quite far from the initial image conceived by Einstein that comes frequently to the physicist's mind; indeed, not a more or less localized entity, but something that may be distributed over the interferometer or the resonant cavity.

However, what we find common to all descriptions of the photon is the coexistence of two key elements: (a) the photon is an individualized structure, integral part of an electromagnetic field, and (b) it is capable of *transferring* a mean energy $\hbar\omega$ (from a stationary single mode or from a traveling wave-packet). It can be identified as an independent excitation of the electromagnetic field. This coincides with the image that emerges from the theory developed in Chap. 3; thus, the expression (3.71) for the fluctuations contains both a particulate and a continuous element.[7] As discussed

---

[7] The *coexistence* of both aspects in quantum behavior has meanwhile become an experimentally verified fact; see Aldemade et al. (1966), Kattke and Ziel (1970). For more recent work on (Footnote 7 continued)
complementarity see e.g. Jaeger et al. (1995), Englert (1996), Engert and Bergou (2000), Liu et al. (2009) and Flores and de Tata (2010).

*Cole, D.C.: Relating work, change in internal energy, and heat radiated for dispersion force situations. In: El-Genk, M.S. (ed.) Proceedings of Space Technology and Applications, International Forum 2000, AIP, vol. 504, p. 960 (2000b)
*Cole, D.C.: Entropy concepts in classical electrodynamics. In: Sheehan, D.P. (ed.) Proceedings of Quantum Limits to the Second Law, First International Conference On Quantum Limits to the Second Law, AIP, vol. 643, p. 960, San Diego (2002)
Cushing, J.T., Fine, A., Goldstein, S. (eds.): Bohmian Mechanics and Quantum Theory: An Appraisal. Boston Studies in the Philosophy of Science, vol. 184. Kluwer, Dordrecht (1996)
Cushing, J.T., McMullin, E. (eds.): Philosophical Consequences of Quantum Theory. University of Notre Dame Press, Indiana (1989)
Czopnik, R., Garbaczewski, P.: Quantum potential and random phase-space dynamics. Phys. Lett. A **299**, 447 (2002)
‡Davidson, M.P.: A dynamical theory of Markovian diffusion. Physica A **96**, 465 (1979a)
‡Davidson, M.P.: A generalization of the Fenyes-Nelson stochastic model of quantum mechanics. Lett. Math. Phys. **3**, 271 (1979b and 2001); arXiv:0112063[quant-ph]
‡Davidson, M.P.: A model for the stochastic origins of Schrodinger's equation. J. Math. Phys. **20**, 1865 (1979c)
‡Davidson, M.P.: Momentum in stochastic quantum mechanics. Lett. Math. Phys. **5**, 523 (1981a); arXiv:0112114v1[quant-ph]
*Davidson, M.P.: Stochastic quantization of the electromagnetic field. J. Math. Phys. **22**, 2588 (1981b)
*Davis, E.W., Teofilo, V.L., Haisch, B., Puthoff, H.E., Nickisch, L.J., Rueda, A., Cole, D.C.: Review of experimental concepts for studying the quantum vacuum field. In: El-Genk, M.S. (ed.) Space Technology and Applications, International Forum-STAIF 2006, AIP Conference Proceedings, vol. 813. http://www.calphysics.org/research.html (2006)
Davies, B., Burkitt, A.N.: On the relationship between quantum, random and semiclassical electrodynamics. J. Phys. **33**, 671 (1980)
de Angelis, G.F., Serva, M.: Jump processes and diffusions in relativistic stochastic mechanics. Ann. Institut Henri Poincaré, Physique Théorique **53**, 301 (1990)
de Broglie, L.: Sur la fréquence propre de l'electron. C. Rend. Acad. Sci. Paris **180**, 498 (1925)
de Broglie, L.: Matière et Lumière. Albin Michel, Paris (1937)
de Broglie, L.: Continu et Discontinu en Physique moderne. Albin Michel, Paris (1941)
de Broglie, L.: La Physique Quantique, restera-t-elle indé terministe?. Gauthier-Villars, Paris (1953)
de Broglie, L.: The Current Interpretation of Wave Mechanics. A Critical Study. Elsevier Publishing Company, Amsterdam (1964)
*Dechoum, K., França, H.M., Malta, C.P.: Gravitation and Cosmology: from the Hubble Radius to the Planck Scale. In: Amoroso, R.L., Hunter, G., Kafatos, M., Vigier, J.-P. (eds.) Proceedings of a Symposium in Honour of the 80th Birthday of Jean-Pierre Vigier, U. C., Berkeley 2000, Kluwer Academic Publishers (2002)
*Dechoum, K., de la Peña, L., Santos, E.: The problem of light detection in the presence of zeropoint: a physical model for the vision of weak signals. Found. Phys. Lett. **13**, 253 (2000)
de la Peña, L.: Stochastic quantum mechanics for particles with spin. Phys. Lett. A **31**, 403 (1970)
de la Peña, L.: Time evolution of the dynamical variables of a stochastic system. Am. J. Phys. **48**, 1080 (1980)
de la Peña, L., Cetto, A.M.: Lagrangian formulation of stochastic equations and quantum theory. Phys. Lett. A **29**, 562 (1969a)
de la Peña, L., Cetto, A.M., New formulation of stochastic theory and quantum mechanics. IV: The two-particle system. Rev. Mex. Fís. **18**, 323 (1969b)
de la Peña, L., Cetto, A.M.: Quantum phenomena and the zeropoint radiation field, I. Found. Phys. **24**, 917 (1994)
de la Peña, L., Cetto, A.M.: Quantum phenomena and the zeropoint radiation field, II. Found. Phys. **25**, 573 (1995)

**de la Peña, L., Cetto, A.M.: Linear Stochastic Electrodynamics: Looking for the Physics Behind Quantum Theory, Abridged Version of Notes XXXI ELAF (1998). In: Hacyan, S., Jáuregui, R., López-Peña, R. (eds.) New Perspectives on Quantum Mechanics (AIP Conference Proceedings 464, New York), p. 151 (1999)

*de la Peña, L., Cetto, A.M.: Contribution from stochastic electrodynamics to the understanding of quantum mechanics. arXiv:0501011[quant-ph] v2 (2005)

*de la Peña, L., Cetto, A.M.: Recent Developments in Linear Stochastic Electrodynamics. In: Adenier, G., Khrennikov, A.Yu., Nieuwenhuizen, T.M. (eds.): Quantum Theory: Reconsideration of Foundations-3, AIP Conference Proceedings, vol. 810, AIP, New York (2006a). An extended version in arXiv:0501011[quant-ph]

*de la Peña, L., Cetto, A.M.: The foundations of linear stochastic electrodynamics. Found. Phys. **36**(2006), 350 (2006)

*de la Peña, L., Cetto, A.M.: On the Ergodic Behaviour of Atomic Systems Under the Action of the Zero-Point Radiation Field. In: Nieuwenhuizen, T.M. et al. (eds.) Beyond the Quantum, World Scientific, Singapore (2007)

*de la Peña, L., Jáuregui, A.: Stochastic electrodynamics for the free-particle. J. Math. Phys. **24**, 2751 (1983)

de la Peña, L., Cetto, A.M., Brody, T.A.: On hidden-variable theories and Bell's inequality. Lett. Nuovo Cim. **5**, 177 (1972)

De Martini, F.: Evidence of first-order coherence of electromagnetic vacuum fluctuations. Phys. Lett. A **115**, 421 (1986)

Dimitrov, V.I.: "Averaged" statistical thermodynamics, energy equipartition and the third law (1997) arXiv:9707007 [quant-ph]

Dohrn, D., Guerra, F., Ruggiero, P.: Spinning particles and relativistic particles in the framework of Nelson's stochastic mechanics. In: Albeverio, A. et al. (eds.) Feynman Path Integrals, Lectures Notes in Physics, vol. 106, Springer, Berlin, p. 165 (1978)

Einstein, A.: Quanten-Mechanik und Wirklichkeit, Dialectica 2, 320 (1948). English translation in Born 1971

Elitzur, A., Doled, S., Kolenda, N. (eds.) Quo Vadis Quantum Mechanics? Springer, Berlin (2005)

Esposito, S.: On the role of spin in quantum mechanics. Found. Phys. Lett. **12**, 165; arXiv:9902019v1 [quant-ph]

*Faria, A.J., França, H.M., Malta, C.P., Sponchiado, R.C.: The Schrödinger picture and the zero-point radiation (2004); arXiv:0409117 [quant-ph] v1

*Faria, A.J., França, H.M., Sponchiado, R.C.: Tunneling as a classical escape rate induced by the vacuum zero-point radiation. Found. Phys. **36**, 307 (2006)

Faye, J.: Copenhagen interpretation of quantum mechanics. In: The Stanford Encyclopedia of Philosophy (2002). http://plato.stanford.edu/archives/sum2002/entries/qm-copenhagen

Ferrero, M., van der Merwe, A. (eds.): Proceedings Oviedo Symposium on Fundamental Problems in Quantum Physics. Kluwer, Amsterdam (1995)

Feshbach, H., Weisskopf, V.: Ask a foolish question. Phys. Today **41**(10), 9 (1988)

Frederick, C.: Stochastic space-time and quantum theory. Phys. Rev. D **13**, 3183 (1976)

*França, H.M., Franco, H., Malta, C.P.: A stochastic electrodynamics interpretation of spontaneous transitions in the hydrogen atom. Eur. J. Phys. **18**, 343 (1997)

*França, H.M., Marshall, T.W.: Excited states in stochastic electrodynamics. Phys. Rev. A **38**, 3258 (1988)

*França, H.M., Santos, R.B.B.: Resonant paramagnetic enhancement of the thermal and zero-point Nyquist noise. Phys. Lett. A **251**, 100 (1999)

Garbaczewski, P.: On the statistical origins of the de Broglie-Bohm quantum potential: Brownian motion in a field of force as Bernstein diffusion. Phys. Lett. A **178**, 7 (1993)

Garbaczewski, P.: Dynamics of coherent states as the diffusion process: Schrödinger's system, Bernstein processes and quantum evolution. In: Feng, D.H., Klauder, J.R., Strayer, M.R. (eds.) Coherent States: Past, Present and Future, World Scientific, Singapore (1994)

Gardner, M.: Guest comment: Is realism a dirty word? Am. J. Phys. **57**, 203 (1989)

earlier, the latter is immediately understood as a manifestation of the Maxwell field. The former, ascribed by Einstein to a *spatial* discreteness of the field, expresses the effects of the zero-point field, according to what we have learned in Chap. 3. Thus the photon is more than a classical Maxwell wave plus the active ZPF: it appears as a packet of the radiation field resulting from its interaction with matter. The field continues to be a field all the time, even if it gains some organization.

As a quantum of the radiation field, the photon appears in the theory, particularly in QED, only interacting (locally) with matter. Indeed, QED is used to describe the mutual effects of matter and field in interaction; this is the standing point from which the present-day notion of photon acquires sense. This standpoint is extended to the free field, assuming that also it requires quantization. However, the free field is an unobserved entity, with properties that are in principle beyond our reach. We have no more ground to assume that the free field *is* quantized, than the heuristic proposal of Einstein for mathematical continuity and physical simplicity of the description. We must be aware that this is an unverifiable extrapolation.

In the present SED account, further to the appearance of the quantized energy of the field in thermal equilibrium with matter (Chap. 3), little contact has been made with the photon. One exception (that addresses the issue only indirectly) is the analysis of radiative corrections presented in Chap. 6, leading to the well-known formulas for the Einstein A and B coefficients for the atomic lifetimes. These formulas clearly imply emission or absorption of a well-defined quantum of energy. In connection with this, we must recall the fact (already discussed above) that in the ergodic regime the atom responds resonantly to field modes of definite frequencies given precisely by the formula $\hbar\omega = \Delta\mathcal{E}$ connecting the energy levels. Therefore, the quantum of photonic energy exchanged between atom and field is in the picture—in relation with radiative processes occurring in matter. An extended SED theory that looks at the entire field-matter system, instead of focusing just on its material part as was done here, should help clarify the nature of the photon.

10.4 Limitations and Extensions of the Theory

To extend the scope of the theory presented in this book, several of the simplifications, reductions and approximations made along its development should be avoided or superseded. An immediate one is our use of a one-dimensional Fourier expansion of the ZPF in both Chaps. 4 and 5, instead of the four-dimensional Fourier expansion proper of QED. This implied averaging over the directions of the wave vectors $\boldsymbol{k}$ for each frequency ω_k, as well as over the states of polarization of the field (except for the analysis of spin, which demanded a separation of the polarization modes). Such averaging limits the scope of the theory from its very beginning and entails the loss of important information, particularly when attempting to construct a more detailed description of the field, to determine higher order corrections, or even to be able to study regions of space, time or matter subject to extreme conditions, as occur in so many situations envisaged by present day cosmology. Similarly, the entire treatment is nonrelativistic, and the Lorentz force has been restricted to its electric component and considered in the dipole (long-wavelength) approximation. Of major significance is the reduction of the description from the entire phase space to the configuration space of the particle; the theory is, consequently, unable to describe very fast processes, since its validity is restricted to small values of the Fourier variable z in Eq. (4.47), such that $pz \leq 1$.

Although the original equations of motion introduced in Chaps. 4 and 5 govern the evolution of the entire system at any given time, their mathematical complexity forced us to focus attention on the solutions near energy balance, when assumptions about ergodicity and stationarity can be reasonably made. This means that the detailed evolution of both the field and the material system during the transient phase from the initial nonequilibrium situation to the quantum regime has not been analyzed. The Heisenberg inequalities, in particular, hold only once the quantum regime is attained. So for extremely short time intervals they can be violated, and although this possibility seems to be difficult to verify at present, it remains open in principle (an example of such possibility is proposed in Sect. 8.1.1 of *The Dice*).

It seems marvelous that after such miscellany of curtailments and approximations, the resulting description, quantum mechanics, manifests itself so powerful and precise as it is known to be. Clearly, each one of these simplifications makes the ensuing theory to depart from the capabilities of a more complete description—and of QED in particular. This explains why, in making contact with QED, the analysis has been restricted to the lowest significative order of approximation (obtaining results that are in full agreement with those of QED). To overcome the limitations of the present theory one should in principle resort to the full phase-space description of the matter-field system. This should help to solve the known problems with the phase space descriptions, of which the Wigner function is the prototype. Of course, the indeterministic description will persist, but the extended theory would offer a finer and fairer description of Nature.

One virtue of the present treatment is that it can help to open opportunities to explore provinces unknown to all three, QM, QED and SED. This is perhaps its most important contribution. Clearly the present theory may—and should—be generalized in several directions. From among the wide range of possibilities let us mention one that seems to be most attractive.

Since the very early days of SED, the possibility of considering space-time metric fluctuations as responsible for the fundamental stochasticity of the quantum world has been entertained. Two immediate advantages are obvious. One is its universality, since all forms of matter and energy are subject to gravitational interaction. This gives an ecumenical answer to the frequent question in SED about neutral particles. An even more significant advantage of such theory would be that it could integratenaturally aspects of both, quantum theory and general relativity, a most desirable property for a theory that should help pave the road to quantum gravity. If such view of nature—intermediate between our present-day quantum theory and the would-be quantum theory of gravity—may be of any help, is something to be explored.

The observation that the vacuum energy fluctuations should give rise to fluctuations of the metric at small distances, has been considered in different contexts and from various points of view. Already in a work by Einstein (1924) stochastic fluctuations of the metric tensor are enterntained as a possible representation of a real, all-pervading material field. Both the mathematical and the physical side of the fluctuating geometry problem have received some attention over the years. The mathematicians have worked on the construction of different types of statistical geometries and their probabilistic topologies, a subject that is of much interest at present. On their part, de Broglie, Rosen, and Blokhintsev, among others, studied time ago the metric fluctuations as a source of the quantum fluctuations; the results appear scattered in the literature, without having led to an accomplished theory (a list of references up to approximately 1980 is given in Vigier 1982). A renewed interest seems to be flourishing, with different and more actual ideas and viewpoints (we refer for example to Carlton 1976; Namsrai 1986; Petroni and Vigier 1984; Bergia et al. 1989; Bergia 1991; Sorkin 1994; Santos 2006; Giovannini 2008; Gasperini 2011) . The connections between a phenomenological description of the fluctuations of the metric and the stochastic theory of quantum mechanics, among other things, have been explored by Roy (1986, 1992) and more recently in Vasudevan et al. (2008), whereas in their formulation of geometrodynamics, Bergia et al. (1989) have considered a model within the Kaluza-Klein framework. Several related aspects of the subject are currently under study, though with emphasis on the relativistic aspect, rather than its quantum implications. It would be premature to draw definitive conclusions about the merits and possibilities of such work; nevertheless, it is clear that they are worth further efforts.

The most ambitious task for present-day theoretical physics is just the development of a unified theory of the forces of nature. The efforts invested, both human and economic, in the search of such theory have been considerable, and the results for physics still quite limited. The two theories that give support to such efforts are general relativity and quantum mechanics. Present-day quantum theory, with all its conundrums, is being used in the search for a theory that pretends to describe the world as it is supposed to be fifteen or more orders of magnitude apart from our present scale of energies and times. This means transferring to the supposedly fundamental theory the problems and confusions that beset quantum theory today—and with an amplified dose of ignorance. Before taking such bold and unwarranted step it is best to clean the house and disclose the rich physics that remains invisible to the usual quantum description. Here we offer a contribution to this task.

References

Aharonov, Y., Albert, D.Z., Vaidman, L.: How the result of a measurement of a component of the spin of a spin-1/2 particle can turn out to be 100. Phys. Rev. Lett. **60**, 1351 (1988)

Aldemade, C.Th.J., Bolwijn, P.T., van der Veer, J.H.C.: Single-beam measurement of Bose-Einstein fluctuations in a natural Gaussian radiation field. Phys. Lett. **22**, 70 (1966)

Bergia, S.: I Fondamenti della Meccanica Quantistica. In: Cattaneo, G., Rossi, A. (eds.) Analisi Storica e Problemi Aperti. EditEl, Commenda di Rende, Cosenza (1991)

Bergia, S., Cannata, F., Pasini, A.: On the possibility of interpreting quantum mechanics in terms of stochastic metric fluctuations. Phys. Lett. A **137**, 21 (1989)

Bohm, D.: Quantum Theory. Prentice-Hall, Englewood Cliffs (1951)

Carlton, P.: Stochastic space-time and quantum theory. Phys. Rev. D **13**, 3183 (1976)

**de la Peña, L., Cetto, A.M.: The Quantum Dice. An Introduction to Stochastic Electrodynamics. Kluwer, Dordrecht (1996) (Referred to in this book as *The Dice*)

Einstein, A.: Über einen die Erzeugung und Verwandlung des Lichtes betreffenden heuristischen Gesichtspunckt. Ann. Phys. **17**, 132 (1905). English translation in The Collected Works of Albert Einstein. Princeton University Press, (1989)

Einstein, A.: Über die Entwicklung unserer Anschauungen über das Wesen und die Konstitution der Strahlung. Zeit. Phys. **10**, 817 (1909). English translation in The Collected Works of Albert Einstein, vol. 2 (Princeton University Press, 1989), 379

Einstein, A.: Über den Äther. Schweiz. Naturforsch. Gesells. Verh. **105**, 85 (1924)

Einstein, A.: In: Spiziale, P. (ed.) Albert Einstein, Correspondence avec Michele Besso 1903–1955. Hermann, Paris (1951)

Englert, B.-G.: Fringe visibility and which-way information: an inequality. Phys. Rev. Lett. **77**, 2154 (1996)

Engert, B.-G., Bergou, J.A.: Quantitative quantum erasure. Opt. Commun. **179**, 337 (2000)

Flores, E.V., de Tata, J.M.: Complementarity paradox solved: surprising consequences. Found. Phys. **40**, 1731 (2010)

Gasperini, M.: Elements of String Cosmology. Cambridge University Press, Cambridge (2011)

Giovannini, M.: A Primer On The Physics Of The Cosmic Microwave Background. World Scientific, Singapore (2008)

Jaeger, G., Shimony, A., Vaidman, L.: Two interferometric complementarities. Phys. Rev. A **51**, 54 (1995)

Jammer, M.: The Conceptual Development of Quantum Mechanics. McGraw-Hill, New York (1966)

Kattke, G.W., van der Ziel, A.: Verification of Einstein's formula for fluctuations in thermal equilibrium. Physica **49**, 461 (1970)

Kuhn, T.S.: Black-Body Theory and the Quantum Discontinuity 1894–1912. Clarendon, Oxford (1978)

Lamb, W.E., Scully, M.O.: The photoelectric effect without photons. In: Polarisation, Matière et Rayonnement. Presses Universitaires de France (1969)

Liu, N.-L., Li, L., Yu, S., Chen, Z.-B.: Duality relation and joint measurement in a Mach-Zehnder interferometer. Phys. Rev. A **79**, 052108 (2009)

Mandel, L.: II The case for and against semiclassical radiation theory. In: Wolf, E. (ed.) Progress in Optics XIII. North-Holland, Amsterdam (1976)

Muthukrishan, A., Scully, M.O., Zubairy, M.S.: The photon wave function. In: Roychoudhuri, Ch., Kracklauer, A.F., Creath, K. (eds.): The Nature of Light. What is a Photon? CRC Press, NY (2008)

Namsrai, K.: Nonlocal Quantum Field Theory and Stochastic Quantum Mechanics. Reidel, Dordrecht (1986)

Newton, T.D., Wigner, E.P.: Localized states for elementary systems. Rev. Mod. Phys. **21**, 400 (1949)

Planck, M.: Über die Begründung das Gesetzes der schwarzen Strahlung. Ann. Phys. **37**, 642 (1912)

Petroni, N.C., Vigier, J.-P.: Random motions at the velocity of light and relativistic quantum mechanics. J. Phys. A **17**, 599 (1984)
Raymer, M.G., Srinivasan, K.: Manipulating the color and shape of single photons. Phys. Today **65**, 32 (2012)
Roy, S.: Stochastic geometry and origin of quantum potential. Phys. Lett. A **115**, 256 (1986)
Roy, S.: Random zero-point field and quantum correction to the metric. Acta Appl. Math. **26**, 209 (1992)
Roychoudhuri, Ch., Kracklauer, A.F., Creath, K. (eds.): The Nature of Light. What is a Photon? CRC Press, New York (2008)
Roychoudhuri, Ch., Roy, R.: Opt. Photonics News **14**, 10 (2003)
Santos, E.: On a heuristic point of view concerning the motion of matter: from random metric to Schrödinger equation. Phys. Lett. A **352**, 49 (2006)
Schiff, L.: Quantum Mechanics. McGraw-Hill, New York (1955)
Sorkin, R.D.: Quantum Mechanics as Quantum Measure Theory. arXiv:gr-qc/9401003 (1994)
Spinoza, B.: Ethics (1677). Penguin Classics, London (2005)
Vasudevan, R., Parthasarathy, K.V., Ramanathan, R.: Quantum Mechanics. A Stochastic Approach. Alpha Science International, Oxford (2008)
Vigier, J.-P.: Non-locality, causality and aether in quantum mechanics. Astron. Nachr. **303**, 55 (1982)
Zajonc, A.: Light reconsidered. In: Roychoudhuri and Roy 2003, vol. 3. (2003)

Ghose, P., Home, D.: Wave-particle duality of single-photon states. Found. Phys. **22**, 1435 (1992)
Ghose, P., Home, D.: The two-prism experiment and wave-particle duality of light. Found. Phys. **26**, 943 (1996)
†Gillespie, D.T.: Why quantum mechanics cannot be formulated as a Markov process. Phys. Rev. A **49**, 1607 (1994)
Golin, S.: Comment on momentum in stochastic mechanics. J. Math. Phys. **27**, 1549 (1986)
*Gómez, B., et al. (eds.) Stochastic Processes Applied to Physics and Other Related Fields. World Scientific, Singapore (1983)
Greenstein, G., Zajonc, A.G.: The Quantum Challenge. Modern Research on the Foundations of Quantum Mechanics. Jones and Bartlett Publications, Boston (1997)
Griffiths, R.B.: Quantum locality. Found. Phys. **41**, 705 (2011)
Grössing, G., Fussy, S., Mesa Pascasio, J., Schwabl, H.: Emergence and collapse of quantum mechanical superposition: orthogonality of reversible dynamics and irreversible diffusion. Physica A **389**, 4473 (2010)
Grössing, G., Mesa Pascasio, J., Schwabl, H.: A classical explanation of quantization. Found. Phys. **41**, 1437 (2011)
Grössing, G., Fussy, S., Mesa Pascasio, J., Schwabl, H.: The quantum as an emergent system. J. Phys. Conf. Series 361, 012008 (2012)
Guerra, F., Loffredo, M.I.: Stochastic quantization of the Maxwell field. Lett. Nuovo Cim. **27**, 41 (1980)
Guerra, F., Ruggiero, P.: New interpretation of the Euclidean-Markov field in the framework of physical Minkowski space-time. Phys. Rev. Lett. **31**, 1022 (1978)
Hacyan, S., Sarmiento, A., Cocho, G., Soto, F.: Zero-point field in curved spaces. Phys. Rev. D **32**, 914 (1985)
*Haisch, B., Rueda, A.: Geometrodynamics, Inertia and the Quantum Vacuum, AIAA Joint Propulsion Conference, Salt Lake, City (2001); arXiv:0106075 [gr-qc] v1
*Hains, B., Rueda, A., Dobyns, Y.: Inertial mass and the quantum vacuum fields. Ann. Phys. (Leipzig) **10**, 393 (2001)
Hanson, N.R.: Copenhagen interpretation of quantum theory. Am. J. Phys. **27**, 1 (1959)
Harré, R.: Varieties of Realism: A Rationale for the Natural Sciences. Blackwell, Oxford (1986)
Hess, K., Michielsen, K., De Raedt, H.: Possible experience: from Boole to Bell, EPL 87 60007 (2009); Reply to the Comment by Leggett, A.J. and Anupam Garg. EPL **91**, 40002 (2010)
Hiley, B.J.: The role of the quantum potential in determining particle trajectories and the resolution of the measurement problem. In: Tarozzi and van der Merwe, p. 237 (1985)
Hooker, C.A.: The nature of quantum mechanical reality: Einstein versus Bohr. In: Philosophical Challenge of the Quantum Domain, vol. 5, Series in the Philosophy of Science, University of Pittsburgh Press, Pittsburgh (1972)
Howard, D.: Who invented the "Copenhagen Interpretation"? a study in mythology. Philos. Sci. **71**, 669 (2004)
*Ibison, M., Haisch, B.: Quantum and classical statistics of the electromagnetic zero-point field. Phys. Rev. A **54**, 2737 (1996)
Kennedy, D., Norman, C.: Do deeper principles underlie quantum uncertainty and nonlocality? Science **309**, 98 (2005)
Khrennikov, A.: Prequantum classical statistical field theory: complex representation, Hamilton-Schrödinger equation, and interpretation of stationary states. Found. Phys. Lett. **19**, 299 (2006)
Khrennikov, A.Yu.: Quantum randomness as a result of random fluctuations at the Planck time scale? Intern. J. Theor. Phys. **47**, 114 (2008a)
Khrennikov, A.Yu.: Born's rule from classical random fields. Phys. Lett. A **372**, 6588 (2008b)
Khrennikov, A.Yu.: Inf. Sci. **179**, 492 (2009a); arXiv:0709.3909 [quant-ph] v2
Khrennikov, A.Yu.: Violation of Bell's Inequality and Non-Kolmogorovness. In: Accardi et al. (2009b)
Khrennikov, A.Yu.: Description of composite quantum systems by means of classical random fields. Found. Phys. **40**, 1051 (2010a)

Khrennikov, A.Yu.: Quantum correlations from classical Gaussian random variables: fundamental role of vacuum noise. Fluctuation Noise Lett. **9**, 331 (2010b)

Khrennikov, A.Yu.: Quantum correlations and dynamics from classical random fields valued in complex Hilbert spaces. J. Math. Phys. **51**, 082106 (2010c)

Khrennikov, A.Yu. Prequantum classical statistical field theory: background field as a source of everything? In: 5th International Workshop DICE2010; JOPCS, vol. 306, p. 012021 (2011). http://iopscience.iop.org/1742-6596/306/1/012021

Kiessling, M.K.-H.: Misleading signposts along the de Broglie-Bohm road to quantum mechanics. Found. Phys. **40**, 418 (2010)

*Kosłowski, M., Marciak-Kosłowska, J.: Zero point field effects in the interaction on the ultra-short laser pulses with matter. Lasers Eng. **12**, 281 (2002)

Kracklauer, A.F.: Nonlocality, Bell's Ansatz and probability. Opt. Spectrosc. **103**, 457 (2007)

Kukla, A.: Scientific realism, Scientific practice and the natural ontological attitude. Brit. J. Phil. Sci. **45**, 955 (1994)

†Lavenda, B.H.: On the equivalence between classical Markov processes and quantum mechanics. Lett. Nuovo Cimento **27**, 433 (1980)

Legget, A.J.: Is quantum mechanics the whole truth?, in Elitzur. Doled Kolenda **2005**, 83 (2005)

Lehr, W.J., Park, J.L.: A stochastic derivation of the Klein-Gordon equation. J. Math. Phys. **18**, 1235 (1977)

Maddox, J.: Licence to slang Copenhagen? Nature **332**, 581 (1988)

*Mareš, J.J., Krištofik, J., Hubik, P., Špička, V.: Weak localization-an experimental tool to investigate electromagnetic vacuum fluctuations. Physica E **29**, 375 (2005)

*Mareš, J.J., Špička, V., Krištofik, J., Hubik, P., Šesták, J.: Filling a cavity with zero-point electromagnetic radiation. Physica E **29**, 213 (2005)

*Marshall, T.W.: O sootnoschenii meschdu klassicheskoi teoriei fluctuatsii i kvantovoi teoriei polei. Isv. Vusov (Physics) **11**, 34 (1968)

*Marshall, T.W.: When is a statistical theory causal?. In: Tarozzi and van der Merwe, p. 257 (1985)

*Marshall, T.W. A causal theory of the Compton effect. preprint 12 January 2000 (2000)

*Marshall, T.W. Engineering the vacuum. In: Amoroso, R.L., Hunter, G., Kafatos, M., Vigier, J-P. (eds.) Gravitation and Cosmology: from the Hubble Radius to the Planck Scale, Proceedings of a Symposium in Honour of the 80th birthday of Jean-Pierre Vigier, U.C., Berkeley 2000, Kluwer Academic Publishers (2002)

*Marshall, T.W.: Are atoms waves or particles? Found. Phys. **36**, 333 (2006)

*Marshall, T.W., Santos, E.: The Present Status of the Quantum Theory of Light. In: Jeffers, S. (ed.), Kluwer Academic, Dordrecht, p. 67 (1997)

Matthews, R.: Inertia: Does empty space put the resistance? Science **263**, 612 (1994)

Matzkin, A., Nurock, V. Classical and Bohmian trajectories in semiclassical systems: mismatch in dynamics, mismatch in reality? Stud. Hist. Philos. Sci. B **39**, 17 (2008)

Milonni, P.W.: Semiclassical and quantum electrodynamical approaches in nonrelativistic radiation theory. Phys. Rep. **25**, 1 (1976)

Milonni, P.W.: Radiation reaction and the nonrelativistic theory of the electron. Phys. Letts. A **82**, 225 (1981)

Mittelstaedt, P.: The problem of interpretation of modern physics. Found. Phys. **41**, 1667 (2011)

†Moore, S.M.: Can stochastic physics be a complete theory of nature? Found. Phys. **9**, 237 (1979a)

Morato, L.M.: On the dynamics of diffusions and the related electromagnetic potentials. J. Math. Phys. **23**, 1020 (1982)

Morato, L.M.: Stochastic derivation of the Klein-Gordon equation from classical relativistic action. Phys. Lett. A **154**, 327 (1991)

Morato, L.M., Ugolini, S.: A connection between quantum dynamics and approximation of Markov diffusions. J. Math. Phys. **35**, 4505 (1994)

Nieuwenhuizen,Th.M.: Classical phase space density for the relativistic hydrogen atom. In: Adenier, G., Khrennikov, Yu., Nieuwenhiuzen, Th. (eds.) Quantum Theory: Reconsideration of Foundations, AIP CP810, vol. 3, p. 198 (2006)

Nieuwenhuizen, ThM: Is the contextuality loophole fatal for the derivation of Bell inequalities? Found. Phys. **41**, 580 (2010)
Nieuwenhuizen, Th.M. et al.: Beyond the Quantum. In: Nieuwenhuizen, T.M., Špička, V., Mehmani, B., Aghdami, M.J., Khrennikov, A.Yu. (eds.). World Scientific, Singapore (2007)
Norsen, T.: The theory of (exclusively) local beables. Found. Phys. **40**, 1858 (2010)
Omnès, R.: Consistent interpretations of quantum mechanics. Rev. Mod. Phys. **64**, 339 (1992)
Onofre, E.: The stochastic interpretation of quantum mechanics: a reply to Ghirardi. Lett. Nuovo Cimento **24**, 253 (1979)
Oriols, X., Martin, F., Suñé, J.: Implications of the noncrossing property of Bohm trajectories in one-dimensional tunneling configurations. Phys. Rev. A **54**, 2594 (1996)
Paty, M.: The nature of Einstein's objections to the Copenhagen interpretation of quantum mechanics. Found. Phys. **25**, 183 (1995)
Petroni, N.C.: A causal fluidodynamical model for the relativistic quantum mechanics. In: Tarozzi, G., van der Merwe, A. (eds.) Open Questions in Quantum Physics, Reidel, Dordrecht (1985)
Petroni, N.C., Vigier, J.-P.: Single-particle trajectories and interferences in quantum mechanics. Found. Phys. **22**, 1 (1992)
Pippard, A.B.: The interpretation of quantum mechanics. Eur. J. Phys. **7**, 43 (1986)
Plotnitsky, A.: Causality and Probability in Quantum Mechanics. In: Accardi et al. (2009)
*Puthoff, H.E.: Ground state of hydrogen as a zero-point-fluctuation-determined state. Phys. Rev. D **35**, 3266 (1987)
Rabinowitz, M.: Challenges to Bohr's wave-particle complementarity principle. Int. J. Theor. Phys. **52**, 668 (2013)
Redhead, M.L.G.: Nonfactorizability, stochastic causality, and passion-at-a-distance. In: Cushing, J.T., McMullin, E. (eds.) Philosophical Consequences of Quantum Theory, University of Notre Dame Press, Indiana (1989)
Riggs, P.J.: Quantum Causality. Conceptual Issues in the Causal Theory of Quantum Mechanics. Springer, Dordrecht (2009)
Ross-Bonney, A.A.: Does god play dice? A discussion of some interpretations of quantum mechanics. Nuovo Cim. B **30**, 55 (1975)
Roychoudhuri, Ch.: Can Photo Sensors Help Us Understand the Intrinsic Differences Between Quantum and Classical Statistical Behaviors? In Accardi et al., p. 167 (2009)
*Rueda, A.: In: Gómez et al. (1983a)
*Rueda, A.: The vacuum electromagnetic field and the energetics of the intergalactic medium. In: Proceedings of 18th International Cosmic Rays Conference, vol. 1, Bangalore (1983b)
*Rueda, A.: Inertia and the vacuum. In: Nieuwenhuizen, Th.M., Špička, V., Mehmani, B., Aghdami, M.J., Khennikov, A.Yu. (eds.) Beyond the Quantum. World Scientific, Singapore (2007)
Rueda, A., Haisch, B.: Gravity and the quantum vacuum inertia hypothesis. Ann. Phys. (Leipzig) **14**, 479 (2005)
Salesi, G., Recami, E.: A velocity field and operator for spinning particles in (nonrelativistic) quantum mechanics. Found. Phys. **28**, 763 (1998)
*Samaniego, F.: Aproximación a las consecuencias filosóficas de la electrodinámica estocástica lineal: realismo y causalidad en la mecánica cuántica. M. Sc Thesis (UNAM, México) (2007)
*Santos, E.: Objectification of classical properties induced by quantum vacuum fluctuations. Phys. Lett. A **188**, 198 (1994)
*Santos, E.: Tests of Bell's inequallities: increasing empirical support for local realism? In: Proceedings Workshop Beyond the Quantum, Leiden University, May–Jun 2006 (2007)
Selleri, F.: La causalitá impossibile. L'interpretazione realisica della fisica del quanti (Jaca Book, Milano) (1987)
Shimony, A.: Bell's theorem. In: Zalta, E.N. (ed.) Stanford Encyclopedia of Philosophy. http://plato.stanford.edu/entries/bell-theorem/ (2004)
Simon, B.: Ed Nelson's work in quantum theory, talk given at Nelsonfest. In: Proceedings of Analysis, Probability and Logic: A Conference in Honor of Edward Nelson. Appl. Math. **171**, p. 411 (17 June 2004)

*Špička, V.: Stochastic Electrodynamics and Experiment. In: Proceedings Workshop Beyond the Quantum, Leiden University (May-Jun 2006) 2007

Špicka, V., Mareš, J.J., Hubík, P., Krištofik, J.: Some Quantum Experiments from the Point of View of Stochastic Electrodynamics, in Nieuwenhuizen et al. 2007, p. 247. (2007)

Stanford Encyclopedia of Philosophy (2006)

Sulcs, S.: The nature of light and twentieth century experimental physics. Found. Sci. **8**, 365 (2003)

Sulcs, S., Gilbert, B.C., Osborne, C.F.: On the interference of fullerenes and other massive particles. Found. Phys. **32**, 1251 (2002)

Sulcs, S., Oppy, G., Gilbert, B.C.: The rejection of local realism is premature. Found. Phys. Lett. **13**, 521 (2000)

*Surdin, M.: Derivation of Schrödinger's equation from stochastic electrodynamics. Int. J. Theor. Phys. **4**, 117 (1971)

Tarozzi, G., van der Merwe, A. (eds.): Open Questions in Quantum Physics. Reidel, Dordrecht (1985)

't Hooft, G.: Equivalence relations between deterministic and quantum mechanical systems. J. Stat. Phys. **53**, 323 (1988)

't Hooft, G.: Quantummechanical behaviour in a deterministic model. Found. Phys. Lett. **10**, 105 (1997). arXiv:9612018

't Hooft, G.: Quantum mechanics and determinism. In: Frampton, P., Ng, J. (eds.) Proceedings of the Eighth International Conference on Particles, Strings and Cosmology. Princeton, Rinton Press, New Jersey, p. 275 (2001a). arXiv:0105105

't Hooft, G.: How does god play dice? Bertlmann, R.A., Zeilinger, A. (eds.) (Pre)determinism at the Planck Scale, p. 307. (2001b). arXiv:0104219v1

't Hooft, G.: Determinism in free bosons. Int. J. Theor. Phys. **42**, 355 (2003). http://arxiv.org/abs/hep-th/0104080hep-th/0104080

't Hooft, G.: The Free-Will Postulate in Quantum Mechanics (2007a). arXiv:0701097v1

't Hooft, G.: A mathematical theory for deterministic quantum mechanics. J. Phys.: Conf. Ser. **67**, 012015 (2007b)

't Hooft, G.: Emergent quantum mechanics and emergent symmetries. AIP Conf. Proc. **957**, 154 (2007c)

't Hooft, G.: Quantum mechanics from classical logic. J. Phys. Conf. Ser. **361**, 012024 (2012)

Towler, M.D., Russell, N.J., Valentini, A.: Timescales for Dynamical Relaxation to the Born Rule. Proc. Roy. Soc. **A 468**, 990 (2011). http://www.tcm.phy.cam.ac.uk/mdt26/papers/born_rule_towler.pdf

van der Merwe, A., Selleri, F., Tarozzi, G. (eds.): Bell's Theorem and the Foundations of Modern Physics. World Scientific, Singapore (1992)

Vink, J.C.: Quantum mechanics in terms of discrete beables. Phys. Rev. A **48**, 1808 (1993)

Whitaker, A.: Einstein, Bohr and the Quantum Dilemma. Cambridge University Press, Cambridge (1996)

Zastawniak, T.: A relativistic version of Nelson's stochastic mechanics. Europhys. Lett. **13**, 13 (1990)

Index

L. de la Peña et al., *The Emerging Quantum*,
DOI: 10.1007/978-3-319-07893-9,

Printed in the USA
CPSIA information can be obtained
at www.ICGtesting.com
CBHW070010191124
17629CB00003B/26